MULTINOMIAL
$(n, p_1, p_2, \ldots, p_k)$

Typical application: numbers of type 1, type 2, . . . , type k outcomes in n independent multinomial trials.

Mass function: $p(n_1, n_2, \ldots, n_k) = \binom{n}{n_1,\ n_2,\ \ldots,\ n_k} p_1^{n_1} p_2^{n_2} \cdots p_k^{n_k}$;
the n_i's are nonnegative integers such that $\sum n_i = n$.

Mean of N_i: np_i.

Variance of N_i: $np_i q_i$.

Covariance of N_i and N_j: $-np_i p_j$.

Moment-generating function: $M(t_1, t_2, \ldots, t_k) = (p_1 e^{t_1} + \cdots + p_k e^{t_k})^n$.

Marginal distributions: binomial (n, p_i).

NEGATIVE BINOMIAL
(k, p)

Typical application: number of failures to the kth success in independent, repeated Bernoulli (p) trials.

Mass function: $p(x) = \binom{-k}{x} p^k (-q)^x$, $x = 0, 1, 2, \ldots$.

Mean: kq/p.

Variance: kq/p^2.

Probability-generating function: $N(z) = p^k/(1 - qz)^k$.

POISSON
(λ)

Typical application: number of arrivals of a Poisson process in a unit of time.

Mass function: $p(x) = e^{-\lambda} \lambda^x/x!$, $x = 0, 1, 2, \ldots$.

Mean: λ.

Variance: λ.

Moment-generating function: $M(t) = e^{\lambda(e^t - 1)}$.

Connection with chi-square: $F(m) = \sum_{i=0}^{m} p(i) = 1 - G(2\lambda)$, where G is the χ^2_{2m+2} CDF.

Characteristics of continuous distributions are given on the inside of the back cover.

Probability

Modeling
Uncertainty

Probability

Modeling
Uncertainty

DONALD R. BARR
Professor of Statistics and Operations Research

PETER W. ZEHNA
Professor of Mathematics and Statistics

Naval Postgraduate School
Monterey, California

ADDISON-WESLEY PUBLISHING COMPANY
Reading, Massachusetts   Menlo Park, California   London
Amsterdam   Don Mills, Ontario   Sydney

This book is in the
ADDISON-WESLEY SERIES IN STATISTICS

Consulting Editor:
Frederick Mosteller

Library of Congress Cataloging in Publication Data

Barr, Donald Roy.
    Probability, modeling uncertainty.

    1. Probabilities. I. Zehna, Peter W. II. Title.
QA273.B2577 1983   519.1   82-11637
ISBN 0-201-10798-8

ISBN 0-201-10798-8
ABCDEFGHIJ-MA-89876543

PREFACE

This book presents a treatment of probability theory and applications that is suitable as background for further work in statistics, stochastic models, and various applications in engineering, science, operations research, and computer science. It utilizes a postcalculus approach and is nonrigorous mathematically. The book has an applied flavor, with emphasis on model building and problem solving. Our goal in writing the book was to provide a treatment of probability more extensive than that found in statistics textbooks yet having the same spirit of immediate applicability. To do so, we have included a large number of relatively simple examples and exercises and have presented discussions of many practical, how-to-do-it methods. We have attempted to keep the discussions at a precisely defined mathematical level, although considerable variation in rigor can be achieved by the instructor's choice of material to be covered, as discussed below. We have also endeavored to weave a compromise path through the numerous specializations within the subject; thus we have made a careful choice (and some reluctant omissions) of materials on combinatorics, conditioning concepts and methods, generating functions, data analysis topics, stochastic processes, limit theorems, and statistical concepts. Applications and examples have been taken from a variety of operations research, science, and engineering contexts. We believe the resulting collection of topics and applications form a good undergraduate introduction to the theory of probability and its applications for students with a wide variety of backgrounds and interests.

The text is organized into seven chapters, each concerned with a major idea in probability. Each chapter is divided into approximately a half dozen sections. The final section in each chapter is a chapter summary, which provides convenient review materials. Most sections are followed by exercises, which are subdivided into two sections. The theory exercises are problems of a more theoretical nature, designed to augment and extend the text material. The application exercises are of a drill nature and are designed to provide experience in applying the ideas introduced in the text. While almost all the exercises at the end of a section are directly related to the material of that section, a few are included that relate to earlier sections or that anticipate following sections. We believe there is some pedagogical merit to assigning such exercises at least occasionally. A few of the exercises are appropriate for computer-aided solution; some of these are labeled in the text as computer-oriented. Answers to selected application exercises are included at the end of the text.

The book can be used for a variety of course lengths, levels, and areas of emphasis by selection of sections to be covered. A number of sections, marked by asterisks, are optional either because of mathematical level or because they relate to more specialized application areas. For example, for a course that introduces stochastic models, the optional sections on occupancy problems (2.4), moment-generating functions (4.4), the Poisson process (6.7), and Markov chains (6.8) should probably be covered. For a course leading to a mathematical statistics course, the optional sections on moment-generating functions (4.4), extensions to higher dimensions (5.7), the bivariate normal distribution (6.5), as well as most of Chapter 7, might be covered. For a course at a lower level of mathematical detail, the following sections might be omitted: multidimensional calculus sections, such as change-of-variable techniques (6.3) and extensions to higher dimensions (5.7), along with moment-generating functions (4.4 and 6.4), and most of Chapters 6 and 7. Sections on the normal (3.5) and bivariate normal (6.5) can be omitted; brief introductions to these families can be gained by appropriate choice of examples and exercises contained elsewhere in the text, if the instructor desires.

The authors wish to express their sincere gratitude to their families, colleagues, and friends for their support and encouragement during the long process of writing this book. We are especially indebted to the reviewers of the manuscript, R. Deane Branstetter (San Diego State University), John P. Lehoczky (Carnegie-Mellon University), Frederick Mosteller (Harvard University), Marion R. Reynolds, Jr. (Virginia Polytechnic Institute and State University), and J. G. Wendel (University of Michigan), for their numerous suggestions; and to the members of the editorial staff of the Addison-Wesley Publishing Company for their assistance in the preparation of the book.

Monterey, California
November 1982

D.R.B.
P.W.Z.

CONTENTS

---

* Optional.

---

* Optional.

Chapter 1

PROBABILITY SPACES

1.1 INTRODUCTION

Historically, probability in its formal sense seems to have had its beginnings in games of chance, around the middle of the seventeenth century. At that time, games of chance were fashionable in France among the noblemen. One of them, de Meré, was concerned about what appeared to be an inconsistency between the basis for betting and the actual outcomes of a certain popular game of the time. Curiosity led de Meré to consult the mathematician Blaise Pascal. In conjunction with a fellow mathematician, Pierre de Fermat, Pascal appears to have developed the first formal analysis of gambling odds in mathematical terms. It is almost certain that, long before that time, people had already been aware that many types of phenomena seemed to be unpredictable and yet displayed a long-run regularity of sorts. This regularity is the heart of probability theory.

From these beginnings, probability theory expanded, slowly at first, into a substantial mathematical theory capable of handling a wide variety of unpredictable phenomena far beyond mere games of chance. Even so, many probability problems are still posed in the language of games of chance. There are generally two attitudes about this situation. On the one hand, there is the feeling that, if this theme is exploited too much, then there is a risk of developing a narrow, and, to some, an uninteresting, view of what probability is about. On the other hand, some feel that games of chance provide an ideal setting in which to discuss probability concepts since they are easily understood in contrast to more complex phenomena like gene structure or electronic switching. Thus game scenarios are often used as prototypes for more complex phenomena. Being divorced from the real setting, such simple applications are not so cluttered with peripheral concepts that may distract from the central issues of the problems and thus solutions. We will attempt to strike a balance between these two points of view in the treatment to follow.

All attempts to define probability mathematically have failed in one way or another. This failure was especially troublesome when twentieth-century ideas like infinite limits were used and the existence of such limits could not even be demonstrated. Rather than attempt to *define* probability, we will use an axiomatic approach. That is, rather than say what probability *is*, we study a short list of assumptions that probabilities must satisfy. In this way, many of the mathematical objections are removed, placing probability theory on a solid foundation, on a par with other mathematical subjects. Such an approach allows a wide variety of methods of probability assignments for a given problem, requiring merely that the axioms be satisfied. At the very least, this approach provides a common ground for comparing diverse methods of assigning probabilities. Later in the book, we will discuss several of the most common methods used in practice to assign probabilities.

The axiomatic system we have adopted was developed first by the Russian mathematician A. Kolmogorov in 1933. Not a lot was done with it, however, until around 1950; indeed, an English translation of the work was not available until then. Since that time, the subject matter of probability has expanded at a rapid rate. A cursory examination of the mathematical literature in a technical library will quickly give you an idea of just how prodigious this development has been. The subject can be studied at many levels, from the very elementary to the very advanced, worthy of doctoral-level research. Our attempt in this book is to offer an intermediate-level treatment with a mix of theory and practice. In this way, you will develop sufficient background to appreciate the mathematical theory of the subject with at least an intuitive understanding of matters that are beyond the scope of the intended level. At the same time, we hope to provide a sufficient basis for applying the subject in a wide variety of settings so that you are left with a respectable working knowledge of probability as it is practiced.

Before turning to our formal development, we point out an essential feature of the structure we plan to follow in the ensuing sections. Almost all sections of the book contain exercises, which are divided into the categories of theory and application. The exercises in the theory section often extend or complement the mathematical developments; they will give you an opportunity to become an active participant in the development. By solving at least some of these exercises, you will expand and deepen your understanding of the subject. The theory exercises at the end of each section should be solved or at least scanned before proceeding to the next section. The applied exercises, on the other hand, are typically specific numerical applications designed to give you practice and to develop your skill in applying the concepts. As in the development of any skill, the quantity of applied exercises required varies from individual to individual. Therefore it is desirable that you undertake at least some practice with the applied exercises.

1.2 MATHEMATICAL MODELS AND SAMPLE SPACES

The ultimate purpose of the scientist is to provide suitable prediction and explanations of phenomena he or she observes in the physical universe. These explanations are often in the form of *models*—that is, systems or structures whose properties are relatively well known. A mathematical model is a mathematical theory (or mathematical system) in which at least some of the statements may be interpreted in terms of a real world phenomenon under investigation. It is hoped that the mathematical system is in some sense analogous to the particular physical process under consideration, so that results in the mathematical system are informative about the process. That is, if the model reflects faithfully, in mathematical terms, the attributes of the physical process, then one may be able to use mathematical methods in the model to arrive at conclusions about the physical process.

For example, we might study a certain type of differential equation in connection with a phenomenon such as the motion of a vibrating string. By applying mathematical techniques to the solution of the differential equation, we may obtain a function that allows us to predict the position of the string at any time and the velocity of the midpoint of the string at any time. But the quality of these predictions will clearly be dependent on how well the differential equation fits the essential attributes of the physical system (the vibrating string). The process of finding such a differential equation is not always simple, and it usually involves a mixture of simplifying assumptions (such as assuming that the string has uniform mass), past experiences, and the analysis of similar or simpler systems (such as vector analysis of the forces on a given string particle).

The final criterion for deciding whether a model is good is whether it yields *useful* information. It should be emphasized that our motivation in considering mathematical models lies in their *utility* rather than in any notion of "correctness." The notion of judging models in terms of their utility gives rise to the possibility of using several different models for the same phenomenon. This situation is not uncommon in science. For example, two models have been used in connection with light phenomena, the wave model and the corpuscular model. Each model is useful in explaining some aspects of light that the other apparently fails to explain; thus both models are used. But notice that using a certain model for light is *not* a claim of what light *is*. It is important to maintain this distinction between a model (or "theory") and the physical process being modeled.

The basic idea of probability theory is to formulate mathematical models for experimental outcomes. The terms "experiment" and "outcome" will be left technically undefined; it is assumed that you are already familiar with these notions in the intuitive sense. The basic aim of an experiment is to produce an outcome of some sort, and we suppose that the various possible outcomes of a given experiment can be identified and distinguished from each other. As a standard example, suppose that a coin is to be tossed into the air and that the upturned face is to be observed after the coin has fallen. The possible outcomes might then be suitably described by the letters H (head) and T (tail). (We will often use the term "outcome" when we mean, more precisely, "outcome description.")

Suppose that the possible outcomes of an experiment have been identified in such a way that performance of the experiment must result in precisely one outcome. Our first task is to associate names or labels with such *distinct* outcomes, as in the case of the coin-tossing experiment. A set of labels of this kind is given a special name:

DEFINITION 1.1   A set S of all possible outcomes of an experiment is called a *sample space* for the experiment.

EXAMPLE 1.1   Suppose a die is rolled and the number of dots on the upturned faced is recorded. A natural sample space is $S = \{1, 2, 3, 4, 5, 6\}$.

EXAMPLE 1.2   For a coin-tossing experiment, we usually take $S = \{H, T\}$

Since the terms "experiment" and "outcome" are left undefined, it is quite possible to use different sample spaces in connection with models for the same phenomenon.

EXAMPLE 1.3   Imagine an experiment that consists of rolling two dice, one red and one green. We might be interested only in the total number of spots showing and would therefore use as our sample space the set $S_1 = \{2, 3, \ldots, 12\}$. Or we might be interested in keeping track of the individual outcomes on the red and green dice. There are several ways of doing so, but one especially convenient method is to use ordered pairs. Thus we might agree to associate the first component of the ordered pair with the number of spots on the red die and the second component with the number of spots on the green die. The resulting sample space would then be the set

$$S_2 = \{(1, 1), (1, 2), \ldots, (2, 1), (2, 2), \ldots, (6, 6)\}.$$

Notice that S_1 has 11 distinct elements, a fact we will denote by $n(S_1) = 11$, while $n(S_2) = 36$. Any set having a definite number of elements (as measured by a positive integer or zero) is called *finite*. Thus $n(A)$ is defined here only for finite sets A, in which case $n(A)$ is a nonnegative integer.

EXAMPLE 1.4   Suppose an experiment consists of tossing a coin repeatedly until "heads" appears for the first time. Suppose the outcome is the number of tosses that was required. It is conceptually possible that any positive integer might be the outcome; moreover, it is possible that the experiment will never terminate, in the sense that there is no positive integer beyond which we can definitely say that no additional tosses would be required. A suitable sample space for this experiment is the set of positive integers, also called the set of natural numbers, denoted $\{1, 2, 3, \ldots\}$. This set, and any set whose elements can be put in a one-to-one correspondence with it, is not finite and is said to be *denumerable* or *countably infinite*. A *countable set*, then, is one that is either finite or denumerable. When we wish to discuss a countable sample space without being specific as to the exact nature of its elements, we will use the notation $S = \{o_1, o_2, \ldots, o_n\}$ for the finite case and $S = \{o_1, o_2, o_2, \ldots\}$ for the denumerable case (o stands for "outcome").

EXAMPLE 1.5   An example of an experiment that cannot be described by either of the foregoing cases is one that consists of observing the amount of time it takes for a light bulb to fail after it is put into use. Conceptually, any nonnegative real number might be used to describe the outcome of this experiment, but it is impossible to list the elements of S or to describe S in the manner used above for denumerable sets because S is *uncountable*. If we let an arbitrary outcome (time to failure) be denoted generically by the letter o, and we let $Q(o)$ be any statement about o, we can use the symbol $\{o : Q(o)\}$ to denote the set of those elements of S that make $Q(o)$ a true statement. Thus for this experiment, a suitable sample space might be denoted $S = \{o : o \geq 0\}$. More simply in this case, we could give S as an interval, $S = [0, \infty)$.

In the previous example, it might be argued that the sample space we selected is really larger than needed. We have no way to physically measure times to the accuracy of an arbitrary real number, since our methods of measuring are limited to discrete values. Nevertheless, it is far more convenient for this type of experiment to allow for more possible labels than might be needed. All that is required of a sample space is that it include labels for all possible outcomes of interest—it isn't necessarily precisely the set of all such descriptions. This is equivalent to saying that one may include in S additional labels, labels that are associated with conceptual outcomes that perhaps may rarely, if ever, actually occur upon performing the experiment. We make use of our freedom from having to pass judgment on precisely what an "outcome" is. One should not, however, attempt to use sample spaces that do not include a sufficient number of labels.

EXAMPLE 1.6   If, in the coin-tossing experiment of Example 1.2, we are also interested in the possibility that the coin will fall on its edge, we would have to expand the sample space to include an additional description, say E, so that $S = \{H, T, E\}$. All such judgments about what constitutes S will have to be made by the experimenter; we will henceforth assume that a suitable choice has been made, although we will continue to indicate spaces that are typically used in some standard situations.

Here are some further examples.

EXAMPLE 1.7   In the die experiment of Example 1.1, we saw that an obvious sample space is given by $S = \{1, 2, 3, 4, 5, 6\}$. But we *could* use other choices, such as $S = \{1, 2, 3, \ldots\}$, $S = \{I, II, III, IV, V, VI\}$, or $S = (-\infty, \infty)$ (the set of all real numbers).

EXAMPLE 1.8   In the light bulb experiment of Example 1.5, we selected $S = [0, \infty)$. But in view of our observations on the freedom of choice of sample spaces, we could also select the set of all real numbers, $S = (-\infty, \infty)$, even though no actual outcome o will be found in the interval $(-\infty, 0)$.

We will make two final comments in this section concerning probability models and sample spaces. The first concerns the idea of *prototype experiments*. It is convenient to use a few simple examples again and again to illustrate various points. Often, these examples involve simple notions from games of chance, such as drawing a card from a deck, spinning a roulette wheel, drawing balls from an urn, or tossing a coin. These simple games, are, in themselves, of little interest, but it is convenient to use them to identify aspects of a real experiment that *are* of interest. Such simple gambling experiments then become important prototypes; a model for drawing balls from an urn may be useful in modeling a quality control application involving the sampling of items from a shipment lot, for example. It is useful to develop models for these prototype situations and to be able to use them "off the shelf" for a wide variety of seemingly different applications.

The second comment concerns the random nature of the experiments we are examining. The outcomes of these experiments cannot be predicted with certainty. All the experimental situations we have been discussing are of this nature, and such experiments are called *random experiments*. It might be argued that, in any of these experiments, if enough information were given, the outcomes could be predicted with certainty. Perhaps this statement is true; but even if such complex information were readily available, it is doubtful that the experimenter would be willing to undertake the complicated task (or perhaps pay the prohibitive cost) of processing it in order to predict the outcome. In some experiments, such as predicting the color of eyes of an unborn child, it may not be possible to predict with certainty because of a lack of sufficient knowledge of the phenomenon being investigated or because of the inadequacy of available measuring devices. The experiment is viewed as being performed *as if* it were not possible to predict the outcome with certainty; hence, the experiment requires a random model. The fact that it might be conceptually possible to remove the element of uncertainty is of no significance, insofar as the model is concerned, if, in fact, the uncertainty is not going to be removed. We are formulating models for experiments in which we cannot (at least at the present state of knowledge) predict exactly which of several alternatives will result. To amplify this point, we will often refer to "a future performance" of the experiment. In fact, the actual physical timing is not of importance. The essential notion is that of being able to "take an observation" on demand, without prior knowledge of the result.

In summary, then, once we have selected a suitable sample space S, each performance of the random experiment will result in exactly one outcome, whose description in S we agree to denote generically by o. By the very nature of randomness, we cannot say, prior to the performance of the experiment, what the particular value of o will be. In the following sections, we ask, "How likely is it that a particular outcome will occur?" It is important to note the *predictive* nature of the question. Any model devised to answer this question will, of necessity, be primarily concerned with conditions *prior to* the performance of the experiment. After the experiment has been performed, the outcome either is or is not described by the value of o in question, and this posterior information cannot answer the question that we posed.

EXERCISES

APPLICATION

1.1   Classify each of the following sets as finite, denumerable, or uncountable: (a) $[0, 1] = \{x : 0 \le x \le 1\}$; (b) $\{x : x$ is a rational number$\}$; (c) $\{i : i$ is an integer$\}$; (d) the set of even numbers; (e) the set of all functions of a real variable; (f) the set of all prime numbers; (g) the set of all subsets of $\{1, 2, 3, 4\}$.

1.2   Suppose our interest in the two-dice-rolling experiment (Example 1.3) is only in whether or not the two dice show the same number of spots. How might S be chosen?

1.3   Suppose that in the one-die-rolling experiment (Example 1.1) we are interested only in whether the outcome is "even" or "odd."
a. What is an acceptable sample space?
b. How does this experiment (and hence model) differ theoretically from that of tossing a coin?

1.4   a. How would you label the outcomes of an experiment that involves the tossing of three coins, a penny, a dime, and a quarter?
b. How would you label the outcomes of an experiment that consists of tossing a dime three times? *Note:* There are several ways to label such outcomes.
c. Determine sample spaces for the experiments in parts (a) and (b).

1.5   A manufactured lot of items consists of n items whose critical dimension is determined by some measuring device. Choose a sample space for the experiment that consists of drawing one item, measuring it, and classifying it as defective or nondefective.

1.6   An experiment consists of forming an arbitrary two-letter "word" from the English alphabet. Describe a suitable sample space S for this experiment. How many elements are in S?

1.7   Suppose a true-false test consisting of n questions is scored as the number right minus the number wrong. Describe a suitable sample space.

1.8   An experiment consists of standing on a certain street corner and observing the number of traffic accidents that occur during the noon hour. Determine a suitable sample space.

1.9   Choose a sample space for an experiment that consists of polling n people for their attitude on an issue, where the responses are classified only as "yes," "no," or "undecided."

1.10 Suppose S consists of the 26 letters of the English alphabet. Let $Q(*)$ denote "* is a vowel." List $\{* : Q(*)\}$.

1.11 A bookshelf contains 14 books. The covers of 4 of them are blue; 3 are green, 2 are orange, and 5 are red. The experiment consists of selecting 2 books from the shelf and noting their colors. Construct a sample space.

1.12 Ten married couples attend a party at which two door prizes are to be given by drawing two names from a hat. Construct a sample space that will describe the identities of the recipients of the prizes.

1.13 An auto repair shop has stocked four spares of a certain type of air conditioner. What is a suitable sample space to describe the number of spares used from its stock this year?

1.14 Write a sample space suitable to describe the number of accidents on all United States highways over the next Labor Day weekend.

1.15 A poker hand of 5 cards is to be dealt from an ordinary deck of 52 cards. Describe a sample space that will account for the cards in one hand.

1.3   EVENTS

Let us return to the one-die experiment (rolling a die and observing the number of dots on the upturned face). We agreed that $S = \{1, 2, 3, 4, 5, 6\}$ was a good choice for a sample space, and it is clear that the experiment is random. The basic idea behind probability is to quantify something about the outcome of the experiment, recognizing that it is not possible to say with certainty what the result will be. In the simple setting of our example, most people would agree that, if the die is fair, any one face is as likely to occur as any other when the die is rolled. With six faces, the chances are $\frac{1}{6}$ for any one of them; that constitutes all the information we have about a future toss of the die. If we were asked to say something about whether the outcome would be fewer than four dots, it would seem reasonable to assert there is an even chance of that happening, since this outcome occurs only if the result is a one, two, or three. There are three out of six ways this result can happen, and the even chance of an outcome less than four can be written as the sum of the chances for each of the three ways, $\frac{1}{6} + \frac{1}{6} + \frac{1}{6}$. Often in probability, one is interested in predicting or assessing the chances of circumstances made up of *collections* of individual outcomes. Assessing the chances of such collections of outcomes may involve evaluating the chances of each outcome individually and adding the results.

To be more formal, we define an *event* to be a collection or set of outcomes, that is, a collection of elements of S. Events are subsets of S typically denoted with capital letters. In the die-rolling experiment, the event described as "the outcome is even" can be written as the set $A = \{2, 4, 6\}$. Similarly, the set $B = \{1, 2, 3, 4\}$ could be described as the event "the outcome is no more than 4", and $C = \{3\}$ is the event "the outcome is 3." All these events are subsets of S, and we let $o \in E$ stand for the fact that the experimental outcome of a trial, o, is a member or element the event E. If the observed outcome is an element of an event E, we say that E occurred.

EXAMPLE 1.9   In the experiment of recording the time (in hours) it takes for a light bulb to burn out, we let $S = [0, \infty)$. The set $[10, 100]$ is a subset of S and represents the event "the time is between 10 and 100 hours, inclusive." The event "the light bulb survives at least 200 hours" would be denoted $[200, \infty)$.

EXAMPLE 1.10   A dime, a nickel, and a penny are tossed and the faces recorded as H or T in a triple, with the first coordinate reserved for the dime, the second for the nickel, and the third for the penny. The event A, "the faces on the dime and penny agree", could be written as $A = \{(H, H, H), (H, T, H), (T, H, T), (T, T, T)\}$, or, more exotically, as $A = \{(x_1, x_2, x_3) : x_1 = x_3\}$.

Since events are subsets of S and since sets enjoy certain special properties, we can exploit those relationships to build more complicated events out of given events. We now review some common terminology relating to events as sets.

Certain subsets of S, because of their special nature, play a special role as events. The *empty set* $\varnothing$, the set having no elements, is a subset of *every* set by default; hence, in particular, $\varnothing \subset S$. The outcome o of an experiment is never a member of $\varnothing$; that is, $o \in \varnothing$ is never true. As an event, $\varnothing$ never occurs and so is referred to as the *impossible event*. On the other hand, $S \subset S$, and thus S qualifies as an event in addition to being the sample space. As an event, S always occurs. Consequently, when S is viewed as an event, we call it the *certain event*. Events like $\{2\}$, $\{(H, T, H)\}$, and so on that contain a single outcome from the corresponding sample

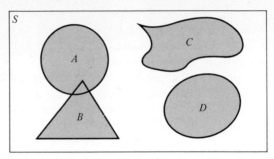

FIGURE 1.1   *Subsets of S Depicted in a Venn Diagram*

space are sometimes called *elementary events*. As we will soon see, they play an important role in making probability assignments.

It is sometimes very helpful to visualize events by means of Venn diagrams, as in Fig. 1.1. If A is a subset of S, so is the set $\overline{A}$ of all points of S not in A; the event $\overline{A}$ is called the *complement* of A. A Venn diagram of the complementary events A and $\overline{A}$ is shown in Fig. 1.2. Now, since $o \in \overline{A}$ if and only if $o \notin A$, we may say that the event $\overline{A}$ occurs if and only if A does not occur. For this reason, $\overline{A}$ is sometimes referred to as the event "not A."

If A and B are events, then so is the *union* of A and B, $A \cup B = \{O : o \in A$ or $o \in B\}$. The event $A \cup B$ occurs if and only if A occurs or B occurs; hence, it is appropriate to name $A \cup B$ the event "A or B." Such an event is depicted as the shaded area in Fig. 1.3. Similarly, the *intersection* of A and B, $A \cap B = \{o : o \in A$ and $o \in B\}$ occurs if and only if both A and B occur simultaneously (i.e., upon a single performance of the underlying experiment), so it is appropriate to name this event "A *and* B." The shaded area in Fig. 1.4 depicts the event "A and B."

EXAMPLE 1.11   Consider the single-die experiment with $S = \{1, 2, 3, 4, 5, 6\}$, $A = \{2, 4, 6\}$, and $B = \{1, 2, 3, 4\}$. Then the event "not even" is $\overline{A} = \{1, 3, 5\}$; $\overline{B} = \{5, 6\}$; the event "even or not more than four" is $A \cup B = \{1, 2, 3, 4, 6\}$; $\overline{A \cup B}$ is the elementary event $\{5\}$. Of course, $\overline{\varnothing} = S$ and $\overline{S} = \varnothing$. Finally, the event "even and not more than four" is $A \cap B = \{2, 4\}$.

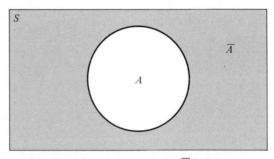

FIGURE 1.2   *Subsets A and $\overline{A}$ of S*

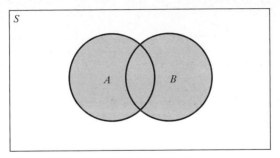

FIGURE 1.3   *A* ∪ *B*

If it should happen that $A \cap B = \emptyset$, as depicted in Fig. 1.5, then the disjoint sets *A* and *B* are called *mutually exclusive events*. It is impossible for such events to occur simultaneously. Note, for example, that complementary events (A and $\overline{A}$) are mutually exclusive. The event described by the sentence "*A* occurs, but *B* does not occur," symbolized by $A \cap \overline{B}$, is sometimes written $A - B$ and is called the *relative complement of B in A*. Note that $\overline{A} = S - A$. Such a set is depicted as the shaded area in Fig. 1.6. Note that, for any events *A* and *B*, the events *A* and $A \cap \overline{B}$ are mutually exclusive. It then follows that

$$A \cup B = A \cup (B \cap \overline{A}) = B \cup (A \cap \overline{B}),$$

which allows us to express the event "*A* or *B*" in terms of mutually exclusive events.

There are many variations of these themes, and some deserve special names. For example, $\overline{A} \cap \overline{B}$ occurs precisely when the experimental outcome *o* is such that $o \notin A$ and $o \notin B$. Hence, it is appropriate to refer to $\overline{A} \cap \overline{B}$ as the event "neither *A* nor *B*." The event "exactly one of *A*, *B*" could be displayed as $(A \cap \overline{B}) \cup (B \cap \overline{A})$. Similarly, $A \cup B$ could be referred to as "at least one of *A*, *B*," since $o \in A \cup B$ iff (if and only if) $o \in A$ or $o \in B$ (possibly both).

EXAMPLE 1.12 Consider once again the die-rolling experiment, and let $A = \{2, 4, 6\}$, $B = \{1, 2, 3, 4\}$, and $C = \{3\}$, as before. Suppose the experiment

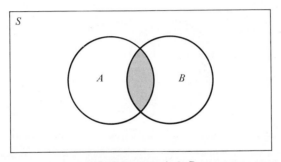

FIGURE 1.4   *A* ∩ *B*

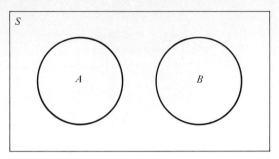

FIGURE 1.5   *Mutually Exclusive Events*

is performed, and the outcome $o = 3$ is observed. Then we can say the following events occurred: B, C, "B or C" (the event $\{1, 2, 3, 4\} \cup \{3\}$), "$A$ or C" (the event $\{2, 3, 4, 6\}$), and "B but not A" (the event $B \cap \bar{A} = \{1, 3\}$).

Some useful identities concerning operations with events are given in Table 1.1.

Most of the above considerations extend in a natural way to more than two events. For example, if A, B, and C are three events, then the union of A, B, and C, $A \cup B \cup C$, is the event "at least one of A, B, C." More generally, if $A_1, A_2, \ldots, A_k$ are k events, their union is the event

$$\bigcup_{i=1}^{k} A_i = \{o : o \in A_i \text{ for } some \ i \in \{1, 2, \ldots, k\}\},$$

also written $A_1 \cup A_2 \cup \cdots \cup A_k$ and described as the event "at least one of $A_1, A_2, \ldots, A_k$." Their intersection is the event

$$\bigcap_{i=1}^{k} A_i = \{o : o \in A_i \text{ for } every \ i \in \{1, 2, \ldots, k\}\},$$

also written $A_1 \cap A_2 \cap \cdots \cap A_k$ and described as the event "each of $A_1, A_2, \ldots, A_k$."

Figure 1.7 shows a general way of representing three sets A, B, and

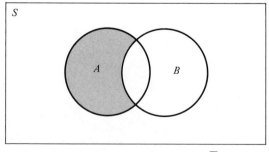

FIGURE 1.6 $A - B = A \cap \bar{B}$

TABLE 1.1   *Summary of Event Operations*

Commutative Laws:	$A \cup B = B \cup A$
	$A \cap B = B \cap A$
Associative Laws:	$A \cup (B \cup C) = (A \cup B) \cup C$
	$A \cap (B \cap C) = (A \cap B) \cap C$
Distributive Laws:	$A \cap (B \cup C) = (A \cap B) \cup (A \cap C)$
	$A \cup (B \cap C) = (A \cup B) \cap (A \cup C)$
De Morgan Laws:	$\overline{A \cup B} = \overline{A} \cap \overline{B}$
	$\overline{A \cap B} = \overline{A} \cup \overline{B}$

C with each pair having points in common. The shaded portion represents $A \cap B \cap C$.

If we have under consideration an event A_i for each positive integer i, the above notation can be extended to define the union and intersection of these events. Thus we have the following notation:

$$\bigcup_{i=1}^{\infty} A_i = \{o : o \in A_i \text{ for some } i\},$$

$$\bigcap_{i=1}^{\infty} A_i = \{o : o \in A_i \text{ for every } i\}.$$

The events $A_1, A_2, A_3, \ldots$ are said to be *mutually exclusive* if it is true that A_i and A_j are mutually exclusive events whenever $i \neq j$; we sometimes say that such a collection of events is *pairwise disjoint*. Sometimes it happens that, in addition to $A_1, A_2, \ldots, A_k$ being mutually exclusive, they are exhaustive, in the sense that $\bigcup_{i=1}^{k} A_i = S$. In this case, we say that the events $A_1, A_2, \ldots, A_k$ form a *partition* of S. A *simple partition* of $S = \{o_1, o_2, \ldots, o_n\}$ is the collection of all elementary events $\{o_1\}, \{o_2\}, \ldots, \{o_n\}$; there are many other possible partitions of S. (Recall our assumption that the members of S are distinct.) Any event A and its complement $\overline{A}$ clearly form a partition of S into two mutually exclusive events.

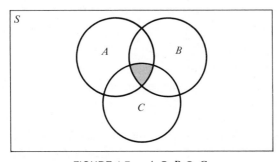

FIGURE 1.7 $A \cap B \cap C$

TABLE 1.2   *Glossary of Event Terminology*

Set Notation	Occurrence Translation
$o \in A \cup B$	At least one of A or B occurs (more simply, A or B occurs)
$o \in A \cap B$	Both A and B occur
$o \in \overline{A} \cap \overline{B} = \overline{A \cup B}$	Neither A nor B occurs
$o \in A \cap \overline{B}$	A occurs but B does not occur
$o \in (A \cap \overline{B}) \cup (\overline{A} \cap B)$	Exactly one of A or B occurs
$o \in A \cup B \cup C$	At least one among A, B, C occurs
$o \in \overline{A} \cap \overline{B} \cap \overline{C}$	None of the three events A, B, C occurs
$o \in (A \cap \overline{B} \cap \overline{C}) \cup (\overline{A} \cap B \cap \overline{C}) \cup (\overline{A} \cap \overline{B} \cap C)$	Exactly one of A, B, C occurs
$o \in (A \cap B \cap \overline{C}) \cup (A \cap \overline{B} \cap C) \cup (\overline{A} \cap B \cap C)$	Exactly two among A, B, C occur
$o \in (A \cap B) \cup (A \cap C) \cup (B \cap C)$	At least two among A, B, C occur
$o \in \overline{A \cap B \cap C}$	At most two among A, B, C occur
$o \in A \cap B \cap C$	All three of A, B, C occur
$o \in (\overline{A} \cap \overline{B}) \cup (\overline{A} \cap \overline{C}) \cup (\overline{B} \cap \overline{C})$	At most one among A, B, C occurs

Occurrences of events are often described in a natural way by phrases like "at least," "at most," and so on. A summary of some common phrases is given in Table 1.2.

EXERCISES

THEORY

1.16 Verify the following by using Venn diagrams or logical arguments:
a. $A \cap (B - A) = \varnothing$; $A \cup (B - A) = A \cup B$.
b. $A \cap B \subset A \subset A \cup B$.
c. $A \cup S = S$; $A \cap S = A$.
d. $A \cup \overline{A} = S$ and $A \cap \overline{A} = \varnothing$.
e. $B = (B \cap A) \cup (B \cap \overline{A})$.
f. $\overline{\overline{A}} = A$, where $\overline{A}$ is the complement of $\overline{A}$.

1.17 Use a Venn diagram to verify that if A, B, and C are events and $B \subset C$, then $A \cap B \subset A \cap C$.

1.18 Use set membership to argue that the following are valid:

a. $\displaystyle\bigcup_{i=1}^{k} A_i = \left(\bigcup_{i=1}^{k-1} A_i\right) \cup A_k$ and $\displaystyle\bigcap_{i=1}^{k} A_i = \left(\bigcap_{i=1}^{k-1} A_i\right) \cap A_k.$

b. $B \cap \left(\bigcup\limits_{i=1}^{k} A_i \right) = \bigcup\limits_{i=1}^{k} (B \cap A_i)$ and

$B \cup \left(\bigcap\limits_{i=1}^{k} A_i \right) = \bigcap\limits_{i=1}^{k} (B \cup A_i).$

c. $\overline{\bigcup\limits_{i=1}^{k} A_i} = \bigcap\limits_{i=1}^{k} \overline{A_i}$ and $\overline{\bigcap\limits_{i=1}^{k} A_i} = \bigcup\limits_{i=1}^{k} \overline{A_i}.$

1.19   Show that $A \subset B$ if and only if $\overline{B} \subset \overline{A}$.

1.20   Show that if $C_1, C_2, \ldots, C_k$ are mutually exclusive and A is any event, then $A \cap C_1, A \cap C_2, \ldots, A \cap C_k$ are also mutually exclusive.

1.21   Show that the last item in Table 1.2 can be equivalently written $o \in (\overline{A} \cup \overline{B})$ $\cap (\overline{A} \cup \overline{C}) \cap (\overline{B} \cup \overline{C}).$

APPLICATION

1.22   Let $S = \{1, 2, 3\}$. Form a list of all the subsets of S.

1.23   Repeat Exercise 1.22 for $S = \{1, 2, 3, 4\}$ and $S = \{1, 2, 3, 4, 5\}$. Can you guess how many subsets there are when $S = \{1, 2, 3, \ldots, n\}$?

1.24   Let $A = \{0, 1, 2, 3\}$ and $B = \{1, 2\}$ be events in $S = \{0, 1, 2, 3, 4, 5\}$. List $A \cap B, A \cap \overline{B}$, and $\overline{A} \cup B$.

1.25   Mark each of the following T if always true or F if false.
_____ a. $A \subset \overline{A} \cup B$.
_____ b. $(A \cup C) \cap B = (A \cap B) \cup (A \cap C)$.
_____ c. If $A \subset B$, then $A \cup \overline{B} \subset \overline{A}$.

1.26   Let A denote the event of throwing an odd total with two dice, and let B denote the event of throwing at least one six. Give a verbal description of each of the following events: (a) $A \cup B$; (b) $A \cap B$; (c) $A \cap \overline{B}$; (d) $\overline{A} \cap B$; (e) $\overline{A} \cap \overline{B}$; (f) $\overline{A} \cup B$.

1.27   For each item in column 1 in the accompanying table, select the item in column 2 that best matches it and place its number in the space provided. *Note:* To say an event C occurs means that the outcome o is in C.

Column 1	Column 2
_____ a. Event "A and B"	1. $\overline{A} \cap \overline{B}$
_____ b. At least one of A or B occurs	2. $o \in A \cap \overline{B} \cap C$
_____ c. Neither A nor B occurs	3. $A \cup B$
_____ d. Not more than one of A or B occur	4. $A \cap B$
_____ e. If A occurs, so does B	5. $A \cap B = \varnothing$
_____ f. A and B are mutually exclusive	6. $A \subset B$
_____ g. At least one of A, B, and C occurs	7. $A \subset \overline{B}$
_____ h. If A occurs, then B does not occur	8. $B \cap \overline{A}$
_____ i. Event "neither A nor B"	9. $o \in A \cup B$
_____ j. Event "B but not A"	10. $o \in \overline{A} \cap \overline{B}$
	11. $o \in \overline{A \cap B}$
	12. $o \in A \cup B \cup C$

1.28 Three high school friends all enlist in the Navy at the same time for identical
        hitches. At the end of his hitch, each man either will or will not reenlist.
        a. Define a sample space for the experiment that consists of observing the
           number among these three that do choose to reenlist.
        b. Let *A* be the event that at least two choose to reenlist, and let *B* be
           the event that at most one chooses to reenlist. Express *A* and *B* as subsets
           of *S*.

1.29 Two cards are to be drawn at once from a standard 52-card bridge deck.
        a. Define a sample space *S* for this experiment.
        b. Let *A* be the event that both cards drawn are red. Express *A* as a subset
           of *S*.

1.30 An automobile supply house has exactly four identical copies of a certain part.
        During a given week, they will supply parts on demand or back-order the part
        if more than four are demanded. Ignoring back orders (they are supplied by
        a different source), the experiment consists of observing how many parts are
        supplied in a given week.
        a. Define a sample space for the experiment.
        b. Let *A* be the event that at least three parts are supplied and *B* the event
           that no more than two are supplied. Express *A* and *B* as subsets of *S*.

1.31 A small company has three automobiles in its car pool. Each automobile may
        break down (only once) or not break down on a given day. The experiment
        consists of counting the number of automobile breakdowns to occur on a given
        day.
        a. Define a sample space for this experiment.
        b. Let *A* be the event that all automobiles break down and *B* the event that
           none break down. Express *A* and *B* as subsets of *S*.

1.4   PROBABILITY AXIOMS

We now consider the problem of measuring the likelihood of occurrence of
each event. One of the earliest approaches to assigning a numerical likelihood to an event *A* was to perform the experiment in question many times,
observing each time whether or not the event *A* occurred. In *n* such trials,
a count of the number of times *A* occurred, say n_A, could then be divided
by the number of trials, say *n*, to form the *relative frequency* n_A/n of the
occurrence of *A*. It was thought that the probability of *A*, *P(A)*, should then
be defined to be the limit of this relative frequency as *n* increased indefinitely:
$P(A) = \lim_{n \to \infty} n_A/n$. On the basis of the fact that many random phenomena,
while unpredictable, seem to display a type of regularity in a large number
of trials, this seems like a reasonable way to proceed. Unfortunately, it is
fraught with mathematical difficulties, such as the existence of the limit in
question. As a means of *defining* probability, it has to be abandoned.

It was mentioned earlier that all attempts to define probability have,
to date, failed for one reason or another. The philosophical and logical details

need not distract us here, but this failure does explain why we take an axiomatic approach. In an axiomatic system such as geometry, for example, we begin with some primitive, undefined objects such as "point," make some assumptions (called axioms) about how those objects are to behave, and proceed to derive properties (called theorems) that logically follow from those axioms. The utility of having such a system is that, if certain real objects are to be modeled by such a system, all that needs to be done is to identify the primitives and check that the axioms are at least reasonably satisfied by the physical nature of the real objects. The theorems then become properties that must be satisfied in the model—and at least reasonably well satisfied by the physical objects. It is not necessary to verify the theorems individually in the real world. Moreover, those properties are inescapable, to within the "fit" of the model. Consider, for example, a plane geometry model for a portion of the earth's surface. Once we identify a square piece of land x feet on a side, the model asserts it is inevitable that the diagonal will measure $\sqrt{2}x$ feet, and we expect a real world measurement would verify this prediction.

To be useful, an axiomatic system should not have too many axioms, for each of those properties must be verified in any real world assignment. The more axioms there are, the more difficult it will be to determine whether various physical systems satisfy the axioms. While theoreticians may enjoy axioms for the sake of axioms, it is desirable to have at hand some system that satisfies the axioms. Axiomatic systems are often invented by having some preconceived ideas of the desired output; probability theory is a good example. From this point of view, developers of an axiomatic system attempt to abstract, and distill to a minimum list, the essential features of a desired system. Individuals applying the system in a real world problem must concern themselves with verifying the axioms' "fit" in the problem context.

With these comments about models in mind, let us return to the notion of relative frequency. Ignoring for the moment the mathematical pitfalls, what would be the essential features of relative frequency regarding events, if probability could be thusly defined? Relative frequencies, by their very nature, are always numbers between 0 and 1 inclusive, that is, $0 \le n_A/n \le 1$. If the limit did exist, it would also be true that $0 \le \lim_{n \to \infty} n_A/n \le 1$. This feature at least provides a convenient number scale on which to compare probabilities of various events. It is clear that $n_\varnothing/n = 0$ and $n_S/n = 1$, establishing the boundary points. Every time A occurs, $\overline{A}$ does not. Therefore, the relative frequency of $\overline{A}$ must be $(n - n_A)/n$, and so $n_{\bar{A}}/n = 1 - n_A/n$. If A and B are mutually exclusive, they never occur at the same time, so it must be true that

$$n_{A \cup B} = n_A + n_B,$$

so, in turn, $$\frac{n_{A \cup B}}{n} = \frac{n_A}{n} + \frac{n_B}{n} \qquad \text{if} \qquad A \cap B = \varnothing,$$

and the same would be true of limits, if they existed.

Instead of listing more properties of relative frequencies, let us turn now to an axiomatic system for probabilities. Since relative frequencies behave the way probabilities should behave, we are anxious to verify that the Kolmogorov axioms about to be introduced do capture this behavior. Briefly consider the games-of-chance setting again. We have already noted that, in situations such as the single-die experiment, we imagine one outcome to be just as likely as any other. One would thus assess the chances of occurrence of an event by dividing the number of favorable possibilities (outcomes in the event) by the total number of outcomes in S. If A is the event "an even number of dots" in the die experiment, the probability of A is $\frac{3}{6}$. Clearly, such a method shares a great deal with the relative frequency approach. Such "probabilities" would all lie between 0 and 1. If k out of m cases are favorable to A, then $m - k$ are not, so the "probability" of $\overline{A}$ is $(m - k)/m = 1 - k/m$, and so on.

We seek a set of axioms that will abstract those properties. The Kolmogorov system is such a set of axioms. The list of axioms is deceptively short, considering the wealth of applications that can be made of the resulting system. For any sample space S with events defined as subsets of S, probability is assumed to be a real-valued function of events, denoted P [or $P(\cdot)$] satisfying the following *Kolmogorov axioms*:

Axiom 1.1 $P(A) \geq 0$ for every event A.
Axiom 1.2 $P(S) = 1$.
Additivity Condition If A and B are mutually exclusive, then $P(A \cup B) = P(A) + P(B)$.

It will usually be easy to verify whether the first two axioms hold for any proposed method of assignment. The additivity condition is usually not so easy to check, for we might have to conceive of every possible pair of mutually exclusive events to verify it. Also, the additivity condition, as stated, is not quite enough if S is infinite. Thus we state the third Kolmogorov axiom in the more general case:

Axiom 1.3.   If $A_1, A_2, A_3, \ldots$ are mutually exclusive, then $P(\bigcup_{i=1}^{\infty} A_i) = \sum_{i=1}^{\infty} P(A_i)$.

If S is finite, the infinite series in Axiom 1.3 isn't necessary; the pairwise additivity condition is equivalent. Henceforth, we refer to Axiom 1.3 as the additivity condition even when S is finite.

Note the axioms are indeed satisfied by equal-likelihood (games-of-chance) assignments. In the equal-likelihood method of assigning probabilities, $P(A) = n(A)/N$ if there are N possibilities (points in S) altogether and $n(A)$ is the number of points in the subset A of S. Also, if we elect to specify $P(A) = n_A/n$, the relative frequency of occurrence of A in n trials, the Kolmogorov axioms are again satisfied. This is true whether $n = 10$, 100, or $1,000,000$ trials were performed.

The beauty of the Kolmogorov system is that it allows one to make probability assignments in a variety of ways and still have a consistent system. One needs only to verify that the axioms are satisfied for a given assignment. Once the axioms are verified, the set of results, theorems, will automatically follow. It is rather surprising that the list of basic theorems we use in most of our applications is so short. In the list of theorems that follows, it is understood that events A, B, A_1, A_2, . . . , A_k are arbitrary events.

Theorem 1.1 $P(\varnothing) = 0$.

Theorem 1.2 $P(A) \leq 1$.

Theorem 1.3 $P(B \cap \overline{A}) = P(B) - P(A \cap B)$.

Theorem 1.4 $P(\overline{A}) = 1 - P(A)$.

Theorem 1.5   If $A \subset B$, then $P(A) \leq P(B)$.

Theorem 1.6 $P(A \cup B) = P(A) + P(B) - P(A \cap B)$.

Theorem 1.7   If $A_1, A_2, . . . , A_k$ are mutually exclusive, then $P(\cup_{i=1}^{k} A_i) = \sum_{i=1}^{k} P(A_i)$.

EXAMPLE 1.13   Suppose S is a square with sides of length one unit, and imagine the events A, B, and so on, are regions within the square. Suppose the probability of an event is taken to be the area of the region. Verify that the axioms are satisfied, and give interpretations of Theorems 1.1–1.7.

SOLUTION   Axiom 1.1 asserts that any area is nonnegative; Axiom 1.2 asserts that the area of the unit square is 1; and Axiom 1.3 asserts that the combined area of several disjoint regions is the sum of individual areas. These interpretations certainly appear reasonable, and they constitute verification that the axioms are satisfied by this proposed assignment of P. Thus Theorems 1.1–1.7 must hold for areas of regions within S. Stated in terms of this application, the theorems state the following:

1. The area of the empty region is zero.

2. The area of a region within the unit square cannot exceed the area of the square itself.

3. The area of $B - A$ is the area in B less the area of $A \cap B$ (draw a Venn diagram).

4. The area (in the square) outside A is 1 minus the area of A.

5. If A is inside B, the area of A cannot exceed that of B.

6. The combined area of (possibly overlapping) regions A and B is the area of A plus that of B, less the area of the overlap (which has been counted twice—once in A and once in B).

7. The combined area of k disjoint regions taken together is the sum of the k individual areas.

We now sketch arguments for proofs of these theorems. In most cases, they are quite simple and will provide you with ideas that can be used in later extensions. A legitimate assignment of mutually exclusive events in Axiom 1.3 is to let $A = \varnothing$ and $B = \varnothing$. Accordingly, $P(\varnothing) = P(\varnothing) + P(\varnothing) = 2P(\varnothing)$. But by Axiom 1.1, $P(\varnothing) \geq 0$, so the only real solution to the equation $2P(\varnothing) = P(\varnothing)$ is $P(\varnothing) = 0$. That takes care of Theorem 1.1. By letting $B = \overline{A}$ in Axiom 1.3 and observing that $A \cup \overline{A} = S$, we obtain, using Axiom 1.2, $1 = P(A) + P(\overline{A})$. From this result, Theorem 1.4 follows immediately; and by observing that $P(\overline{A}) \geq 0$ by Axiom 1.1, we see that Theorem 1.2 also follows from this result. If we note that $A \cap B$ and $B \cap \overline{A}$ are mutually exclusive events, with $(A \cap B) \cup (B \cap \overline{A}) = B$ [see Exercise 1.16(e)], then a direct substitution in Axiom 1.3 yields $P(B) = P(A \cap B) + P(B \cap \overline{A})$, and Theorem 1.3 follows. Using the result of Exercise 1.16(a), together with Axiom 1.3, we have $P(A \cup B) = P(A) + P(B \cap \overline{A}) = P(A) + P(B) - P(A \cap B)$ from the result just established, which verifies Theorem 1.6. If $A \subset B$, as stated in Theorem 1.5, then $A \cap B = A$, and Theorem 1.3 becomes $P(B \cap \overline{A}) = P(B) - P(A)$. But $P(B \cap \overline{A}) \geq 0$, so $P(B) \geq P(A)$, establishing Theorem 1.5.

The proof of Theorem 1.7 employs mathematical induction. Utilizing Exercise 1.18(a), we may "split" $\cup_{i=1}^{k} A_i$ into the union of *two* sets, $(\cup_{i=1}^{k-1} A_i) \cup (A_k)$. It is easy to argue that A_k and $(\cup_{i=1}^{k-1} A_i)$ are two mutually exclusive sets given that $A_1, A_2, \ldots, A_k$ are mutually exclusive. Then Axiom 1.3 applies and yields

$$P\left(\bigcup_{i=1}^{k} A_i \right) = P\left(\bigcup_{i=1}^{k-1} A_i \right) + P(A_k).$$

Starting with $k = 2$, the theorem is a reaffirmation of Axiom 1.3. Using the inductive assumption on $P(\cup_{i=1}^{k-1} A_i)$, we see that the result follows.

We close with one more theorem of interest in the material to follow. It has been asserted that if $C_1, C_2, \ldots, C_k$ constitutes a partition of S, then for any event A, $A \cap C_1, A \cap C_2, \ldots, A \cap C_k$ are mutually exclusive (see Exercise 1.20). In Fig. 1.8, we show a case where $k = 9$ and it happens that $A \cap C_1$ as well as $A \cap C_9$ are empty. Using the result of Exercise 1.18(b) and Theorem 1.7, we have

$$A = A \cap S = A \cap \left(\bigcup_{i=1}^{k} C_i \right) = \bigcup_{i=1}^{k} (A \cap C_i),$$

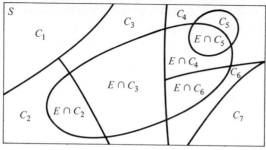

FIGURE 1.8   *Partition of S*

from which we obtain the *law of total probability* that follows:

Theorem 1.8   If $C_1, C_2, \ldots, C_k$ is a partition of S, then for any event A, $P(A) = \sum_{i=1}^{k} P(A \cap C_i)$.

EXAMPLE 1.14   For the assignment of P in terms of areas discussed in Example 1.13, imagine that $C_1, C_2, \ldots, C_k$ are maps of k states that make up the nation S. Now for any arbitrary region A (say a national forest) within the nation, the area of A can be obtained by summing the areas of the portions of A within each state.

In addition to being theoretical consequences of the axioms, these theorems provide formulas that may allow computation of probabilities of some events, given the probabilities of others. This application is illustrated in the exercises that follow.

EXERCISES

THEORY

1.32 a. Prove that if A, B, and C are events, then

$$P(A \cup B \cup C) = P(A) + P(B) + P(C) - P(A \cap B) - P(A \cap C)$$
$$- P(B \cap C) + P(A \cap B \cap C).$$

b. State and prove a similar result for events A, B, C, and D.

1.33 Assume A and B are events and $P(A)$, $P(B)$, and $P(A \cap B)$ are given. Find formulas for the probabilities of the following events: (a) exactly one of A, B; (b) neither A nor B; (c) at least one of A, B; (d) A but not B.

1.34 Show that if $P(A) = \frac{3}{4}$ and $P(B) = \frac{3}{8}$, then A and B cannot be mutually exclusive. Formulate a general statement in this direction (a good guess is appropriate).

1.35 Show that if A and B are events, then $P(A \cap B) \leq P(A)$ and $P(A \cap B) \leq P(B)$; hence, $P(A \cap B) \leq \min\{P(A), P(B)\}$.

1.36 Show that if A and B are events, then $P(A \cup B) \geq P(A)$ and $P(A \cup B) \geq P(B)$; hence, $P(A \cup B) \geq \max\{P(A), P(B)\}$.

1.37 Show that if A and B are events, then $P(A \cap B) \geq P(A) + P(B) - 1$; hence, $P(A \cap B) \geq \max\{0, P(A) + P(B) - 1\}$.

1.38 If A and B are events, show that $P(A - B) \geq P(A) - P(B)$; hence, $P(A - B) \geq \max\{0, P(A) - P(B)\}$.

1.39 Show that if A and B are events, $A \subseteq B$, and $P(B) = 0$, then $P(A) = 0$.

1.40 If A and B are events and either $P(A) = 0$ or $P(B) = 0$, show that $P(A \cap B) = 0$.

1.41 Suppose A, B, and C are events. What is an expression for the probability that none of the events occurs?

1.42 If A and B are events, show that $P(A) \geq P(A - B)$.

1.43 If $A_1, A_2, \ldots, A_k$ are events, show that $P(\cup_{j=1}^{k} A_j) \leq \Sigma_{j=1}^{k} P(A_j)$.

1.44 Suppose a sample space consists of n points, $S = \{a_1, \ldots, a_n\}$. Moreover, suppose probabilities have been assigned to just the elementary events $\{a_1\}$, $\{a_2\}, \ldots, \{a_n\}$, where, say, $P(\{a_i\}) = p_i$.
   a. Show that $\Sigma_{i=1}^{n} P_i = 1$. Hint: $\{a_1\} \cup \{a_2\} \cup \cdots \cup \{a_n\} = S$.
   b. Show that for any nonempty event E, the probability assigned to E is determined through the p_i's assigned to the events $\{a_i\}$ by the formula $P(E) = P(\{a_{j_1}\}) + P(\{a_{j_2}\}) + \cdots P(\{a_{j_m}\})$, where $E = \{a_{j_1}\} \cup \{a_{j_2}\} \cup \cdots \cup \{a_{j_m}\}$ for some $m \leq n$.

1.45 Prove the following extension of Theorem 1.8: Suppose A and B are events such that $A \subseteq B$ and suppose, moreover, that $B = B_1 \cup B_2 \cup \cdots \cup B_k$, where the B_i's are mutually exclusive. Then $P(A) = \Sigma_{i=1}^{k} P(A \cap B_i)$. Hint: $B_1, B_2, \ldots, B_k, \overline{\cup_{j=1}^{k} B_j}$, is a partition of S.

1.46 Show that, in the notation of Axiom 1.3, the sequence $\Sigma_{j=1}^{n} P(A_j)$ is nondecreasing as n increases. Note: This result shows that the series expressed in Axiom 1.3 either converges or else diverges to ∞.

APPLICATION

1.47 Suppose $P(A) = \frac{1}{2}$, $P(B) = \frac{1}{8}$, $P(C) = \frac{1}{4}$, where A, B, and C are mutually exclusive. Determine the values of the following:
   a. $P(A \cup B)$.   b. $P(A \cup B \cup C)$.   c. $P(A - B)$.
   d. $P(B \cap \overline{A})$.   e. $P(\overline{A} \cap \overline{B})$.
   f. The probability that exactly one of the events A, B, and C will occur.
   g. The probability that at least one of the events, A, B, and C will occur.
   h. The probability that none of the events A, B, and C will occur.

1.48 Evaluate the probabilities in Exercise 1.47 when $P(A) = \frac{1}{2}$, $P(B) = \frac{1}{8}$, and $P(C) = \frac{1}{4}$, but A and B are not mutually exclusive, say $P(A \cap B) = \frac{1}{16}$.

1.49 The sample space is $S = \{1, 2, 3\}$. Is it possible to have a probability function with values $P(\{1, 2\}) = \frac{1}{2}$, $P(\{2, 3\}) = \frac{1}{3}$, and $P(\{1\}) = \frac{1}{2}$? Explain your answer.

1.50 Given an experiment in which $P(A) = P(B) = \frac{1}{3}$ and $P(A \cap B) = \frac{1}{10}$. Compute the following: (a) $P(\overline{A})$; (b) $P(\overline{A} \cup B)$; (c) $P(A \cap \overline{B})$; (d) $P(\overline{A} \cup \overline{B})$.

1.51 A bowl contains several slips of paper. Each slip has either a 1, 2, or 3 printed on it (exactly one digit on each slip). The probability is $\frac{1}{2}$ that the number drawn is at least 2 and the probability is $\frac{5}{6}$ that the number drawn is at most 2. Compute the probability that the number drawn is the following: (a) 1, (b) 2, and (c) 3.

1.52 A clerk is sent out to buy a light bulb. The wattage of the light bulb the clerk returns with will be 40, 60, or 75. Let A be the event that the wattage of the bulb he returns with is at least 60 and B the event that it is not 60. Suppose $P(A) = \frac{3}{5}$ and $P(B) = \frac{2}{5}$. Find the probability that he returns with a bulb whose wattage is (a) 75, (b) 40, and (c) 60.

1.53 Given $P(A) = \frac{1}{4}$, $P(B) = \frac{1}{3}$, and $P(A \cup B) = \frac{1}{2}$, determine the following: (a) $P(A \cap B)$; (b) $P(\overline{A} \cap \overline{B})$; (c) $P(A \cup \overline{B})$; (d) $P(B \cap \overline{A})$.

1.54 A sample space for a random experiment is $S = \{a, b, c\}$. A proposed model assigns values $P(\{a, b\}) = .9$, $P(\{b, c\}) = .9$, and $P(\{a, c\}) = .1$. Are these consistent with the axioms?

1.55 A memo addressed to two people is circulated by the originator. Assume it will eventually be received by 0, 1, or 2 of the addressees. The probability that it will be received by at least one addressee is .9; the probability that it will be received by at most one is .8. What is the probability that the number of addressees to receive the memo is (a) 0, (b) 1, and (c) 2?

1.56 A family drives a camper across the United States. Assume that the number of flat tires they will have during the trip is 0, 1, or 2. Suppose the probability that they will have at most one flat is .9 and that .2 is the probability that they will have at least one flat. Calculate the probability that the number of flat tires they will have is (a) 0, (b) 1, and (c) 2.

1.57 Suppose that a study of 900 college graduates 25 years after graduation revealed that 300 were "successes," 300 had studied probability theory in college, and 100 were both "successes" and students of probability theory. Find, for $k = 0, 1, 2$, the number of persons in the group who were, of these two things ("students of probability" and "successes"), (a) exactly k, (b) at least k, and (c) at most k.

1.58 a. Assume that 60% of the employees of a certain firm are women and 70% are married. Determine the percentage *at least* that are married women.
   b. Suppose 75% of the employees of the same firm have blond hair. What percentage *at least* of the employees are blond women?
   c. What percentage *at least* are married blond women?

1.59 The mayor of a village had a feeling that some of the residents were cheating the village of tax payments. According to the laws of the village, a tax had to be paid by any homeowner whose home satisfied three criteria: at least 1500 square feet of space ("large"), at least two bathrooms ("convenient"), and at least three bedrooms ("comfortable"). The mayor knew that no resident would admit guilt, and so she decided on a trick scheme. She inquired of the 2000 homeowners in the village how many had houses that were large; convenient; comfortable; large and convenient; large and comfortable; convenient and comfortable. She knew that she could expect honest answers to these questions, the results of which are given in the accompanying table. It was observed that every homeowner admitted owning a house having at least one of the properties. How many homeowners should be paying taxes? *Hint:* Assume that the mayor knew a little about the algebra of sets.

Property	Number of Homeowners Owning Property
Large	1100
Convenient	1200
Comfortable	1600
Large and comfortable	900
Convenient and comfortable	900
Large and convenient	700

1.60 Suppose the sample space is $S = \{1, 2, 3, \ldots, 25\}$, and P satisfies

$$P(\{i\}) = \begin{cases} 0 \text{ if } i \text{ is odd,} \\ .1 \text{ if } i \text{ is even and less than 10,} \\ .05 \text{ if } i \text{ is even and } 10 \le i \le 22, \\ ? \text{ if } i = 24. \end{cases}$$

Find the probability of each of the following events: (a) $\{24\}$; (b) $\{1, 2, 5, 7, 10\}$; (c) $\{i \in S : i^2 > 4\}$; (d) $\{i : i \text{ is prime}\}$; (e) $\{i : i \text{ is divisible by 3}\}$; (f) $\{i : i > 0\}$; (g) $\{i : i < 0\}$.

1.61 Is it possible to have $P(A) = 0$ even if $A \ne \emptyset$? *Hint:* Consider the event $\{1, 3, 5\}$ in Exercise 1.60.

1.5 CONDITIONAL PROBABILITY; BAYES'S THEOREM

Often, the likelihood of occurrence of an event need not be computed in the complete absence of information about the outcome o; rather, it may be known that one or more other events did occur (did contain o, whatever it was). For example, the probability that an individual sampled from some population favors a particular social issue might materially be affected by the *additional* information that the person being polled is over 35 years of age. On the other hand, age may not tell us anything new, in which case we would tend to say attitude and age are "independent." There is a way to make these ideas formal within a given probability structure.

To keep things simple, consider the die experiment with $S = \{1, 2, 3, 4, 5, 6\}$. Guided by Exercise 1.44 and the assumed fairness of the die, we suppose each of the six elementary events has probability $\frac{1}{6}$. Let $B = \{2, 4,$

6} be the event the die is even and $A = \{2, 4, 5\}$ the outcome is a 2, 4, or 5. Then, $P(B) = \frac{1}{2}$ and $P(A) = \frac{1}{2}$. Now, given that the event B occurs, should the probability that A occurs be changed from the value $\frac{1}{2}$? It would seem so, for given B occurs, it is not even possible for the outcome to be 5. Thus, given B occurs, A will only occur if 2 or 4 is the outcome. This result is two out of the three possibilities, so one would change the probability for A from $\frac{1}{2}$ to $\frac{2}{3}$. It is as though $B = \{2, 4, 6\}$ is a new sample space and all the other events are evaluated according to their joint occurrence with B. Thus, for example, given B occurs, the event $D = \{1, 3\}$ should have new probability zero since $D \cap E = \phi$, while the new probability of B should be 1.

What is needed is a new concept, consistent with our present formulation, that will provide a satisfactory answer to the question, "What is the probability A occurs given B occurs?" For a fixed event B, we want to be able to answer the question for each and every event A. To do so, we need to define a new probability function that depends on B as well as P. The standard way of defining such a new probability function is given in the following definition:

Definition 1.2   If S is a sample space, P is a probability, and B is an event in S with $P(B) > 0$, then for every event A in S, the *conditional probability* of A *given* B is defined by

$$P(A|B) = \frac{P(A \cap B)}{P(B)}.\qquad(1.1)$$

It is important to note that B is *fixed* and that $P(B) > 0$. If $P(B) = 0$, then $P(A|B)$ is not defined. The event A is allowed to be any event contained in the sample space. As a function of A, with B fixed, it is then possible to verify that this new probability does in fact satisfy the Kolmogorov axioms. The first axiom is automatic, and, since $S \cap B = B$, we have $P(S|B) = 1$. Also, if E and F are mutually exclusive events, then $(E \cup F) \cap B = (E \cap B) \cup (F \cap B)$, with $(E \cap B)$ and $(F \cap B)$ also mutually exclusive according to Exercise 1.20. Accordingly,

$$P(E \cup F|B) = \frac{P(E \cap B) + P(F \cap B)}{P(B)} = P(E|B) + P(F|B).$$

Thus, all the probability theorems apply as well to conditional probabilities.

EXAMPLE 1.15   In terms of Venn diagrams, this definition can be viewed as follows. Suppose B and A are subsets of S as shown in Fig. 1.9, and

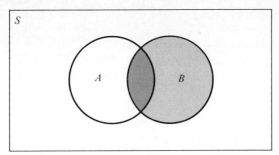

FIGURE 1.9   *Conditional Probability of A, Given B, Viewed as That*
              Proportion of the Area of B Belonging to A

suppose we imagine that the original probabilities of B and A are given by the areas of the corresponding regions. Then the conditional probability of A, given B, is the ratio of the area of A that is contained in B to the area of B.

To distinguish between $P(A|B)$ and $P(A)$, we sometimes refer to the latter as the *unconditional probability* of the event A. Even though Definition 1.2 provides a formal definition, very often the entire sample space and P are not given; rather, just the appropriate probabilities for $A \cap B$ and B are given. In that case, Eq. (1.1) may be viewed as a computational formula (as may any of the theorems) provided the event B, called the *conditioning event*, is held fixed throughout. Or $P(A|B)$ and $P(B)$ may be given when it is required to calculate the probability of the joint occurrence of A and B. According to Eq. (1.1), this probability would be the product $P(A \cap B) = P(A|B) \cdot P(B)$.

EXAMPLE 1.16 In a small community, 30% of the voting population is Republican and 80% of these voters favor a tax reduction. What proportion of this community is both Republican and in favor of the reduction?

SOLUTION Interpreting proportions as probabilities and letting R stand for Republican and F for favoring reduction, we could translate the given conditions into $P(R) = .3$ and $P(F|R) = .80$. Accordingly, $P(F \cap R) = (.3)(.80) = .24$ answers the question.

EXAMPLE 1.17 They say bond values rise whenever interest rates fall, not with certainty, but with probability .8. If there is a 70% chance that interest rates will fall tomorrow, how likely is it that bonds will rise and interest rates will fall?

SOLUTION Let B denote the event "bond values rise" and F the event "interest rates fall." Then the question is one of evaluating $P(B \cap F)$. But from the given information, $P(F) = .70$ and $P(B|F) = .8$. Hence, $P(B \cap F) = P(B|F) \cdot P(F) = .56$.

EXAMPLE 1.18   In the context of Example 1.17, what is the probability that bond prices will drop tomorrow?

SOLUTION   Since this event is just $\overline{B}$, we may use Theorem 1.4 with F fixed to calculate $P(\overline{B}|F) = 1 - P(B|F) = .2$.

The way Eq. (1.1) is used in Examples 1.16 and 1.17 to calculate the probability of intersections can be generalized to three or more events. Suppose we start with three events, A, B, and C. If you think of $B \cap A$ as a single event to begin with, $P(C \cap B \cap A) = P(C|B \cap A)P(B \cap A)$ would be one application of the formula. Now break $P(B \cap A)$ into $P(B|A)P(A)$ as a second application of the formula to obtain

$$P(A \cap B \cap C) = P(A)P(B|A)P(C|A \cap B).$$

This formula can be generalized by induction to provide a formula for n events, sometimes referred to as the *general multiplication rule*:

$$P(A_1 \cap A_2 \cap \cdots \cap A_n)$$
$$= P(A_1)P(A_2|A_1)P(A_3|A_1 \cap A_2) \cdots P(A_n|A_1 \cap A_2 \cdots \cap A_{n-1}). \quad (1.2)$$

EXAMPLE 1.19   Suppose five cards are drawn one at a time from an ordinary deck. What is the probability that all are red?

SOLUTION   Let R_i denote the event of selecting a red card on the ith draw, so $P(R_1) = \frac{26}{52} = \frac{1}{2}$, under the usual assumptions about shuffling the cards such that each card in the deck has an equal likelihood of being drawn. Now, given A_1, there are 51 cards left, of which 25 are red, so it is reasonable to write $P(A_2|A_1) = \frac{25}{51}$. Likewise, $P(A_3|A_1 \cap A_2) = \frac{24}{50}$, and so on. By the multiplication rule, with $n = 5$,

$$P(A_1 \cap A_2 \cap A_3 \cap A_4 \cap A_5)$$
$$= P(A_1)P(A_2|A_1)P(A_3|A_1 \cap A_2)P(A_4|A_1 \cap A_2 \cap A_3)$$
$$\cdot P(A_5|A_1 \cap A_2 \cap A_3 \cap A_4)$$
$$= \tfrac{26}{52} \cdot \tfrac{25}{51} \cdot \tfrac{24}{50} \cdot \tfrac{23}{49} \cdot \tfrac{22}{48} = .0253.$$

Because probabilities of events of the form $A \cap C_i$ may be written now in terms of conditional probabilities, the law of total probability [$P(A) = \Sigma P(A \cap C_i)$] may be written

$$P(A) = \sum_{i=1}^{k} P(A|C_i)P(C_i), \quad (1.3)$$

where $C_1, C_2, \ldots, C_k$ is a partition of S. Naturally, which form of this law is more useful depends on what is known. An interesting application known as Bayes's theorem can be used when $P(A)$ can be computed by using

Eq. (1.3). Suppose we wish to calculate $P(C_j|A)$ for any member C_j of the partition, when $P(C_j \cap A)$ is not given. By definition, $P(C_j|A) = P(C_j \cap A)/P(A)$. We know that $P(A \cap C_j) = P(A|C_j)P(C_j)$ is the same as $P(C_j \cap A)$, so we may write

$$P(C_j|A) = \frac{P(A|C_j)P(C_j)}{\sum_{i=1}^{k} P(A|C_i)P(C_i)} \qquad \text{for each } j = 1, 2, \ldots, k. \qquad (1.4)$$

This result is known as Bayes's theorem. In this context, the members of the partition are sometimes referred to as *causes* or *hypotheses,* and $P(C_j|A)$ is called the *inverse* or *a posteriori probability* (being inverse to the given, or *a priori,* conditional probabilities $P(A|C_j)$).

EXAMPLE 1.20   Two manufacturing companies, M_1 and M_2, produce a certain unit used in an assembly plant. Company M_1 is larger than M_2, and it supplies the plant with twice as many units per day as M_2. Unfortunately, M_1 also produces more defects than M_2. Because of past experience with these suppliers, it is felt that 10% of M_1's units have some defect, whereas only 5% of M_2's units are defective. Suppose that a randomly selected unit has been observed to be defective. What is the probability that it came from M_1?

SOLUTION   If we let D denote the event of observing a defective unit, and if we let C_1 and C_2 denote the events that the selected unit was produced by M_1 and M_2, respectively, the question requires an evaluation of $P(C_1|D)$. If we assume that M_1 and M_2 are the only manufacturers of this type of unit, then it is safe to assume $P(C_1) = \frac{2}{3}$ and $P(C_2) = \frac{1}{3}$. Also, $P(D|C_1) = .10$ and $P(D|C_2) = .05$. We then have all the required inputs for Bayes's theorem, with C_1 and C_2 constituting the partition into causes. Accordingly, $P(D) = P(D|C_1)P(C_1) + P(D|C_2)P(C_2) = \frac{1}{12}$ is the denominator, and $P(C_1|D) = (.10)(\frac{2}{3})(12) = .80$ is the answer.

It would make life easier if the value of $P(C_j|B)$ were insensitive to what the a priori assumptions are. That this is not so in general can be seen by returning to Example 1.20. In our solution, we argued that, with no additional information, the a priori probability of C_1 was $\frac{2}{3}$. But suppose it happens that the assembly plant is located near M_2, so even though M_2 is smaller, it usually supplies far more units to the particular warehouse at the assembly plant where the experiment is being performed. Relying on discussions with plant personnel, suppose we determine that now $P(C_1) = .10$ instead of $\frac{2}{3}$. The solution to the question posed in Example 1.20 is now $P(C_1|D) = .18$. This result would tend to reverse our judgment on the source of the defective unit.

Our example thus shows that any inference made on the basis of the a posteriori probabilities in Bayes's theorem is only as valid as the information that determined the given probabilities. Moreover, nothing within the theory will determine these probabilities; consequently, the burden of supplying valid numerical quantities rests on the practitioner. Unfortunately, in some cases in which Bayes's theorem is applied, not enough attention is paid to this point, and a "garbage in, garbage out" situation ensues.

EXERCISES

THEORY

(Assume that all conditioning events have positive probability.)

1.62 Prove that if $B \subset A$, then $P(A|B) = 1$, while if $A \subset B$, then $P(A|B) = P(A)/P(B)$; hence, $P(A|A) = 1$.

1.63 If A, B, and C are three events, show that $P(C|A \cap B)P(A|B) = P(A \cap C|B)$.

1.64 If A, B, and C are events, show that $P(C|A \cap B)P(A|B) = P(A \cap C|B)$.

1.65 a. Show that if A and B are disjoint events, then $P(A|B) = 0$.
      b. Is the converse of the statement in part (a) true? *Hint:* Do not confuse "zero probability" with "having no elements."

1.66 Verify that $P(A|S) = P(A)$ for every event A.

1.67 Prove that $P(A|B) = P(A)$ if and only if $P(\overline{A}|\overline{B}) = P(\overline{A})$.

1.68 Suppose that $P(B) = 1$. Show that $P(A|B) = P(A)$ for every event A.

1.69 Construct an example to show that it is not necessarily true that $P(A|B) = 1 - P(A|\overline{B})$.

1.70 Suppose A and B are mutually exclusive. Determine expressions for $P(A|A \cup B)$ and $P(B|A \cup B)$.

1.71 In the context of Bayes's theorem, show that $\sum_{j=1}^{k} P(C_j|A) = 1$.

APPLICATION

1.72 In a drawing of five cards, one at a time, from an ordinary deck, what is the probability that all are spades?

1.73 A course grade is based on the results of three tests. A student figures that his chance of getting an A on the first test is 75%, but that if he does, his chance of an A on the second test increase to 90%. Finally, he believes that if he does get an A on the first two tests, he will almost surely (99%) get an A on the third test. What is the probability that this student gets all A's in the course?

1.74 In the game of craps, what is the probability of 7 or 11, given that an odd sum is rolled? (Assume that two fair dice are rolled and that the outcome is the sum of the spots showing.)

1.75 In the manufacturing companies example (Example 1.20), what must $P(C_2)$ be in order for $P(C_2|D)$ to be exactly $\frac{1}{2}$ when $P(D) = .005$?

1.76 In a certain locality, the conditional probability that a clear day (C_2) follows a clear day (C_1) is given by $P(C_2|C_1) = .9$, so the conditional probability that a foul day follows a clear day is given by $P(\overline{C}_2|C_1) = .1$. Similarly, the probability that a clear day follows a foul day is assumed to be $P(C_2|\overline{C}_1) = .5$. Now suppose that, at this time of year, the probability of a clear day is $P(C_1) = .8$. What is the a posteriori probability that yesterday was clear, given that today is clear?

1.77 Girls tend to outscore boys on a certain aptitude test given at college. In fact, the probability of a grade of A for girls (G) is given by $P(A|G) = .75$, while for boys (B) the result is $P(A|B) = .6$. Boys outnumber girls three to two at this particular school. A test is turned in without a name and is graded A. What is the probability that the test belongs to a boy?

1.78 Three bolt machines operate with equal capacity and feed a common stockpile. The machines are known to produce defective items on a given day with the respective conditional probabilities of .1, .15, and .2. A bolt is inspected from the stockpile and observed to be defective. What are each of the respective probabilities that it came from a particular one of the three machines?

1.79 Three prisoners, A, B, and C, with apparently equally good records have applied for parole. The parole board has decided to release two of the three, and the prisoners know this, but they do not know which two. A guard friend of prisoner A knows who are to be released. Prisoner A realizes that it would be unethical to ask the guard if he, A, is to be released but thinks of asking for the name of *one prisoner other than himself* who is to be released. He thinks that, before he asks, his chances of release are $\frac{2}{3}$. He reasons that if the guard says, "B will be released," his own chances will then go to $\frac{1}{2}$, because then either A and B or B and C are to be released. And so A decides not to reduce his chances by asking. However, A is mistaken in his calculations. Explain.

1.80 Assume that a family will have three children and that each has equal likelihood of being a boy or a girl.
    a. What is the probability that the second child will be a boy given that the first and third will be boys?
    b. What is the probability that the second child will be a boy given that two of the three will be boys?

1.81 What is the probability that a team will win four games in a row if the probability of winning the first game is .6 and the conditional probabilities of winning each succeeding game, given that it got there, are .6, .8, and .5?

1.82 A manufacturer of hand calculators has three different assembly plants, A, B, C. The proportions of defective calculators produced by these three plants are, respectively, .01, .02, .04. Now plant A produces 50% of the calculators, while plants B and C, respectively, produce 30% and 20%. A customer purchases a calculator and finds it to be defective.
    a. What is the probability of this event?
    b. What is the conditional probability that the defective calculator was produced at plant A? At plant B? At plant C?

1.83 A shipment of 50 television sets is received by a dealer. Four of the sets are defective. What is the probability that the first three in a row that he checks are all nondefective?

1.84 The probability is 1 that a fisherman will say he had a successful day when, in fact, he did, but the probability is only .6 that he will say he had a successful day when, in fact, he did not. Assuming half his days are successful, what is the probability that he had a successful day if he says he had a successful day? Now suppose only 25% of his days are successful. What would the answer then be?

1.85 An artist just finished a painting. She has observed in the past that clients respond to her works somewhat like her husband does. She feels that, when clients like a painting, about 85% of the time her husband likes it also, but even when a client does not like her painting, her husband still likes it about 85% of the time. She is a popular artist, and at showings, about 90% of the clients respond favorably to her work. Her husband does not like the painting she just finished. How likely is it that the clients will like it anyway?

1.86 An oil wildcatter figures that there is a 50–50 chance of oil on a piece of property he purchased. He has a test he performs that is 80% reliable; that is, if there is oil, it indicates that there is with probability .8; and if there is none, it so indicates with the same probability. His test just indicated oil. What are the chances that there really is oil? What would the answer be if the test indicated no oil?

1.87 A life insurance company gives an aptitude test to sales applicants in order to predict sales ability. No such test is perfect, however, and the company decides to experiment on the test's reliability by giving the test to a random group of 100 salespeople, 80 of whom have good sales records and 20 of whom have poor sales records. Of those who were good salespeople, only 60% passed the test; 20% of those salespeople with poor records also passed the test. Later, the test is administered to an applicant and he passes. How likely is it that he will be a good salesman? A poor salesman?

1.6 INDEPENDENCE

If $P(A|B)$ is found to be the same as $P(A)$, that would imply that the additional information concerning the occurrence of B does not change the probabilistic information about the occurrence of A. One might then be tempted to say that A is independent of B, since the occurrence of A in no way *depends* on the occurrence of B. But we should be careful about using the word "independent" because of its common use to imply that one thing has nothing to do with another, thereby suggesting mutual exclusiveness. As we will see, independence and mutual exclusiveness are only loosely related ideas.

Suppose that A and B are two events. As mentioned above, if the information that B occurs is not to affect the assignment of probability to the occurrence of A, it would seem reasonable that $P(A|B) = P(A)$. But if $P(A|B) = P(A)$ and, by definition,

$$P(A|B) = \frac{P(A \cap B)}{P(B)},$$

then we have $P(A \cap B) = P(A)P(B)$ for the particular events A and B. It would also then follow that

$$P(B|A) = \frac{P(B \cap A)}{P(A)} = \frac{P(A)P(B)}{P(A)} = P(B),$$

which is a desirable consequence. That is, if A and B are independent in the sense that $P(A|B) = P(A)$, it should follow that B and A are independent in the sense that $P(B|A) = P(B)$, so the occurrence of A does not affect the likelihood of the occurrence of B either.

DEFINITION 1.3   Two events A and B are (*statistically*) *independent* if $P(A \cap B) = P(A)P(B)$.

This definition embodies both the conditional probability properties discussed above. It may even be used when the conditional probabilities are not defined. When it is necessary to be careful about the technical use of the term "independent," you will sometimes see the term "stochastically independent." Normally, however, in absence of the qualifier, it is understood that the technical meaning is intended in any probabilistic context.

Although the concept of independence is motivated by intuitive considerations, the only test of independence *in the theory* is the multiplicative property of the definition. In applications, independence is often interpreted in keeping with the intuitive ideas we have been discussing. Often, in a given application, it is *assumed* that certain events are independent because of the analyst's intuitive feeling about the occurrence of such events in their real world context. Thus the assumption by the analyst that particular events are independent may be important in building a probability model for an experiment. Again, judgment about whether such assumptions are warranted must await the use of the model and comparisons of results in the model ("theoretical" results) with actual results that we experience in the real world (actual "data").

EXAMPLE 1.21   Suppose, in a human study, 1000 randomly selected subjects were cross-classified according to gender and smoking behavior (smokers and nonsmokers). Suppose the results were as given in the accompanying table. Make probability assignments suggested by these data and investigate the independence of gender and smoking behavior.

	M	F
S	420	280
N	180	120

SOLUTION   With relative frequencies used as probabilities assigned for a model, it would follow that $P(M \cap S) = .42$ and $P(F \cap S) = .28$, so the probability that a randomly selected person is a smoker is $P(S) = .70$ by the law of total probability. Similarly, $P(M) = .60$, $P(F) = .40$, and $P(N) = .30$. Now, since $P(M) \cdot P(S) = P(M \cap S)$, it may be concluded that the two events "male" and "smoker" are independent in this model, whether or not this result complies with our intuitive ideas of their independence.

EXAMPLE 1.22   Suppose that, in Example 1.21, it is found someone rounded the original results and, in fact, the complete tabulation was as shown in the accompanying table.

	M	F
S	423	279
N	181	117

Suppose the analyst now assigns $P(S) = .702$ and $P(M) = .604$, only slightly different from before. However, $P(S) \cdot P(M) = .424 \neq P(S \cap M)$. Under this slightly revised model, M and S would not be independent. (There are statistical procedures to test whether such data are consistent with an independence hypothesis.)

EXAMPLE 1.23   A fair coin is to be tossed twice and the results recorded in a sample space $S = \{(H, H), (H, T), (T, H), (T, T)\}$. Let A be the event $\{(H, H), (H, T)\}$ (which can be described as "heads on the first toss"), and let B denote "heads on the second toss," $B = \{(H, H), (T, H)\}$. So strongly do we feel, from the nature of the experiment, that A and B should be independent that we would probably reject any model in which this is not the case. Guided by that fact, we might *assume* as a model that each elementary event should have probability $\frac{1}{4} = \frac{1}{2} \cdot \frac{1}{2}$. In accordance with Exercise 1.44, $P(A) = \frac{1}{2}, P(B) = \frac{1}{2}$, and $P(A \cap B) = P(\{(H, H)\}) = \frac{1}{4}$. Not surprisingly, A and B are independent by fiat.

It is a mistake to confuse mutually exclusive events with independent events, but that is easy to do unless the technical meanings of the two are kept in mind. You have seen how independence is dictated by the choice of P. This is not true of mutual exclusiveness. Two events A and B are mutually exclusive iff $A \cap B = \varnothing$. That is purely a set property, a question of whether or not A and B have points in common. No choice of P is going to alter that position. Indeed, if $A \cap B = \varnothing$, then $P(A \cap B) = 0$ regardless of the choice of P; and if $P(A) > 0$ and $P(B) > 0$, then A and B could not be independent.

EXAMPLE 1.24   Let's return to our die experiment with $S = \{1, 2, 3, 4, 5, 6\}$ and let each elementary event have probability $\frac{1}{6}$ as a model. Let $A = \{1, 2\}$, $B = \{2, 4, 6\}$, and $C = \{1, 3, 5\}$. The events B and C are mutually exclusive and certainly not independent. However, $A \cap B = \{2\}$ and $P(A) \cdot P(B) = \frac{2}{6} \cdot \frac{3}{6} = \frac{1}{6} = P(A \cap B)$. Thus A and B are independent, with no apparent intuitive reason why they should be.

Once it has been established that A and B are independent events, it follows immediately that so are the pairs of events $\overline{A}$ and B; A and $\overline{B}$; $\overline{A}$ and $\overline{B}$. For example, if A and B are independent, then

$$P(A \cap \overline{B}) = P(A) - P(A \cap B) = P(A) - P(A)P(B)$$
$$= P(A)[1 - P(B)] = P(A)P(\overline{B}),$$

so that A and $\overline{B}$ are also independent. Other relationships are left as exercises.

These ideas can be extended to the consideration of independence for more than two events. It seems reasonable that any concept of independence of the events A, B, and C should impose, in particular, the assumption that the events are *pairwise independent* in the sense that each of the pairs of events A and B, A and C, and B and C is independent. Thus, in particular, it is desired that

$$P(A \cap B) = P(A)P(B), \qquad P(A \cap C) = P(A)P(C),$$

and
$$P(B \cap C) = P(B)P(C).$$

In addition, it would seem natural that the multiplicative property should hold for all three events; that is, $P(A \cap B \cap C) = P(A)P(B)P(C)$. For if we think of $A \cap B$ as a single event, say D, then, intuitively speaking, if A

and B are both independent of C, we would want D and C to be independent. (Recall, for example, that if A and C are independent, the related events $\overline{A}$ and C are also independent.) Thus $P(D \cap C) = P(D)P(C)$ is desired, and since A and B are independent, then $P(D) = P(A)P(B)$. The condition $P(A \cap B \cap C) = P(D \cap C) = P(A)P(B)P(C)$ holds in addition to the pairwise statements. This condition will not necessarily follow from pairwise independence, as can be seen by examining the model in Example 1.23 again.

EXAMPLE 1.25   In the situation of Example 1.23, let A and B be as before and let $C = \{(H, H), (T, T)\}$ be the event that both faces are alike. Then $P(A) = P(B) = P(C) = \frac{1}{2}$. We have already observed that A and B are independent. Since $A \cap C = \{(H, H)\}$ and $P(A \cap C) = \frac{1}{4} = P(A)P(C)$, we see that A and C are also independent. Also, $B \cap C = \{(T, T)\}$, with $P(B \cap C) = \frac{1}{4} = P(B)P(C)$, so B and C are independent. But $A \cap B \cap C = \varnothing$; and so

$$P(A \cap B \cap C) = 0 \neq P(A)P(B)P(C);$$

that is, the multiplicative property does not hold for all three sets.

On the other hand, it is easy to construct examples of sets for which $P(A \cap B \cap C) = P(A)P(B)P(C)$ but A, B, and C are not pairwise independent.

EXAMPLE 1.26   Let $S = \{1, 2, 3, \ldots, 16\}$, and suppose all elementary events have probability $\frac{1}{16}$ as a model. Let

$$A = \{1, 2, 3, 4, 5, 6, 7, 8\}, \qquad B = \{1, 2, 3, 4\},$$

and $C = \{1, 9, 10, 11, 12, 13, 14, 15\}.$

Since $A \cap B \cap C = \{1\}$, we have

$$P(A \cap B \cap C) = \tfrac{1}{16} = \tfrac{1}{2} \cdot \tfrac{1}{4} \cdot \tfrac{1}{2} = P(A)P(B)P(C).$$

But $A \cap B = B$, so $P(A \cap B) = \frac{1}{4} \neq P(A)P(B)$; and $A \cap C = \{1\}$, so that

$$P(A \cap C) = \tfrac{1}{16} \neq P(A)P(C)$$

and $$P(B \cap C) = \tfrac{1}{16} \neq P(B)P(C).$$

Thus no two of the events A, B, and C are independent.

This example provides a clue as to how a general definition of independence should be stated. A total of n events should be called independent if the probability of the joint occurrence of any 2, 3, 4, and, in general, k of them, is the product of the probabilities of their individual occurrence.

DEFINITION 1.4   Events $A_1, A_2, \ldots, A_n$ are *independent* if for any subcollection $A_{i_1}, A_{i_2}, \ldots, A_{i_k}, 2 \le k \le n,$

$$P\left(\bigcap_{j=1}^{k} A_{i_j}\right) = \prod_{j=1}^{k} P(A_{i_j}),$$

where Π stands for product.

EXERCISES

THEORY

1.88 If A and B are independent, show that $\overline{A}$ and B are independent and also that $\overline{A}$ and $\overline{B}$ are independent.

1.89 Suppose A, B, and C are (mutually) independent. Show that the following triples of events are also independent: $A, \overline{B}, C; A, \overline{B}, \overline{C}; \overline{A}, \overline{B}, \overline{C}$.

1.90 a. Show that S and any event E are independent.
   b. Show that $\varnothing$ and any event E are independent.

1.91 If A, B, and C are independent, show that $P(C|A \cup B) = P(C)$. Are C and $A \cup B$ independent?

1.92 If A, B, and C are independent, show that A and $B - C$ are (two) independent events.

1.93 Under what circumstances can we say that A is independent of itself?

1.94 If $A \subseteq B$ and A and B are independent, show that $P(A) = 0$ or $P(B) = 1$.

1.95 Suppose A_1, A_2, and A_3 are independent events, each having probability p. Find the probabilities of the following events:
   a. Exactly m of the A_i's occur, $m = 0, 1, 2, 3$.
   b. At least m of the A_i's occur, $m = 0, 1, 2, 3$.
   c. At most m of the A_i's occur, $m = 0, 1, 2, 3$.

APPLICATION

1.96 Two contractors are asked to design a particular system "independent" of each other. The probability that contractor A will evolve a satisfactory design is .6 and that probability for contractor B is .7.
   a. What is the probability of obtaining a satisfactory design?
   b. Given a satisfactory design, what is the probability that it was furnished by B and not A?

1.97 The probability that any fleet exercise will be concluded with no major personal injuries is .9. Suppose five exercises are to be conducted in a year. Compute the probability that there will be no major injuries in any of them.

1.98 It has been determined that a certain kind of rocket engine will misfire 5% of the time. What is the probability that there will be at least one misfire if 10 of them are tested independently? *Hint:* "At least one misfire" is the complement of "no misfires."

1.99 The probability that a 35-year-old United States citizen will live to age 65 is .725 according to recent mortality tables.
   a. Suppose you and your friend are both 35. What is the probability that both of you will live to age 65?
   b. What is the probability that neither of you will live to age 65?

1.100 Three missiles are fired at the same target. Their respective probabilities of hitting it are .6, .7, and .8. What is the probability that the target is hit?

1.101 Whenever an experiment is performed, the probability of occurrence of an event A equals .2. The experiment is repeated until A occurs. What is the probability that it will be necessary to carry out a fourth experiment?

1.102 Two machines, A and B, are operated "independently" of each other. The probability of a breakdown in a given shift is .04 for A and .07 for B. Compute the probabilities of the following events for a given shift: (a) both A and B break down; (b) A breaks down but B does not; (c) A breaks down given that there is a breakdown.

1.103 Let A and B be two events with $P(A) = .4$ and $P(A \cup B) = .7$.
   a. What must be the value of $P(B)$ if A and B are independent?
   b. What must be the value of $P(B)$ if A and B are mutually exclusive?
   c. Can $P(B)$ be chosen so that both are satisfied?

1.104 An electronic assembly consists of two subsystems, I and II. Let A be the event that system I fails (regardless of what occurs for II), and let B be the event that system II fails (regardless of what occurs for I). From past records, it is assumed that $P(A) = .02$, $P(B \cap \overline{A}) = .01$, and $P(B \cap A) = .015$. Are A and B independent? What is the value of $P(A \cap \overline{B})$?

1.105 In a survey of viewing habits, a television station classified the returns according to age, with the relative frequencies as shown in the accompanying table. Did viewing the special depend on the age category?

	Viewing Habit	
Age	*Watched Special*	*Did Not Watch Special*
Over 40	.14	.17
Under 40	.27	.42

1.106 The use of Bayes's theorem can result in counterintuitive results. An example occurs when the prior probabilities are "extreme" (close to zero or close to one). Suppose there is an extremely reliable medical screening test for a certain rare disease; the test detects a person having the disease with probability .90, and it causes false alarms with probability .08. Let D denote the event that a randomly selected person has the disease, and let I denote the event that the test of a randomly selected person is positive (indicates the person has the disease). Our suppositions about the reliability of the test can be written $P(I|D) = .90$ and $P(I|\overline{D}) = .08$. Now, suppose the rarity of the disease is given by $P(D) = .001$.

a. Suppose you have just been tested for the disease and the test was positive. What is the probability you have the disease?

b. Are you surprised that, even when a very reliable test is positive, the probability that you actually have the disease is so low?

c. Rework part (a) for the reliability of the test increased to $P(I|D) = .99$ and $P(I|\overline{D}) = .05$.

d. What result would you have gotten in part (c) if the disease was extremely rare, say afflicting only one person in 100,000 on the average?

1.7 APPROACHES TO ASSIGNING PROBABILITIES

We have taken the position that there is no inherent definition of probabilities in a given situation. Rather, it is the obligation of the analyst (model builder) to *assign* probabilities to certain key events in the model he or she intends to apply. Then other probabilities can be computed, using the theorems. If the analyst's assignment is good, the model should be useful. But how does an analyst go about making a good assignment of probabilities? A few of the approaches used in practice are discussed in what follows. Some of our examples anticipate work to be done later; you may wish to just scan over some of these now, returning to them later when you have the necessary background.

As you have seen in many of our examples involving simple physical experiments, apparently satisfactory assignments can be made by using a premise of symmetry or other geometric considerations.

EXAMPLE 1.27   In the die-rolling experiment, most people believe the cubical solid has symmetry properties that make each face equally likely to land up—hence the usual equal-likelihood model. It is possible that a particular die has irregularities, such as inhomogeneous material, so this model is not "good" enough. Below, we discuss how the analyst might make a different assignment, at the cost of additional analyses of the die.

EXAMPLE 1.28   —*Buffon's Needle*   A hardwood floor is made of boards 4 inches wide. A needle of length 2 inches is tossed in the air and lands on the floor. What is the probability the needle comes to rest crossing a joint between the floorboards?

SOLUTION   As a model, imagine the joints are parallel vertical lines on a plane and the needle's resting position is a line segment on the plane. For reference, imagine an x-axis drawn perpendicular to the parallel lines, through the center C of the needle. Let x_0 denote the distance from C to the nearest joint and let θ denote the acute angle between the x-axis and the needle. It is assumed the center C is equally likely to fall anywhere between the two parallel lines to which it is closest, and the angle θ is equally likely to be any value between 0 and $\pi/2$ radians. The needle intersects a joint, provided x_0 is less than $\cos \theta$; see Fig. 1.10(a). This event may be envisioned as the shaded region in Fig. 1.10(b), which is plotted in the "x_0, θ space." Under the assumptions of the random toss of the needle, the actual outcome of the experiment is represented by a random point in the rectangle of Fig. 1.10(b). The probability of intersection is thus the relative area of the region that satisfies the condition for intersection, $x_0 < \cos \theta$. This probability is

$$\frac{(\text{area under } x_0 = \cos \theta)}{(\text{area of rectangle})} = \frac{(\int_0^{\pi/2} \cos \theta \, d\theta)}{(\pi/2 \cdot 2)} = \frac{1}{\pi}.$$

A second method often used in assigning probabilities is related to the observation of relative frequencies. If the experiment can be repeated a number n of times, the resulting collection of outcomes contains statistical information about the underlying probabilities governing each experiment. The analyst is aware that the observed relative frequency of occurrence of a given event will fluctuate, as n increases, due to the nature of randomness. But with a preplanned sample of adequate size, the observed relative frequency should provide useful assignments. The problem of how to design the series of experiments, including sample size requirements, is one falling in the area of statistics.

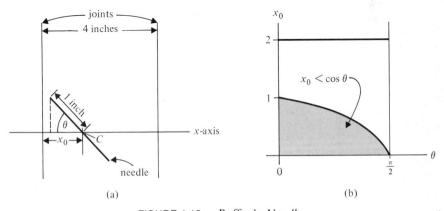

FIGURE 1.10   *Buffon's Needle*

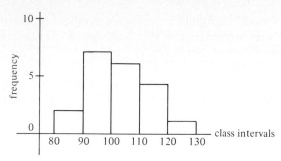

FIGURE 1.11   *Histogram for Heart Rate Data*

EXAMPLE 1.29   If the analyst suspected a die might not be fair, a series of, say, 1000 tosses of the die might be performed. Suppose this is done and the results are as tabulated in the accompanying table. The analyst might believe (again, on the basis of statistical considerations) these outcomes are not consistent with a sample of 1000 outcomes on a fair die, and an assignment of probabilities equal to the observed relative frequencies might be made. These relative frequencies might be viewed as estimates of the biased die probabilities.

	Outcome					
Frequency	*1*	*2*	*3*	*4*	*5*	*6*
Observed	201	165	143	151	177	163
Relative	.201	.165	.143	.151	.177	.163

EXAMPLE 1.30   —*Histograms*   If the experiment can have any real number as an outcome (at least in principle), it is useful to categorize the observed data into class intervals and to consider the relative frequencies within these intervals. Suppose the analyst wishes to model outcomes on experiments consisting of measuring the heart rates, after exercise, of randomly selected schoolchildren. To get an idea of the shape of the distribution of heart rates, the analyst might plan to measure a sample of 20 children and tally the results into five class intervals, say 80–89, 90–99, 100–109, 110–119, 120–129. Suppose the measurement is made on each of 20 randomly selected children, and the following data result: 99, 93, 86, 94, 103, 101, 121, 106, 115, 97, 119, 105, 111, 108, 93, 88, 111, 96, 107, 94. Then a count of the frequencies of outcomes within each class interval is as shown in the next table.

Class Interval	80–89	90–99	100–109	110–119	120–129
Frequency	2	7	6	4	1

A bar graph showing class intervals along the *x*-axis and bars with heights proportional to the frequencies in each interval is called a histogram. A histogram for the heart rate data, shown in Fig. 1.11, gives an idea of how the population of heart rates of all children are distributed. In turn, this result gives the analyst an idea of how probabilities might be assigned in a model for the experiment of selecting a child at random and measuring the heart rate.

A third method of assigning probabilities involves using model types that experience has shown to be useful in situations of certain kinds. The idea is somewhat like our discussion of prototype experiments, with applications to seemingly different situations. The analyst may have learned, through experience, that, in experiments involving measuring times required for experimental subjects to complete tasks (such as to detect a target or change a flat tire), there is a certain type of model that is likely to be useful. The task then becomes one of specifying parameter values in the model type, to specialize it to the particular application at hand. This activity often involves an area of statistics known as *parametric inference*.

EXAMPLE 1.31   From experience, the analyst may have reason to believe that measurements of line voltage, made at random times at a certain point in a power transmission system, will be distributed according to the bell-shaped normal curve, but the location of the highest point of the bell may be unknown. This location is a parameter, which might be estimated by using a sample of data collected for the purpose. Suppose five measurements yield the data 119, 122, 117, 116, 117. The analyst would typically calculate the average of the data, 118.2 in this case, as an estimate of the unknown location parameter. A discussion of why the average of the data provides the best estimate in this case again falls in the realm of statistics.

A fourth method of assigning probabilities concerns situations where the analyst is trying to model a decision maker's subjective probabilities that certain events will occur in the future. In this case, the probabilities in question represent the decision maker's degree of belief that the events will occur. However, the decision maker may not feel comfortable in trying to assign numerical values to the probabilities, even though he may have equivalent information in mind. The decision maker takes actions, buying or selling a certain stock, for example, based on his belief in the likelihoods that the stock will have various prices at a certain future time. Decision theorists have devised various methods of eliciting, from the decision maker, information about actions that should be taken in hypothetical situations and using this information to determine the decision maker's subjective probability assignment. The decision maker is acting as if the probability assignment were the one derived.

EXAMPLE 1.32   Consider a very simple case in which the analyst wants to determine the likelihood a horse-racing expert places on a certain horse winning in tomorrow's race. The expert can't (or won't) assert directly what he believes the probability in question to be. But he will answer questions of the following form, with a yes or no answer: "Would you be willing to bet $a against my $1 that the horse wins?" (In such a wager, if the horse wins, the expert gets my dollar; if the horse loses, I get his $a.)

Suppose this question is asked, in various forms and in a randomized order, and the results are as shown in the accompanying table (after reordering on the value of a). It then follows that the expert believes a bet of $0.30 against $1 is, approximately, a fair bet. That is, in a long series of such bets over many races (with this same threshold of $0.30), the expert would about break even. The proportion p of such bets he would win is the (subjective) probability the expert places on tomorrow's race. For a break-even position, long-run wins equal long-run losses. The winnings for n races at $a = \$0.30$ are projected to be $\$1 \cdot p \cdot n$; the losses for these races are $\$0.30 \cdot (1 - p) \cdot n$. So at the break-even point, $1 \cdot p \cdot n = .3(1 - p) \cdot n$, or $p = .23$. This value is the expert's subjective probability that the horse will win tomorrow's race.

a	0.10	0.20	0.30	0.31	0.5	1
Answer	Yes	Yes	Yes	No	No	No

Once probabilities are assigned to certain of the events in a model, whatever the approach used, the probabilities of *all* events are determined, and such determinations often involve computations involving the probability theorems we have developed (and will develop). They often also involve mathematical arguments, utilizing special features of the model adopted. Some of these, such as combinatoric arguments and the use of calculus methods, form a major part of this book. Others, such as simulation methods, are just briefly touched on here; treatments of specialized methods such as simulation can be found elsewhere. The role of statistical analyses of observed data in assigning probabilities is a very important one; this is an important topic in treatments of statistics. For the most part, we will not concentrate on how specific numerical probabilities are assigned. We will concern ourselves with the framework in which this activity takes place. We want to show how one can handle, mathematically, the characterization of probability functions, and what some commonly used probability models are. We concentrate on the theory of probability, illustrating it with rather simple examples that suggest applications.

EXERCISES

APPLICATION

1.107 Suppose, in the Buffon needle problem (Example 1.28), the parallel lines are w units apart, and the needle is $n < w$ units long. Show that the probability of intersection is $2n/w\pi$.

1.108 In the subjective probability situation (Example 1.32), suppose the expert is just willing to wager $a = \$0.50$. What is his subjective probability?

1.109 a. In the subjective probability situation (Example 1.32), suppose the expert is willing to give you odds of 8–5 that the horse will win. (That is, he puts up \$8 and you put up \$5. If the horse wins, he takes the pot; otherwise, you take the pot.) What is his subjective probability that the horse will win?
  b. Repeat part (a) for odds of a to b.

1.110 Suppose in 1000 replications of an experiment, the analyst keeps track of the frequencies with which two events A and B occur, in combination with one another. Suppose the results are cross-tabulated as shown in the accompanying table. Thus, for example, $A \cap B$ occurred 208 times in the 1000 trials.

	B	$\overline{B}$
A	208	386
$\overline{A}$	351	55

  a. Estimate $P(A)$.
  b. Estimate $P(A \cap B)$.
  c. Estimate $P(A \cup B)$.
  d. Do you think A and B are independent?

1.111 a. Construct a histogram for the following data, using five class intervals: 9, 14, 6, 46, 22, 41, 34, 21, 26, 11, 15, 40, 19, 35, 28, 21.
  b. Do you believe the distribution underlying the process that generated these data is symmetric?
  c. Estimate the probability that a future observation on this process will be less than 20.

1.112 Suppose you believe the data in Exercise 1.111 were from a population with a normal (bell-shaped) distribution. Estimate the location at which the bell is highest.

1.113 A train arrives at a station at a random time between 7:00 and 7:15, and it leaves the station exactly 2 minutes later. You plan to arrive at the station at a random time between 7:00 and 7:15, and you will leave the station if the train isn't available within 2 minutes of your arrival. What is the probability that you take the train? *Hint:* Construct a two-dimensional (train arrival, your arrival) space, and identify in it the event "you catch the train."

1.114   In a common carnival game, the player tosses a dime at a table with squares ruled on its surface. The player gets the dime back and wins a prize worth $0.25 if the dime comes to rest within a square; otherwise, the player loses the dime.

    a. Assuming the dime is $\frac{1}{2}$ inch in diameter, the squares are $\frac{3}{4}$ inch on a side, and the dime lands on the table, what is the probability the player wins?

    b. For what size squares would the game be fair (see Exercise 1.109)?

1.8   SUMMARY

A probability model for an experiment consists of three things: a sample space S consisting of the set of possible outcomes o of the experiment; a collection of subsets of S, called events; a probability function P, which assigns numerical likelihoods of occurrence of events.

An event is said to occur if the outcome o of the experiment is an element of the event. Thus when the experiment is performed, each of the events will have occurred or not occurred.

The events $A_1, A_2, A_3, \ldots$ are mutually exclusive if any two of the A's are disjoint. S is the certain event; $\varnothing$ is the impossible event. A and $\overline{A}$ are called complementary events. A collection of events $C_1, C_2, C_3, \ldots$ such that the C's are mutually exclusive and exhaustive (i.e., $\cup_{i=1}^{\infty} C_i = S$) is a *partition* of S.

A probability function satisfies the Kolmogorov axioms:

1. For any event A, $P(A) \geq 0$.

2. $P(S) = 1$.

3. If $A_1, A_2, A_3, \ldots$ are mutually exclusive events, then $P(\cup_{i=1}^{\infty} A_i) = \sum_{i=1}^{\infty} P(A_i)$.

Some important consequences of the Kolmogorov axioms are as follows: $P(\overline{A}) = 1 - P(A)$ (probabilities of complementary events); if $C_1, C_2, \ldots, C_n$ is a partition of S, then for any event B,

$$P(B) = \sum_{i=1}^{n} P(B \cap C_i) \qquad \text{(law of total probability)};$$

and $\qquad P(C_i|B) = \dfrac{P(B|C_i) \cdot P(C_i)}{\sum_{j=1}^{n} P(B|C_j) \cdot P(C_j)} \qquad \text{(Bayes's theorem)};$

and

$$P(A_1 \cap A_2 \cap \cdots \cap A_n) = P(A_1) \cdot P(A_2|A_1) \cdots P(A_n|A_1 \cap A_2 \cap \cdots \cap A_{n-1})$$

(general multiplication rule), where the conditional probability of A_i given B is $P(A_i|B) = P(A_i \cap B)/P(B)$.

In Bayes's theorem, $P(C_i)$ is called the a priori probability of the cause C_i, and $P(C_i|B)$ is called the a posteriori probability of C_i, given the effect B.

The events $A_1, A_2, \ldots, A_n$ are independent provided that, for any collection of the A's, say $A_{1*}, A_{2*}, \ldots, A_{k*}$,

$$P(A_{1*} \cap A_{2*} \cap \cdots \cap A_{k*}) = P(A_{1*}) \cdot P(A_{2*}) \cdots P(A_{k*}).$$

The concepts of independence and mutual exclusiveness are quite different and should not be confused with one another.

Chapter 2

FINITE SAMPLE SPACES

2.1   PROBABILITY STRUCTURE

When the sample space is finite, the characterization of a probability model is a relatively simple task. If the probability of each elementary event $\{o_i\}$ is known, then the probability of other events can be calculated from them. Thus when S is finite, that is all that needs to be specified. For suppose $S = \{o_1, o_2, \ldots, o_n\}$ and $P(\{o_i\}) = p_i$ is known for $i = 1, 2, \ldots, n$. For A as any nonempty event, let $P(A) = \Sigma_{o \in A} P(\{o\})$. [If $A = \emptyset$, then, of course, $P(A) = 0$.] The Kolmogorov axioms are satisfied, since $P(A) \geq 0$ for any event A, $P(S) = \Sigma_{o \in S} P(\{o\}) = \Sigma_{i=1}^{n} p_i = 1$, and if A and B are mutually exclusive, then

$$P(A \cup B) = \sum_{o \in A \cup B} P(\{o\}) = \sum_{o \in A} P(\{o\}) + \sum_{o \in B} P(\{o\}) = P(A) + P(B).$$

(The second step follows from the fact that A and B have no sample points in common.)

Thus all we have to do to determine a model in this situation is to specify n nonnegative numbers that sum to unity, each one being selected as the probability of an elementary event. If n is very large, that specification might not be an easy task to actually carry out, but for many of the common cases of interest, this characterization of P can be implemented with a reasonable amount of effort.

There is a technical distinction between an elementary event $\{o_i\}$ and the single outcome o_i from which it is constructed. Practically, however, there is little harm in using the simpler notation for both. Thus we may speak of the probability of the outcome 6, for example, writing ''$P(6)$'' when we really mean ''$P(\{6\})$.'' The value $p_i = P(o_i)$ is called the *point probability* of o_i. A probability model with a sample space $S = \{o_1, o_2, \ldots, o_n\}$ of n descriptions is determined by any choice of point probabilities p_1, $p_2, \ldots, p_n$, where the p's are nonnegative numbers that sum to unity.

There are infinitely many ways to assign such p's. This feature allows us a great deal of freedom (and responsibility) in constructing models for particular situations.

EXAMPLE 2.1   Among the many possibilities for selecting point probabilities for $S = \{o_1, o_2, \ldots, o_n\}$ is the choice where all the p_i's are the same. Since the p_i's must add to unity, it follows, in this case, that $p_i = 1/n$ for each i. This model is called the *equal-likelihood model*. It is a common model for most games of chance, and we used it in Chapter 1 in examples such as the die experiment, where $S = \{1, 2, 3, 4, 5, 6\}$ and $p_i = \frac{1}{6}$.

Although the equal-likelihood model is very special and limited, nevertheless, it is often used in practice, and it is convenient for illustrating basic concepts.

EXAMPLE 2.2   In the toss of a single coin with $S = \{H, T\}$, the assumption of the equal-likelihood model, $p_1 = p_2 = \frac{1}{2}$, is often described as assuming that the coin is fair. Other possibilities such as $p_1 = \frac{3}{4}$ and $p_2 = \frac{1}{4}$, or $p_1 = \frac{1}{3}$ and $p_2 = \frac{2}{3}$ are legitimate models and reflect different ideas on the biasedness of the coin. In general, p_1 may be any number between 0 and 1 inclusive; then automatically, $p_2 = 1 - p_1$ is determined.

The model for this coin-tossing experiment is often referred to as a *Bernoulli model* or a *Bernoulli trial*. It can be used as a prototype for situations in which there are only two possible outcomes of the experiment. For example, the actual experiment might be classifying the gender of an unborn child, where H stands for "boy" and T stands for "girl." Current statistics show that these are not equally likely events, the odds being slightly in favor of boys. A better model is provided by $p_1 = .51$ and $p_2 = .49$. As another example, the experiment might be one of firing a highly reliable missile, where H stands for "success" and T for "failure." Experience with the missile system might suggest a choice of a reliability figure such as $p_1 = .99$. Even the choice $p_1 = 1$ (and, therefore, $p_2 = 0$) is admissible, though maybe not realistic; it reflects a belief that the missile simply will not fail.

This last example brings up a point alluded to before, namely, that an event may be nonempty and still have probability zero. In the case of the perfectly reliable missile, $\{T\}$ is not empty, and yet $P(T) = 0$. At the opposite end of the spectrum, $\{H\} \neq S$, yet $P(H) = 1$. The event $\{H\}$ is not certain, because it fails to include all the points of S; yet we are convinced it will occur according to the choice of P. Just as with the distinction between independence and mutual exclusiveness, an event A is empty or not empty according to whether or not it has elements from S; that fact does not depend on how P is assigned. One choice of P can make $P(A) = 0$ and another model could have $P(A) \neq 0$.

Some of the best examples of equal-likelihood models are associated with the so-called *urn problems*, in which we imagine an urn containing balls that are numbered from 1 through n but are otherwise indistinguishable. An experiment of this kind usually consists of drawing a ball from the urn and noting the number on the ball drawn. Indeed, a single trial for almost any experiment involving a finite sample space can be paraphrased as an urn problem. That is, urn problems provide us with a large class of prototype experiments. To apply the equal-likelihood probability model to such an experiment, imagine shaking the balls in the urn until they are well mixed, in the sense that any one ball is just as likely to be drawn as any other. Then draw a ball "at random" from the urn. Generally speaking, the phrase *at random*, when applied to experiments with finite sample spaces, will be interpreted to mean that the equal-likelihood model is assumed.

EXAMPLE 2.3   One special example of an urn problem is that of drawing a digit at random; the result is called a *random digit*. A sample space for this experiment is the set $S = \{0, 1, 2, 3, 4, 5, 6, 7, 8, 9\}$. The equal-likelihood model then assigns point probabilities of .1 to each of these 10 possible outcomes. This model is the basis for all random number generators, which we discuss later.

EXAMPLE 2.4   A variation of the basic urn problem is as follows. Suppose an urn contains five white balls, four red balls and seven green balls. If a ball is drawn at random, what is the probability that it is red?

SOLUTION   Most people would answer $\frac{4}{16} = \frac{1}{4}$ even without a course in probability. To model the problem as an urn problem, imagine the white balls numbered 1 through 5, the red balls numbered 6 through 9, and the green balls numbered 10 through 16. The equal-likelihood model then selects $p_i \equiv \frac{1}{16}$, and since the event "red ball drawn" is the set $A = \{6, 7, 8, 9\}$, then

$$P(A) = \tfrac{1}{16} + \tfrac{1}{16} + \tfrac{1}{16} + \tfrac{1}{16} = \tfrac{4}{16} = \tfrac{1}{4}.$$

EXAMPLE 2.5   A game involving a toss of two dice may take the sample space $S = \{(1, 1), (1, 2), \ldots, (6, 6)\}$, a set consisting of all possible pairs of outcomes on the two dice. The equal-likelihood model assigns $p_i = \frac{1}{36}$ for each i. The event "the sum of spots on the two dice is 7" may be written $A = \{(1, 6), (2, 5), (3, 4), (4, 3), (5, 2), (6, 1)\}$, a set consisting of six points. By our rule,

$$P(A) = \sum_{o \in A} P(o) = \underbrace{\tfrac{1}{36} + \cdots + \tfrac{1}{36}}_{\text{6 terms}} = \tfrac{6}{36} = \tfrac{1}{6}.$$

It is no mere coincidence in the last two examples that $P(A) = \frac{6}{36}$ is simply the number of points in A, $n(A)$, divided by n, the number of points in S. Indeed, in any equal-likelihood model with $P(A) = \Sigma_{o \in A} P(o)$ and $P(o) \equiv 1/n$, $P(A) = n(A)/n$. This feature makes the determination of $P(A)$ merely one of counting the number of elements in A, which seems like a trivial task. But, in fact, counting the elements in A, or in S, for that matter, can be a difficult task in itself. For example, in dealing five cards from an ordinary deck of cards for a poker hand, a reasonable sample space is the set of all possible five-card hands. It turns out that this number of possibilities is in the millions. The manner in which the numbers of such possibilities are computed is a study in itself. This topic is treated more extensively in the next section.

EXERCISES

THEORY

2.1   Suppose the equal-likelihood model is assumed for an arbitrary probability space in which $n(S) = n$. Show that $P(A|B) = n(A \cap B)/n(B)$.

2.2   Show that in any equal-likelihood model, no two distinct elementary events can be independent.

2.3   Show that, if the causes have equal likelihood in Bayes's theorem, then the formula becomes

$$P(C_j|A) = \frac{P(A|C_j)}{\Sigma_{i=1}^{k} P(A|C_i)} .$$

2.4   Suppose, in the three-color problem (Example 2.4), the analyst had used the sample space $S' = \{w, r, g\}$. Then the event "red ball drawn" is an elementary event, $\{r\}$, and the probability of this event would seem to be $\frac{1}{3}$ since there are three elementary events. Yet in Example 2.4, it was claimed that the probability of drawing a red ball is only $\frac{1}{4}$. Explain.

APPLICATION

2.5   A die is loaded so that the probability of the event $\{o\}$ is proportional to o, where $o \in S = \{1, 2, 3, 4, 5, 6\}$. Describe the probability function thereby determined. *Hint*: List the point probabilities.

2.6   A person is to be selected at random from a group and asked his preference concerning a certain issue. Suppose that twice as many people favor "yes" as "no," and that three times as many are undecided as favor "yes." Establish a probability model under these assumptions.

2.7   Each of two persons tosses three fair coins. What is the probability that they obtain the same number of heads?

2.8   A child has three bags of marbles. Bag 1 has five white and five blue marbles, bag 2 has two white and six blue, bag 3 has four white and ten blue. If a bag is selected at random and then a marble is selected, what is the probability that it is white? That it is blue?

2.9   In Exercise 2.8, suppose it is given that a blue marble was selected. Which bag is most likely to be the one that was selected?

2.10 Apply Bayes's theorem to the following experiment. One urn (I) contains six white balls and two green balls. Another urn (II) contains four white balls and three green balls. An urn is selected *at random,* and a ball is drawn *at random* from that urn. If the ball drawn is white, what is the probability that it came from urn II?

2.11 Two coins look the same, but one is fair while the other is biased, with the probability of {H} being .7. A coin is selected at random and is tossed.
      a. What is the probability that the selected coin is biased?
      b. What is the probability that the event {H} occurs?
      c. What is the probability that the coin is the fair one if the outcome on the toss is tails?
      d. What is the probability that the coin is the fair one if, in tossing it twice, we observe tails both times?
      e. Same as part (d), except that we observe heads once and tails once. Does it matter whether heads was tossed first or second?

2.12 A chest contains three drawers; drawer 1 contains two gold coins, drawer 2 contains one gold and one silver coin, and drawer 3 contains two silver coins. The experiment consists of tossing an unbiased coin twice and selecting drawer 1 if both tosses result in heads, drawer 3 if both are tails, and drawer 2 otherwise; then a coin is selected at random from the drawer. Given the information that the resulting coin is silver, what is the conditional probability that it came from drawer 2? Drawer 3? What is the probability of getting a silver coin?

2.13 Suppose that, in answering a question on a multiple-choice test, where each question has five possible answers, a student either knows the answer or she guesses. Let p denote the probability that she will know the answer, so that $1 - p$ is the probability that she guesses. Assume that the student selects an answer at random if she does not know the answer. Find the probability that the student knew the answer, given that she correctly answered it.

2.14 A number N is chosen at random from the set {1, 2, 3}, and a second number M is chosen from the set {0, 1, . . . , N}.
      a. Find the probability that M is j, given that N was k; $j = 0, 1, 2, 3$; $k = 1, 2, 3$.
      b. Find the probability that N was k, given that M is j; $k = 1, 2, 3$; $j = 0, 1, 2, 3$.
      c. Find the probability that M is 2.
      d. Find the probability that M is at least 2.
      e. Find the probability that M is at most 2.

2.15 Two cards are drawn at random from among the four jacks of a deck of playing cards.
      a. What is the probability that they are both red?
      b. What is the probability that they are both red, given that one of them is red?

    c. What is the probability that they are both red, given that one of them is the jack of hearts?

2.16 A recreational gear issue room has five footballs it can issue; one of the five has a slow leak. Three people are standing in line to check out a ball. Each time a ball is issued, it is selected at random from the balls available.
    a. What is the probability that the ith person gets the leaky ball?
    b. What is the probability that none of the three get the leaky ball?

2.2   COMBINATORIAL PROBLEMS

Problems of counting the various ways of grouping objects together, such as the poker hand problem, furnish a context for many kinds of classical probability models based on the equal-likelihood assumption. The experiments involved can often be described as taking place sequentially, making an n-tuple representation of points in S desirable.

EXAMPLE 2.6 Ten horses are entered in a race. The experiment consists of observing the top three winners as win, place, and show (assume no ties). A typical outcome may be conveniently represented as a 3-tuple, $o = (x_1, x_2, x_3)$, in which x_1 is the name of the horse that wins, x_2 the name of the horse that shows, and x_3 the name of the horse that places. It is convenient to let the horses be named by the numbers 1, 2, . . . , 10, and to take

$$S = \{(x_1, x_2, x_3); x_i \in \{1, 2, \ldots, 10\}, x_i \neq x_j \text{ if } i \neq j\}.$$

    To count the number of elements of such a sample space, we employ what is often referred to as a *fundamental principle of counting*. If a first task can be performed in α_1 ways and after that a second task in α_2 ways, and so on until a kth task can be performed in α_k ways, then all k tasks can be performed in sequence in $\alpha_1 \cdot \alpha_2 \cdot \alpha_3 \cdots \alpha_k$ ways.

EXAMPLE 2.7 In Example 2.6, imagine selecting numbers to "build" 3-tuples. Now, x_1 can be selected in any of ten different ways, and, once x_1 is selected, x_2 can be chosen in any of the nine remaining ways, and, finally, x_3 can be selected in any of the eight ways then remaining. Consequently, the total number of ways of building (x_1, x_2, x_3) is $10 \cdot 9 \cdot 8 = 720$. It follows that $n(S) = 720$.

    In the counting principle, when α_1 is some integer n, $\alpha_2 = n - 1, \ldots, \alpha_k = n - k + 1$, then the product $\alpha_1 \alpha_2 \cdots \alpha_k =$

$n(n - 1) \cdots (n - k + 1)$ is known as the *number of permutations of n things k at a time* and is denoted $(n)_k$. In our example, $n(S) = (10)_3$ is the permutation of 10 things (names) taken 3 at a time. The permutation of n things n at a time is the product $n(n - 1) \cdots (3)(2)(1)$ and is denoted $n!$, read n *factorial*. This is the number of ways of arranging all n things into an ordered n-tuple. An alternative formula for $(n)_k$, for $k \leq n$, is $(n)_k = n!/(n - k)!$. In our example, $720 = 10!/7! = 3,628,800/5040 = 10 \cdot 9 \cdot 8$. With this expression, observe that $(n)_0 = 1$.

EXAMPLE 2.8 What is the probability of getting all clubs in a five-card poker hand?

SOLUTION To solve this problem, imagine the deck of cards numbered from 1 through 52 to constitute balls in an urn, with clubs numbered 1 through 13. Then

$$S = \{(x_1, x_2, x_3, x_4, x_5); x_i \in \{1, 2, \ldots, 52\}, x_i \neq x_j \text{ if } i \neq j\}$$

and $n(S) = (52)_5 = \dfrac{52!}{47!} = 52 \cdot 51 \cdot 50 \cdot 49 \cdot 48 = 311,875,200.$

If A is the event in question, then $A = \{(x_1, x_2, x_3, x_4, x_5); x_i \in \{1, 2, \ldots, 13\}$, and $n(A) = (13)_5 = 13!/8! = 154,440$. Consequently, $P(A) = 154,440/311,875,200 = .0005$, if we assume all five-card hands are equally likely [so each particular one has probability $1/(52)_5$].

The solution to Example 2.8, while legitimate, accounts for more than is necessary. For example, in that solution, the hand (1, 7, 15, 39, 4) is distinguished from the hand (7, 1, 15, 39, 4) among all other possible rearrangements of the integers involved. By the counting principle, in all, there are $(5)_5 = 5! = 120$ such hands that are, from the point of view of the game, unnecessarily made distinct. If all 120 ordered hands were represented by the set {1, 7, 15, 39, 4}, then the whole sample space could be redefined as the set of all subsets of size 5 taken from a set of 52 objects, that is, all unordered five-card hands. Since there are $(52)_5$ *ordered* 5-tuple representations of a hand, every 5! of these can be replaced by an *unordered* set representation of five objects in a subset, for a total of $(52)_5/5! = 2,598,960$ unordered sets. If the sample space is taken to be this collection of sets of 5 cards, then the event "all clubs," as a subset of this new sample space, would be the set of all subsets of 5 from 13 objects. There are $(13)_5/5! = 1287$ unordered club hands. Then the probability of drawing an all-club hand is $1287/2,598,960 = .0005$, the same numerical value as in Example 2.8. The unordered-hand approach represents another way to model the same problem. Note, however, that the sample space associated with unordered arrangements would not allow you to form an event representing "first card drawn was an ace," so choice of S depends on what questions you want to ask (events you wish to form).

In general, $(n)_k/k!$ is denoted $\binom{n}{k}$ and is called the *number of combinations of n things k at a time*. It is used to measure the number of subsets of size k taken from a universe of n objects, so we assume $k \le n$. If $k > n$ (or $k < 0$), it is convenient to define $\binom{n}{k} = 0$. An alternative formula for $\binom{n}{k}$ is

$$\binom{n}{k} = \frac{n!}{k!(n-k)!}. \tag{2.1}$$

Since 0! is defined to be 1, it follows that $\binom{n}{0} = 1$. In Example 2.8, $\binom{52}{5} = 2{,}598{,}960$.

EXAMPLE 2.9   In Example 2.6, suppose five of the ten horses are grey. What is the probability that the three leaders of the race are all grey?

SOLUTION   Using S as the set of subsets of size 3, we have $n(S) = \binom{10}{3} = 120$. The event in question, G, then has $n(G) = \binom{5}{3} = 10$ elements, so that $P(G) = \frac{1}{12}$ under the equal-likelihood model. Alternatively, with the ordered-triple sample space of Example 2.6, $n(S) = (10)_3 = 720$, and, with the grey horses numbered one through five, $n(G) = (5)_3 = 60$. Again, $P(G) = \frac{1}{12}$. *Note:* The equal-likelihood model would usually not be appropriate in an actual horse race application.

Both of the example situations may be viewed as drawing balls successively from an urn; such situations are often called *sampling without replacement*. In the k-tuple formulation of S, this situation is exemplified by the condition $x_i \ne x_j$ when $i \ne j$. This condition forces coordinates of each k-tuple to be distinct. In the subset formulation of S, it is understood that elements of a set are not to be duplicated, so distinctness is built in. Many classical problems may be formulated as urn problems, but, as Example 2.9 brings out, some attention should be paid to whether or not the equal-likelihood assumption is reasonable.

This basic experiment can be modified slightly by allowing the ball to be replaced in the urn after each drawing. Such situations are called *sampling with replacement*. In sampling with replacement, a number may occur more than once in the k-tuple. Indeed, it is possible that the same ball is drawn every time. Again, S has a representation as a set of k-tuples, $S = \{(x_1, x_2, \dots, x_k); x_i \in \{1, 2, \dots, n\}\}$, only this time there is no restriction that $x_i \ne x_j$. With the fundamental counting principle, $n(S) = n^k$ in this case.

EXAMPLE 2.10   A three-letter word is to be formed at random from the English alphabet, and any three letters in juxtaposition constitutes a word. What is the probability that the letters are all from A to M inclusive and are distinct?

SOLUTION Since any three letters, alike or not, constitute a word, $S = \{(x_1, x_2, x_3); x_i \in \{1, 2, \ldots, 26\}\}$, where we imagine the letters numbered successively starting with A and ending with Z. Then, $n(S) = 26^3 = 17{,}576$. The event in question, say E, can be represented as $E = \{(x_1, x_2, x_3); x_i \in \{1, \ldots, 12\}, x_i \neq x_j$, where $i \neq j\}$, so $n(E) = (13)_3 = 1716$. Finally, since a word is to be formed at random, the equal-likelihood assumption is valid, and we may compute $P(E) = 1716/17{,}576 = .0976$.

A classical application of sampling from an urn with replacement is the so-called *birthday problem*.

EXAMPLE 2.11 Suppose a room contains m people. What is the probability that two or more persons in the room have the same birthday (i.e., they were born on the same day of the year but not necessarily in the same year)? To be specific, suppose there are 25 people in the room. Do you think the odds of a common birthday are very high? (You will shortly see these odds are better than one to one!)

SOLUTION The general problem can be solved in a context of drawing balls from an urn. Suppose the days of the year are numbered systematically from 1 through 365 (ignore leap years), and imagine an urn with $n = 365$ balls from which a sample of size m is drawn with replacement. (If sampling were without replacement, there would be no problem.) It is easier to compute the probability of the event A that no two people have the same birthday and then calculate $P(\overline{A}) = 1 - P(A)$. The event A is the set of m-tuples (of elements from $\{1, 2, \ldots, 365\}$) whose coordinates are all distinct, so $n(A) = (365)_m$. Under the equal-likelihood model, $P(A) = (365)_m/(365)^m$. Consequently, the event we are interested in, $\overline{A}$, has probability $P(\overline{A}) = 1 - (365)_m/(365)^m$. For $m = 50$, the probability is almost unity, and it is just over $\frac{1}{2}$ when $n = 23$; even for $m = 10$, this probability is .12, which most people would generally consider high. Of course, this result depends on some strong assumptions, which could easily be challenged for this example. (You might find it fairly easy to argue that the probabilities of common birthdays are even higher for actual groups of people.)

Not all combinatorial problems can be answered so simply with permutations, combinations, or powers. Occasionally, some innovative analysis is required, perhaps using the fundamental counting principle.

EXAMPLE 2.12 The state of California uses automobile license plates that consist of three letters followed by three digits or three digits followed by three letters. The total number of possibilities in the first case, since replacement is allowed, is $26^3 \cdot 10^3 = 17{,}576{,}000$. This result is the same as the number of possibilities in the second case, so the total number is 35,152,000. If certain letter combinations, say 100 of them, are not allowed on the grounds of being offensive, then we must substract 100 from the 26^3 possibilities. The fundamental counting principle gives a revised count of the number of possible license plates of $2 \cdot (26^3 - 100) \cdot 10^3 = 34,952,000$.

It should be noted that the basic distinction between using permutations and using combinations in finding the number of ways of selecting m distinct objects from a total of n is whether the *order* in which the objects are selected is to be taken into account. If the order in which the objects are selected is taken into account (different orders of the same objects being counted), then permutations should be used. If the order is not taken into account, combinations should be used. Again, note that the events that can be formed will depend on this choice.

EXAMPLE 2.13   a. In how many ways can a committee of three people be formed, if they are to be selected from a group of seven people?

b. In how many ways can a committee consisting of a president, vice-president, and secretary be formed, if they are to be selected from a group of seven people and no person can hold more than one office?

SOLUTION   a. Here, the order in which the selection is made is not considered. (That is, a committee consisting of Joe, Bill, and Dick is the same as that consisting of Bill, Dick, and Joe, and we do not anticipate need for events such as "Bill was selected first.") We therefore use combinations, and the answer is $\binom{7}{3} = 35$.

b. In this case, it is different to have Joe as president, Bill as vice-president, and Dick as secretary than to have the same three people in different offices. Thus the order of selection is counted. Imagine that the president is selected first, the vice-president next, and the secretary last. Then the number of committees is calculated by using permutations, and the solution is $(7)_3 = 210$.

An interesting and useful connection between combinations and permutations can be established by using the fundamental counting principle. Consider the process of selecting an ordered k-tuple with distinct components drawn from a set of n objects. The number of ways of making such a selection is $(n)_k$ (since order counts). But imagine the selection taking place in two steps: first, select the k objects to be used as components; then assign each of the k objects to a particular component position. The first step may be done in $\binom{n}{k}$ ways (since order does not count in the first step). Once the k items are selected, they can be assigned in $k! = (k)_k$ ways. Hence, by the fundamental counting principle, the overall task can be completed in $\binom{n}{k}$ $(k)_k$ ways. But this result must be the same number as that given above, $(n)_k$. Thus $(n)_k = \binom{n}{k}$ $(k)_k$. (You are asked to verify this directly in Exercise 2.21.)

If the group from which the selections are made consists of various subgroups that are alike, or if the selections are to form subgroups that are alike, then the combination-permutation counting procedures must be modified. The following examples illustrate how this modification can be done.

EXAMPLE 2.14   A 3-flag signal is formed by selecting 3 flags, placing one on top, one in the middle, and one on the bottom. The available supply of 16 flags consists of 2 yellow, 3 red, 5 blue, and 6 green flags. How many signals can be formed?

SOLUTION   If the 16 flags are all distinguishable, the number would be $(16)_3$. However, assuming that the flags of one color are alike, the number of distinct signals is much smaller than $(16)_3$. If there were at least 3 flags of each color, we could argue as follows: The top flag can be any of the four colors, and, after it is chosen, the middle flag can be chosen in four "ways" (colors). Finally, after the first two colors are chosen, the bottom flag can be chosen in any of four ways. Thus the number of signals is $4 \cdot 4 \cdot 4 = 4^3$. Noting that there cannot be a signal with all yellow flags, we conclude that the correct number is $4^3 - 1$.

EXAMPLE 2.15   A group of 15 people is to be divided into three "working groups" of 5 people each. In how many ways can this division be done?

SOLUTION   Imagine at first that the working groups are to be labeled, say, according to the room to which they are assigned. Then we may argue as follows: Room 1 may be occupied by any 5 of the 15 people, and this selection may be made in $\binom{15}{5}$ ways. Once Room 1 is assigned, Room 2 may be filled with 5 people in $\binom{10}{5}$ ways. Finally, Room 3 may be filled in $\binom{5}{5}$ ways, so that the total number of distinguishable ways in which to fill Rooms 1, 2, and 3 is $\binom{15}{5}$ $\binom{10}{5}$ $\binom{5}{5}$. However, if we are concerned only with the membership of the groups and not with the room to which they are assigned, the number $\binom{15}{5}$ $\binom{10}{5}$ $\binom{5}{5}$ is too large, inasmuch as it includes the number of different ways in which to assign three groups to three rooms (equal to 3!). Thus the number of different groups is given by

$$\frac{\binom{15}{5} \binom{10}{5} \binom{5}{5}}{3!} = \frac{15!}{10!5!} \cdot \frac{10!}{5!5!} \cdot \frac{5!}{5!0!} \cdot \frac{1}{3!} = \frac{15!}{5!5!5!} \cdot \frac{1}{3!} \; .$$

Ratios of factorials of the type shown above are commonly encountered in counting problems, and a special notation is used in connection with them.

We will use the following notation conventions:

$$\binom{n}{m_1, m_2, \ldots, m_k} = \frac{n!}{m_1! m_2! \cdots m_k!},$$

where for all i, $m_i \in \{0, 1, \ldots, n\}$ and $\Sigma_{i=1}^{k} m_i = n$. Thus

$$\frac{15!}{5!5!5!} = \binom{15}{5, 5, 5}.$$

Remark: In selecting m items from a group of n, we also automatically determine another group—those items not selected. Thus we should have

$$\binom{n}{m} = \binom{n}{m, n - m},$$

which is easily verified.

The combination symbols we have discussed are extremely useful in connection with the expansion of powers of binomial and multinomial terms. You should recall the binomial theorem:

Binomial Theorem   For every positive integer n,

$$(a + b)^n = \binom{n}{0} b^n + \binom{n}{1} ab^{n-1} + \cdots$$

$$+ \binom{n}{n-1} a^{n-1}b + \binom{n}{n} a^n$$

$$= \sum_{j=0}^{n} \binom{n}{j} a^j b^{n-j}.$$

EXERCISES

THEORY

2.17   Show that the number of subsets of size m from a set having $n \geq m$ objects is $\binom{n}{m}$.

2.18   Use of binomial theorem to prove that $\Sigma_{j=0}^{n} \binom{n}{j} = 2^n$.

2.19   For $S = \{o_1, o_2, \ldots, o_n\}$, use Exercises 2.17 and 2.18 to show that the total number of events in S is 2^n.

2.20 Show the following:

a. $\dbinom{n}{m}\dbinom{n-m}{s} = \dfrac{n!}{m!s!\,(n-m-s)\,!}$.

b. $\dbinom{n}{m-1} + \dbinom{n}{m} = \dbinom{n+1}{m}$.

2.21 Show that $(n)_k = \binom{n}{k}(k)_k$ by using the factorial representations of these combinations and permutations.

2.22 Show that $\binom{n}{m} = \binom{n}{n-m}$.

2.23 If $k \neq 0$, show that $\binom{n}{k}/\binom{n-1}{k-1} = n/k$.

2.24 Use the binomial expansion of $(1 - 1)^n$ to show that if n is even,

$$\binom{n}{1} + \binom{n}{3} + \cdots + \binom{n}{n-1} = \binom{n}{0} + \binom{n}{2} + \cdots + \binom{n}{n}.$$

2.25 Expand $(x + 1/x)^5$ by using the binomial theorem. Express the answer as the ratio of two polynomials in x.

2.26 How many distinct partitions of the set $\{1, 2, 3, 4\}$ are there? Can you generalize this result to a set of n elements?

2.27 Show that $n\binom{n}{k} = k\binom{n+1}{k+1} + \binom{n}{k+1}$.

2.28 Verify that $(n - 1)_{(r-1)} = (n)_r \div n$.

APPLICATION

2.29 Three children are selected in succession, without replacement, from a group of three boys and three girls.
 a. What is the probability that the second child will be a boy, given that the first and third will be boys?
 b. What is the probability that the second child will be a boy, given that two of the three will be boys?

2.30 An appliance store carries a large stock of three sizes of refrigerators. Of all refrigerators sold, 20% are small, 50% are medium, and the rest are large. Assume the next three customers make independent purchases of refrigerators.
 a. What is the probability that each of the next three refrigerators sold will be large?
 b. What is the probability that, in the next three refrigerators sold, one will be of each size?

2.31 An urn contains 100 balls, 6 of which are black and 94 of which are red. Three balls are to be drawn. Let A be the event that all 3 sampled balls are black and B the event that 2 are red and 1 is black. Compute $P(A)$ and $P(B)$ if the sampling is conducted (a) without replacement and (b) with replacement.

2.32 An urn contains six red and eight blue marbles. Five marbles are drawn at random without replacement. Compute the following probabilities: (a) three are red and two are blue; (b) two are red and three are blue; (c) none are red; (d) at least one is red; (e) at most two are red.

2.33 A box contains eight red, three white, and nine blue balls. Three balls are to be drawn at random without replacement. Determine the probability of these events: (a) all three are red; (b) all three are white; (c) two are red and one is blue; (d) at least one is white; (e) the first drawn is red, the second white, and the third blue; (f) one of each color is drawn.

2.34 If a positive integer from 1 through 100 is selected at random, what is the probability that it will be a multiple of 3? That it will contain a 3 in its name? That it will be the square of a digit? That it will be the square of a prime?

2.35 a. What is the probability of any particular poker hand of 5 cards being dealt from an ordinary well-shuffled deck of 52 cards?
   b. What is the probability of a flush (all cards of one suit)?
   c. What is the probability that the queen of spades is one of 5 cards dealt?
   d. What is the probability that two queens are dealt?
   e. What is the probability of getting a pair?

2.36 What is the probability that a bridge hand is all of one suit? What is the probability of any *particular* bridge hand?

2.37 A card is drawn at random from an ordinary deck. What is the probability that it will be red? That it will be a face card? That it will be a red face card?

2.38 A letter is drawn at random from the English alphabet. What is the probability that it will be a vowel? That it will be a letter appearing in this sentence?

2.39 If an event A has probability m/n, then the *odds in favor of* A are said to be m to $n - m$ and the *odds against* A are said to be $n - m$ to m. What are the odds in favor of drawing an ace from an ordinary deck of cards? What are the odds against throwing a seven in a game of craps?

2.40 A total of 50 chances are to be sold on a new television set. You purchase one ticket. What are the odds against winning?

2.41 In the toss of two fair dice, what is the probability of a seven or an eleven?

2.42 a. A change purse contains five nickels, eight dimes, and three quarters. If a coin is selected at random, what is the probability of selecting a dime?
   b. How could you arrange this experiment so that each type of coin has an equal chance of being selected?

2.43 The name of a month is drawn at random. What is the probability that it begins with the letter M?

2.44 Continuing Example 2.14, suppose that a signal is chosen by selecting 3 flags at random from the 16 flags. Find the probability of each of the following events: (a) there is one red flag in the signal; (b) the topmost flag is red; (c) there are no yellow flags; (d) a blue flag appears with a yellow flag; (e) a blue flag appears above a yellow flag; (f) the signal consists of all one color.

2.45 How many four-flag signals can be formed in the signal flag example?

2.46 In how many ways can a set of 19 elements be partitioned into four subsets consisting of 2, 3, 4, and 10 elements, respectively? *Note:* It makes a difference whether you consider the subsets to be ordered (say, the 2-element subset first, the 3-element subset second, and so on) or not ordered. Work this problem both ways.

2.47 An ordinary die is to be rolled, with some outcome, say x. Then x dice are rolled, and the total number of spots showing is considered to be the outcome.
   a. Find a sample space for this experiment.
   b. Find the probability that the outcome does not exceed five.

2.48   For what value of n is $3 \cdot \binom{n+1}{3} = 7 \cdot \binom{n}{2}$?

2.49   A committee of three people is to be chosen from four married couples.
   a. How many different committees are possible?
   b. How many committees contain two women and one man?
   c. How many committees are there such that no two committee members are married to each other?

2.50   In how many ways can seven people be seated at a round table if (a) they can sit anywhere; (b) two particular people must not sit next to each other?

2.51   From five statisticians and six economists, a committee consisting of three statisticians and two economists is to be formed. How many different committees can be formed if (a) there are no additional restrictions; (b) two particular statisticians must be on the committee; (c) one particular economist refuses to be on the committee?

2.52   Three men wearing identical hats, except for their names inside on a label, have lunch together. The hatcheck person places their hats on a shelf side by side. If, after lunch, the checker randomly selects the hats as the men are leaving, what is the probability that (a) each will get his own hat; (b) at least one will get his own hat; (c) exactly one will get his own hat?

2.53   In how many ways can two men, four women, three boys, and three girls be selected from six men, eight women, four boys and five girls if (a) there are no restrictions; (b) a particular man and woman must be selected; (c) one particular woman cannot be selected?

2.54   Ten people go salmon fishing on a commercial boat. When the boat returns, it is observed that three salmon were caught. Assuming no differences in fishing ability, compute the probability that (a) the three fish were caught by three different people; (b) all three fish were caught by the same person; (c) all three fish were caught by two people.

2.55   In the game of Keno, 100 Ping-Pong balls are in a container. Each ball has a different pair of digits on it, ranging over 00, 01, 02, . . . , 99. Two balls are selected at random without replacement. Compute the following probabilities:
   a. The first ball selected has its first digit equal to 1.
   b. The first ball has a one in either position.
   c. Both balls have a one in either position.
   d. Neither ball has a one in either position.

2.3 REPEATED BERNOULLI TRIALS

Many times the balls in an urn problem may be thought of as being of two types (colors, characteristics). Indeed, in some of the urn problems in the last section, the objects were imagined to be numbered just for convenience. There are many applications for urn problems in which the final result of sampling balls from the urn is to note how many of the two types of balls were obtained. For example, one might have a lot consisting of defective and nondefective items. It is expensive, and perhaps even destructive, to sample each and every item in the lot to determine whether it is defective. An alternative is to sample from the lot and observe how many defectives are obtained and then perhaps make some judgment about the entire lot on the basis of the results in the sample.

Imagine an urn with balls numbered 1, 2, . . . , N, with the first M balls of one type (color, defective, female, pass inspection, etc.) and the last $N - M$ of the other type in the two categories. The end result of drawing a ball and observing whether or not it has a number from 1 through M or a number from $M + 1$ to N is called a *Bernoulli trial*. We will record an H if the number 1 through M is observed and a T otherwise. Consider the result of n successive draws, or trials, in which a ball is not replaced, once drawn, and the contents of the urn are thoroughly mixed after each successive draw. This experiment constitutes drawing a sample of size n *without replacement*.

The result of such an experiment is an n-tuple $(x_1, x_2, \ldots, x_n)$ of distinct coordinates chosen from the set $\{1, 2, \ldots, N\}$. The value of x_1 can be any one of the numbers in this set, with probability $1/N$. Let us denote this probability $p(x_1)$. What the value of x_2 will be depends on the value of x_1, but whatever x_1 is, there are $N - 1$ balls left in the urn, and under the mixing requirement, each has probability $1/(N - 1)$ of being drawn on the second draw. But this probability is a conditional probability, which we will denote $p(x_2|x_1)$. Similarly, the conditional probability of x_3, given whatever x_1 and x_2 are, is $1/(N - 3)$, or $p(x_3|x_1, x_2) = 1/(N - 3)$. The conditional probability of any coordinate x_i, given the values of the preceding x's, is $p(x_i|x_1, x_2, \ldots, x_{i-1}) = 1/(N - i + 1)$. From the multiplication rule for conditional probabilities (Eq. 1.2), the point probability of any n-tuple ($n \le N$) must be

$$\frac{1}{N} \cdot \frac{1}{N - 1} \cdots \frac{1}{N - n + 1} = \frac{1}{(N)_n}.$$

This is a model we used in these circumstances in the last section. In the end, however, each x_i is replaced with an H or a T in accordance with the earlier agreement, so the final result will be an n-tuple of H's and T's. Thus the final experiment will be referred to as one constituting *repeated, dependent Bernoulli trials*, the dependence stemming from the fact that the result of each trial depends on the previous trials in a very specific probabilistic way. Repeated, dependent Bernoulli trials can arise in situations other than sampling from an urn and with different dependence structures.

Let B_k denote the event of obtaining exactly k H's [and $(n - k)$ T's] in the sample, where $k \le n$ and $k \le M$. To evaluate the probability of B_k, observe that B_k occurs if and only if the outcome is an n-tuple $(x_1, x_2, \ldots, x_n)$ in which exactly k of the x's are numbers from 1 through M and $n - k$ are numbers from $M + 1$ through N. To account for all these possibilities, imagine first selecting the k coordinates that are to be H's. This selection can be done in $\binom{n}{k}$ ways. For each of these choices, the k selected coordinates can be filled with distinct numbers from 1 through M (hence, H's) in $(M)_k$ ways *and* the remaining $n - k$ coordinates with distinct numbers from $M + 1$ through M in $(N - M)_{(n-k)}$ ways. By the fundamental counting principle, the total number of such coordinates is then $\binom{n}{k}(M)_k(N - M)_{(n-k)}$, and, since each has point probability $(N)_n$, it follows that

$$P(B_k) = \binom{n}{k}\left[\frac{(M)_k(N - M)_{(n-k)}}{(N)_n}\right] \qquad 0 \le k \le \min(n, M). \quad (2.2)$$

EXAMPLE 2.16   An urn contains five red balls and seven green ones. In a sample of size three without replacement, what is the probability of obtaining two red balls? At most two red balls?

SOLUTION   To answer the first question, observe that the problem can be modeled by letting H stand for red, in which case $M = 5$, $N = 12$, and $n = 3$. Now $P(B_2)$ is, by Eq. (2.2),

$$\binom{3}{2}\left[\frac{(5)_2(7)_1}{(12)_3}\right] = 3 \cdot \frac{(20)(7)}{1320} = .3182.$$

To answer the second question, observe that the event in question is $B = B_0 \cup B_1 \cup B_2$ and that these are mutually exclusive events. But

$$P(B_0) = \binom{3}{0}\left[\frac{(5)_0(7)_3}{(12)_3}\right] = \frac{(7)_3}{(12)_3} = \frac{210}{1320} = .1591$$

and $$P(B_1) = \binom{3}{1}\left[\frac{(5)_1(7)_2}{(12)_3}\right] = 3 \cdot \frac{(5)(42)}{1320} = .4773.$$

Hence, $P(B) = P(B_0) + P(B_1) + P(B_2) = .9546$.

EXAMPLE 2.17   A manufacturer regularly ships items in lots of size 100 and has a record of 5% defective items. How likely is it that a sample of five without replacement from a particular lot would have no defectives? At least two defectives?

SOLUTION   As a model, take the lot to be an urn with $N = 100$, $M = 5 = .05(100)$, and $n = 5$. The answer to the first question is

$$P(B_0) = \binom{5}{0}\left[\frac{(5)_0(95)_5}{(100)_5}\right] = .7696.$$

To answer the second question, we need to compute the probability of $B_2 \cup B_3 \cup B_4 \cup B_5$. But this is the complement of $B_0 \cup B_1$, which has a probability that is easier to compute. So, with

$$P(B_1) = \binom{5}{1} \left[\frac{(5)_1(95)_4}{(100)_5} \right] = .2114,$$

the answer is $1 - P(B_0) - P(B_1) = .0190$.

In Example 2.17, M is not given directly, but a proportion p is given, so that $M = Np$. With $q = 1 - p$, Eq. (2.2) is often written

$$P(B_k) = \binom{n}{k} \left[\frac{(Np)_k(Nq)_{(n-k)}}{(N)_n} \right]. \qquad (2.3)$$

You will recall that, with an unordered sample space, the outcomes are expressed as subsets of $\{1, 2, \ldots, N\}$ rather than n-tuples. For such a sample space and sampling without replacement, B_k occurs if and only if a set $\{x_1, x_2, \ldots, x_n\}$ is obtained in which exactly k of the numbers 1 through M, and $n - k$ of the numbers $M + 1$ through N, are present. The first of these may be chosen in $\binom{M}{k}$ ways and the second in $\binom{N-M}{n-k}$ ways. Among the $\binom{N}{n}$ possible samples, then, there are $\binom{M}{k} \binom{N-M}{n-k}$ that meet the requirements of B_k. With the equal-likelihood assumption,

$$P(B_k) = \frac{\binom{M}{k}\binom{N-M}{n-k}}{\binom{N}{n}} \quad \text{or} \quad P(B_k) = \frac{\binom{Np}{k}\binom{Nq}{n-k}}{\binom{N}{n}}. \qquad (2.4)$$

The second equation applies if proportions are given. You will be asked in the exercises to show that, in fact, this formula will always give the same answer as Eqs. (2.2) and (2.3). Both formulas are often found under the name *hypergeometric distribution*; we will consider them again in the following chapters.

EXAMPLE 2.18   In a lottery of 100 tickets, there are 10 winning ones. You buy 5 tickets. What is the probability that you will win a prize?

SOLUTION   With an urn model with $N = 100$, $M = 10$, and $n = 5$, the probability that you will *not* win is

$$P(B_0) = \frac{\binom{10}{0}\binom{90}{5}}{\binom{100}{5}} = \frac{43,949,268}{75,287,528} = .5838.$$

So the probability that you will win is .4162.

In this last example, it is clear that you cannot sample with replacement. But if you could, as in Example 2.16, you would expect to get a different answer, since the effect of a preceding set of trials on a given trial would now be nil. Thus in the discussion leading up to Eq. (2.2), in the n-tuple outcome $(x_1, x_2, \ldots, x_n)$, each coordinate can be any of the numbers from $\{1, 2, \ldots, N\}$ with probability $1/N$, and there may now be repetitions. Accordingly, the point probability of any n-tuple (now we need not even restrict n to be less than or equal to N) would be $1/N^n$ instead of $1/(N)_n$. In classifying each coordinate as H or T, we would now be dealing with repeated, *independent* Bernoulli trials.

To count the number of points in the event "exactly k H's," again denoted B_k, imagine choosing the k coordinates for H's. This choice can be made in $\binom{n}{k}$ ways. For each of these choices, there are M choices for a number classified as an H and $N - M$ choices for a number classified as a T. In all, there are $\binom{n}{k} M^k (N - M)^{n-k}$ points in B_k, each with probability $1/N^n$, so that

$$P(B_k) = \binom{n}{k} \left[\frac{M^k (N - M)^{n-k}}{N^n} \right]$$

$$= \binom{n}{k} \left(\frac{M}{N} \right)^k \left(1 - \frac{M}{N} \right)^{n-k}, \qquad k = 0, 1, 2, \ldots, n. \tag{2.5}$$

EXAMPLE 2.19 Evaluate the answers to Example 2.16 if sampling is conducted with replacement.

SOLUTION From Eq. (2.5),

$$P(B_0) = \binom{3}{0} \left(\frac{5^0 7^3}{12^3} \right) = \frac{343}{1728} = .1985,$$

$$P(B_1) = \binom{3}{1} \left(\frac{5^1 7^2}{12^3} \right) = \frac{(3)(5)(49)}{1728} = .4253,$$

and $\qquad P(B_2) = \binom{3}{2} \left(\frac{5^2 \cdot 7}{12^3} \right) = \frac{(3)(25)(7)}{1728} = .3038.$

Thus $P(B) = .9276$.

EXAMPLE 2.20 Suppose, in Example 2.17, the items supplied by the manufacturer are not destroyed in the testing procedure, and sampling is conducted with replacement. In this case,

$$P(B_0) = \binom{5}{0} \left(\frac{5^0 95^5}{100^5} \right) = \frac{7,737,809,375}{1 \times 10^{10}} = .7738,$$

$$P(B_1) = \binom{5}{1} \left(\frac{5 \cdot 95^4}{100^5}\right) = \frac{2,036,265,625}{1 \times 10^{10}} = .2036,$$

so the answer is .0226. Contrast this answer with .0190, obtained when sampling is conducted without replacement.

When, as in this last example, M is expressed as Np for some proportion p, then $p = M/N$ and $q = 1 - M/N$, and Eq. (2.5) may be rewritten as

$$P(B_k) = \binom{n}{k} p^k q^{n-k}, \qquad k = 0, 1, 2, \ldots, n. \qquad (2.6)$$

This equation allows us to generalize as follows: Suppose p is any number between 0 and 1 (a ratio M/N or not) and imagine a Bernoulli trial (result is H or T) having probability p for H and q for T. If these trials are performed n times in such a way that the trials may be assumed independent, then Eq. (2.6) measures the probability of obtaining exactly k H's in the n trials. Since $P(B_k)$ is the $(k + 1)$st term in the binomial expansion of $(p + q)^n$, this expression has come to be known as the *binomial distribution*. With $p + q = 1$, you see that

$$1 = (p + q)^n = \sum_{k=0}^{n} \binom{n}{k} p^k q^{n-k},$$

as should be the case since $B_0, B_1, \ldots, B_n$ constitute a partition of the sample space (the n-tuple outcomes) into $n + 1$ events. We will return to study this distribution more extensively in the following chapters.

In summary, a model for the problem of drawing n items from a batch containing M items of type I and $N - M$ items of type II and classifying the sample of n items depends on whether sampling is conducted with or without replacement. In the former case, the sampling may be viewed as performing n repeated, independent Bernoulli trials; in the latter case, the trials are dependent. In each case, the probability of the event "exactly k items of type I are drawn" may be computed by either the hypergeometric distribution in the dependent case or the binomial distribution in the independent case. Unless the type of sampling is specified or implied in a problem, the answers with both types of sampling should be given for the sake of clarity. If the population sizes are large relative to the sample size, it doesn't make much difference which model is used. Cases where sampling without replacement is obviously implied are examples where items are destroyed or otherwise cannot possibly occur in the sample again.

EXERCISES

THEORY

2.56 Show mathematically (i.e., using the definitions of permutations and combinations), that

$$\binom{n}{k}\left[\frac{(M)_k(N-M)_{(n-k)}}{(N)_n}\right] = \frac{\binom{M}{k}\binom{N-M}{n-k}}{\binom{N}{n}}$$

so that Eqs. (2.2) and (2.4) are equivalent, quite apart from any probabilistic arguments.

2.57 Show that $P(B_k)$ in Eq. (2.4) is zero if $k > M$ or $k > n$.

2.58 Argue that the events $B_0, B_1, \ldots, B_n$ constitute a partition of the sample space of n-tuples as asserted. Therefore, show that the sum of the terms in Eq. (2.4), and, hence, in Eq. (2.2), must be unity, and thus $\Sigma_{k=0}^n \binom{M}{k}\binom{N-M}{n-k} = \binom{N}{n}$.

2.59 If a sample of size n is drawn with replacement from an urn, $\{1, 2, \ldots, N\}$, and A is the event "no ball is drawn more than once," show that

$$P(A) = \frac{(N)_n}{N^n} = \left(1 - \frac{1}{N}\right)\left(1 - \frac{2}{N}\right) \cdots \left(1 - \frac{n-1}{N}\right).$$

APPLICATION

2.60 Suppose that the gender of a newborn child is viewed as a Bernoulli experiment with equally likely outcomes. Assuming independent trials, what is the probability that a family with four children has all girls? Suppose that the probability of a boy is higher than $\frac{1}{2}$, say .52. What then is the probability of the above event?

2.61 a. A projectile is fired at a target with probability .2 of a hit. In seven independent firings, what is the probability of seven hits? What is the probability of at least one hit?

b. Verify that a reasonable mass function for this experiment is the binomial, with parameters $n = 7$ and $p = .2$.

2.62 In the game of craps, what is the probability of rolling a five and then making that point (rolling another five) on the next trial? Set this experiment up as two independent Bernoulli trials. Compare it with the experiment of rolling a single die four times.

2.63 A die is suspected of being loaded and will be so labeled if an even number shows more than 50 times in 100 independent rolls. What is the probability that the die will be called loaded when it is not?

2.64 a. Suppose that team I is considered better than team II in the sense that, on a given day, the probability that team I will win over team II is $\frac{2}{3}$. Assuming that the games constitute independent trials, what is the probability that team II will win the next two games in a five-game series?

b. What is the probability that team II will win the series?

2.65 A manufacturer claims that only 5% of the items he produces are defective. A sample of size 5 is drawn from a lot containing 60 such items, and the items are tested according to some standard. Each item is destroyed when tested. What is the probability of finding more than 5% defective items in the sample if the lot, in fact, contains 5% defectives?

2.66 Suppose that the probability of hitting a target in a single trial is .4. Assume independent trials.
   a. At least how many trials are required so that the probability of at least three hits is at least .6?
   b. What is the probability that no more than three shots are required to hit the target for the first time?

2.67 What is the probability, correct to two decimal places, that in 20 tosses of a biased coin $[P(H) = .1]$, you will observe at most two heads?

2.68 In sampling without replacement from an urn containing five white and five green balls, what is the probability that, in five draws, each of the odd-numbered draws results in a white ball?

2.69 Two tubes out of the 16 in an electronic component are known to be defective. If 4 are selected at random and tested, what is the probability that neither of the defective tubes will be discovered?

2.70 A lottery consists of tickets numbered 1 through 100. The tickets whose numbers are multiples of 10 are to be winners. What is the probability of at least one win with the purchase of two tickets? Of exactly one win? Of a win on the first ticket purchased?

2.71 A pencil holder contains 10 pencils, of which 3 are sharp. What is the probability of drawing 2 sharp pencils in a sample of size 5?

2.72 Suppose a lot consisting of 100 bolts is known to be 2% defective. What is the probability of drawing 2 defective bolts in a sample of 2 drawn without replacement?

2.73 In Exercise 2.70, what is the conditional probability of drawing a winning ticket on the second draw, given that the first draw was the winning ticket numbered 10?

2.74 A box contains 12 items. We are asked to determine whether the box contains intolerably many defective items, but we are allowed to remove and test only 3 of the items. (Assume that the items are picked at random and are not replaced after testing.) We have decided to reject the lot (box of 12) if we find one or more defectives in the three tests. What is the probability that we accept the lot, under the hypothesis that it actually contains (a) no defectives; (b) two defectives; (c) five defectives; (d) eight defectives; (e) eleven defectives? (f) How will the above answers be modified if 5 items are sampled rather than 3?

2.75 A fair coin is tossed five times.
   a. What is the probability of exactly three heads?
   b. Find the probability that at least three heads are tossed.
   c. What is the most likely number of heads to be tossed?
   d. What is the probability that heads will occur on each of the last two tosses?

2.76 A shipment of 50 television sets is received by a dealer. Four of the sets are defective. What is the probability that, in a random sample of 4, there are one or more defectives?

2.77   Three students, G, H, and I, each have probability .9 of passing a course in probability. Let A be the event that G passes, B the event that H and I pass, and C the event that exactly two of the three pass. Use conditional probabilities to show that A and B are independent but A and C are not.

2.78   A shipment of 100 stoves is received; suppose, unknown to the dealer, 5 of the stoves in the shipment are defective. The dealer decides to take a random sample of 3, and if one or more defectives are found in the sample, the shipment will be rejected. What is the probability that the shipment will be rejected (a) if sampling is done without replacement; (b) if sampling is done with replacement?

2.79   A young couple is planning to have three children. Suppose that the probability of having a boy is $\frac{1}{2}$. Let A be the event that the second child is a boy, B the event that the first and third children are boys, and C the event that two of the three children are boys. Show that A and B are independent but A and C are not, using conditional probabilities. Assume that the three births constitute independent Bernoulli trials.

2.4   OCCUPANCY PROBLEMS

Various prototype situations involving drawing balls from an urn and assigning them to various categories, or cells, have been introduced. In this section, we will discuss a set of such problems, called "occupancy problems" in the probability literature, leading to several models used in physics related to states occupied by physical particles. A large variety of seemingly different applications are abstractly equivalent to problems involving placing balls in cells. A good discussion of combinatorial methods and their use in assigning probabilities is given in Feller's classic work.*

The number of ways of assigning n distinguishable balls to k cells is k^n, since the cell number to which each given ball will go can be assigned in k ways and this assignment is done for each of the n balls. Put in another way, there are k^n different n-tuples in which any of the integers 1, 2, . . . , k can occur as each component.

Suppose each of n balls in an urn is assigned at random to one of n cells. What is the probability all cells are occupied (have a ball assigned)? First, there are n^n ways to assign the balls; under random assignment, each has probability $1/n^n$. There are $(n)_n = n!$ assignments that leave all cells occupied. Thus the probability in question is $n!/n^n$.

EXAMPLE 2.21   Five salesmen each visit an important client on randomly chosen days of the workweek. The probability that one salesman visits each day is $5!/5^5 \approx .0384$, which is surprisingly small. It is very likely two or more

---

* William Feller, *An Introduction to Probability Theory and Its Applications*, vol. I, 2nd ed. (New York: Wiley, 1957).

salesmen call on the same day and that, on other days, no salesman calls. The expression $n!/n^n$ gets small very rapidly as n increases; with $n = 7$, it is .0061, for example. Thus with random assignments to the cells, it is very unlikely that a uniform assignment will result.

Any assignment of n indistinguishable balls to k cells is characterized by the *cell counts,* say $n_1, n_2, \ldots, n_k$, where n_i is the number of balls assigned to the ith cell. The number of distinguishable assignments of balls to the cells is thus the number of integer solutions to the equation $n_1 + n_2 + \cdots + n_k = n$, $n_i \geq 0$. This number can be obtained by a "barrier argument," as follows: Imagine the n balls are in a groove in a rack, and k cells are to be formed by inserting $k - 1$ spacers between the balls. The total number of positions for objects in the groove is $n + k - 1$, and the number of ways of choosing positions for the $k - 1$ spacers is $\binom{n+k-1}{k-1}$. Alternatively, the number of ways of choosing n positions for the balls is $\binom{n+k-1}{n}$, which is the same number.

EXAMPLE 2.22   With $n = 5$ balls and $k = 2$ cells, the number of assignments is $\binom{6}{1} = 6$, which amounts to choosing a position for the spacer from among 6 positions in the groove.

If $n \geq k$, the number of assignments for which no cell is empty is $\binom{n-1}{k-1}$, since, in this case, no two spacers can be adjacent. Thus if you imagine a potential location ("gap") for a *single* spacer between any two balls, there are $n - 1$ such gaps; the number of ways of assigning $k - 1$ spacers to the $n - 1$ gaps (so no gap receives more than one spacer) is $\binom{n-1}{k-1}$ by the multiplication principle.

Occupancy problems occur in statistical mechanics, where a physical system with n indistinguishable particles is modeled as occupying, at any point in time, some selection from a large number k of cells. These cells represent subdivisions of space into discrete categories. There are three contending models, stated in terms of assumptions of the probability distributions associated with the various occupancy configurations. One model is the Maxwell-Boltzmann model, in which the k^n possible arrangements are assumed to have equal likelihoods. In this case, the probability of a certain set of cell counts, $n_1, n_2, \ldots, n_k$, is

$$\frac{\binom{n}{n_1, n_2, \ldots, n_k}}{k^n},$$

since there are $n!/n_1! n_2! \cdots n_k!$ ways in which n balls can be divided into k ordered cells with n_1 balls in the first cell, n_2 balls in the second cell, and so on (see Exercise 2.83).

In the Bose-Einstein model, only distinguishable arrangements are considered, each having the probability $1/(^{n+k-1}_n)$. In a third model, the Fermi-Dirac model, an equal likelihood is ascribed to all arrangements for which no two particles occupy the same cell. Such an arrangement is determined once the choice is made of which n of the k cells are occupied. This choice can be made in (^k_n) ways, so each of these configurations is assigned probability $1/(^k_n)$.

There is physical evidence that certain types of particles "follow" (are modeled well by) the Bose-Einstein model, while other particles appear to follow the Fermi-Dirac model. There is no evidence supporting the Maxwell-Boltzmann model.

EXAMPLE 2.23 Suppose $n = 3$ and $k = 4$. The cell counts $n_1 = 1$, $n_2 = 0$, $n_3 = 2$, and $n_4 = 0$ are a configuration that has the probability shown next under each model.

Maxwell-Boltzmann: $3/4^3 = .047$.

Bose-Einstein: $1/(^6_3) = .05$.

Fermi-Dirac: $1/(^4_3) = .25$.

EXERCISES

THEORY

2.80 If n balls are randomly placed into n cells, what is the probability that at least one cell is empty?

2.81 If n balls are randomly placed into r cells, what is the probability a specified cell contains exactly k balls?

2.82 Suppose n balls are randomly placed into $k > n$ cells. Show that the probability that no cell contains more than one ball is $(k)_n/k^n$.

2.83 Show that the number of ways of assigning n balls to k ordered cells, so that there are n_i balls in the ith cell for $i = 1, 2, \ldots, k$, is $n!/n_1!n_2! \cdots n_k!$. *Hint:* Use the multiplication principle to show that the number is

$$\binom{n}{n_1} \cdot \binom{n - n_1}{n_2} \cdots \binom{n - n_1 - \cdots - n_{k-2}}{n_{k-1}},$$

then simplify.

APPLICATION

2.84 Suppose numbers four digits long are constructed by generating individual digits at random and putting them in ordered strings of four digits. What is the probability that in a random sample of 25 such four-digit numbers, there are no 2 numbers the same?

2.85 Six passengers on a bus are traveling independently.
  a. What is the probability that exactly one traveler departs at each of the remaining six stops on the bus route?
  b. What is the probability that each traveler departs at a different stop if ten stops remain on the route? *Hint:* See Exercise 2.82.

2.86 How many partial derivatives of order four could there be of a function of three variables?

2.87 Twelve independent calls will be received by an individual next week (seven days). What is the probability at least one call is received each day?

2.88 Four typographical errors occur in a page of typed manuscript. Suppose the page has 28 lines, each 60 characters long.
  a. If each of the $60 \cdot 28$ character positions is a "cell" and each of the 4 typographical errors is a "particle," which of the three models of statistical mechanics is most reasonable for calculating the probability of any set of cell counts?
  b. What is the probability no line contains more than one typographical error?
  c. What is the probability the fourth line contains no typographical error?

2.89 a. What is the probability that no six occurs in ten tosses of a die?
  b. What is the probability that all faces occur at least once in ten tosses?

2.90 Use an occupancy argument to show there are 2^n ways to assign n balls to two cells, and relate this result to the same result found earlier with the binomial theorem.

2.91 Suppose a random sample of 5 balls is taken, with replacement, from an urn containing 15 distinguishable balls. What is the probability that the sampled balls are all different?

2.5 SUMMARY

When S is finite, say $S = \{o_1, o_2, \ldots, o_n\}$, the probabilities of all 2^n possible events are determined once the point probabilities $p_i = P(o_i)$ are assigned. In equal-likelihood models, the point probabilities are all the same, $p_i = 1/n$.

The permutation of n items r at a time is $(n)_r = n \cdot (n - 1) \cdots (n - r + 1)$, which represents the number of ordered arrangements of r objects selected from a set of n distinguishable objects.

The combination of n items r at a time is

$$\binom{n}{r} = \frac{n!}{r!(n - r)!},$$

which represents the number of unordered subsets of r items selected from a set of n distinguishable objects.

For a population of n objects, there are n^r different samples of size r with replacement; there are $(n)_r$ different samples of size r without replacement.

An assignment of n indistinguishable balls to k cells is described by the cell counts, $n_1, n_2, \ldots, n_k$. The number of possible assignments is $\binom{n+k-1}{n}$. If $n \geq k$, the number of assignments for which no cell count is zero is $\binom{n-1}{k-1}$.

Chapter 3

RANDOM VARIABLES

3.1 BASIC CONCEPTS

A problem with many of the sample spaces introduced so far is that their elements are general labels, not necessarily having numerical properties. This feature can be a shortcoming, for it is desirable to analyze experimental outcomes using such processes as averaging, taking absolute values, and squaring results. This problem is resolved with the concept of a random variable. What we have in mind is a relabeling function that always results in a numerical sample space.

Basically, a *random variable* is a mapping (function) of the sample space into the real line [which we will abbreviate as Re or $(-\infty, \infty)$], thereby converting or relabeling abstract outcomes in S to real number labels in Re. But instead of using the familiar symbols for ordinary functions of real variables (such as f and g), for historical reasons, probabilists use capital letters, usually near the end of the alphabet, such as X or Y, as symbols for such functions. Then instead of the familiar $y = f(x)$ to display y as the *value* of f at x, probabilists use the lowercase letter corresponding to the function to display $x = X(o)$ so that x is the value of the function X at the point o of the sample space S. The probability analogues to the more familiar notation in elementary analysis is displayed in the accompanying table.

	Typical Function	Independent Variable	Dependent Variable
Analysis	f	x	$y = f(x)$
Probability	X	o	$x = X(o)$

Whereas ordinary functions f have the real line, or some subset of it, as its domain, the domain of the function X is always S. The usual method

73

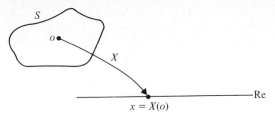

FIGURE 3.1   *X as a Mapping from S to* R_e

of graphing is not available in probability theory because of the nature of the domain S. One can display the function X as in Fig. 3.1, which conveys the nature of X as a mapping carrying the point o into the real number x.

EXAMPLE 3.1   We have often discussed the prototype experiment of tossing two coins, using $S = \{(H, H), (H, T), (T, H), (T, T)\}$. If interest lies in the total number of heads resulting from the toss, a random variable X can be defined explicitly by using a table like the accompanying table. Such a table does as well in this simple case as the graph shown in Fig. 3.2 to display the function X, with range $R_X = \{0, 1, 2\}$.

o	(H, H)	(H, T)	(T, H)	(T, T)
$x = X(o)$	2	1	1	0

EXAMPLE 3.2   When two dice are rolled, a suitable sample space is $S = \{(1, 1), (2, 2), \ldots, (6, 6)\}$, which can be represented as a set of points in an $x_1 x_2$-plane as in Fig. 3.3. The game of craps, however, is concerned with the sum of the coordinates of the outcome. We can represent this sum as the function $Y((x_1, x_2)) = x_1 + x_2$. Here, the function Y has a two-dimensional domain and hence could be graphed in three-dimensional space. We depict it in Fig. 3.3 as a mapping. We could also have used a table like the accompanying table.

o	(1, 1)	(1, 2)	(1, 3)	(1, 4)	(1, 5)	(1, 6)	(2, 1)	(2, 2)	(2, 3)	(2, 4)	(2, 5)	(2, 6)
$y = X(o)$	2	3	4	5	6	7	3	4	5	6	7	8
o	(3, 1)	(3, 2)	(3, 3)	(3, 4)	(3, 5)	(3, 6)	(4, 1)	(4, 2)	(4, 3)	(4, 4)	(4, 5)	(4, 6)
$y = X(o)$	4	5	6	7	8	9	5	6	7	8	9	10
o	(5, 1)	(5, 2)	(5, 3)	(5, 4)	(5, 5)	(5, 6)	(6, 1)	(6, 2)	(6, 3)	(6, 4)	(6, 5)	(6, 6)
$y = X(o)$	6	7	8	9	10	11	7	8	9	10	11	12

An alternative to Fig. 3.3 is Fig. 3.4, where the numerical outcome y is appended to the point representation of o in S. However Y is viewed, $R_Y = \{2, 3, 4, 5, 6, 7, 8, 9, 10, 11, 12\}$.

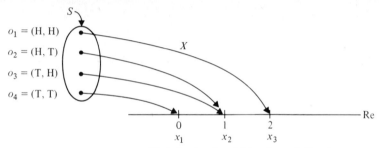

FIGURE 3.2 $x = X(o)$ *Is the Total Number of Heads*

EXAMPLE 3.3   In the die experiment, $S = \{1, 2, 3, 4, 5, 6\}$ already consists of a set of real numbers and as such could be viewed as the range of a random variable. To be consistent in this case, define a random variable X to be the identity function, $x = X(o) = o$. This definition amounts to denoting o-values as x. The situation can easily be depicted as in Fig. 3.5. Naturally, $R_X = \{1, 2, 3, 4, 5, 6\}$.

EXAMPLE 3.4   A Bernoulli trial has often been represented by $S = \{H, T\}$. But it is just as instructive to represent heads by 1 and tails by 0. We could have used the numerical labels in the first place, but having failed to do so, we can simply define a random variable W by the rule $W(H) = 1$ and $W(T) = 0$, so $R_W = \{0, 1\}$. If we announce that the *value* of W was 0, we know that T was the original outcome. We will often use this numerical replacement of H and T when discussing Bernoulli trials.

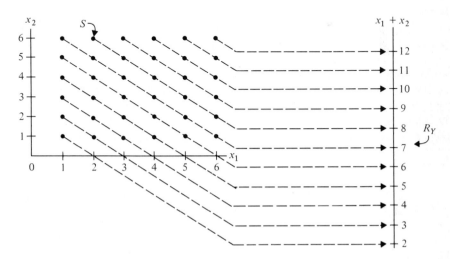

FIGURE 3.3   *Game of Craps as the Function Y*

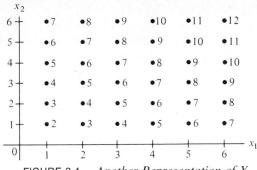

FIGURE 3.4   *Another Representation of Y*

DEFINITION 3.1   Let S be the sample space of a random experiment. A real-valued function X with domain S is called a *random variable*.

Two remarks are in order. First, if X is a random variable (a mapping from S to Re), then such functions as X^2, sin X, and e^X must also be random variables. We will need to explore how such functions of X are related to X probabilistically, for they are often of as much interest to us as X. Second, whether we start with an experiment where S is already a set of real numbers and use the identity map, as in Example 3.3, or have a nonnumerical sample space whose labels o are mapped into points x by an explicit rule, in the end, our attention is focused on the range R_X of X as the set of outcomes of interest to us. In that sense, R_X becomes a new sample space that has replaced the old sample space S.

It is necessary to determine a probability structure on R_X; in doing so, we encounter two situations. If X is the identity map, then the task is exactly as it was in the preceding chapters. For finite R_X, we need only specify point probabilities for elements of R_X (real numbers this time) and continue as before. But if S already has a probability structure, we are not free to specify point probabilities on R_X in any manner we please. It is necessary to carry over the probability structure from S to R_X. For instance, in the two-dice game (Example 3.2), the point 8 in R_X can only result from the

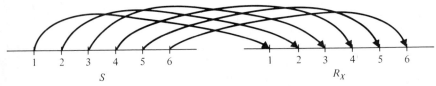

FIGURE 3.5   *Identity Mapping*

dice roll being one of the outcomes (2, 6), (3, 5), (4, 4), (5, 3), or (6, 2). Under the usual assumptions, the probability of this event happening is $\frac{5}{36}$, so it seems logical that the point probability of 8 in R_X should be $\frac{5}{36}$.

In any case, if R_X is the new desired sample space, then X may be thought of as a *placeholder* for a value in its range R_X. As such, it represents the *random* status *prior* to the performance of the experiment; hence the name *random variable*. After the experiment is performed, the variable X is then replaced by exactly one point x in R_X. We will want to make predictions or likelihood assessments of what the value of X will be (the number it will be replaced by) before the experiment is performed. *After* the experiment is performed, x has occurred and there is no prediction to be made.

In applications, the random variable (or variables) is (are) usually identified early in the model-building process. Once this identification is done, the underlying experiment with outcomes o in S is often essentially ignored. For modeling purposes, all efforts are applied to the probability space induced by the variable(s). Thus if the experiment involved performance of a chemical analysis, with an outcome involving the color of an indicator, this outcome might be transformed immediately to a numerical value X, perhaps by measuring the amount of light absorbed by the indicator. It is the value of X that is recorded in the chemist's notebook and is later analyzed. From the probabilist's point of view, it is as if the random variable X were a projector. The probabilist will see a number projected on the wall, and he will deal with it as an outcome x. The amount and nature of the chemists' activity that gave rise to the slide put in the projector is, from a purely theoretical point of view, not of concern to the probabilist. (Actually, from a practical point of view, it is usually a good idea for the analyst to have at least some understanding of the underlying experiment.) You will notice that most of our discussions from now on will begin with the assumption that a random variable has been defined, and attention will be concentrated on the new probability space, involving R_X, rather than the original space involving S.

3.2 INDUCED PROBABILITY

How should the probability structure on R_X be defined when a probability function P has already been defined for events in S? If S is finite, then no matter what the functional form of X, the new sample space R_X is finite. Consider a point $x \in R_X$. By definition, x must be the image of at least one point o in S. Temporarily designate the set of such points o in S as $A(x)$. Thus $A(x) = \{o : X(o) = x\} \subset S$. Start with an $x \in R_X$ and search through S to find all o-points that are mapped into x, as depicted in Fig. 3.6. For a given x, there may be just one o, or there may be several different ones. For instance, in the game of craps (Example 3.2), $A(8) = \{(2, 6), (3, 5),$

(4, 4), (5, 3), (6, 2)}, while $A(12) = \{(6, 6)\}$. Define the point probability for x as $P[A(x)]$, the probability of the event $A(x)$ in S.

DEFINITION 3.2   Let X be a random variable having range R_X. If S is finite, let $p(x) = P[A(x)]$ for each $x \in R_X$, where $A(x) = \{o : X(o) = x\}$. Then $p(x)$ is called the (*probability*) *mass function* for X evaluated at x. $p(x)$ is also called the *probability function* for X, or the probability *induced* by X, or the *distribution* of X.

The reason for calling $p(x)$ a mass function is that, since R_X is a finite set of isolated points on the real line, these points, together with the associated point probabilities, may be viewed as a physical mass system. In this system, a total mass of 1 unit is divided up in such a way that the portion $p(x)$ is placed at x. Figure 3.7 depicts such a mass system with four mass points and the mass values depicted as heights, or "spikes," above the points.

Let us verify that a mass function, as defined, does indeed provide a set of point probabilities for R_X (which, in turn, determines the probabilities of all events in Re). First, since the value of a function is unique, it must be true that $A(x_1) \cap A(x_2) = \phi$ for any distinct points x_1, x_2 in R_X. Next, since R_X is the range of X, it contains every possible value of X. Then the union of all the $A(x)$-sets, again as subsets of S, must be S. That is, if $R_X = \{x_1, x_2, \ldots, x_m\}$, then $\bigcup_{i=1}^{m} A(x_i) = S$. Thus $A(x_1), A(x_2), \ldots, A(x_m)$ is a partition of S. Accordingly,

$$\sum_{i=1}^{m} p(x_i) = \sum_{i=1}^{m} P(A(x_i)) = P\left[\bigcup_{i=1}^{m} A(x_i) \right] = P(S) = 1.$$

Since $p(x_i) \geq 0$, the set of numbers $p(x_1), p(x_2), \ldots, p(x_m)$ is a set of point probabilities for R_X.

Once we have established a method of inducing a probability function on R_X (given P and X), the problem of calculating probabilities for events

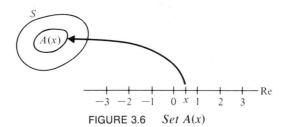

FIGURE 3.6   *Set $A(x)$*

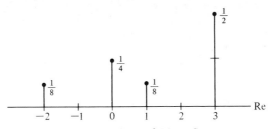

FIGURE 3.7   *Physical Mass System*

in R_X proceeds as before. Thus if $B \subset R_X$, then the probability of B is given by $P(B) = \Sigma_{x \in B}\ p(x)$. Also, it is conventional to denote $P[A(x)]$ by $P(X = x)$, read as "the probability the *value* of X will be x" or, more briefly, "the probability that X equals x." Then $p(x) = P(X = x)$ is a numerical index of how likely it is that X will be replaced by x, once the experiment is performed. This is how we will interpret a mass function value.

It is convenient to extend Definition 3.2 so that x can be any real number. Thus if $x \notin R_X$, surely $A(x) = \phi$, and, accordingly, $p(x) = P(X = x) = 0$. Recall, however, that even if x is in R_X, it may be that $p(x) = 0$. There is a subtle and not too important distinction in interpreting $p(x) = 0$ when $x \in R_X$ and when $x \notin R_X$. In the latter case, it is not possible for x to even occur as a value of X, a replacement for X; in the former case, it is possible for x to arise but so extremely unlikely that it is viewed as impossible for all practical purposes. With Definition 3.2 so extended, $p(x)$ is defined over all of the real line. With the agreement that $\Sigma_{x \in B}\ p(x)$ means $\Sigma_{x \in B \cap R_X}\ p(x)$, that is, those points in B where p is positive, that sum may be interpreted as the probability that the value of X will be found in the set $B \subset \text{Re}$; we denote this probability by $P(X \in B)$. For special B-sets, such as intervals $B = [a, b], (a, b), (a, b], [a, b)$, the notation is abbreviated further to $P(A \leq X \leq b), P(a < X < b), P(a < X \leq b)$, and $P(a \leq X < b)$, respectively. The values $P(a \leq X < \infty), P(-\infty < X < b), P(-\infty < X \leq b)$, and $P(a < X < \infty)$ are usually written $P(a \leq X), P(X < b), P(X \leq b)$, and $P(a < X)$, respectively.

EXAMPLE 3.5   Let us complete the determination of the induced probability mass function for the game of craps. By Fig. 3.4 or the table in Example 3.2, you can see that

$$A(2) = \{(1, 1)\}, \qquad A(3) = \{(1, 2), (2, 1)\},$$

$$A(4) = \{(1, 3), (2, 2), (3, 1)\},$$

$$A(5) = \{(1, 4), (2, 3), (3, 2), (4, 1)\},$$

$$A(6) = \{(1, 5), (2, 4), (3, 3), (4, 2), (5, 1)\},$$

$$A(7) = \{(1, 6), (2, 5), (3, 4), (4, 3), (5, 2), (6, 1)\},$$

$$A(8) = \{(2, 6), (3, 5), (4, 4), (5, 3), (6, 2)\},$$

$$A(9) = \{(3, 6), (4, 5), (5, 4), (6, 3)\},$$

$$A(10) = \{(4, 6), (5, 5), (6, 4)\},$$

$$A(11) = \{(5, 6), (6, 5)\}, \qquad A(12) = \{(6, 6)\}.$$

Accordingly, assuming fair dice, the accompanying table determines the positive values of $p(x)$.

x	2	3	4	5	6	7	8	9	10	11	12
$p(x)$	$\frac{1}{36}$	$\frac{2}{36}$	$\frac{3}{36}$	$\frac{4}{36}$	$\frac{5}{36}$	$\frac{6}{36}$	$\frac{5}{36}$	$\frac{4}{36}$	$\frac{3}{36}$	$\frac{2}{36}$	$\frac{1}{36}$

It follows that $p(x) = 0$ for all values x not in $R_X = \{2, 3, 4, 5, 6, 7, 8, 9, 10, 11, 12\}$. It is understood that $p(x) = 0$ for $x \notin R_X$, even when not explicitly stated. To compute the probability that X will be odd, $P(\{3, 5, 7, 9, 11\})$, add the point probabilities given by $p(x)$ over the points $x = 3, 5, 7, 9, 11$. Also, $P(4 \le X \le 8) = \frac{23}{36}$, $P(X \le 3) = p(2) + p(3) = \frac{3}{36} = \frac{1}{12}$, and $P(X > 12) = 0$. A graph of the mass function is shown in Fig. 3.8.

EXAMPLE 3.6   A local drugstore stocks five copies of the town's morning paper for sale each morning it is open. The number of papers sold on a given day varies, however. The possible sales $\{0, 1, 2, 3, 4, 5\}$ may be thought of as the range of a random variable X described loosely by writing $X =$ (the number of papers sold on a given day). What is the distribution of X?

SOLUTION   There is no basis for a priori assumptions about probabilities, so, for example, the equal-likelihood model does not seem appropriate.

Suppose we are able to observe sales over 100 days and find the accompanying results.

x	0	1	2	3	4	5
Frequency	1	0	5	3	67	24

As a model, then, we might *define* $p(x)$ as estimated by the relative frequencies given in the next table.

x	0	1	2	3	4	5
$p(x)$	.01	0	.05	.03	.67	.24

The probability of selling more than three papers on a given day would be, according to this model, $P(X > 3) = p(4) + p(5) = .91$. In this case, it happens that $p(1) = P(X = 1) = 0$, but we would not infer from this result that selling only one newspaper is *impossible*. On the other hand, $p(6) = P(X = 6) = 0$, and selling six newspapers is impossible.

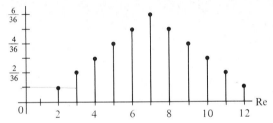

FIGURE 3.8   *Graph of the Induced Mass Function*
for the Game of Craps

To summarize, whether X is a random variable with a specified range of possible values at the outset or one that was derived through an explicit mapping of a structured sample space S, the probability mass function for X is a function satisfying two properties:

1. $p(x) \geq 0$.
2. $\Sigma_{x \in R_X} p(x) = 1$.

If X is specified at the outset, $p(x)$ is determined by the experimenter on one basis or another and assigned as a model; if a probability model with a structured sample space S has been assigned, $p(x)$ is induced by the structure on S and the subsets $A(x) = \{o : X(o) = x\}$. It is often convenient to redefine the range of X to be $R_X = \{x : p(x) > 0\}$. Whenever it is necessary to distinguish the mass function for X from other functions in a given context, we subscript p, writing p_X. Then for any subset B of Re, $P(X \in B) = \Sigma_{x \in B} p_X(x)$, where $P(X \in B)$ is an abbreviation for "the probability that the value of X will be an element of B."

It is of special interest to evaluate $P(X \leq x)$ as x is allowed to vary over all the real numbers. This result creates a function of x, which we denote $F(x)$ [or $F_X(x)$ when necessary to avoid confusion]. Since the value of F at x is formed by adding, or cumulating, all the mass values up to and including x, F is called the *cumulative distribution,* or the *distribution function,* or the CDF, for short. Formally, we have the following definition.

DEFINITION 3.3   Let X be a random variable. The *cumulative distribution function* for X is a real-valued function F, defined for each real x by $F(x) = P(X \leq x)$.

EXAMPLE 3.7   For Example 3.6, the newspaper sales example, $F(x)$ would be computed by $\Sigma_{t \leq x} p(t)$, where p is the mass function assigned on the basis of the observed frequencies. Accordingly, for $x < 0$, $F(x) = 0$, for there are no mass points to the left of 0. Just at 0, however, there is mass .01, so $F(0)$

= .01. $F(x)$ remains .01 up to the value $x = 2$. For all $2 \leq x < 3$, $F(x) =$ .01 + .05 = .06, and so on. Finally, from $x = 5$ on, $F(x) = 1$ since all the mass values cumulate to 1. A formula for F is

$$F(x) = \begin{cases} .00 \text{ if } x < 0, \\ .01 \text{ if } 0 \leq x < 2, \\ .06 \text{ if } 2 \leq x < 3, \\ .09 \text{ if } 3 \leq x < 4, \\ .76 \text{ if } 4 \leq x < 5, \\ 1.00 \text{ if } 5 \leq x. \end{cases}$$

As a function of a real variable, $y = F(x)$ has the graph shown in Fig. 3.9. It is a step function with jumps at each mass point. The height of each jump is equal to the mass value at the corresponding point. All the random variables treated so far have step function CDFs, since probabilities are concentrated at isolated points.

EXAMPLE 3.8   For the two-dice game (Example 3.5), the CDF has value 0 for $x < 2$; $F(x) = \frac{1}{36}$ for $2 \leq x < 3$; $F(x) = \frac{3}{36}$ for $3 \leq x < 4$; $F(x) = \frac{6}{36}$ for $4 < x \leq 5$; and so on, until $F(x) = 1$ for $12 \leq x$. A graph of this CDF is shown in Fig. 3.10.

In each of these cases, the distribution function F is bounded below by zero and above by unity. That is, for all x, $0 \leq F(x) \leq 1$, and, in fact,

$$\lim_{x \to -\infty} F(x) = 0 \qquad \text{while} \qquad \lim_{x \to +\infty} F(x) = 1.$$

These results are intuitively appealing, since $F(x) = P(X \leq x)$ is a probability. Also, as $x \to -\infty$, the event $(X \leq x)$ becomes empty, while as $x \to +\infty$, the event becomes Re. Another important property that F possesses in each example above is that it is a monotone increasing function; that is, if $x \leq y$, then $F(x) \leq F(y)$. A third property that F possesses is that of right continuity, which is to say that the right-hand limit $\lim_{y \to x+} F(y) = F(x)$

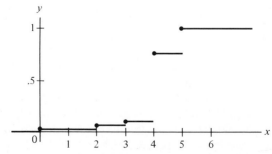

FIGURE 3.9   *CDF for the Newspaper Sales Example*

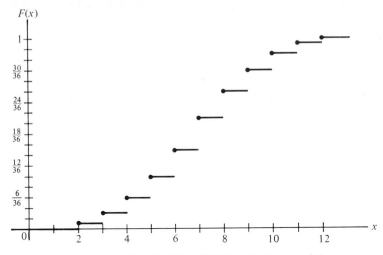

FIGURE 3.10   *Graph of the CDF for the Game of Craps*

for each $x \in$ Re. Basically, this property is so because $F(x + \epsilon) - F(x)$ $= P(x < X \le x + \epsilon)$, and as $\epsilon \to 0$, the interval $(x, x + \epsilon]$ becomes empty. However, we see from Fig. 3.9 and 3.10 that F need not be continuous.

As an example, consider the function F defined by

$$F(x) = \begin{cases} 0 \text{ if } x \le 0, \\ x \text{ if } 0 < x < 1, \\ 1 \text{ if } 1 \le x. \end{cases}$$

It is easy to see that this function possesses the foregoing properties and is continuous (so it is, in particular, left continuous). We are interested in knowing when a function may be considered to be a CDF so that, in the construction of a model, once a CDF is defined, a probability function may be defined by $P(X \le x) = F(x)$. It turns out that any function with the three above properties (goes to 0 on the left and to 1 on the right; is monotone increasing; is right continuous) can be considered a CDF.

Finally, another connection between the CDF and the mass function is $p(x) = F(x) - \lim_{t \to x-} F(t)$, expressing the value of p at x as the CDF at x minus the left-hand limit at x. This expression is just the jump of F at x. From this expression, it follows that $p(x) = 0$ at every point where F is continuous. At every point of discontinuity of F, the difference in heights of the CDF graph is precisely the positive mass value at that point.

The foregoing suggests that, in fact, a CDF contains all the probabilistic information about a random variable. Thus if X and Y have the same CDF [i.e., $F_X(z) = F_Y(z)$ for each z], then, for any event E, $P(X \in E) = P(Y \in E)$. This result means that, from a probabilistic point of view, one cannot distinguish between X and Y even though they may be different functions (see Exercises 3.7 and 3.8). When this situation prevails, X and Y are said to be *identically distributed*.

EXERCISES

THEORY

3.1 Suppose S is a sample space and a probability function has been defined for events in S. Let A be such an event, and define a function with domain S as follows:

$$I_A(o) = \begin{cases} 1 \text{ if } o \in A, \\ 0 \text{ if } o \notin A. \end{cases}$$

Show that I_A is a random variable for any event A. It is called the *indicator function* for A.

3.2 In Exercise 3.1, determine the mass function for I_A and also its CDF.

3.3 In Exercise 3.1, suppose E is any event in Re. Show that $P(I_A \in E) = 1$ if $\{0, 1\} \subset E$. Evaluate this probability for the conditions (a) $1 \in E, 0 \notin E$; (b) $1 \notin E; 0 \in E$; (c) $1 \notin E, 0 \notin E$.

3.4 Let I_A and I_B be indicator functions of two events $A, B \subset S$. For a and b real numbers with $a < b$, let $Y = aI_A + bI_B$. Show that Y is a random variable on S, and determine the range of Y.

3.5 If A and B are mutually exclusive events, show that $I_{A \cup B} = I_A + I_B$.

3.6 If A and B are two arbitrary events, show that $I_{A \cap B} = I_A I_B$, $I_A = 1 - I_{\bar{A}}$, and $I_{A \cup B} = I_A + I_B - I_{A \cap B}$.

3.7 Let $S = \{(H\ H), (H, T), (T, H), (T, T)\}$ be the sample space of the two-coin experiment. Let X be the function that counts the number of heads in an outcome, and let Y be the function that counts the number of tails. List X and Y explicitly. Show that X and Y are different functions but are identically distributed.

3.8 For the sample space $S = \{H, T\}$, let $P(\{H\}) = \frac{1}{2} = P(\{T\})$. Let X be defined by $X(H) = 1, X(T) = 0$; and let Y be defined by $Y(H) = 0, Y(T) = 1$. Show that $X \neq Y$ but $F_X(z) = F_Y(z)$ for all z, so X and Y are identically distributed.

3.9 Let X be a random variable and suppose $R_X \subseteq [a, b]$, where $a < b$. Show that $F_X(x) = 0$ if $x < a$ and $F_X(x) = 1$ if $x > b$.

3.10 Let $F(x) = \frac{1}{2} + (1/\pi)\tan^{-1}(x)$ for each $x \in$ Re. Show that F is a distribution function.

3.11 Establish each of the following formulas, assuming $a < b$. In each case, use only the fact that $F_X(x) = P(X \leq x)$ and $p_X(x) = P(X = x)$.
   a. $P([a < X \leq b]) = F_X(b) - F_X(a)$.
   b. $P([a < X < b]) = F_X(b) - F_X(a) - p_X(b)$.
   c. $P([a \leq X < b]) = F_X(b) - F_X(a) - p_X(b) + p_X(a)$.
   d. $P([a \leq X \leq b]) = F_X(b) - F_X(a) + p_X(a)$.
   e. $P([X > a]) = 1 - F_X(a)$.
   f. $P([X \geq a]) = 1 - F_X(a) + p_X(a)$.
   g. $P([X < b]) = F_X(b) - p_X(b)$.

3.12 If $b > 0$, evaluate $P([|X| \leq b])$ and $P([|X| \geq b])$ in terms of F_X and p_X, as in Exercise 3.11.

3.13 In terms of the convention adopted for $P(X \in C)$ for $C \subset \text{Re}$, it should be clear that, if $D \subset \text{Re}$ also, then

$$P(X \in C | X \in D) = \frac{P(X \in C \cap D)}{P(X \in D)} .$$

Verify this result and evaluate this expression for discrete X with probability mass function p_X. *Note:* Often, you will see $P(X \in C \cap D)$ written as $P(X \in C, X \in D)$.

APPLICATION

3.14 Let X be a random variable for the experiment of rolling an unbiased die (X is the identity map). Describe the cumulative distribution function for X and sketch its graph. Also determine the mass function.

3.15 In a roll of two fair dice, a player wins \$1 if he rolls 7 or 11 and loses \$1 if he rolls 2 or 12; otherwise, he neither wins nor loses. Define a random variable describing the game and determine the induced probability mass function.

3.16 Describe, by means of a random variable, the experiment of drawing a digit at random.

3.17 Suppose that X induces the mass function given by

$$p_X(x) = \begin{cases} \frac{2}{3} \text{ if } x = 0, \\ \frac{1}{3} \text{ if } x = 1, \\ 0 \text{ otherwise.} \end{cases}$$

Graph the CDF of X and verify that it has the required properties.

3.18 An experiment is described as follows: A box contains five transistors. The experimenter knows four of them are good and one is defective. Transistors are selected at random and removed for testing, the tested items not being returned to the box. Let X denote the number of items required to be withdrawn in order to locate the defective transistor. *Hint:* $S = \{1, 2, 3, 4\}$.
   a. Describe the range of X.
   b. Find the mass function p_X of X.
   c. Find the CDF of X.

3.19 Repeat Exercise 3.18 for a box that contains four good and three defective transistors. Sampling continues until the three defective transistors have been located.

3.20 Let X be the number of heads in two independent tosses of a fair coin. Calculate $P(X = 2 | X \geq 1)$ by using the formula of Exercise 3.13.

3.21 Let the experiment consist of tossing five fair coins, and let X count the number of heads. Determine the following: (a) R_X; (b) $p_X(x)$ for each $x \in R_X$; (c) $P(X = 3)$; (d) $P(X \leq 4)$; (e) $P(X = 3 | X \geq 4)$.

3.22 A rifleman fires at a target three times, and the probability is .8 that he hits the X ring each time. Let Y denote the number of times he misses the X ring. Determine the probability mass function for Y and its CDF.

3.23 Suppose $f(x) = (\frac{1}{2})^x$, $x = 1, 2, 3, 4, 5$; $f(6) = \frac{1}{32}$ and $f(x) = 0$ otherwise.
   a. Graph $f(x)$.
   b. Can $f(x)$ be used as a probability mass function?

3.24 A boxcar contains 50 packaged appliances, 4 of which are defective. Four are sampled without replacement for inspection. Let X denote the number of defectives found.
a. What is the probability mass function for X?
b. Calculate $P(X \leq 1)$ and, in general, the CDF for X.

3.25 A receiving depot receives a shipment of 100 generators, 5 of which are defective. Four are selected at random without replacement for inspection. Let Y be the number of defectives.
a. What is the probability mass function for Y?
b. Calculate $P(1 \leq Y \leq 3)$.

3.26 Three fair dice are rolled one time. Let X denote the number of sixes that occur.
a. What is the probability mass function for X?
b. Calculate $P(0 \leq X \leq 2)$.

3.27 Suppose $f(x) = c/(x + 1)$, $x = 1, 2, 3, 4$.
a. What must the value of c be in order for f to be a probability mass function?
b. Calculate the probability that the value of X is odd if it has this probability function.

3.28 Three fair dice are rolled one time. Let Y denote the number of odd dots that occur. What is the probability mass function and the CDF for Y? Graph both functions.

3.29 One bulb in a package of six is defective. Bulbs are tested, one after another, until the defective one is found. Let X be the number of tests.
a. What is the probability mass function and the CDF for X?
b. Calculate $P(X > 1)$.

3.30 Suppose $h(x) = c/x^2$ for $x = -2, -3, 4, 5$ and is zero otherwise. What would the value of c have to be in order for h to be a mass function?

3.31 Suppose Z is discrete with probability function

$$p_Z(z) = \begin{cases} \frac{1}{5} \text{ if } z = 1, 2, 4, \\ \frac{2}{5} \text{ if } z = 3, \\ 0 \text{ otherwise.} \end{cases}$$

a. What is R_Z?
b. Determine a formula for the CDF F_Z.
c Sketch the graph of F_Z.

3.32 Suppose Y is a random variable with CDF

$$F_Y(t) = \begin{cases} 0 \text{ if } t < 1, \\ \frac{1}{3} \text{ if } 1 \leq t < 4, \\ \frac{1}{2} \text{ if } 4 \leq t < 6, \\ \frac{5}{8} \text{ if } 6 \leq t < 10, \\ 1 \text{ if } 10 \leq t. \end{cases}$$

a. Verify that Y is discrete.
b. What is the probability mass function for Y?
c. Calculate $P(1 \leq Y \leq 5)$, $P(1 < Y \leq 5)$, $P(1 \leq Y < 5)$, and $P(1 < Y < 5)$, using the formulas of Exercise 3.11.

3.33 Suppose Z is a random variable with probability mass function

$$p_Z(z) = \begin{cases} \frac{1}{6} \text{ if } z = 6, 7, 8, 10, \\ \frac{1}{3} \text{ if } z = 9, \\ 0 \text{ otherwise.} \end{cases}$$

a. Verify that Z must be discrete.
b. Determine the CDF for Z.
c. Sketch the graph of F_Z.
d. Calculate $P(7 \le Z \le 9)$, $P(7 < Z \le 9)$, $P(7 \le Z < 9)$, and $P(7 < Z < 9)$, using the formulas of Exercise 3.11.

3.3 DISCRETE FAMILIES

The random variables dealt with so far have had finite ranges. Before we proceed to random variables with infinite ranges, it is useful to summarize and catalog several standard distributions and settings in which they are commonly encountered.

EXAMPLE 3.9   The simplest random variable is one whose range consists of a single point. If $R_X = \{c\}$ for some constant c, $p_X(c) = 1$ and $p_X(x) = 0$ for all $x \ne c$. For any event $E \subset \text{Re}$,

$$p(X \in E) = \begin{cases} 1 \text{ if } c \in E, \\ 0 \text{ if } c \notin E. \end{cases}$$

A trivial random variable, such as that in the preceding example, or its distribution,* is said to be *degenerate,* or, more precisely, *degenerate at c.* Such trivial random variables are introduced because they arise in practice, and they are simple cases for illustrating various probabilistic concepts. For example, they provide a simple context in which to discuss the notion of a "distribution family." Since there is a distinct degenerate random variable (and associated distribution) for each choice of c, the latter serves to index an infinite set of such distributions, a family of distributions. The constant c is then referred to as a *parameter* of the family. Each member of the family resembles every other (e.g., the functional form of the mass function is the same), and yet there are differences between members of the family [the parameter value at which $p(x) = 1$ is different, for example]. The concept of families of distributions will be useful in the work ahead.

---

* In what follows, the term "distribution" refers to either the mass function or the CDF, since either characterizes the probability model.

EXAMPLE 3.10 —*The Bernoulli Family*   A Bernoulli trial can be modeled with a random variable X having range $\{1, 0\}$. The outcome 1 on X might be identified with "newborn child is a boy," "outcome is a success," or "item meets specifications." The mass function of X involves the point probability $p = P(X = 1)$:

$$p_X(x) = p^x q^{1-x}, \qquad x = 0, 1. \tag{3.1}$$

This formula expresses the fact that $p_X(1) = p$ and $p_X(0) = q = 1 - p$. There is an infinite class of Bernoulli mass functions, one for each choice of p between 0 and 1. The variable p constitutes a parameter for this family of mass functions. The distribution of the random variable X is called the *Bernoulli distribution*; the class of Bernoulli distributions is a one-parameter family.

EXAMPLE 3.11 —*The Binomial Family*   The binomial distribution was introduced in Chapter 2 as a model for an experiment consisting of performing n independent Bernoulli trials and counting the number of "successes." Suppose each trial can result in a 1 with probability p and a 0 with probability $q = 1 - p$. Let Y be a count of the total number of 1s. Then $R_Y = \{0, 1, 2, \ldots, n\}$, and a formula for the mass function is

$$p_Y(y) = \binom{n}{y} p^y q^{n-y}, \qquad y = 0, 1, 2, \ldots, n. \tag{3.2}$$

The random variable Y is said to be a binomial random variable, and p_Y is called a *binomial distribution*. The binomial distribution has two parameters: n and p. The first parameter is restricted to be a positive integer, and the second parameter must be a real number between 0 and 1. Thus every member of the binomial family has two names, n and p; the binomial is a two-parameter family of distributions. A specific member of the family is sometimes denoted $b(n, p)$.

The first clue suggesting the use of a binomial distribution as a model is that the experiment consists of a number of trials and each trial has only two possible outcomes. One of these is arbitrarily labeled 1 and the other 0. Independence of the trials is required and is often difficult to assure. Sometimes it is simply (and dangerously) assumed, unless there is compelling evidence to the contrary. The trials must also all have the same "one-shot probability of success," p. A special case is sampling with replacement from an urn whose contents have been dichotomized, so $p = M/N$ for some choice of $M < N$. When $n = 1$, the binomial formula reduces to the Bernoulli formula, so the Bernoulli family is a *subfamily* of the binomial family.

EXAMPLE 3.12 As a particular case, consider a couple planning four children. Arbitrarily identify 1 with a girl and 0 with a boy, and view the sex of each child as a Bernoulli trial with $p = .5$. Whether the trials are independent can be debated on several grounds, but, in terms of current knowledge, it may be a reasonable assumption. Given all of that, Y, the number of girls that will be born, has a binomial distribution with $n = 4$ and $p = .5$ (.49 might be a slightly better choice). Then the values $P(Y = 3)$, $P(Y > 2)$, $P(1 < Y < 3)$, and so on, can be found by using Eq. (3.2).

The binomial distribution is so widely used that extensive tables are available giving values of $p_Y(Y)$ and the CDF, $F_Y(y)$, for various choices of n and p. Most computers and many calculators have the capability of evaluating the binomial mass function and CDF, for any reasonable choices of n and p. To assist you with computations for the exercises in the absence of such calculating capability, we provide a brief binomial table in the Appendix (Table 1). Values for $p > .5$ are not given in the table because

$$\binom{n}{y} p^y q^{n-y} = \binom{n}{n-y} q^{n-y} p^{n-(n-y)}. \tag{3.3}$$

Every problem with $p > .5$ can be converted to one with $p < .5$. For example, if $n = 5$ and $p = .7$, $p_Y(3)$ can be evaluated as follows:

$$p_Y(3) = \binom{5}{3} (.7)^3(.3)^2 = \binom{5}{2} (.3)^2(.7)^3 = p_{Y'}(2),$$

where Y' is binomial with $n = 5$ and $p = .3$. The value of $p_{Y'}(2)$ from the binomial table is .3087, so $p_Y(3) = .3087$.

EXAMPLE 3.13 —*The Hypergeometric Family*   The hypergeometric distribution arises in connection with sampling without replacement from an urn having M balls of one type and $N - M$ of another. If Y counts the number of balls of the first type in a sample of size $n \leq N$, then $R_y = \{0, 1, 2, \ldots, n\}$, just as with the binomial. The formula for the mass function is

$$p_Y(y) = \frac{\binom{M}{y}\binom{N-M}{n-y}}{\binom{N}{n}}, \qquad y = 0, 1, 2, \ldots, n. \tag{3.4}$$

Such a random variable (and its distribution) is called *hypergeometric*. The hypergeometric is a three-parameter family of distributions, with parameters n, M, and N, where $M \leq N$ and $n \leq N$. In the formula, $\binom{M}{y} = 0$ whenever $y > M$, so it is not necessary to restrict y to be less than or equal to M; the formula automatically takes care of that contingency.

EXAMPLE 3.14 —*The Geometric Family*   Suppose independent, repeated Bernoulli trials are to be performed, but there are not a fixed number of trials; rather, the experiment continues until a trial outcome of 1 is observed for the first time. In this case, the number of 1s that will be observed is controlled, but there is no control over the sample size. The number of trials X is random in this setting. The experiment might terminate on the first trial, maybe the second, possibly the third, and so on. Thus the range of possible values of X is $R_X = \{1, 2, 3, \ldots\}$, a countably infinite set. When R_X is countable, probabilities are still characterized by mass functions or point probabilities. Now, a sum of mass values may actually be a series. The only way the value of X can be the integer k is if, in k trials, there are exactly $k - 1$ 0s followed by exactly one 1. From the independence of the trials, such an event has probability pq^{k-1}. Hence,

$$p_X(x) = pq^{x-1}, \qquad x = 1, 2, 3, \ldots, \tag{3.5}$$

which defines a probability mass function. The geometric is a one-parameter family, with one member for each choice of $0 < p < 1$. To verify that Eq. (3.5) defines a mass function, first note that $p_X(x) \geq 0$ for each choice of x. All that remains is to check $\Sigma_{x \in R_X} p_X(x)$, and this sum will involve an infinite series. But the series $\Sigma_{k=1}^{\infty} r^{k-1}$, known as the geometric series, converges any time $|r| < 1$, to the value $1/(r - 1)$. Thus

$$\sum_{x=1}^{\infty} q^{x-1} = \frac{1}{1 - q} = \frac{1}{p},$$

since $0 < q < 1$, and, hence, $\Sigma_{k=1}^{\infty} pq^{x-1} = p \cdot 1/p = 1$. This distribution is called *geometric*. A formula for the nth partial sum of a geometric series is

$$\sum_{i=1}^{n} r^{i-1} = \frac{1 - r^n}{1 - r}.$$

Accordingly, for any positive integer k,

$$F_X(k) = \sum_{i=1}^{k} pq^{x-1} = p\left(\frac{1 - q^k}{1 - q}\right) = 1 - q^k, \qquad k = 1, 2, 3, \ldots . \tag{3.6}$$

One important application of the geometric distribution is as follows: Imagine a switch on a piece of electronic equipment. It is desired to model the number of times the switch can be turned on, then off, until, finally, it fails. The number of on-off *cycles* of the switch until it fails is a random variable X, with possible values 1, 2, 3, ... Each cycle of the switch is modeled as a Bernoulli trial, with the same probability p of failure (identified as the outcome 1). Assuming independent cycles, x is geometric with parameter p. For example, if $p = .0005$ (a fairly reliable switch), $p_X(6) = (.0005)(.9995)^5 = .0005$; there is only a small probability that the switch will fail on the sixth cycle and not before.

In this context, $P(X > k)$ measures the probability that the switch will survive at least k cycles. With k fixed by the analyst, this expression is referred to as the *reliability* of the switch at k, denoted $R(k)$. Since $P(X > k) = 1 - P(X \le k) = 1 - F_X(k)$, this right-hand "tail" probability is called the *complementary cumulative distribution function* (CCDF, for short) and is denoted $G_X(k)$. That is,

$$G_X(k) = P(X > k) = 1 - F_X(k). \tag{3.7}$$

The reliability at k is given by $R(k) = G_X(k) = q^k$. In the switch example, with $p = .0005$, the reliability at 50 cycles is $R(50) = (.9995)^{50} = .975$. Thus there is a good chance that the switch will survive at least 50 cycles. However, 50 cycles are not many power on-off's for some systems. The chances of surviving 1000 cycles is $R(1000) = (.9995)^{1000} = .61$. Indeed, a value of k such that $R(k) = .50$ is found as follows: $(.9995)^k = .50$ provided $k = (\ln .50)/(\ln .9995) \approx 1386$. If it is desired that the system have a switch that performs reliably in the 1000-to-1500-cycle range, this switch is not acceptable in spite of its low probability of failure on a given trial.

Before leaving this cycles-to-failure example, we point out one more feature of the geometric distribution. For any k, $m > 0$, the quantity $P(X > k + m | X > k)$ would measure the likelihood that, given that the switch has already survived k cycles, it will survive m additional cycles. A curious property of the geometric distribution is that, as will be seen shortly, this conditional probability is the same as $P(X > m)$. It is as though, once the switch has survived k cycles, there is no memory of having survived those k cycles. In this model, there is no wear-out factor in the switch (reflected, in a sense, by the assumption that the probability of failure on each trial never increases from p). To show the memoryless property of a geometric distribution, we observed that, since $k + m > k$, $p(X > k + m, X > k) = P(X > k + m)$. That is, if the value of X is greater than $k + m$, it is already greater than k, so the event $(X > k + m)$ is a subset of $(X > k)$, and $(X > k + m) \cap (X > k) = (X > k + m)$. From conditional probability definitions,

$$P(X > k + m | X > k) = \frac{P(X > k + m, x > k)}{P(X > k)} = \frac{P(X > k + m)}{P(X > k)} \tag{3.8}$$

$$= \frac{q^{k+m}}{q^k} = q^m = P(X > m).$$

EXAMPLE 3.15   —*The Negative Binomial Family*   In the performance of n independent Bernoulli trials until we observe a 1 for the first time, the number of 0s is random. If Y is that random number, then the possible values of Y include 0 along with the positive integers, and $p_y(y) = pq^y$.

Now, suppose Y measures the random number of 0s until 1 is observed for the rth time in those same circumstances. Still, $R_Y = \{0, 1, 2, \ldots\}$, and 1 must be observed on the yth trial. But the other $r - 1$ observations of 1 may occur on any of the preceding $y + r - 1$ trials and, hence, in any of the $\binom{y+r-1}{r-1} = \binom{y+r-1}{y}$ possible arrangements. For each such choice, the point probability is $p^{r-1} \cdot q^y \cdot p = p^r q^y$. Thus the number of trials until the rth success has the distribution

$$p_y(y) = \binom{y + r - 1}{y} p^r q^y, \qquad y = 0, 1, 2, \ldots . \tag{3.9}$$

This distribution is called the *negative binomial*, or *Pascal*. The formula defines a two-parameter family with parameters $r \geq 1$ and $0 < p < 1$. In fact, Eq. (3.9) defines a mass function even when r is not an integer (see Exercise 3.38). This model is currently used by the Navy, for example, to model the number of demands on an inventory system when the average annual demand is moderate (2 to 20).

We could take any convergent series of positive terms and normalize it to form a probability mass function. Whether such a procedure would lead to a useful model is a question. One series that, when normalized, does provide useful models is the series $e^z = \Sigma_{k=0}^{\infty} z^k/k!$. This is another countably infinite case, and it has been named in honor of the French mathematician S. D. Poisson.

EXAMPLE 3.16   —*The Poisson Family*   For any $\lambda > 0$, the series with general term $e^{-\lambda}(\lambda^k/k!)$ must converge to 1 since $e^{-\lambda} \Sigma_{k=0}^{\infty} \lambda^k/k! = e^{-\lambda} \cdot e^{\lambda} = e^0 = 1$. Therefore,

$$p_X(x) = e^{-\lambda} \left(\frac{\lambda^x}{x!} \right), \qquad x = 0, 1, 2, \ldots \tag{3.10}$$

defines a one-parameter family of probability distributions indexed by the single parameter $\lambda > 0$. This family of mass functions is called *Poisson*. Poisson mass functions are often used in applications, and they have been tabulated for a wide range of values of λ. A brief table of Poisson mass values is given in the Appendix (Table 2). It is quite easy to develop a program for computing values of Poisson mass functions for more extensive choices of λ. It may be observed from Table 2 that a Poisson mass function is increasing from 0 up to a value of x near λ; then it decreases (all for $x \in R_X = \{0, 1, 2, \ldots\}$).

The Poisson family has a long history of application in modeling counts of the number of occurrences of an event over time when the *rate* of occurrence is known. For example, suppose demand for a certain automobile part at an automotive supply store occurs randomly, say at a rate of about one every five years. In stocking the shelves of the parts department, the store manager might wish to know the probability that a demand for the part will occur within the next year, the next six months, or the next quarter. Other typical applications are counting the number of accidents on a freeway overpass in some time period; counting the number of air bubbles in a cross section of solid rocket propellant (here, units are not measured in time but space, such as cubic centimeters); counting the number of incoming telephone calls to a certain switchboard in the next 20 minutes.

In all these applications, a rate of occurence r is given as so many per unit of measurement, and a time span of t units of measurement is considered, over which the counting will take place. Under a set of assumptions that will be explored in Chapter 6, the random count of occurrences, X, is Poisson, with $\lambda = rt$. In the parts department situation, the analyst could use five years as the time unit; then $t = .1$ (six months is one-tenth of five years), in which case the Poisson parameter is $\lambda = .1$. Or the analyst could choose one year as a time unit, in which case $r = .2$ and $t = 5$, whence $\lambda = .1$ again. The point is that λ is the same regardless of how r is selected, as long as consistency is maintained with the desired time horizon. If the analyst adopts $r = 1$ every five years and $t = .1$, then the probability of no demands in the next six months would be $p_X(0) = e^{-.1}[(.1)^0/0!] = e^{-.1} = .90$. To find the probability of one or more demands in the next year, change t from .1 to .2, so that $\lambda = .2$. Then $P(X \geq 1) = 1 - P(X = 0) = 1 - e^{-.2} = .18$. For a probability such as $P(3 \leq X \leq 7) = \sum_{x=3}^{7} e^{-.2}(.2)^x/x! = F_X(7) - F_X(2)$, a Poisson table can be used to find an answer of .0012.

A Poisson mass function can be used to provide an approximation to a binomial mass function in which n is large and p is small. The Poisson parameter used for this purpose is $\lambda = np$. For example, on a certain section of a freeway, the probability of an accident for a given vehicle may be very small, say .0001, so this is a rare event from the point of view of each vehicle. But during rush hours, the opportunities for an accident are large, because there are many vehicles at risk. If the occurrence of an accident is thought of as a Bernoulli trial with $p = .0001$, and if 3000 vehicles cross the section during the time period in question, one can imagine these constitute $n = 3000$ independent trials. Then the number of accidents during the time period Y would be binomial. The probability of more than one accident would be given by the CCDF at 1, that is, $P(Y > 1) = 1 - F_y(1)$. With the parameter values $n = 3000$ and $p = .0001$, we could experience difficulty calculating this value. With even modest computing capability available, we would have no problem calculating this value (.0369, to four decimal places). But the Poisson approximation to the binomial can also be used to compute the answers. To see this, let X be Poisson with $\lambda = np = .3$. Then $P(X > 1) = 1 - e^{-.3} - .3e^{-.3} = .0369$, to four decimal places.

To show this approximation result, we need only demonstrate that each binomial mass value is approximately the corresponding Poisson mass value. For any $k \in \{0, 1, 2, \ldots, n\}$,

$$p_Y(k) = \binom{n}{k} p^k q^{n-k} = \frac{n!}{k!(n-k)!} p^k (1-p)^{n-k}$$

$$= \frac{n(n-1)(n-2) \cdots (n-k+1)}{k!} \left(\frac{\lambda}{n}\right)^k \left(1 - \frac{\lambda}{n}\right)^{n-k}$$

$$= \frac{n(n-1)(n-2) \cdots (n-k+1)}{k!} \left(\frac{\lambda^k}{n^k}\right) \left(1 - \frac{\lambda}{n}\right)^{-k} \left(1 - \frac{\lambda}{n}\right)^n$$

$$= \frac{\lambda^k}{k!} (1) \left(1 - \frac{1}{n}\right) \left(1 - \frac{2}{n}\right) \cdots \left(1 - \frac{k-1}{n}\right) \left(1 - \frac{\lambda}{n}\right)^{-k} \left(1 - \frac{\lambda}{n}\right)^n$$

But with k fixed,

$$\lim_{n \to \infty} \left(1 - \frac{\lambda}{n}\right)^{-k} = 1^{-k} = 1, \qquad \lim_{n \to \infty} \left(1 - \frac{\lambda}{n}\right)^n = e^{-\lambda},$$

$$\text{and} \qquad \lim_{n \to \infty} \left(1 - \frac{x}{n}\right) = 1.$$

Hence, for large values of n, $p_Y(k) \approx e^{-\lambda}(\lambda^k/k!)$, which is the Poisson mass at k, with $\lambda = np$. There is no hard and fast rule regarding how large n should be and how small p should be in order to make the Poisson approximation to the binomial sufficiently accurate. Roughly speaking, the results are satisfactory when $np \le 10$.

The analyst is often faced with a situation requiring a discrete model that does not fall within any of the families we have cataloged above. All that may be known is the set $\{x_1, x_2, \ldots, x_N\}$ of possible values (expressed numerically) that can occur as a result of performing the experiment. Faced with this situation, the analyst has several options. One is to set a model on a priori grounds, perhaps the equal-likelihood model as in games of chance. In that case the random variable X is said to have a *discrete uniform distribution*. Naturally, some care has to be exercised in the use of such an a priori choice. A second option is to rely on familiarity with similar kinds of experiments or perhaps expert advice to define a mass function for X. One problem with such subjective selections is that such a choice, however valid, may not appear compelling to a second analyst. Nevertheless, expert knowledge about a phenomenon should not be discounted. A third option is to perform the experiment many times and then rely on relative frequency as a method of determining the mass function for X. This option is a common one to employ. One problem with it is that there are almost always costs associated with such trials (people have to be polled, missiles have to be fired), and such costs often prohibit performing a number of trials sufficiently large that the resulting relative frequencies can be relied on. Such choices, whether or not they are based on a large number of trials, are valid ways of selecting a model and allow the objectivity of experimental evidence as an alternative to the subjectivity of expert guidance. Even when a standard family is selected as a model, the parameters in that family may have to be chosen, and these same options are available for such choices.

EXERCISES

THEORY

3.34 Show that, if X is binomial and $p_k = p_X(k)$, then

$$p_k = \frac{n - k + 1}{k} \cdot \frac{p}{q} \cdot p_{k-1}, \qquad k = 1, 2, 3, \ldots, n.$$

Note: This recursive relationship can be used with initial condition $p_0 = q^n$ to program the mass function for calculating mass values.

3.35 Show that, if X is hypergeometric and $p_k = p_X(k)$, then

$$p_k = \frac{M - k + 1}{k} \cdot \frac{n - k + 1}{N - M - n + k} \cdot p_{k-1}, \qquad k = 1, 2, \ldots, n;$$

$$p_0 = \frac{(N - M)_n}{(N)_n}.$$

3.36 Show that, if X is Poisson and $p_k = p_X(k)$, then $p_k = (\lambda/k)p_{k-1}$, $k = 1, 2, 3,$ $\ldots$; $p_0 = e^{-\lambda}$.

3.37 Show that, if X has a negative binomial distribution and $p_k = p_Y(k)$, then $p_k = [(k + r - 1)/k] \cdot q \cdot p_{k-1}$, $k = 1, 2, 3, \ldots$; $p_0 = p^r$.

3.38 For any positive real number r and nonnegative integer n, let $\binom{-r}{n}$ be defined as

$$\binom{-r}{n} = \frac{(-r)(-r - 1)(-r - 2) \cdots (-r - n + 1)}{n!}.$$

Show that, for each integer k, $\binom{k+r-1}{k} = (-1)^k \binom{-r}{k}$. Hence, the negative binomial mass function may be written as $p_Y(y) = \binom{-r}{y}p^r(-q)^y$, $y = 0, 1,$ $2, \ldots$. The family gets its name from the fact that $\binom{-r}{y}$ is called a negative binomial coefficient. The parameter r need not be an integer.

3.39 Let X denote the number of Bernoulli trials until 1 is observed for the rth time. Using arguments similar to those preceding Eq. (3.9), show that $R_X = \{r, r + 1, \ldots\}$ and

$$p_X(x) = \binom{x - 1}{r - 1} p^r q^{x-r}, \qquad x = r, r + 1, r + 2, \ldots.$$

Sometimes this family is called negative binomial also. The geometric family is a subfamily, with $r = 1$.

3.40 Suppose independent Bernoulli trials are performed until 1 is observed for the *r*th time. Let *y* be any fixed nonnegative integer, and suppose *Y* is negative binomial with parameters *r* and *p*, while *X* is binomial with parameters *r* + *y* and *p*.

    a. Argue that the events $(Y \leq y)$ and $(X \geq r)$ are equivalent (i.e., one occurs if and only if the other occurs) for proper identification of *X* and *Y*.

    b. Show that the CDF for *Y* at *y* may be written as

$$F_Y(y) = \sum_{j=r}^{r+y} \binom{r+y}{j} p^j q^{r+y-j}.$$

3.41 The probability that a hit will be accomplished is *p* each time a shot is fired from a large-caliber gun. Shots are to be fired until a hit is accomplished. Let *v* denote the probability that an odd number of shots is required. What must the value of *p* be in order to make $v = \frac{5}{8}$?

3.42 For what value of *c* would the function given by $f(k) = c \cdot 2^{-|k|}$, $k = 0, \pm 1, \pm 2, \ldots$, serve as a probability function of a discrete random variable *Y*? What is the range? Calculate $P(-3 < Y < 3)$.

APPLICATION

3.43 The probability that a riflewoman hits a target is estimated to be .8 for each shot. She is to keep firing until she hits the target. What is the probability that it will take more than six shots?

3.44 A pair of fair dice are rolled until a sum of ten occurs.

    a. What is the probability that at least two rolls are required?

    b. What is the probability that at least six rolls are required?

3.45 Five people each fire one round, simultaneously, at the same target. The probability is .8 that each person hits the target.

    a. What is the distribution of *X*, the number of hits on the target?

    b. What is the probability that the target is hit at least twice?

3.46 A population of 1000 insects infests a rose garden. The gardener sprays the garden with insecticide. Suppose the probability is .995 that an insect will be killed.

    a. What is the probability distribution of *X*, the number of surviving insects?

    b. Approximate the probability that at least 997 insects are killed.

3.47 An auditor is hired to inspect the financial accounts of a firm. Rather than inspect every individual record, he will inspect a random sample of 20% of the records. Assume the firm has 100 accounts so that 20 will be selected for inspection. Suppose that there is at least one error in 15 of the 100 accounts. Let *X* be the number of erroneous accounts among the sampled ones.

    a. What is the probability distribution of *X*?

    b. Calculate the probability that exactly 3 sampled accounts are erroneous.

3.48 A game at a local carnival consists of tossing a ring at a peg. A player with unlimited wealth decides to play the game until he gets a ring successfully on a peg. Suppose the probability is .2 for each toss.

    a. What is the probability that he rings the peg on the first toss?

    b. What is the probability that he rings the peg within 5 tosses?

c. Calculate the probability that, if he continues to play until he rings the peg twice, he will end the game in ten or fewer tosses.

3.49 Orders for a specific part arrive at a supply depot at a rate of two every three years ($\frac{1}{6}$ per quarter). Let X denote the quarterly demand for the item.
a. What is the probability that there will be no orders next quarter?
b. What is the probability that two or more orders will be received next quarter?

3.50 In the game of Keno, 100 Ping-Pong balls are in a container, and each one has a two-digit number from 00, 01, to 99. In playing the game, you select three different two-digit numbers. Then 20 balls are selected at random from the container. Let X denote the number of balls in the sample that correspond to the three numbers you pick.
a. What is the probability distribution for X?
b. What is the probability that X has the value 0?

3.51 A person wishing to be admitted to the California Bar Association is allowed to take a qualifying exam as many times as desired until passing. Suppose a candidate's median number of attempts is estimated to be two. (That is, the candidate has equal likelihood of requiring two or less, and two or more, attempts.)
a. What is the probability that the candidate will pass in two attempts?
b. What is the probability that it takes the candidate three or more attempts before passing?

3.52 When items are purchased in lots, frequently the acceptance or rejection of a lot by a quality control inspector is based on the results of a sample taken from the lot. Suppose a lot consists of 200 items, and a sample of 5 is selected without replacement. Also, suppose the lot actually has 10 defectives and the sampling plan is to accept the lot if and only if there are no defectives in the sample.
a. What is the probability that the lot will be accepted?
b. What is the probability of finding two defectives in the sample?

3.53 Suppose one baby in 1000 is born with a certain defect. In a large city, 2000 babies are born in a year. Calculate the probability that at least 3 babies with the defect are born this year.

3.54 A supermarket dairy case contains 50 half gallons of milk, 20 of which exceed the recommended date for sale. You buy 4 half-gallon cartons.
a. What is the probability that all 4 of your purchases are fresh, that is, do not exceed the recommended date?
b. Calculate the probability that 2 are fresh and 2 are not.

3.55 A service station has ten tires in stock of the brand and size you want to buy. Two of the tires have blemishes. You buy four tires and they are randomly selected from the service stations' stock.
a. What is the probability that none of the tires you buy are blemished?
b. What is the probability that half the tires you buy have blemishes?

3.56 Five percent of the light bulbs made by a manufacturer are defective. These bulbs are sold in packages of four.
a. Calculate the probability that all the bulbs in a package are nondefective.
b. You buy three packages of these bulbs. Let Y denote the number of packages that have no defective bulbs. Calculate $P(Y = 2)$.

3.57 A keypunch operator has probability .005 of making an error in a single key-stroke. Suppose a program requires 600 keystrokes.
a. What is the probability that the operator makes no errors?
b. Calculate the probability of two or more errors.

3.58 The probability that a hand-blown glass is defective is .01. In one week, a worker turns out 200 glasses.
a. What is the distribution of the number of defective glasses produced by the worker in one week?
b. Compute the probability that the worker produces no more than one defective glass in a week.

3.59 A bowl contains 40 chips, 20 of which are red, 10 green, and 10 white. Ten chips are selected without replacement from the bowl.
a. What is the distribution of the number of red chips in the sample?
b. Compute the probability that there will be 5 or fewer red chips in the sample.

3.60 A professional golfer hits golf balls repeatedly from a certain spot until one stays on the green. Suppose the probability is .6 that each ball will stay on the green.
a. What is the distribution of Y, the number of balls the golfer will hit?
b. Calculate $P(Y \le 3)$.
c. The golfer continues to chip until balls stay on the green for the third time. What is the probability distribution of the number of trials?

3.61 A certain switch is supposed to have a reliability of .95. In cyclings of the switch, it failed on the third cycle. How rare is this event? How many cycles might be typical?

3.62 It has been observed that cars pass a certain point on a rural road at the average rate of three per hour. Assume that the instants at which the cars pass are independent, and let X be the number that pass this point in a 20-minute interval. Compute $P(X = 0)$, $P(X \ge 2)$.

3.63 It has been observed empirically that deaths due to traffic accidents occur at a rate of eight per hour on long holiday weekends in the United States. Assuming that these deaths occur independently, compute the probability that a 1-hour period would pass with no deaths. That a 15-minute period would pass with one death. That four consecutive, nonoverlapping 15-minute periods would pass with one death in each. Compare the latter with the probability of four deaths in a 1-hour period.

3.64 An ice cream company sells chocolate-covered ice cream bars on sticks for 25 cents each. Suppose they put a star on every 50th stick; anyone who buys a bar with a starred stick gets a free ice cream bar. You decide to buy ice cream bars until you get a free one. How much would you expect to spend before getting a free bar?

3.65 Assume the sales made by a used car salesman occur as events in a Poisson process with parameter $r = 1$ per week.
a. What is the probability that the salesman makes (exactly) three sales in a two-week period?
b. What is the probability of three two-week periods in a row with no sales?

3.66 Assume a printed page in a book contains 40 lines, and each line contains 75 positions (each of which may be left blank or filled with some symbol). Thus

each page has 3000 positions to be set. Assume a particular typesetter makes one error per 6000 positions, on the average.
a. What is the distribution of X, the number of errors per page?
b. Compute the probability that a page contains no errors.
c. What is the probability that a 16-page chapter contains no errors?

3.67 In a given semester, a large university will process 100,000 grades. In the past, .1% of all grades have been erroneously reported. Assume you are taking five courses at this university in one semester. What is the probability that all your grades are correctly reported?

3.4 CONTINUOUS DISTRIBUTIONS

If a piece of electronic equipment (such as a transistor) were put on test by operating until it failed, then the operating time to failure would be a random variable X. Even under the most controlled manufacturing conditions, different transistors would fail at different times. What would be the range of such a random variable? If time is measured in whole numbers of hours, the countable set $\{0, 1, 2, \ldots\}$ would suffice. Usually, it would be preferred to allow for measurements in tenths of an hour, hundredths of an hour, and possibly even finer time divisions. Although there may not be a measuring device capable of reading it, conceptually, the true time to failure could be any positive real number, rational or irrational. It is thus reasonable to use the infinite half line $[0, \infty)$ as a range of possible values for X.

But now the probability distribution of X can no longer be characterized by a mass function. For one thing, there are too many points in such a set to add over, even in an infinite series sense. This condition is true of all intervals on the real line whether or not they have infinite endpoints. (Such uncountable sets are called *continuous*.) The way out of this dilemma is through differential and integral calculus. Imagine a total mass of one unit spread continuously over the real line (or part of it), not necessarily in a uniform manner. In physics, such a mass system is characterized by a density function, a function whose values determine the amount of mass over any interval by computing the integral of the mass function over that interval. The situation is depicted in Fig. 3.11. What are the requirements of such a function? For one thing, $f(x)$ should be nonnegative for each x (there is no physical meaning to a negative amount of mass), and for another, the entire area over the real line should be the total mass (which we deliberately set at 1 for use in probability models); that is,

$$f(x) \geq 0 \quad \text{and} \quad \int_{-\infty}^{\infty} f(x)\, dx = 1.$$

Any function satisfying these two requirements will be called a (*probability*) *density function*. Instead of physical mass, the application concerns probability mass.

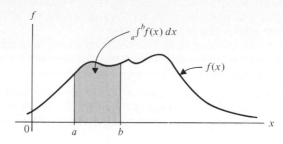

FIGURE 3.11   *Density $f(x)$ with Mass in (a, b) Calculated as an Integral*

EXAMPLE 3.17   Suppose $f(x) = e^{-x}$ for $x > 0$ and $f(x) = 0$ otherwise. Now $f(x) \geq 0$ and $\int_{-\infty}^{\infty} f(x)\, dx = \int_{0}^{\infty} e^{-x}\, dx = -e^{-x}\big|_{0}^{\infty} = 0 - (-1) = 1$. Thus the function e^{-x} qualifies as a probability density function.

EXAMPLE 3.18   Let $f(x) = 3x^2$ for $0 < x < 1$ and $f(x) = 0$ otherwise. Again, $f(x) \geq 0$ and $\int_{-\infty}^{\infty} f(x)\, dx = \int_{0}^{1} 3x^2\, dx = x^3\big|_{0}^{1} = 1$, so the function $3x^2$ is a probability density function.

Given such a function, an area interpretation of probability is adopted. For example, the CDF for X may be written as

$$P(X \leq x) = F_X(x) = \int_{-\infty}^{x} f(t)\, dt; \qquad (3.11)$$

this number is the area under the density curve from $-\infty$ to x. To verify that this expression does indeed yield a CDF, check the three conditions studied earlier. If $x < y$,

$$F(y) = \int_{-\infty}^{y} f(t)\, dt = \int_{-\infty}^{x} f(t)\, dt + \int_{x}^{y} f(t)\, dt = F(x) + \int_{x}^{y} f(t)\, dt.$$

Since f is a nonnegative function, $\int_{x}^{y} f(t)\, dt \geq 0$, and, hence, $F(y) \geq F(x)$, so F is monotone increasing. It is shown in calculus that the integral F is continuous (and therefore right continuous). Finally, applying limits in Eq. (3.11), we have $\lim_{x \to -\infty} F(x) = 0$, while $\lim_{x \to +\infty} F(x) = 1$.

The relationship between f and F is graphically portrayed in Fig. 3.12. The *value* of F at a point x_0 may be graphically interpreted as the total area under f to the left of x_0. As x_0 increases, the monotonicity of F reflects the fact that more area is being accumulated to the left of x_0. From the elementary properties of integrals, it is clear that the value of the density f can be changed at a finite or denumerable number of points and not affect the distribution function F; the area under a curve is not affected by changes in the curve at a countable number of points. It also follows that F is differentiable at each point x where f is continuous, and at each such point, $F'(x) = f(x)$.

DEFINITION 3.4   Let X be a random variable having CDF F. Then X is a *continuous random variable* (and X is said to have a *continuous distribution*) if there exists a density function f such that F is given by

$$F(x) = \int_{-\infty}^{x} f(t)\, dt. \qquad (3.12)$$

Remark: The density function f in Eq. (3.12) will often be denoted by f_x to show the dependence on X and will be referred to as the *density function* of X. A model for an experiment to be described by a continuous distribution can thus be completed by choosing a density function f. As with the mass function, a density function is defined over the entire real line, even when (as in the time-to-failure case) certain intervals do not even contain possible values of X. Whenever a formula is given for $f_X(x)$, it will be understood that $f_X(x) = 0$ for any portion of the real line not specified in the formula. It is then convenient to redefine the range of a continuous random variable to be $R_X = \{x : f_X(x) > 0\}$.

   If X is a continuously distributed random variable, then F_X is a continuous function, so the mass function p_X is identically zero. In particular, this condition means that $P(X = x) = 0$ for every real number x; if $f_X(x) > 0$, then $f_X(x)$ cannot be $P(X = x)$. Thus while the density function f_X is *analogous* to the mass function for a discrete random variable, with integration replacing the summation, f_X must be *interpreted* differently. At first, the fact that all point probabilities in a continuous distribution are zero may seem counterintuitive. For we know that when the experiment is performed, *some* point in R_X will be obtained. And yet for any such x, the probability

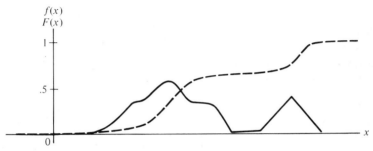

FIGURE 3.12   *Graphs of a Hypothetical Density Function* (Solid) *and Corresponding CDF* (Dotted)

that the outcome will be that particular x is zero. Keep in mind that probability theory is concerned with predicting something about outcomes *prior* to performance of the experiment. From that point of view, it seems entirely reasonable that, if you choose an x beforehand, it would be extremely unlikely—indeed, almost impossible—that an experiment with a continuous distribution would result in that specific x.

If a small enough interval is taken about x, say $(x - h/2, x + h/2)$, then $f(x)$ times the length h of the interval is approximately the area (probability) under the curve and over the interval. However, $f(x)$ is not itself that probability. To be more precise, from the definition of F_X and Eq. (3.12), it follows that if $h > 0$,

$$P(x < X \le x + h) = F_X(x + h) - F_X(x) = \int_x^{x+h} f_X(t)\, dt \quad (3.13)$$

for any fixed x. Geometrically, the area under the density curve between the points x and $x + h$ (and above the x-axis) is the probability that X will assume a value in the interval $(x, x + h]$. From the mean value theorem for integrals, recall that, if f is continuous on $[a, b]$, where $a < b$, then there exists a point $\xi \in (a, b)$ such that $\int_a^b f(t)\, dt = f(\xi)(b - a)$. Hence, *if f_X is continuous in a neighborhood of x*, for h sufficiently small, there is a point $\xi \in (x, x + h)$ such that $P(x < X \le X + h) = f_X(\xi)h$. But since f_X is continuous at x, $\lim_{h \to 0} f_X(\xi) = f_X(x)$. For this reason, one may say that, if h is "sufficiently small," then $h \cdot f_X(x)$ is "approximately" the probability $P(x < X \le x + h)$. This expression also agrees with the fact that $F_x'(x) = f_X(x)$ for

$$f_X(x) = \lim_{h \to 0} \frac{P(x < x \le x + h)}{h} = \lim_{h \to 0} \frac{F_X(x + h) - F_X(x)}{h} = F_X'(x).$$

In general, $\quad P(a < X \le b) = \int_a^b f(t)\, dt \quad (3.14)$

for any real numbers a and b with $a < b$. Since the mass function is identically zero in this continuous case, the integral in Eq. (3.14) also yields the values of $P(a < X < b)$, $P(a \le X < b)$, and $P(a \le X \le b)$ (compare this result with Exercise 3.11). Finally, if B is any event in Re, then $P(X \in B) = \int_B f(t)\, dt$.

EXAMPLE 3.19 Let the density function e^{-x} of Example 3.17 serve as a model for a continuous random variable X. (Exponential densities like this are often used as models for continuous time to failure of electronic devices, for example.) Then $R_X = [0, \infty)$, and the CDF for X is given by

$$F_X(x) = \begin{cases} 0 \text{ if } x < 0, \\ 1 - e^{-x} \text{ if } x \ge 0, \end{cases}$$

since, for $x > 0$, $\int_{-\infty}^x f(t)\, dt = \int_0^x e^{-t}\, dt = -e^{-t}\big|_0^x = 1 - e^{-x}$. The graph

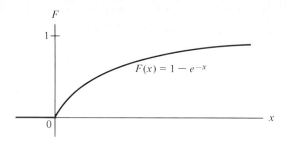

FIGURE 3.13   *Exponential CDF*

of F_X in Fig. 3.13 displays typical characteristics of a CDF. In this case, $F_X(x)$ never quite reaches the value 1, although the latter is a limiting value. The CCDF for this example is $G_X(x) = 1 - F_X(x) = e^{-x}$.

EXAMPLE 3.20   —*The Uniform Family*   Perhaps the simplest density function is one that is constant over some interval (a, b) and is zero elsewhere. If the function is to be a probability density function, the constant height c over (a, b) must be positive, and since the area in this case is just c times $b - a$, c must be $1/(b - a)$ (see Fig. 3.14). Thus the function $f(x) = 1/(b - a)$, $a < x < b$, is a density function, called the *uniform density*. The uniform is a two-parameter family with parameters a and $b > a$. It is easily verified (Exercise 3.68) that the CDF is given by the formula

$$F_X(x) = \begin{cases} 0 \text{ if } x \le a, \\ \dfrac{x - a}{b - a} \text{ if } a < x < b, \\ 1 \text{ if } b \le x. \end{cases} \tag{3.15}$$

Thus F_X is linear in $R_X = (a, b)$, as shown in Fig. 3.15.

The particular choice with $a = 0$, $b = 1$, the uniform distribution over the unit interval, is of special interest and is the basis for generating random

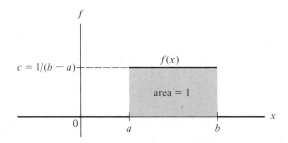

FIGURE 3.14   *Uniform Density Function*

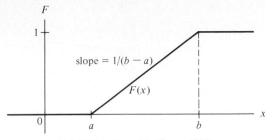

FIGURE 3.15   *Uniform CDF*

numbers. Indeed, the phrase "choosing a number at random in the interval (a, b)" will be taken to mean that such a number is to be the observed value of X, where X has a uniform density over (a, b). The reason is that, with this model, subintervals in (a, b) of equal length have the same probability. For example, if $a = 0$, $b = 1$, then $P(.1 < X < .2) = .2 - .1 = .1$, $P(.3 < X < .4) = .4 - .3 = .1$, and $P(.8 < X < .9) = .9 - .8 = .1$. In a sense, this model is the continuous analogue of the equal-likelihood model. The phrases "equal likelihood" and "at random," applied in the context of a continuous interval, are taken to mean the assumption of a model with uniform density.

EXAMPLE 3.21   —*The Exponential Family*   The density function e^{-x} of Example 3.19 can be generalized by introducing a parameter. For any $\lambda > 0$, let

$$f(x) = \lambda e^{-\lambda x}, \qquad x > 0. \tag{3.16}$$

Since $\int e^{-\lambda x}\, dx = -(1/\lambda)e^{-\lambda x}$, this formula defines a density function for each $\lambda > 0$; hence, it is a one-parameter family. The CDF for a random variable having this density function is

$$F(x) = \begin{cases} 0 \text{ if } x < 0, \\ 1 - e^{-\lambda x} \text{ if } x \geq 0, \end{cases} \tag{3.17}$$

and the CCDF is $G(x) = e^{-\lambda x}$ for $x \geq 0$.

Exponential densities are often used to model random times to failure of various systems. In this context, λ is called the *failure rate* (see Fig. 3.16). Exponential distributions also have the memoryless property possessed by geometric distributions. Thus if t and u are positive times,

$$P(X > t + u \,|\, X > u) = \frac{P(X > t + u, X > u)}{P(X > u)} = \frac{P(X > t + u)}{P(X > u)} = \frac{G(t + u)}{G(u)}$$

$$= \frac{e^{-\lambda t}e^{-\lambda u}}{e^{-\lambda u}} = e^{-\lambda t} = P(X > t) \tag{3.18}$$

Thus if the item has survived (not failed) for u units of time, the conditional probability that it will survive t additional units of time is the same as the probability of surviving t units of time to begin with. It may be shown that the exponential family is the only continuous family having this property.

As a time-to-failure model, $G(t)$ is often called the reliability at time t and is denoted $R(t)$. If time to failure is exponential, the reliability at time t is given by the simple expression $R(t) = e^{-\lambda t}$. For example, if the failure rate is 1 per 1000 hours, then the reliability at 2000 hours is $R(2000) = e^{-2} = .14$. That is, the probability that the system will survive for at least 2000 hours is .14. If the analyst wanted to know the value of t that would yield a reliability of 95%, the analyst would solve $.95 = e^{-t/1000}$ to obtain $t = 51$ hours. Or the analyst might ask what the failure rate would have to be to have 95% reliability at 1000 hours, in which case the analyst would solve $.95 = e^{-1000\lambda}$. The answer, $\lambda = .00005$, might have equipment design applications, for example.

Technically, any integrable, nonnegative function can be converted to a density function for a random variable. Whether such a density function is useful as a model depends on the application. One such density that is of theoretical interest (it actually arises in practice as well) is named Cauchy in honor of the French mathematician.

EXAMPLE 3.22   —*The Cauchy Family*   Suppose $h(x) = 1/(1 + x^2)$ over the entire real line. Then $h(x) > 0$, and

$$\int_{-\infty}^{\infty} \frac{dx}{1 + x^2} = \text{Arctan } x \Big|_{-\infty}^{\infty} = \frac{\pi}{2} - \left(-\frac{\pi}{2}\right) = \pi.$$

Consequently, the function $f(x) = 1/\pi(1 + x^2)$ is a probability density function. Its graph is shown in Fig. 3.17. The corresponding CDF may be written as $F(x) = (1/\pi) \text{Arctan } x + \frac{1}{2}$, a formula valid over the entire real line. In this case, $F(x)$ is never 0 or 1 and is strictly monotone increasing

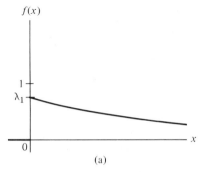

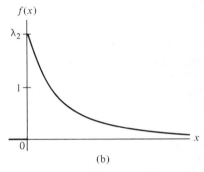

FIGURE 3.16   *Two Exponential Densities with Different Failure Rates*

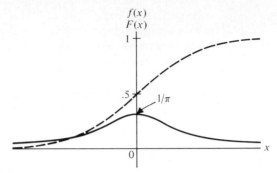

FIGURE 3.17   *Graphs of the Cauchy Density Function*
                (Solid) and CDF (Dotted)

(see Fig. 3.17). This basic density can be expanded to a family by introducing a *location* parameter θ. If $f(x) = 1/\pi[1 + (x - \theta)^2]$, this is the same density curve shifted θ units to the right, so its maximum is at θ instead of 0 (Exercise 3.69). Here, θ can be any real number and indexes a family of density functions called the *Cauchy family*. This family will be examined in more detail in the next chapter.

Many other continuous distributions will be taken up in the work ahead. Analogous to specifying a mass function to define a model induced by a discrete random variable, an approach for a continuous random variable is to specify a density function. A typical starting point would then be stated, "Let X be a continuous random variable with density f_X." There are, of course, random variables that are neither discrete nor continuous, since a distribution function need not be either a step function or a continuous function. Although random variables having such mixed distributions can arise in practice, they cannot easily be analyzed without more advanced mathematical tools (Lebesgue-Stieltjes integration). Hence, in this book, attention is focused mainly on the two special cases of discrete and continuous random variables. Fortunately, these cases include the vast majority of models used in applications.

EXERCISES

THEORY

3.68 Verify that Eq. (3.15) for the uniform CDF is correct.

3.69 Use calculus to verify the sketch of the Cauchy density in Fig. 3.17. Where are the points of inflection? Show by the change of variable $y = x - \theta$ that the general Cauchy density has the same characteristics as the case $\theta = 0$ (Fig. 3.17) with its unique maximum located at θ. Find a formula for the CDF in the general case (arbitrary θ).

3.70 The *Weibull family* of density functions is a two-parameter family with densities given by

$$f(x) = (\alpha\beta)x^{\beta-1} \exp(-\alpha x^\beta), \qquad x > 0, \tag{3.19}$$

where $\exp(z)$ is e^z, $\beta > 0$, and $\alpha > 0$.

a. Verify that Eq. (3.19) defines a density function for each admissible choice of parameters.

b. For what choices of the parameters is Eq. (3.19) exponential?

c. Verify that the CDF, for $x > 0$, is given by the formula $F(x) = 1 - \exp(-\alpha x^\beta)$ so that the CCDF is $G(x) = \exp(-\alpha x^\beta)$. *Note:* This density is often used to model time to failure when the *failure rate*, defined to be $f(x)/G(x)$, is not necessarily reasonably assumed constant. The constant–failure rate case corresponds to the exponential family, which is a subfamily of the Weibull family.

3.71 Show that the uniform density does not have the memoryless property.

3.72 a. Show that the function f defined by $f(x) = 2x$, $0 < x < 1$, defines a density function. Generalize the result to the function f_n defined by $f_n(x) = nx^{n-1}$, $0 < x < 1$, where n is any positive integer.

b. For the case $n = 2$, find $P(X < \tfrac{1}{2})$.

3.78 How must the constant k be chosen so the function f defined by $f(x) = ke^{-|x|}$ will be a density function: *Note:*

APPLICATION

3.74 a. For $f(x) = c \sin x$ for $0 < x < \pi/3$, determine c so that f will be a density function.

b. If Y has this density, find $P(-\pi/6 < Y < \pi/6)$.

3.75 Suppose f is defined by

$$f(x) = \begin{cases} x + 1 \text{ if } -1 \le x \le 0, \\ 2x - 6 \text{ if } 3 < x < c, \\ 0 \text{ otherwise,} \end{cases}$$

where $c > 3$.

a. Determine c so that f is a density function.

b. Determine the distribution function corresponding to f.

c. Find the probability that the next outcome is greater than $c/2$.

3.76 Suppose X has distribution function defined by

$$F_X(x) = \begin{cases} 0 \text{ if } x < 0, \\ x - \tfrac{1}{4}x^2 \text{ if } 0 \le x < 2, \\ 1 \text{ if } 2 \le x. \end{cases}$$

Show that X is a continuous random variable and determine its density function. Verify that F_X satisfies the conditions of a CDF.

3.77 Let f be defined by

$$f(x) = \begin{cases} x \text{ if } 0 < x \le 1, \\ 2 - x \text{ if } 1 < x \le 2, \\ 0 \text{ otherwise.} \end{cases}$$

Show that f is a density function. When X is a continuous random variable with density $f_X = f$, evaluate $P(\frac{1}{2} < X < \frac{5}{4})$, $P(X < a)$.

3.78 How must the constant k be chosen so the function f defined by $f(x) = ke^{-|x|}$ will be a density function: *Note:*

$$e^{-|x|} = \begin{cases} e^{-x} \text{ if } x \geq 0, \\ e^{x} \text{ if } x < 0. \end{cases}$$

This distribution is called the *Laplace distribution*.

3.79 Suppose a random variable has a density function as graphed in the accompanying figure. What is $P(X < \frac{1}{2})$? What is $P(X^2 > 1)$? *Hint:* First, find b.

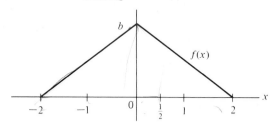

3.80 An electronic device has a life length that is assumed to be exponential, and reliability engineers have determined its reliability at 100 hours to be 90%. At how many hours is the reliability 95%?

3.81 The life length of a satellite is exponentially distributed with a failure rate of $\frac{2}{3}$ per year. Three such satellites are launched simultaneously. What is the probability that at least two will be in orbit after two years?

3.82 Suppose time to failure (in hours) is exponential with $\lambda = .01$. The reliability is $R(t) = .90$. What is the value of t? Suppose 100 such components are put into operation. What is the distribution of the number still operating after t hours?

3.83 Suppose that quarter-pound bars of butter are cut from larger slabs by a machine. The larger slabs are quite uniform in density; if the length of the bar is exactly $3\frac{3}{8}$ inches, then the bar wieghs one-quarter pound. Suppose the true length X of a bar cut by this machine is equally likely to lie in the interval from 3.35 inches to 3.45 inches. Assuming the lengths are independent, what is the probability that all four bars in a particular pound package of butter will weigh at least one-quarter pound? That exactly three will weigh at least one-quarter pound?

3.84 Assume the time of occurrence of the next major earthquake is modeled to be random between 0 and 120 hours.
a. What is the probability that the next earthquake will occur in the next 24 hours?
b. What is the probability that it will be at least three days before it occurs?
c. What is the conditional probability that it will occur tomorrow given that it will take at least that long? Assume time 0 is at 12:00 midnight.

3.85 A test load on a 10-foot wire cable is increased until the cable breaks. Assume the cable does not stretch and that the break occurs at random along its length.
a. What is the probability that the longer piece of cable is exactly three times as long as the shorter piece?

b. What is the probability that the bottom piece is at least three times as long as the top piece?

3.86 A point is chosen at random between 0 and 1. A circle is then drawn with a radius equal to the value of the point. Compute the probability that the area of the circle is (a) equal to $\frac{1}{2}$; (b) less than $\frac{1}{2}$; and (c) between $\frac{1}{3}$ and $\frac{1}{2}$.

3.87 The winning time for a horse is believed to be equally likely to lie in an interval from 1.30 to 1.33 minutes.
   a. What is the probability that the winning time exceeds 1.32 minutes?
   b. What is the probability that the winning time is less than 1.31 minutes?
   c. What is the probability that the winning time will be one of the values 1.30, 1.31, 1.32, 1.33 minutes?

3.88 Assume X is a random variable with distribution function

$$F_X(t) = \begin{cases} 0 \text{ if } t < -2, \\ \frac{1}{4}(t + 2) \text{ if } -2 \le t \le 2, \\ 1 \text{ if } 2 < t. \end{cases}$$

   a. Establish that X is continuous.
   b. Derive the density function for X.
   c. Sketch the graphs of F_X and f_X.
   d. Compute $P(-1 < X \le 1)$ and $P(X > 0 | X \le 1)$.
   e. Identify the distribution of X.

3.89 Let Z be continuous with density function $f_Z(z) = 2/z^2$, $1 < z < 2$.
   a. Determine the CDF of Z.
   b. Sketch the graphs of f_Z and F_Z.
   c. Calculate $P(Z > 1.3)$, $P(1.3 < Z < 1.7 | Z < 1.5)$.

3.90 Write the formulas for f_X and F_X if X is uniform over $[.5, 16.5]$. Determine $P(1 < X < 16)$.

3.91 Assume Y is a random variable whose CDF is

$$F_Y(t) = \begin{cases} 0 \text{ if } t > 1, \\ \log_e t \text{ if } 1 \le t \le e, \\ 1 \text{ if } e \le t. \end{cases}$$

   a. Verify that Y is continuous.
   b. What is the density function for Y?
   c. Sketch the graphs of f_Y and F_Y.

3.92 Suppose Z is a continuous random variable whose density function is $f_Z(z) = c(z + z^2)$ for $1 \le z \le 4$.
   a. What must be the value of c?
   b. Write a general formula for the CDF F_Z.
   c. Calculate $P(2 \le Z \le 3)$, $P(Z > 2)$, $P(Z < 3 | Z > 2)$.

3.93 Suppose W is a continuous random variable with density function $f_W(w) = k(w + 1)$ for $-1 < w < 1$.
   a. Determine the value of k.
   b. Find a formula for F_W.
   c. Calculate $P(W < \frac{1}{2})$, $P(-\frac{1}{2} < W < \frac{1}{2})$, $P(W < \frac{1}{2} | -\frac{1}{2} < W < \frac{1}{2})$.

3.94 In a calm sea, the heights X of waves above, and troughs below, the average depth have a triangular distribution similar to that shown in Exercise 3.79, except $R_X = (-5, 5)$.
a. What is the density function of X?
b. What proportion of waves exceed 3 feet above average depth?

3.95 Computations made with a digital computer must necessarily be done with a finite number of digits. So that a finite number are obtained, a number with an infinite decimal expansion is rounded off at the appropriate point. Assume that, in a particular program, all numbers used are rounded to the nearest integer. Thus 5.76 is rounded to 6, 5.49 is rounded to 5, π (3.14159 . . .) is rounded to 3, e (2.71828 . . .) is rounded to 3, and so on. The error made in this rule for rounding any number must lie between $-\frac{1}{2}$ and $\frac{1}{2}$. Assume all possible values in this interval have equal likelihoods of occurring, and let W be the error made in rounding one number.
a. What is the density function for W?
b. What is the probability that W, in absolute value, exceeds .3?

3.96 The length of time a commercial airliner must hold before being given landing clearance at a busy airport is an exponential random variable H with $\lambda = \frac{1}{5}$ per minute.
a. What is the density function of H?
b. Assume the airliner has 30 minutes worth of fuel on board, and if the pilot is not given clearance within 20 minutes, he will request (and receive) emergency clearance. What is the probability that the pilot must request emergency clearance?

3.97 Z is an exponential random variable with $\lambda = 3$. Evaluate (a) $P(Z > 3)$ and (b) $P(1 < Z < 6)$.

3.5   NORMAL FAMILY

The next family of continuous probability distributions is so important that it is singled out and treated in a separate section. The importance of this family, called *normal,* or *Gaussian,* derives from a number of sources. First, it has been discovered through experience that normal probability functions lead to good models for a wide variety of experiments, ranging from errors in reading meters to scores on IQ tests. Second, it can be shown that, under certain conditions, sums of independent random variables will tend to be normally distributed, regardless of the distribution of each individual random variable. We will return to this important theorem later, in the discussion of the central limit theorem. Finally, normal distributions may be used as approximations to certain discrete distributions, as we will see.

Before introducing the normal distributions, we will find it convenient to review the gamma function and some of its properties. The gamma function, denoted by Γ, is defined by

$$\Gamma(x) = \int_0^\infty t^{x-1} e^{-t} \, dt, \qquad x > 0. \tag{3.20}$$

An important property of Γ is the recursive property $\Gamma(x + 1) = x\Gamma(x)$. To see this, use integration by parts with $\Gamma(x + 1)$ to obtain

$$\Gamma(x + 1) = \int_0^\infty t^x e^{-t}\, dt = -t^x e^{-t} \bigg|_0^\infty + x \int_0^\infty t^{x-1} e^{-t}\, dt, \quad (3.21)$$

letting $y = t^x$ and $dv = e^{-t}\, dt$. By repeated applications of L'Hôpital's rule, it is easy to verify that $\lim_{t\to\infty} t^x e^{-t} = 0$, so the second term in Eq. (3.21) is $x\Gamma(x)$. Hence, for every $x > 0$,

$$\Gamma(x + 1) = x\Gamma(x). \quad (3.22)$$

This basic recursion relation provides some important properties of Γ. By definition,

$$\Gamma(1) = \int_0^\infty e^{-t}\, dt = 1,$$

since the integrand is simply an exponential density function. By Eq. (3.22), $\Gamma(2) = 1 \cdot \Gamma(1) = 1$, $\Gamma(3) = 2\Gamma(2) = 2 \cdot 1$, $\Gamma(4) = 3\Gamma(3) = 3 \cdot 2 \cdot 1$, and, in general,

$$\Gamma(n) = (n - 1)!. \quad (3.23)$$

There is also use for gamma function values at half integers. In order to establish these values, we must show that

$$\int_{-\infty}^\infty e^{-(1/2)z^2}\, dz = \sqrt{2\pi},$$

as follows. Let

$$I = \int_{-\infty}^\infty e^{-(1/2)z^2}\, dz.$$

Then

$$I^2 = \int_{-\infty}^\infty e^{-(1/2)z^2}\, dz \int_{-\infty}^\infty e^{-(1/2)y^2}\, dy = \int_{-\infty}^\infty \int_{-\infty}^\infty e^{-(1/2)(y^2 + z^2)}\, dy\, dz.$$

Changing to polar coordinates (recall that $dy\, dz$ is replaced by $\rho\, d\rho\, d\theta$), and using the symmetry of the integrand, we obtain

$$I^2 = 4 \int_0^{\pi/2} \int_0^\infty \rho e^{-(1/2)\rho^2}\, d\rho\, d\theta = 4 \int_0^{\pi/2} -e^{-(1/2)\rho^2} \bigg|_0^\infty d\theta$$

$$= 4 \int_0^{\pi/2} d\theta = 4\left(\frac{\pi}{2}\right) = 2\pi.$$

Since the integrand in I is nonnegative, $I = \sqrt{2\pi}$. It thus also follows that

$$\int_0^\infty e^{-(1/2)z^2}\, dz = \sqrt{\frac{\pi}{2}}. \quad (3.24)$$

Now consider $\Gamma(\frac{1}{2})$. By definition, $\Gamma(\frac{1}{2}) = \int_0^\infty t^{-(1/2)} e^{-t} dt$. Letting $t = (\frac{1}{2})z^2$, so that $t^{-(1/2)} = \sqrt{2}\, z^{-1}$ and $dt = z\, dz$, we have

$$\Gamma(\tfrac{1}{2}) = \int_0^\infty \sqrt{2}\, e^{-(1/2)z^2}\, dz = \sqrt{2} \cdot \sqrt{\frac{\pi}{2}} = \sqrt{\pi}. \tag{3.25}$$

From Eq. (3.23), it follows that

$$\Gamma(n + \tfrac{1}{2}) = \frac{(2n - 1)(2n - 3) \cdots 5 \cdot 3 \cdot 1}{2^n} \sqrt{\pi} \tag{3.26}$$

Consequently, if k is an even integer, say $k = 2n$ for some n, then $\Gamma(k/2) = \Gamma(n) = (n - 1)!$, while if k is odd, so that $k = 2n + 1$ for some n, $\Gamma(k/2) = \Gamma(n + \frac{1}{2})$ is given by Eq. (3.26). Using a change of variables in the integral, we can easily verify the following (Exercise 3.98): For $\alpha > 0$ and $\beta > 0$,

$$\int_0^\infty t^{\alpha - 1} e^{-(t/\beta)}\, dt = \beta^\alpha \int_0^\infty z^{\alpha - 1} e^{-z}\, dz = \beta^\alpha \Gamma(\alpha). \tag{3.27}$$

We begin our discussion of normal distributions with a special case, the so-called *standard normal density function*. In establishing that $\Gamma(\frac{1}{2}) = \sqrt{\pi}$, we demonstrated that

$$\int_{-\infty}^\infty e^{-(1/2)z^2}\, dz = \sqrt{2\pi}.$$

It thus follows that the function

$$\phi(z) = \frac{1}{\sqrt{2\pi}} e^{-(1/2)z^2}$$

defines a density function on Re. Let us investigate the graph of this function. Since the derivative

$$\phi'(z) = -\frac{z}{\sqrt{2\pi}} e^{-(1/2)z^2}$$

exists, ϕ is everywhere continuous. Also, $\phi'(z) = 0$ only at $z = 0$, so ϕ has a single critical value. Since

$$\phi''(z) = \frac{1}{\sqrt{2\pi}} (z^2 - 1)e^{-(1/2)z^2},$$

$\phi''(0) < 0$, so ϕ has a single maximum at the origin. Points of inflection occur at $z = \pm 1$, where the curve changes from concave downward to concave upward. Since ϕ is an even function, its graph is symmetric about the vertical axis $z = 0$. Finally, $\phi(z) \to 0$ as either $z \to +\infty$ or $z \to -\infty$. With these facts in mind, we can easily sketch the graph of ϕ, as shown in Fig. 3.18. This bell-shaped curve is called the *standard normal curve.*

If X has the standard normal distribution, the CDF of X is

$$\Phi(x) = \int_{-\infty}^{x} \frac{1}{\sqrt{2\pi}} e^{-(1/2)z^2} \, dz.$$

The fundamental theorem of calculus guarantees the *existence* of an antiderivative for the integrand (the standard normal density), but it cannot be given in closed form in terms of familiar functions. However, numerical integration can be used to find the value of $\Phi(x)$ for any selected value of x. For convenience in calculating probabilties, a table of values of Φ is given in the Appendix (Table 3).

The standard normal distribution is a member of the normal family. The general normal distribution is defined in terms of two parameters.

DEFINITION 3.5   A random variable X is *normally distributed* with parameters $\mu \in \text{Re}$ and $\sigma^2 \in (0, \infty)$, written "X is $N(\mu, \sigma^2)$," provided X has the density

$$f(x) = \frac{1}{\sqrt{2\pi\sigma^2}} e^{-(1/2\sigma^2)(x-\mu)^2}. \tag{3.28}$$

The term "normal" is used here in a technical sense. It must not be assumed that the normal is the "usual" distribution or that a nonnormal distribution is "abnormal." The standard normal distribution is simply the $N(0, 1)$ distribution. The normal family quite often provides good models for a wide variety of random phenomena. Among these are scores on achievement tests, tensile strengths of materials, and measurement errors when using various devices to measure some critical dimension or quantity. In many of these examples, negative values are not physically realizable.

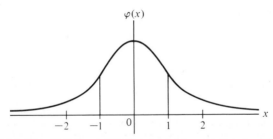

FIGURE 3.18   *Graph of the Standard Normal Density Function*

In others (such as measurement errors), both positive and negative values are possible, but we hardly believe that their magnitude is limitless. And yet if X is a normal random variable, $R_X = (-\infty, \infty)$. Both of these issues can be resolved by the appropriate choice of parameter values. From the normal table, we see that for $x > 3$, $\Phi(x) > .999$; hence, if $x < -3$, $\Phi(x) < .001$. Consequently, as we will shortly see, if X is $N(\mu, \sigma^2)$ and $\mu > 3\sigma$, then $P(X < 0) = \Phi(\mu/\sigma) < .001$. We can make this bound as small as we please for fixed σ by choosing μ sufficiently large. From the point of view of actual applications, the event $(X < 0)$ can be made nearly impossible, and so the normal model may fit well even in an application where X cannot be negative. In the measurement errors case, in which it is often assumed that the random variable is $N(0, \sigma^2)$, the degree to which the analyst believes the error must be bounded is controlled by choosing σ^2 sufficiently small. It will be shown below that linear transformations may be made to render problems involving general normal distributions solvable by using the *standard* normal distribution. For this reason, it is not necessary to tabulate CDF functions for various normal distributions; one can always transform to standard normality and use the tabulated $N(0, 1)$ CDF, Φ.

EXAMPLE 3.23   Suppose X is $N(0, 1)$. Find (a) $P(X < 2)$; (b) $P(X > 2)$; (c) $P(-1 < X < \frac{1}{2})$; (d) $P(-.6 < X < -.1)$.

SOLUTIONS   a. $\Phi(2) = P(X \le 2) = P(X < 2)$, since X is continuous, so $P(X < 2) = .9772$ (to four decimal places) by the Table 3 in the Appendix.
   b. $P(X > 2) = 1 - P(X \le 2) = 1 - .9772 = .0228$.
   c. $P(-1 < X < \frac{1}{2}) = P(X < \frac{1}{2}) - P(X \le -1)$. See Fig. 3.19. Now $P(X < \frac{1}{2}) = .6915$, whereas $P(X \le -1) = P(X \ge 1)$ (by symmetry), and the latter is given by $1 - P(X \le 1) = 1 - .8413 = .1587$. Thus $P(-1 < X < \frac{1}{2}) = .6915 - .1587 = .5328$.
   d. $P(-.6 < X < -.1) = P(.1 < X < .6) = .7257 - .5398 = .1859$.

If X is $N(\mu, \sigma^2)$, a transformation of variables in the CDF integral can be used to obtain the standard normal integral.

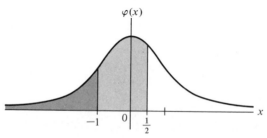

FIGURE 3.19 $P(-1 < X < \frac{1}{2})$ *May Be Given as* $P(X \le \frac{1}{2})$ *Minus* $P(X \le -1)$

EXAMPLE 3.24   Suppose V is $N(-1, 4)$. Find $P(-2 < V < 1)$.

SOLUTION   Since V has density

$$f_V(t) = \frac{1}{2\sqrt{2\pi}} e^{[-1/2(4)](t+1)^2},$$

then

$$P(-2 < V < 1) = F_V(1) - F_V(-2) = \int_{-2}^{1} \frac{1}{\sqrt{2\pi}} e^{-(1/2)[(t+1)/2]^2} \frac{dt}{2}.$$

Let $Z = (t + 1)/2$ (so $dz = dt/2$). Then $t \in (-2, 1)$ if and only if $z \in (-\frac{1}{2}, 1)$, so

$$\int_{-2}^{1} f_V(t)\, dt = \int_{-(1/2)}^{1} \frac{1}{\sqrt{2\pi}} e^{-(1/2)z^2}\, dz = \Phi(1) - \Phi(-\tfrac{1}{2})$$

$$= \Phi(1) + \Phi(\tfrac{1}{2}) - 1 = .5328.$$

A general method is now clear. If Y is $N(\mu, \sigma^2)$ and it is desired to calculate

$$P(a < Y < b) = F_Y(b) - F_Y(a) = \int_{a}^{b} \frac{1}{\sqrt{2\pi}} e^{-(1/2)(x-\mu/\sigma)^2} \frac{dx}{\sigma},$$

make the change of variable $z = (x - \mu)/\sigma$ in the integral to obtain $dx = \sigma\, dz$, and

$$P(a < Y < b) = \int_{(a-\mu)/\sigma}^{(b-\mu)/\sigma} \frac{1}{\sqrt{2\pi}} e^{-(1/2)z^2}\, dz$$

$$= \Phi\left(\frac{b-\mu}{\sigma}\right) - \Phi\left(\frac{a-\mu}{\sigma}\right). \qquad (3.29)$$

This equation may be viewed as a computing formula. A special case is

$$F_Y(y) = \Phi\left(\frac{y-\mu}{\sigma}\right). \qquad (3.30)$$

Another reason for the importance of the normal distribution is the fact that it can be used to approximate binomial probabilities when n is large. We have seen the use of the Poisson in this regard but have noted that it shouldn't be applied when n is small or $np > 10$. It is proved in

Chapter 7 that, if X is binomial with large n, then for any integer k in R_X,

$$F_X(k) \approx \Phi\left(\frac{k - np}{\sqrt{npq}}\right),\tag{3.31}$$

that is, F_X is approximated by a normal distribution having parameters $\mu = np$ and $\sigma^2 = npq$.

EXAMPLE 3.25 Find the probability of at most 230 successes in 500 independent Bernoulli trials with $p = \frac{1}{2}$.

SOLUTION   Here, $np = 100$ is not close to meeting the requirement guideline for the Poisson approximation, and the prospect of directly computing

$$P(X \le 230) = \sum_{k=0}^{230} \binom{500}{k} (.5)^{500}$$

is not appealing whatever method is employed. The normal approximation, Eq. (3.31), gives

$$F_X(230) \approx \Phi\left(\frac{230 - 250}{\sqrt{125}}\right) = \Phi(-1.79) = .04.$$

When using an approximation, we are, of course, interested in how close the approximate value is to the true value. Although this information is not readily available, we can obtain bounds on the maximum error that might be involved. The approximation is best for p-values near .5, where two-place accuracy is attained, even for n as small as 4 or 5 (see Exercise 3.111). As a general rule of thumb, the approximation is good for any combination of parameter values such that $npq \ge 3$. Compare this result with the guideline $np \le 10$ for the Poisson approximation. It is possible that both $np \le 10$ and $npq \ge 3$. In such a case, either approximation may be used with good results. This result also suggests that, under such circumstances, the normal and Poisson distributions involved are nearly the same. Thus a normal distribution may be used to approximate a Poisson distribution when the latter has a "large" parameter λ. In such a case, the $N(\lambda, \lambda)$ distribution closely approximates the Poisson (λ) distribution, as will be shown in Chapter 7.

EXAMPLE 3.26 Suppose X is binomial with $n = 100$ and $p = .1$. Find $P(X \le 10)$.

SOLUTION   Using the Poisson distribution with $\lambda = (100)(.1) = 10$, we have

$P(X \le 10) \approx .58$. On the other hand, using a normal approximation with $\mu = (100)(.1) = 10$ and $\sigma^2 = (100)(.1)(.9) = 9$, we have

$$P(X \le 10) \approx \Phi \left(\frac{10 - 10}{\sqrt{9}} \right) = \Phi(0) = .50.$$

The correct value, obtained from a calculator routine, is about .5832. (The normal approximation can be improved with a "continuity correction," which is discussed later.)

EXAMPLE 3.27   Imperfections in tires produced at a certain plant occur in such a way that it is reasonable to model the presence of an imperfection as a Bernoulli trial having probability $p = .005$. What is the probability that, in a lot of 500 tires, not more than two tires are imperfect?

SOLUTION   Let Z denote the number of defective tires in a (random) lot of 500 tires, so Z is binomial with $n = 500$ and $p = .005$. For a Poisson approximation, take $\lambda = np = 2.5$, so

$$P(Z \le 2) \approx \sum_{i=0}^{2} \frac{e^{-2.5}(2.5)^i}{i!} = .54.$$

(Use the Poisson table, Table 2, in the Appendix.) The normal approximation is shaky for this example since $npq = 2.49 < 3$. Applying this approximation would give the value $\Phi(2 - 2.5/\sqrt{2.49}) = \Phi(-.32) = .38$, a poor approximation, even though n is rather large. The correct value is .5435, to four decimal places.

EXERCISES

THEORY

3.98   Verify Eq. (3.27).

3.99   Verify that for $x > 0$, $\lim_{t \to \infty} t^x e^{-t} = 0$ as claimed in Eq. (3.21).

3.100   Prove that $\Gamma(n + \frac{1}{2}) = [(2n - 1)(2n - 3) \cdots 5 \cdot 3 \cdot 1] \sqrt{\pi}/2^n$.

3.101   a. Discuss the graph of the Cauchy density with respect to symmetry, concavity, extrema, and asymptotes, and compare with the $N(0, 1)$ graph.

b. Sketch, on a single coordinate system, the normal densities for the parameter pairs shown.

$$\mu = -1, \quad \sigma^2 = 1; \quad \mu = 0, \quad \sigma^2 = 2;$$

$$\mu = 0, \quad \sigma^2 = 1; \quad \mu = 0, \quad \sigma^2 = \tfrac{1}{2};$$

$$\mu = +1, \quad \sigma^2 = 1.$$

c. Discuss the effects of changing the parameters μ and σ^2 in terms of the graphs of the normal densities.

3.102 a. Assume X is $N(\mu, \sigma^2)$. Find parameters μ and σ^2 so that $P(X < 1) = .5$.
       b. Is the pair (μ, σ^2) given in part (a) unique?
       c. Do there exist parameters μ and σ^2 for any $\alpha \in (0, 1)$ such that
          $P(-1 < X < .2) = \alpha$?

3.103 A random variable U has a standard normal distribution.
       a. Find the real number α such that $P(\alpha < U < \alpha + .1)$ is maximized.
       b. Show that the maximum probability given in part (a) is $2\Phi(.05) - 1$.
       c. Compare the value in part (b) with $(.1)\phi(0)$ and suggest graphically why
          the latter is slightly larger.

3.104 Generalize Exercise 3.103 so that $\alpha(\beta)$ is found that maximizes
       $P(\alpha < U < \alpha + \beta)$.

3.105 Verify directly (by evaluating the integral) that $\Gamma(1) = 1$.

3.106 For $r > 0$ and $\lambda > 0$, show that the function f defined by

$$f(x) = \frac{\lambda^r x^{r-1} e^{-\lambda x}}{\Gamma(r)}, \qquad x > 0,$$

       is a density function. Such a density is called *gamma*, and if X has this
       density, we say X is distributed gamma (r, λ).

3.107 Verify Eq. (3.30).

APPLICATION

3.108 Assume X is a standard normal random variable. Find each of the following:
       (a) $P(X > .2)$; (b) $P[X \notin (.7, .9)]$; (c) $P(X$ is an integer);
       (d) $P(-\pi/2 < X < 1/e)$; (e) $P(X < -1$ or $X > .68)$; (f) $P(|X| < 4.6)$.

3.109 Assume Z is $N(\mu, \sigma^2)$ and a and b are real numbers, $a < b$.
       a. Show that $P(a < Z < b) = \Phi((b - \mu)/\sigma) - \Phi((a - \mu)/\sigma)$.
       b. Show that $P(a < Z < b) = \Phi((b - \mu)/\sigma) + \Phi((\mu - a)/\sigma) - 1$.

3.110 Suppose V is $N(.5, 9)$. Find (a) $P(V > 1)$; (b) $P(-2 < V < 1)$,
       (c) $P(V \le .5)$, and (d) $P(V < -8)$.

3.111 Calculate $P(X \le 3)$, where X is binomial with the parameter values listed
       below, using the binomial tables and using the normal approximations to the
       binomial. (The cases for which $npq \ge 3$ appear to result in "reasonable"
       normal approximations to the binomial.)

       a. $n = 20, p = .5$.   b. $n = 20, p = .4$.
       c. $n = 20, p = .3$.   d. $n = 20, p = .05$.
       e. $n = 10, p = .5$.   f. $n = 5, p = .5$.
       g. $n = 3, p = .5$.

3.112 Suppose that printing errors in a book occur on each page with probability
       .05.
       a. Find the probability that a 500-page book contains no more than 25 errors.
       b. Find the probability that an error occurs on page 15.
       c. Find the probability that the first 40 pages are error-free.
       d. Discuss the "random experiment" aspect of this exercise. (That is, what
          is random, and what is the random variable involved?)

3.113   Each customer entering a certain store will make a purchase with probability .4. What is the probability that, among 1000 customers, (a) 423 customers make purchases; (b) at least 423 customers make purchases; (c) at most 423 customers make purchases?

3.114   Assume that, on each day, each person in a certain community with a population of 10,000 requires a hospital bed with probability 1/2000. At least how many beds should the hospital have so that it can accommodate all people requiring a bed on a given day with a probability of at least .95?

3.115   In Exercise 3.114, find the number of beds required so that all people requiring a bed during a given three-day period will be accommodated with a probability of at least .95, assuming independence between days.

3.116   The time required in a final assembly of a production line item is a normal random variable Y, with $\mu = 120$ minutes and $\sigma = 10$ minutes.
   a. What proportion of these items require less than 135 minutes in final assembly? *Hint:* Interpret "proportion" as probability.
   b. What is the probability that two successive items require at least 135 minutes in final assembly?

3.117   Assume that weights of Valencia oranges, in a good year, are described by a normal distribution with $\mu = 16$, $\sigma = 2$ (ounces).
   a. One orange is selected at random. What is the probability that its weight exceeds 17 ounces?
   b. Three oranges are selected at random. What is the probability that the weight of exactly one of them exceeds 17 ounces (and the other two do not)?

3.118   A normal random variable has the properties $P(|V - \mu| < 1) = .95$ and $P(V > 0) = .6$.
   a. Find μ and σ^2.
   b. Evaluate $P(0 < V < \frac{1}{2})$.

3.6 SUMMARY

A random variable X is a function mapping the underlying sample space S to the induced sample space $(-\infty, \infty)$. If $o \in S$ is the outcome to the experiment, the outcome on the random variable is the number $X(o)$.

The random variable is discrete if there is a mass function $p(x)$ giving the point probabilities $P(X = a) = p(a)$. Any function that is zero outside a discrete set, and that has positive values on the discrete set that sum to 1, can be considered a mass function. Once a mass function for a discrete random variable X is assigned, the probability of any induced event B, $P(X \in B)$, can be found by summing the discrete masses in B.

Several families of discrete distributions are composed of mass functions involving one or more parameters. These include the following:

The distribution degenerate at c; $p(c) = 1$.

The Bernoulli distribution with success parameter p; $p(x) = p^x(1 - p)^{1-x}$, where $x \in \{0, 1\}$.

The binomial distribution with parameters n and p; $p(x) = \binom{n}{x}p^x(1 - p)^{n-x}$, $x = 0, 1, \ldots, n$.

The hypergeometric distribution with parameters n, M, N; $p(x) = \binom{M}{x}\binom{N-M}{n-x}/\binom{N}{n}$, $x = 0, 1, \ldots, n$.

The geometric distribution with parameter p; $p(x) = pq^{x-1}$, $x = 1, 2, 3, \ldots$.

The negative binomial distribution with parameters r and p; $p(x) = \binom{x+r-1}{x}p^r(1 - p)^x$, $x = 0, 1, 2, \ldots$.

The Poisson distribution with parameter λ; $p(x) = e^{-\lambda}\lambda^x/x!$, $x = 0, 1, 2, \ldots$.

The random variable X is continuous if there is a density function $f(x)$ such that, for any event B, $P(X \in B) = \int_B f(x)\, dx$. In this case, the probability of any point $[P(X = a)]$ is zero. A density function, defined over the real numbers, must be nonnegative and integrate to 1 over the reals, $\int_{-\infty}^{\infty} f(x)\, dx = 1$. While the density value $f(x)$ is not a probability, a useful approximation relating it to the probability of an interval of width Δx is $P(x \le X \le x + \Delta x) \approx f(x) \cdot \Delta x$.

Several continuous families are composed of the following distributions:

The exponential distribution with parameter $\lambda > 0$; $f(x) = \lambda e^{-\lambda x}$, $x \ge 0$.

The Cauchy distribution with parameter θ; $f(x) = 1/\pi[1 + (x - \theta)^2]$.

The Weibull distribution with parameters $\alpha > 0$ and $\beta > 0$; $f(x) = \alpha\beta x^{\beta - 1}e^{-\alpha x^\beta}$, $x > 0$.

The uniform distribution with parameters a and b; $f(x) = 1/(b - a)$, $a < x < b$.

The normal distribution with parameters μ and $\sigma^2 > 0$;

$$f(x) = \frac{1}{\sqrt{2\pi\sigma^2}} e^{-(1/2\sigma^2)(x - \mu)^2}.$$

The normal distribution with $\mu = 0$ and $\sigma^2 = 1$, denoted $N(0, 1)$, has density denoted by $\phi(x)$ and is called the standard normal.

The CDF F of a random variable X is defined, for each real x, as $F(x) = P(X \le x)$. CDFs are monotone increasing, tend to 0 on the left and 1 on the right, and are right continuous. CDFs associated with discrete distributions are step functions; those associated with continuous distributions are continuous functions. The standard normal CDF is denoted by Φ.

If X is $N(\mu, \sigma^2)$, $P(a < X < b)$ can be calculated in terms of the standard normal distribution as

$$P(a < X < b) = \Phi\left(\frac{b - \mu}{\sigma}\right) - \Phi\left(\frac{a - \mu}{\sigma}\right).$$

Values of Φ may be obtained from the normal table, Table 3, in the Appendix.

Probabilities for binomial events may be approximated by Poisson and normal distributions in various circumstances. As a rule of thumb, the Poisson distribution with $\lambda = np$ may be used when n is large and $np \leq 10$; the normal distribution with $\mu = np$ and $\sigma^2 = np(1 - p)$ may be used when $np(1 - p) \geq 3$.

Chapter 4

EXPECTATION

4.1 EXPECTED VALUE OF A RANDOM VARIABLE

We have discussed how the distribution of a random variable can be characterized in terms of its CDF, and how in certain cases the distribution can also be characterized by the mass function (for discrete random variables) or density function (for continuous random variables). In this chapter, we will investigate a third method of characterizing distributions, under the heading of "generating functions." Before doing so, however, we must consider two very important topics: the central concept of *mathematical expectation* and the technical concept of *functions of a random variable*. Each of these topics is of general interest in itself, and each can be useful both in the development of probabilistic models and in the interpretation of the results obtained in such models.

The concept of mathematical expectation concerns a notion of the center of a distribution. In terms of a random variable, it concerns an "average" outcome. Consider a discrete random variable X, with masses $\frac{1}{4}$ at each point in $R_X = \{1, 2, 3, 4\}$. If it were possible to repeat the experiment a number of times, one would "expect" about $\frac{1}{4}$ of the outcomes to be 1s, about a quarter of the outcomes to be 2s, and so on. Of course, there would be variations in these frequencies; it is not likely that exactly $\frac{1}{4}$ of the outcomes would fall in each category. Imagine it is desired to predict the *average* of the outcomes in the repeated experiment. Since about $\frac{1}{4}$ of the outcomes are "expected" in each category, the average outcome is expected to be about 2.5. Indeed, if the relative frequencies *did* turn out to be the same as the point probabilities, the average would be exactly 2.5. What we have in mind is to measure this theoretical average, calculated with the masses, as a theoretical, long-term expected value of the random variable. While the expected value is not a value that can be observed (2.5 $\notin R_X$) in this case, it has useful applications. For example, if X is the number of dollars spent by a random customer at a lunch counter, knowing the expected expenditure is $2.50 makes it easy to predict that 100 customers will spend about $250.

Now, consider a discrete random variable X having range $\{x_1, x_2, \ldots, x_n\}$ with point probabilities $p(x_i) \equiv 1/n$. The possible outcomes of the experiment in R_X have average $\Sigma_{i=1}^{n} x_i/n$, which may be written as $\Sigma_{i=1}^{n} x_i \cdot p(x_i)$. If some discrete model other than the equal-likelihood model applies, some values of X are more likely than others. Then the preceding equation may be viewed as defining a weighted average of the outcomes, where each x_i is weighted by its point probability $p(x_i)$. This weighted average is called the *mean* of X or the *expected value* of X. When R_X is countably infinite or X is continuous, the notion of an ordinary average must be extended. The concept of a mean or expected value is defined by analogy with the finite case, replacing the sum by a series or integral as appropriate.

DEFINITION 4.1   The *expected value* (or *mean*) of a random variable X, denoted $E(X)$, is defined by

$$E(X) = \sum_{x \in R_X} x p_X(x), \qquad (4.1)$$

if X is discrete, provided $\Sigma_{R_X} |x| p(x) < \infty$. If X is continuous,

$$E(X) = \int_{-\infty}^{\infty} x f_X(x) \, dx, \qquad (4.2)$$

provided $\int_{-\infty}^{\infty} |x| f(x) \, dx < \infty$.

EXAMPLE 4.1   Let X be the first digit that results from a call to the random number generator in a computer system. Then $R_X = \{0, 1, \ldots, 9\}$ and $p_X(x_i) \equiv .1$, so

$$E(X) = \sum_{x=0}^{9} \frac{x}{10} = \frac{1}{10} \sum_{i=0}^{9} x = \frac{1}{10} \cdot \frac{10 \cdot 9}{2} = 4.5.$$

Again, the expected value of X is not an outcome to be expected since an outcome for this experiment can never be 4.5.

EXAMPLE 4.2   Suppose time to failure of a motorcycle, X, is exponential with parameter $\lambda = .01$ per week. Then

$$E(X) = \int_{0}^{\infty} x(.01e^{-.01x}) \, dx = .01 \int_{0}^{\infty} x e^{-x/100} \, dx$$

$$= (.01)(100)^2 \Gamma(2) \qquad \text{[by Eq. (3.27) related to the gamma function]}$$

$$= 100.$$

Since X measures time to failure (in weeks), then *mean time to failure* is 100 weeks. In general, for exponential distributions, $E(X) = 1/\lambda$. (see Exercise 4.3.)

EXAMPLE 4.3   Let X denote the number of cycles to failure of a complex system with one-cycle reliability .9. Then X is geometric with $p = .1$, and

$$E(X) = \sum_{x=1}^{\infty} x(.1)(.9)^{x-1} = (.1) \sum_{x=1}^{\infty} x(.9)^{x-1}.$$

The value of the series may be found by observing the following facts about a geometric series $\sum_{y=1}^{\infty} r^{y-1}$, for $|r| < 1$. If this function is denoted $f(r)$, then, since $1/(1 - r)$ is the "sum,"

$$\frac{1}{1-r} = f(r) = \sum_{y=1}^{\infty} r^{y-1} = 1 + \sum_{y=2}^{\infty} r^{y-1}.$$

Differentiating both sides of this equation with respect to r yields $1/(1 - r)^2 = \sum_{y=2}^{\infty} (y - 1)r^{y-2}$. Changing the dummy index from y to $x = y - 1$, we may write this expression as $1/(1 - r)^2 = \sum_{x=1}^{\infty} xr^{x-1}$. Substituting this expression with $r = .9$ in $E(X)$ yields $E(X) = (.1)/(1 - .9)^2 = 1/.1 = 10$. Thus the expected number of cycles to failure is 10. It may be shown, similarly, that, for geometric distributions, $E(X) = 1/p$ (Exercise 4.4).

EXAMPLE 4.4   —*Mean of a Poisson Distribution*   Suppose X is Poisson with parameter λ. Then

$$E(X) = \sum_{x=0}^{\infty} xe^{-\lambda} \frac{\lambda^x}{x!} = e^{-\lambda} \sum_{x=1}^{\infty} \frac{\lambda^x}{(x - 1)!}$$

(dropping the term with $x = 0$). In this series, change from x to $y = x - 1$ and obtain

$$\sum_{x=1}^{\infty} \frac{\lambda^x}{(x - 1)!} = \sum_{y=0}^{\infty} \frac{\lambda^{y+1}}{y!} = \lambda \sum_{y=0}^{\infty} \frac{\lambda^y}{y!} = \lambda e^{\lambda}.$$

Accordingly, $E(X) = e^{-\lambda} \lambda e^{\lambda} = \lambda$. For a Poisson distribution, the mean is just the parameter.

EXAMPLE 4.5   There is an interesting connection between the probability of an event B and the expected value of the associated indicator random variable I_B. [For each event B, I_B is a Bernoulli random variable defined by $I_B(x) = 1$ if $x \in B$; otherwise, $I_B(x) = 0$.] Accordingly, the mass function of I_B is $p(1) = P(I_B = 1) = P(B); p(0) = P(I_B = 0) = 1 - P(B) = P(\bar{B})$. Then $E(I_B) = 0p(0) + 1p(1) = P(B)$. Some axiomatic approaches to probability theory first define expectation and then derive probabilities as expected values of indicator random variables.

EXAMPLE 4.6   —*Mean of a Binomial Distribution*   Suppose X is binomial. Then

$$E(X) = \sum_{x=0}^{n} x \binom{n}{x} p^x q^{n-x} = \sum_{x=1}^{n} x \frac{n!}{x!(n-x)!} p^x q^{n-x}$$

$$= \sum_{x=1}^{n} \frac{n!}{(x-1)!(n-x)!} p^x q^{n-x}$$

$$= np \sum_{x=1}^{n} \frac{(n-1)!}{(x-1)![(n-1)-(x-1)]!} p^{x-1} q^{(n-1)-(x-1)}.$$

Now let $y = x - 1$, and obtain

$$E(X) = np \sum_{y=0}^{n-1} \binom{n-1}{y} p^y q^{(n-1)-y} = np(p+q)^{n-1} = np.$$

In all these cases, it is useful to provide some additional interpretation of $E(X)$. Two particular interpretations are of interest at this point. One stems from the analogy of a probability distribution with a physical mass system. The center of gravity of a mass system is just the position at which a fulcrum must be placed to balance the system. When the total mass being distributed is 1 (as it is for probability mass systems), the formula for calculating the value of the center of gravity, relative to its location on a number line, is exactly the formula given in the definition of $E(X)$, Eq. (4.1) or (4.2), depending on whether the mass system is discrete or continuous. For example, if there are just two mass points at -1 and 3, as shown in Fig. 4.1, then the center of gravity is closer to 3 than -1 because the weight at 3 is more than that at -1.

Since $E(X)$ may be interpreted physically as the center of mass of the distribution induced by X, it is reasonable to refer to $E(X)$ as the mean of

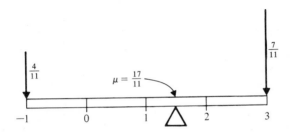

FIGURE 4.1   *Balance Point μ of a Two-Point System*

the distribution of X as well as the mean of X. It has traditionally been denoted by μ_X (the Greek letter μ standing for "mean"), or simply by μ if there is no danger of confusion.

A second interpretation commonly used is that μ_X represents a long-run average value of X. That is to say, if the experiment that X represents is performed a large number of times and the resulting outcomes are averaged (arithmetic average) over the number of trials, that average value will be very close to $E(X)$. The precise manner in which this result is true is known as the law of large numbers, and this result will be derived in Chapter 7. Meanwhile, we will use this interpretation as the occasion arises. In the random digit example (Example 4.1), and in many games of chance, this interpretation is particularly suitable. Thus if a die were rolled a million times and those million numbers were averaged, it would be very surprising if the average were very different from 3.5. In the time-to-failure example (Example 4.2), we would not really expect a given motorcycle to last exactly 100 weeks. Some would last longer, others not so long. But in the long run, we can expect the average time to failure of a large number of such machines to be close to 100 weeks. Similar remarks apply to the cycles to failure of equipment, reflected by the geometric assumptions of Example 4.3.

We have been tacitly assuming that the expressions in the definitions of expected values are meaningful. Thus when R_X is countably infinite, we have assumed that the corresponding series expressed in Eq. (4.1) is absolutely convergent, and hence, the order of summation is immaterial. Similarly, when X is continuous, we have supposed that the integral in Eq. (4.2) converges absolutely, in the sense that $\int_{-\infty}^{\infty} |x| f_X(x)\, dx$ exists and is *finite*. With very few exceptions, soon to be discussed, the series and integrals encountered in this book will possess this property, and we will say no more about it save for the exceptional cases.

EXAMPLE 4.7   —*A Distribution Having No Mean*   A classic example of a distribution in which the mean fails to exist is the Cauchy distribution. Recall that a random variable X is said to be Cauchy-distributed if it has the density function $f(x) = 1/\pi(1 + x^2)$. In such a case, the application of the definition for $E(X)$ (Definition 4.1) would yield

$$E(X) = \frac{1}{\pi} \int_{-\infty}^{\infty} \frac{x\, dx}{1 + x^2} = \frac{1}{2\pi} \int_{-\infty}^{\infty} \frac{2x\, dx}{1 + x^2}. \tag{4.3}$$

But the integral

$$\int_0^{\infty} \frac{2x\, dx}{1 + x^2} = \log(1 + x^2) \Big|_0^{\infty} = \infty$$

diverges; consequently, the integral in Eq. (4.3) does not converge absolutely, so $E(X)$ does not exist. The Cauchy density is a bona fide density function; it is just that when this density function is "weighted" by multiplying $f(x)$ by x itself, the resulting function does not converge to zero fast enough to be integrable.

EXERCISES

THEORY

4.1   Show that, if X is degenerate at c, then $E(X) = c$. For this reason, we sometimes deliberately confuse a degenerate random variable and its (essentially) constant value, and write $E(c) = c$.

4.2   Let X have a hypergeometric distribution with parameters N, M, and n. Show that $E(X) = n(M/N)$. Compare this result with that for a binomial distribution having parameters n and M/N. Discuss the significance for sampling with and without replacement from an urn.

4.3   Find the mean of an exponential distribution having parameter λ.

4.4   Find the mean of a geometric distribution having parameter p. *Hint:* Use the results of Example 4.3.

4.5   Show that the mean of a negative binomial distribution having parameters r and p is rq/p, where $q = 1 - p$.

4.6   Show that, if a density function has a *point of symmetry* c, that is, $f(c + x) = f(c - x)$ for all x, and if $E(X)$ exists, then $E(X) = c$. *Hint:* Write $\mu_x = \int_{-\infty}^{c} xf(x)\, dx + \int_{c}^{\infty} xf(x)\, dx$ and make the change of variable $y = x - c$ in the first integral and $y = c - x$ in the second. Then use the fact that $\int_{-\infty}^{c} f(x)\, dx = \int_{c}^{\infty} f(x)\, dx = \frac{1}{2}$.

4.7   Establish that 0 is a point of symmetry for the Cauchy density of Example 4.7. Why doesn't Exercise 4.6 apply to this case?

4.8   Let X be uniform over the interval (a, b). Apply the result of Exercise 4.6 to verify that $E(X) = (a + b)/2$. Verify this result directly.

4.9   Let X be normal with parameters μ and σ^2. Show that $E(X) = \mu$. (This result is the motivation for denoting this parameter as μ in the first place.) *Hint:* Make the change of variable $z = (x - \mu)/\sigma$ in the integral expression for $E(X)$.

4.10   Let X be continuous with density function f and distribution function F. If $E(X)$ exists, show that

$$E(X) = \int_{0}^{\infty} [1 - F(x)]\, dx - \int_{-\infty}^{0} F(x)\, dx.$$

Hence, if $f(x) = 0$ for $x \le 0$, then $E(X) = \int_{0}^{\infty} [1 - F(x)]\, dx$. See Fig. 4.2 for a graphical interpretation. *Hint:*

$$\int_{0}^{\infty} [1 - F(x)]\, dx = \int_{0}^{\infty} \left[\int_{x}^{\infty} f(t)\, dt \right] dx = \int_{0}^{\infty} \int_{0}^{t} f(t)\, dx\, dt = \int_{0}^{\infty} tf(t)\, dt$$

and, similarly, $\int_{-\infty}^{0} F(x)\, dx = -\int_{-\infty}^{0} tf(t)\, dt$.

APPLICATION

4.11   Use Exercise 4.10 to verify that $E(X) = \lambda$ for the general exponential random variable X.

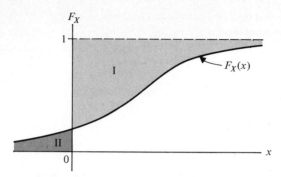

FIGURE 4.2   *Graphical Interpretation of* $\mu_X = I - II$

4.12 What is the expected outcome of the toss of a fair coin if we let heads correspond to 1 and tails correspond to 0?

4.13 Find the mean for the experiment consisting of drawing one transistor from a box containing four good transistors and one defective transistor. (Let "good" correspond to 1 and "defective" correspond to 0.) Does it matter whether sampling is done with or without replacement?

4.14 Find μ_X for the experiment in which X denotes the total thrown on two dice.

4.15 Let $R_X = \{-1, 1\}$ and $p_X(1) = \frac{1}{3}$. Find $E(X)$. What must p_X be in order that $E(X) = 0$?

4.16 Suppose that the birth of a baby boy versus a baby girl is modeled as a Bernoulli trial with probability $\frac{1}{2}$. A couple decides to continue having children until a boy is born. What is the expected number of children for this family?

4.17 The number of bacteria occurring in a small square of a microscopic field for a certain smear is estimated from past data to be $\frac{1}{5}$ per square. What is the probability of observing ten such bacteria in an area of 50 squares? What is the expected number in an area of 50 squares?

4.18 The number of particles emitted per minute from a radioactive source is often assumed to be a Poisson random variable with rate r per minute. What is the expected number of such particles that will be emitted in 1 hour?

4.19 If X is Poisson and $m_X(0) = m_X(1)$, what is $P(X \geq 2)$? What is $E(X)$?

4.20 A telephone operator normally places 23 long distance calls per hour. What is the probability that he or she will not have to place a single call in a given 5-minute period? What is the expected number of 5-minute periods in an hour in which no call is placed?

4.2   FUNCTIONS OF RANDOM VARIABLES

If X is a random variable, it defines a mapping from a sample space S to the real line. If $g(x)$ is a real-valued function defined over the real line, then the composition (function of a function) $g(X)$ is also a mapping from the set S

to the real line. Thus $g(X)$ is itself a random variable. All the usual functions from ordinary analysis such as $g(x) = x^2$, $g(x) = \sin x$, and $g(x) = ax + b$, when composed with a random variable X, are themselves random variables, which would be denoted X^2, $\sin X$, and $aX + b$. Since X has a probability distribution, the probability distribution of $g(X)$ will depend on that assignment in a direct way.

EXAMPLE 4.8 Suppose X is Bernoulli with parameter p. Let $g(x) = 1 - x$ for $x = 0, 1$, and represent the random variable $g(X)$ by Y so that $Y = 1 - X$. First, it is clear that $R_Y = \{0, 1\}$, and the distribution of Y will be determined by its mass function. But $p_Y(1) = P(Y = 1) = P(1 - X = 1) = P(X = 0) = 1 - p = q$, and hence, $p_Y(0) = p$. Thus we may say that Y is Bernoulli with parameter q. If $p = \frac{1}{2}$, then $q = \frac{1}{2}$ also, so X and Y are identically distributed although they are not the same function [indeed, $P(X = Y) = 0$].

EXAMPLE 4.9 Suppose X is a discrete random variable characterized by the distribution shown in the accompanying table.

x	-2	-1	0	$\frac{2}{3}$	2	5
$p_X(x)$	.1	.3	.1	.1	.2	.2

Let $g(x) = x^2$, and denote $g(X) = X^2$ by Z. Then the range of possible values of Z is the set $\{0, \frac{4}{9}, 1, 4, 25\}$. The function g is not 1 to 1 since both -2 and 2 are mapped into 4. That result makes it reasonable and proper to assign point probability .3 to 4 in R_Z since, literally, the value of Z will be 4 when the value of X is either -2 or $+2$. That is, the event $(Z = 4)$ may be written as the union of the disjoint events $(X = -2) \cup (X = +2)$, and so $P(Z = 4) = P(X = -2) + P(X = +2) = .1 + .2 = .3$. Otherwise, each possible value of Z results from just one value of X, and so the distribution of Z must be as shown in the accompanying table.

z	0	$\frac{4}{9}$	1	4	25
$p_Z(z)$	.1	.1	.3	.3	.2

If we adopt the notation $g^{-1}(A)$ to denote $\{x : g(x) \in A\}$ for any set A on the real line, so that, in particular, $g^{-1}(\{y\}) = \{x : g(x) = y\}$, then we can write a general formula for describing the distribution of $g(X)$ when that for a discrete random variable X is given. The first task is to find the range of possible values of $Y = g(X)$. But that is just the set of all images obtained when x from R_X is substituted into the formula $g(x)$. This is to say, $R_Y = \{y : y = g(x) \text{ for some } x \in R_X\}$. The value of Y will be y precisely when the value of x is some point in $g^{-1}(\{y\})$. The second task is to identify the sets $g^{-1}(\{y\})$. The sets $g^{-1}(\{y\})$ partition R_X, and the event $[Y = y]$ occurs precisely when the event $[X \in g^{-1}(\{y\})]$ occurs. Thus

$$p_Y(y) = P(Y = y) = P[X \in g^{-1}(\{y\})] = \sum_{x \in g^{-1}(\{y\})} p_X(x).$$

In summary, the method of obtaining the mass function for $Y = g(X)$, when the mass function for a discrete random variable X is known, involves two steps:

$$R_Y = \{y : y = g(x) \text{ for some } x \in R_X\}, \qquad (4.4)$$

$$p_Y(y) = \sum_{x \in g^{-1}(\{y\})} p_X(x) \qquad \text{for each } y \in R_Y.$$

In many practical problems, performing these steps amounts to careful book-keeping. From our preceding remarks regarding the partitioning of R_X, you can see that $p_Y(y) \geq 0$ and $\sum_{y \in R_Y} p_Y(y) = 1$.

EXAMPLE 4.10 Suppose a fair die is rolled (X) and you win one dollar if the outcome is even and lose one dollar if the outcome is odd. Consider the distribution of your gain. Your gain can be described by the function $g(x)$ $= 1$ if $x = 2, 4, 6$ and $g(x) = -1$ if $x = 1, 3, 5$. The model is $p_X(x) \equiv \frac{1}{6}$ for $x \in \{1, 2, 3, 4, 5, 6\}$. Now, $R_Y = \{-1, +1\}$ and $g^{-1}(\{-1\}) = \{1, 3, 5\}$, while $g^{-1}(\{1\}) = \{2, 4, 6\}$. Consequently, $p_Y(1) = p_X(2) + p_X(4) + p_X(6)$ $= \frac{1}{2}$, and, of course, $p_Y(-1) = \frac{1}{2}$ also. For this case, $E(Y) = 0$ so that, in the long run, there is no gain (or loss) in playing this game provided you pay nothing to play it.

EXAMPLE 4.11 Suppose X is Poisson with $\lambda = 2$ and $g(x) = 2x$. Then $Y = 2X$ has range $R_Y = \{0, 2, 4, 6, \ldots\}$. The mapping from R_X to R_Y is 1 to 1, and so, for each y in R_Y, $g^{-1}(\{y\}) = \{y/2\}$ and $p_Y(y) = p_X(y/2) = e^{-2}[2^{y/2}/(y/2)!]; y = 0, 2, 4, \ldots$.

When X is continuous, $g(X)$ could be discrete or continuous. We illustrate the former case first.

EXAMPLE 4.12 At a particular university, letter grades are given grade point values as follows: A $= 4.0$, B $= 3.0$, C $= 2.0$, D $= 1.0$, and F $= 0$. A test used to decide whether or not a student can waive a certain course is supposed to generate scores that are normally distributed with $\mu = 70$ and $\sigma = 10$, and letter grades are assigned by the scheme A $= 95+$, B $= 85$–95, C $= 70$–85, D $= 60$–70, and F $=$ below 60. What is the probability distribution of the grade point value of a test administered to a randomly selected student?

SOLUTION   We can define the grade point value of a score as the function

$$g(x) = \begin{cases} 4 \text{ if } x > 95, \\ 3 \text{ if } 85 < x < 95, \\ 2 \text{ if } 70 < x < 85, \\ 1 \text{ if } 60 < x < 70, \\ 0 \text{ if } x < 60. \end{cases}$$

(Since X is continuous, we need not worry about inclusive inequalities.) Clearly, $Y = g(X)$ is a discrete random variable with range $R_Y = \{0, 1, 2, 3, 4\}$. Now, however, $g^{-1}(\{y\})$ is a continuous subset of the real line. For example, $g^{-1}(\{3\})$ is the x-interval $(85, 95)$; that is, $(Y = 3) = (85 < X < 95)$. The value $p_Y(3)$ is thus given by $p_Y(3) = P(85 < X < 95) = F_X(95) - F_X(85) = .0606$. The mass function of Y can similarly be determined for each possible value in R_Y and is given by the accompanying table.

y	4	3	2	1	0
$p_Y(y)$	.0062	.0606	.4332	.3413	.1587

If a grade of B or better is required to waive the course, this scheme will only allow about 7% waivers. Indeed, the expected grade point value of the test is given by $E(Y) = \Sigma_y \, y p_Y(y) = 1.4143$.

The general method is clear. If $Y = g(X)$ is discrete with countable range R_Y when X is continuous, then

$$p_Y(y) = \int_{g^{-1}(\{y\})} f_X(x) \, dx \quad \text{for each } y \in R_Y.$$

When X is continuous and the range of $Y = g(X)$ is not countable, matters are not quite so simple, since Y is not discrete and, hence, its distribution cannot be characterized by a mass function. But that distribution is always characterized by a CDF, so we may attempt to construct F_Y. Once that is accomplished, we may be able to retrieve the density function for Y by differentiation.

EXAMPLE 4.13   Suppose X is uniform on $(0, 1)$ and $g(x) = 2x + 1$. Now (see Fig. 4.3), the range of X is mapped into the interval $R_Y = (1, 3)$. Restricting our attention to a value $y \in (1, 3)$, we see that the value of Y will be less than y if and only if the value of X is less than $(y - 1)/2$. Thus

$$F_Y(y) = P\left(X \le \frac{y-1}{2}\right) = F_X\left(\frac{y-1}{2}\right) = \frac{y-1}{2},$$

using the fact that $F_X(x) = x$ for any $0 < x < 1$. Differentiating F_Y yields

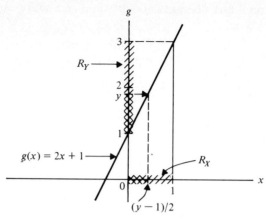

FIGURE 4.3   *A Linear Transformation*

$f_Y(y) = \frac{1}{2}$ for $1 < y < 3$. The distribution of Y is found to be uniform over $(1, 3)$.

The example is easily generalized. In our adopted notation for g^{-1},

$$F_Y(y) = P(Y \le y) = P[g(X) \le y] = P[X \in g^{-1}((-\infty, y])]$$
$$= \int_{g^{-1}((-\infty, y])} f_X(x)\, dx, \tag{4.5}$$

for each y in $R_Y = \{y : y = g(x), x \in R_X\}$. Then $f_Y(y) = F'_Y(y)$ for each $y \in R_Y$. The success of this approach clearly depends on our ability to integrate over the set $g^{-1}((-\infty, y])$. The resulting density function f_Y may not be a member of any of the families of distributions we have discussed.

EXAMPLE 4.14   Suppose X is uniform over the interval $(-2, 2)$, and $Y = X^2 = g(X)$. Then $R_Y = (0, 4)$ (see Fig. 4.4). In this case, if y is arbitrarily

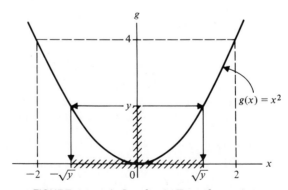

FIGURE 4.4   *A Quadratic Transformation*

taken in R_Y (the only points of interest), then $g^{-1}((-\infty, y]) =$ $(-\sqrt{y}, \sqrt{y})$ and, hence,

$$F_Y(y) = \int_{-\sqrt{y}}^{\sqrt{y}} \tfrac{1}{4}\, dx = \tfrac{1}{4}(\sqrt{y} + \sqrt{y}) = \tfrac{1}{2}\sqrt{y},$$

so $f_Y(y) = 1/4\sqrt{y}$, $0 < y < 4$. Additionally,

$$E(Y) = \int_0^4 \frac{y}{4\sqrt{y}}\, dy = \tfrac{1}{4} \cdot \tfrac{2}{3}y^{3/2} \Big|_0^4 = 1\tfrac{1}{3}.$$

This last example can be generalized to produce a formula valid for any given density f_x. Clearly, if $Y = X^2$, then $R_Y \subset (0, \infty)$, and for any $y \in R_{Y,}$ $(Y \le y) = (-\sqrt{y} < X < \sqrt{y})$, so that $F_Y(y) = F_X(\sqrt{y}) - F_X(-\sqrt{y})$. It then follows that

$$f_Y(y) = f_X(\sqrt{y}) \cdot \frac{1}{2\sqrt{y}} - f_X(-\sqrt{y})\left(-\frac{1}{2\sqrt{y}}\right)$$

$$= \frac{f_X(\sqrt{y}) + f_X(-\sqrt{y})}{2\sqrt{y}}, \qquad y \in R_Y. \qquad (4.6)$$

It is important to note that the formula is valid only for $y \in R_Y$, and care must be exercised in applying f_X to both $\sqrt{y}$ and $-\sqrt{y}$.

EXAMPLE 4.15 Suppose X is uniform over the interval $(-1, 9)$. The range of $Y = X^2$ is the interval $(0, 81)$, and so its density, by Eq. (4.6), is

$$f_Y(y) = \frac{f_X(\sqrt{y}) + f_X(-\sqrt{y})}{2\sqrt{y}}, \qquad 0 < y < 81.$$

Now, for any choice of $y > 1$, $f_X(-\sqrt{y}) = 0$ (see Fig. 4.5), and it would not do at all to substitute .1 at such a point. Hence, a final formula for f_Y

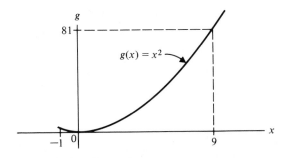

FIGURE 4.5   *Quadratic Transformation over an Asymmetric Domain*

will have to be given in two parts:

$$
f_Y(y) = \begin{cases} \dfrac{.10}{\sqrt{y}} & \text{if } 0 < y < 1, \\[2ex] \dfrac{.05}{\sqrt{y}} & \text{if } 1 < y < 81. \end{cases}
$$

You may verify that f_Y is a density, and $E(Y) = \frac{73}{2}$.

EXAMPLE 4.16   Let X be $N(0, 1)$ and $Y = X^2$. Here, $R_Y = [0, \infty)$, so there are no restrictions on y to worry about. According to Eq. (4.6),

$$
f_Y(y) = \frac{(1/\sqrt{2\pi})e^{-(1/2)y} + (1/\sqrt{2\pi})e^{-(1/2)y}}{2\sqrt{y}} = \frac{e^{-(1/2)y}}{\sqrt{2\pi y}}, \qquad y > 0.
$$

This CDF technique may be applied to find a general formula when g is linear. The range of $Y = aX + b$ may easily be determined from the fact that g is a 1-to-1 transformation. Since $(Y \le y) = (aX + b \le y) = (aX \le y - b)$, two cases arise according to whether $a > 0$ or $a < 0$. (If $a = 0$, then Y is degenerate at b and the distribution of Y is immediate.)

If $a > 0$,

$$
F_Y(y) = P\left(X \le \frac{y - b}{a}\right) = F_X\left(\frac{y - b}{a}\right), \quad \text{so } f_Y(y) = \frac{1}{a} f_X\left(\frac{y - b}{a}\right).
$$

If $a < 0$,

$$
F_Y(y) = P\left(X \ge \frac{y - b}{a}\right)
$$

$$
= 1 - F_X\left(\frac{y - b}{a}\right), \quad \text{so } f_Y(y) = -\frac{1}{a} f_X\left(\frac{y - b}{a}\right).
$$

These expressions can be written as one formula since $|a| = a$ if $a > 0$, while $|a| = -a$ if $a < 0$. Thus

$$
f_Y(y) = \frac{1}{|a|} f_X\left(\frac{y - b}{a}\right), \qquad y \in R_Y.
$$

EXAMPLE 4.17   As in Example 4.16, let X be $N(0, 1)$, and let $Y = 2X + 3$. Then $R_Y = (-\infty, \infty)$ and, for any y,

$$f_Y(y) = \frac{1}{2} f_X \left(\frac{y - 3}{2} \right) = \frac{1}{2 \sqrt{2\pi}} \exp \left[-\frac{1}{2} \left(\frac{y - 3}{2} \right) \right]^2$$

$$= \frac{1}{2 \sqrt{2\pi}} \exp \left[-\frac{1}{2 \cdot 4} (y - 3) \right]^2.$$

Thus Y is $N(3, 4)$. In general, if X is $N(0, 1)$ and $Y = aX + b$, then Y is $N(b, a^2)$ (Exercise 4.21).

It was the monotonicity of g in the discussion above that made it possible to write the event $(g(X) \leq y)$ in terms of an inequality on X. A slightly more general formula can be derived, as follows. Suppose $g(x)$ defines a strictly monotone increasing function. Then g is 1-to-1, and g^{-1} is also an increasing function on R_Y. Consequently, $g(x) \leq y$ if and only if $x \leq g^{-1}(y)$, and so $(g(X) \leq y) = (X \leq g^{-1}(y))$. By definition,

$$F_Y(y) = F_X(g^{-1}(y)). \qquad (4.7)$$

With g^{-1} denoted by ψ for convenience, the chain rule for differentiation can be applied to obtain the density function

$$f_Y(y) = f_X(g^{-1}(y))\psi'(y), \qquad y \in R_Y, \qquad (4.8)$$

assuming ψ is differentiable (so $\psi' = d[g^{-1}(y)]/dy > 0$ since g^{-1} is strictly increasing).

If $g(x)$ defines a strictly monotone decreasing function, then g^{-1} is strictly decreasing and its derivative ψ' will be everywhere negative. Thus

$$(g(X) \leq y) = (X \geq g^{-1}(y)),$$

$$F_Y(y) = 1 - F_X(g^{-1}(y)),$$

and $$f_Y(y) = -f_X(g^{-1}(y))\psi'(y) \qquad \text{for each } y \in R_Y. \qquad (4.9)$$

Again, Eqs. (4.8) and (4.9) may be combined into a single equation:

$$f_Y(y) = f_X(g^{-1}(y))|\psi'(y)|, \qquad y \in R_Y. \qquad (4.10)$$

It should be recalled that, when X is discrete, the formula for the mass function of $g(X)$ has no term corresponding to $|\psi'(y)|$. From elementary calculus, the mechanics of finding $\psi'(y)$, that is, $dx/dy = d[g^{-1}(y)]/dy$, is to take the reciprocal of $dy/dx = dg(x)/dx$, where $x = g^{-1}(y)$ is expressed in terms of y. For example, if $y = g(x) = x^3$, then $g^{-1}(y) = y^{1/3}$. Then $dg^{-1}(y)/dy = \frac{1}{3} y^{-2/3}$ directly. But also, $g(x) = x^3$ yields $dg^{-1}(x)/dy = 3x^2$, and with $x = y^{1/3}$,

$$\frac{1}{dg(x)/dx} = \frac{1}{3x^2} = \frac{1}{3y^{2/3}} = \frac{1}{3}y^{-(2/3)}$$

indirectly.

EXAMPLE 4.18 Suppose X is exponential with $\lambda = 3$ and $Y = X^3$. The range for Y is $[0, \infty)$, and according to Eq. (4.10), with $x = y^{1/3} = g^{-1}(y)$,

$$f_Y(y) = f_X(y^{1/3}) \cdot \tfrac{1}{3}y^{-2/3} = 3e^{-3y^{1/3}} \cdot \tfrac{1}{3}y^{-2/3} = y^{-2/3}e^{-3y^{1/3}}, \qquad y \geq 0.$$

EXAMPLE 4.19 Let X be uniform over the interval $(-\pi/2, \pi/2)$ and $g(x) = \tan x$. From a knowledge of the tangent function, $R_Y = (-\infty, \infty)$, and with $x = \mathrm{Tan}^{-1}(y)$, $dx/dy = 1/(1 + y^2)$, and $f_Y(y) = (1/\pi)(dx/dy) = 1/\pi(1 + y^2)$. But this expression can be recognized as the Cauchy density!

Among the various properties possessed by a CDF for a continuous random variable X is the fact that it is a continuous monotone increasing function. As such, it qualifies as a choice for $g(x)$ in the present context. When $g(x)$ is so chosen, a surprising result is obtained, namely, that $g(X)$ is always uniform over $(0, 1)$. We first illustrate with the exponential case.

EXAMPLE 4.20 Suppose X is exponential with parameter λ. Then the CDF for X is $F_X(x) = 1 - e^{-\lambda x}$ for $x > 0$. For $Y = g(X)$, where $g(x) = F_X(x)$, it follows that $R_Y = (0, 1)$, and for any y in that interval,

$$F_Y(y) = P(Y \leq y) = P(1 - e^{-\lambda X} \leq y) = P(e^{-\lambda X} \geq 1 - y)$$

$$= P[-\lambda X \geq \ln(1 - y)] = P\left[X \leq \frac{-\ln(1 - y)}{\lambda}\right]$$

$$= F_X\left(\frac{-\ln(1 - y)}{\lambda}\right) = 1 - \exp\left\{-\lambda\left[\frac{-\ln(1 - y)}{\lambda}\right]\right\}$$

$$= 1 - e^{\ln(1 - y)} = 1 - (1 - y) = y.$$

Accordingly, $f_Y(y) = 1$, $0 < y < 1$, the uniform density.

The example is not a special case. For if $Y = F_X(X)$, where X is *any* continuous random variable, $R_Y = (0, 1)^*$ and $F_Y(y) = P(Y \leq y) = P[F_X(X) \leq y]$, $0 < y < 1$. Now if F_X is strictly monotone, F_X^{-1} is a well-defined function with the properties that it, too, is strictly increasing, and

---

* For various cases, one or both endpoints might be included; since Y has a continuous distribution, we may ignore these points.

$F_X(F_X^{-1}(y)) = y$ for all $y \in R_Y$, and $F_X^{-1}(F_X(x)) = x$ for all $x \in R_X$. Thus

$$F_Y(y) = P[F_X(X) \leq y] = P[X \leq F_X^{-1}(y)] = F_X(F_X^{-1}(y)) = y,$$

so $f_Y(y) = 1$ for $0 < y < 1$. If F_X is not strictly monotone, then we only have to define $F_X^{-1}(y) = \min\{x : F_X(x) = y\}$ whenever F_X is constant at y to obtain the same result.

Theorem 4.1   —*The Probability Integral Transformation*   If X is continuous with CDF F_X, then the random variable $Y = F_X(X)$ is uniform over $(0, 1)$.

This theorem has many important applications in computer simulation and Monte Carlo methods. If $Y = F_X(X)$ is solved for X, the solution is $X = F_X^{-1}(Y)$, where Y is uniform over $(0, 1)$. Consequently, if y is an observed value of Y in the unit interval, then $x = F_X^{-1}(y)$ is a unique value of x in R_X, which may be considered to be an outcome on X. Most computer centers have programs called random number generators that are devised to simulate the experiment of observing an outcome on a random variable Y that is uniform over $(0, 1)$. Using such a program, one can simulate or imitate the performance of the X-experiment (which might be a costly venture) by mapping a generated uniform observation through F^{-1}. If the successive generated uniform outcomes are independent, this approach will produce the equivalent of independent trials of the X-experiment.

EXAMPLE 4.21   A table of random numbers was used to find successive random numbers (uniform observations) of $y_1 = .29$, $y_2 = .68$, $y_3 = .69$, $y_4 = .81$. An electronic device has a life length represented by an exponential random variable X with mean time to failure of 100 hours, so $F_X(x) = 1 - e^{-.01x}$, $x \geq 0$. Accordingly, $x = F_X^{-1}(y) = -\ln(1 - y)/.01$. Substituting the successive values of y into this formula produces $x_1 = 34$, $x_2 = 114$, $x_3 = 117$, $x_4 = 166$. These may be viewed as simulated times to failure of four independently tested devices having this life distribution.

EXERCISES

THEORY

4.21   Suppose X is $N(0, 1)$. Show that $Y = aX + b$ is $N(b, a^2)$.

4.22   Suppose X is $N(\mu, \sigma^2)$. Let $Z = (X - \mu)/\sigma = (1/\sigma)X - (\mu/\sigma)$. Show that z is $N(0, 1)$, and hence, validate the formula

$$F_X(b) - F_X(a) = \Phi\left(\frac{b - \mu}{\sigma}\right) - \Phi\left(\frac{a - \mu}{\sigma}\right)$$

for calculating probabilities in a general normal model.

4.23 Suppose X is standard normal. Find the density of $Y = |X|$.

4.24 Let X be uniform over (c, d). Show that $Y = aX + b$ is also uniformly distributed.

4.25 Let X be binomial with parameters n and p. Show that $Y = n - X$ is also binomial with parameters n and q.

4.26 Let $X = I_A$ (indicator function). Show that $g(X)$ is discrete for any g, and characterize the mass function.

APPLICATION

4.27 The game of roulette is equivalent to drawing a number from the set $\{00, 0, 1, 2, \ldots, 36\}$ at random. You pay \$1 to play once and bet on an odd outcome. Let X denote your gain. (You receive \$2 if the outcome is odd; otherwise, you receive nothing.) Describe the distribution of X and compute $E(X)$.

4.28 A small drugstore buys five copies of the local daily newspaper for resale each day. It pays 10¢ for each copy and sells them for 25¢ apiece. Let X be the number of papers sold and suppose its probability mass function is as shown in the accompanying table.

x	0	1	2	3	4	5
$p_X(x)$	.01	.01	.02	.18	.41	.37

Any papers not sold are scrapped. Let Z be the daily profit. Derive the probability function for Z. Find $E(Z)$.

4.29 Suppose X is discrete with range $R_X = \{-2, -1, 0, 1, 2, 3\}$ with mass function

$$p_X(x) = \begin{cases} \frac{1}{12} & \text{if } x = -2, 0, \\ \frac{1}{6} & \text{if } x = -1, 1, \\ \frac{1}{4} & \text{if } x = 2, 3. \end{cases}$$

Find the distribution of $|X| - 2$.

4.30 In many war games, a procedure like the following is used to decide whether or not a hit is made when a round is fired. A random digit X is selected; score a miss (0) if that digit exceeds 2 and a hit (1) otherwise. Let H represent the random outcome. Find the probability distribution of H.

4.31 Suppose X has the geometric mass function $p_X(x) = \frac{1}{3}(\frac{2}{3})^{x-1}$, $x = 1, 2, 3, \ldots$. Determine the probability distribution of $Y = X/(X + 1)$.

4.32 Let X have an exponential density and let $g(x) = 1/x$.
a. Show that $R_Y = (0, \infty)$, and find the density function for $Y = g(X)$.
b. Find $E(Y)$, and compare the result with $1/E(X)$.

4.33 Let X be exponential with $\lambda = 1$. Find the density function for $Y = \ln(X)$.

4.34 Suppose X is uniform over $(0, 1)$. For any pair of real numbers a and b, with $a < b$, let $g(x) = a + (b - a)x$ for each $x \in (0, 1)$. Show that $Y = g(X)$ has a uniform density over (a, b). Thus a random number generator is easily modified to simulate outcomes on a uniform-over-(a, b) random variable.

4.35 Let X be uniform over $(-1, 1)$. Find the density function of $g(X)$, where $g(x) = |x|$ for each $x \in (-1, 1)$.

4.36 a. Let Y be uniform over $(0, \pi)$. Find the density function for $g(Y)$, where
       $g(y) = \sin y$ for each $y \in [0, \pi]$.
   b. Find $E[g(Y)]$.
   c. Find $F_Y(y)$.

4.37 Suppose that an electronic device has lifetime denoted by T and that it has
   "value" 5 if it fails before $t = 3$; otherwise, its value is $2T$. Assume that T
   is exponential (1.5), that is, exponential with parameter $\lambda = 1.5$.
   a. Find the distribution of the "value" of the device.
   b. Find the expected value of the device.

4.38 Suppose time to failure X of a machine has an exponential density $f_X(x) =$
   e^{-x}, $x > 0$. The cost of production of a certain unit on the machine is \$500
   if the time to machine failure is less than one day, \$200 if it is between one
   and two days, and \$100 if it exceeds two days. Determine the probability
   distribution for the cost Y of producing a unit.

4.39 Suppose W is a random variable with CDF

$$F_W(w) = \begin{cases} 0 \text{ if } w < 0, \\ \sqrt{w} \text{ if } 0 \le w \le 1, \\ 1 \text{ if } 1 \le w. \end{cases}$$

   Let $V = \sqrt{W}$. Find F_V and f_V.

4.40 Suppose X is an angle, measured in radians, with a probability density that
   is uniform over $[-\pi/2, \pi/2]$. Determine the density function of $Y = \cos X$.

4.41 Suppose a random number generator yielded five successive values, .03, .53,
   .42, .78, and .17. What would be the corresponding sample values of Y in the
   preceding exercise?

4.42 Suppose X has the exponential density $f_X(x) = e^{-x}$, $x > 0$, and let $Y = e^X$.
   Find f_Y, F_Y, and $E(Y)$. Compare these results with $\exp[E(X)]$.

4.43 Suppose X has density function $f(x) = \frac{1}{18}(x + 2)$ if $-2 \le x \le 4$. Find the
   density function for $Y = X^2$ and an expression for F_Y.

4.44 Suppose X is continuous with density function $f_X(x) = \frac{1}{12}(x^2 + 1)$ for $0 \le$
   $x \le 3$. Find the density function for the random variable $Y = 2X^2$. Determine
   the CDF F_Y.

4.3 DISTRIBUTION CHARACTERISTICS

We have seen that a function $Y = g(X)$ of a random variable X is itself a
random variable and admits a probability distribution of its own related to
that of X. The $E(Y)$, if it exists, may be computed from that distribution.
That $E[g(X)]$ may fail to exist even when $E(X)$ exists is exemplified by
example 4.19 of the preceding section where X is uniform over $(-\pi/2, \pi/2)$,
so $E(X) = 0$, but $Y = \tan X$ is Cauchy and hence has no mean. On the
other hand, since the distribution of Y is so intimately related to that of X,
it seems reasonable that $E(Y)$, when it exists, might be computed directly

from the distribution of X. That this can be done is a theorem in advanced calculus.

EXAMPLE 4.22 Suppose $R_X = \{-2, -1, 0, 1, 2, 3\}$ and $p_X(x) \equiv \frac{1}{6}$. Let $g(x) = |x| + 5$, so the range of $Y = g(X)$ is $R_Y = \{5, 6, 7, 8\}$ with $p_Y(5) = \frac{1}{6} = p_Y(8)$, while $p_Y(6) = p_Y(7) = \frac{1}{3}$. Then $E(Y) = \Sigma_y\, y p_Y(y) = \frac{5}{6} + \frac{6}{3} + \frac{7}{3} + \frac{8}{6} = \frac{39}{6}$. But, in turn, this sum may be written (with $y = |x| + 5$) as

$$\frac{1}{6}(|0| + 5) + [\frac{1}{6}(|1| + 5) + \frac{1}{6}(|-1| + 5)] + [\frac{1}{6}(|2| + 5) + \frac{1}{6}(|-2| + 5)]$$

$$+ \frac{1}{6}(|3| + 5) = \sum_x (|x| + 5)p_X(x).$$

When X is discrete, it is relatively easy to see that the computation in Example 4.22 is not a special case. Since for each $y \in R_Y$, $p_Y(y) = \Sigma_{x \in g^{-1}(\{y\})}\, p_X(x)$, we may write, substituting $y = g(x)$,

$$E(Y) = \sum_y y p_Y(y) = \sum_y y \left[\sum_{x \in g^{-1}(\{y\})} p_X(x) \right] = \sum_y \sum_{x \in g^{-1}(\{y\})} g(x) p_X(x).$$

The sets $g^{-1}(\{y\})$ constitute a partition of R_X, so $\Sigma_y \Sigma_{x \in g^{-1}(\{y\})}$ is equivalent to the single sum $\Sigma_{x \in R_X}$. In Example 4.22, for instance, R_X is partitioned into four sets, $g^{-1}(\{5\}) = \{0\}$, $g^{-1}(\{6\}) = \{-1, 1\}$, $g^{-1}(\{7\}) = \{-2, 2\}$, and $g^{-1}(\{8\}) = \{3\}$. Thus $\Sigma_{x \in g^{-1}(\{0\})}$ amounts to substituting $x = 0$; $\Sigma_{x \in g^{-1}(\{6\})}$ amounts to adding in the substitutions for $x = -1$ and $x = 1$; and so on. In the end, all x-values from R_X will have been substituted and summed over. That is,

$$E(Y) = \sum_{x \in R_X} g(x) p_X(x). \tag{4.11}$$

When X is continuous, the analogue of Eq. (4.11) is true, with integration replacing summation. If g happens to be differentiable and strictly increasing, then [see Eq. (4.8)] $f_Y(y) = f_X(g^{-1}(y))\psi'(y)$, $y \in R_Y$, where $\psi'(y)$ is the derivative of $g^{-1}(y)$, an increasing function. Hence, if the change of variable $y = g(x)$ is made in the integral

$$\int_{-\infty}^{\infty} y f_Y(y)\, dy = \int_{-\infty}^{\infty} y f_X(g^{-1}(y))\psi'(y)\, dy,$$

the result is

$$E(Y) = \int_{-\infty}^{\infty} g(x) f_X(x)\, dx, \tag{4.12}$$

where $x = g^{-1}(y)$ and $dx = d[g^{-1}(y)] = \psi'(y)\, dy$. This result makes it at least plausible that the result in Eq. (4.12) is true in general. (See Exercise 4.45.)

EXAMPLE 4.23   Return to Example 4.15, where X is uniform over the interval $(-1, 9)$ and $Y = X^2$. We found that Y had density $f_Y(y) = .1/\sqrt{y}$ if $0 < y < 1$ and $f_Y(y) = .05/\sqrt{y}$ if $1 < y < 81$. From that density function, we found that $E(Y) = \frac{73}{3}$. But for this case,

$$\int_{-\infty}^{\infty} g(x) f_X(x)\, dx = \int_{-1}^{9} x^2 \cdot \frac{1}{10}\, dx = \frac{x^3}{30}\bigg|_{-1}^{9} = \frac{729 + 1}{30} = \frac{73}{3},$$

the same value.

The significance of Eqs. (4.11) and (4.12) is that if only the expected value of $Y = g(X)$ is of interest, it is not necessary to derive the actual distribution of Y, which, in itself, can be an unpleasant, or even impossible, task. The fact that the mean of Y may be found by integrating $g(x)$ with respect to the density for X is sometimes called "the rule of the unconscious statistician," or RUS for short. This name presumably derives from the idea that a statistician would naturally (unconsciously) integrate $g(x)$ to find $E[g(X)]$. One immediate application of RUS shows that expectation may be viewed as a linear operator.

Let X be a random variable, and suppose $Y = aX + b$. According to Eq. (4.11), if X is discrete,

$$E(aX + b) = \sum_{x \in R_X} (ax + b) p_X(x)$$

$$= a \sum_{x \in R_X} x p_X(x) + b \sum_{x \in R_X} p_X(x) = aE(X) + b$$

since $\sum_{x \in R_X} x p_X(x)$ defines $E(X)$ and $\sum_{x \in R_X} p_X(x) = 1$. If X is continuous, the integral enjoys the same linearity properties as do series, so

$$E(aX + b) = \int_{-\infty}^{\infty} (ax + b) f_X(x)\, dx = a \int_{-\infty}^{\infty} x f_X(x)\, dx + b \int_{-\infty}^{\infty} f_X(x)\, dx$$

$$= aE(X) + b.$$

In both cases, we have the general result

$$E(aX + b) = aE(X) + b, \tag{4.13}$$

displaying E as a linear operator.

An interesting application of RUS leads to a discrete random variable having no mean (compare with the Cauchy distribution), which, because of its interpretation in games of chance, is sometimes viewed as paradoxical.

EXAMPLE 4.24 —*The St. Petersburg Paradox*   Suppose a game is as follows: A player bets one dollar against the house and a fair coin is tossed until a head appears for the first time. The payoff is 2^X dollars, where X is the random number of tosses. Then X is geometric with mean $\mu = 1/p = 2$, and the payoff function is the function of X defined by $g(X) = 2^X$. According to RUS, the expected payoff would be given by

$$E(2^X) = \sum_{x=1}^{\infty} 2^x \cdot \frac{1}{2} \left(\frac{1}{2}\right)^{x-1} = \sum_{x=1}^{\infty} 2^x \cdot \frac{1}{2^x} = \sum_{x=1}^{\infty} (1) \to \infty.$$

Since the series defining $E(2^X)$ diverges, $Y = 2^X$ has no mean. The reason this result is sometimes believed to be paradoxical is that this game is not fair; indeed, it favors the player. (A game is said to be *fair* if the amount you pay to play is equal to the expected gain, or payoff. Thus in the long run, a player can expect to break even on a fair game.) In spite of the fact that the present game terminates, on the average, in two tosses, there isn't enough money in the world to accommodate the expected payoff to the player. Put another way, in terms of expected winnings, you should be willing to pay any amount of money you can accumulate to play such a game; yet few rational people would actually be willing to do so.

    If the game is modified slightly by imposing a house limit on payoffs, there will be a finite payoff. For example, if the game is limited to five plays, so you win 2^X dollars if the value of X is 1, 2, 3, 4 and \$32 if it takes five or more plays, the expected payoff now would be

$$\sum_{x=1}^{4} (1) + 32 \sum_{x=5}^{\infty} \frac{1}{2^x} = 4 + 32 \left(1 - \sum_{x=1}^{4} \frac{1}{2^x}\right) = 4 + 32(\tfrac{1}{16}) = 6.$$

You should be willing to pay up to \$6 to play the game; at \$6, this game would be a fair game.

    RUS can be used to obtain various characteristics of a random variable that are extremely important for applications in statistics. The simplest among these are the *moments* of X (or of the distribution of X).

> **DEFINITION 4.2**   For each positive integer n, the *n*th *moment of X*, denoted μ_n', is defined by $\mu_n' = E(X^n)$.

    The first moment of X, μ_1', is just the mean of X, which is often denoted by μ rather than by μ_1'. A more useful notion results when $g(x) = (x - \mu)^n$.

DEFINITION 4.3   For each positive integer n, the nth *central moment* of X, denoted μ_n, is defined by $\mu_n = E[(X - \mu)^n]$.

For $n = 1$, $\mu_1 = E(X - \mu) = E(X) - \mu = \mu - \mu = 0$, using Eq. (4.13). The second central moment, μ_2, is of special interest. It is called the *variance* of X. In the physical mass system analogy, μ_2 corresponds to the moment of inertia of the system. The second central moment of X is typically denoted σ^2 or σ_X^2 or sometimes $V(X)$ if there is a need to emphasize the random variable. $V(X)$ is a measure of the spread or variability of the values of X, in a sense to be discussed shortly.

DEFINITION 4.4   The *variance* of a random variable X is the value $V(X) = \sigma^2 = E[(X - \mu)^2]$.

EXAMPLE 4.25   —*Variance of a Normal Distribution*   Since the symbol σ^2 appears as a parameter in the $N(\mu, \sigma^2)$ distribution, it is desirable to justify its use by verifying that the variance of the distribution is indeed this second parameter value. The variance of an $N(\mu, \sigma^2)$ distribution is, according to Definition 4.4 (and using RUS),

$$\int_{-\infty}^{\infty} (x - \mu)^2 \frac{1}{\sqrt{2\pi}\,\sigma} \exp\left[-\frac{1}{2\sigma^2}(x - \mu)^2 \right] dx$$

$$= 2\sigma^2 \int_{\mu}^{\infty} \left(\frac{x - \mu}{\sigma} \right)^2 \frac{1}{\sqrt{2\pi}\,\sigma} \exp\left[-\frac{1}{2\sigma^2}(x - \mu)^2 \right] dx,$$

since the integrand is symmetric about $x = \mu$. Now let $y = \frac{1}{2}[(x - \mu)/\sigma]^2$ so that $dy = [(x - \mu)/\sigma]\,dx/\sigma$; the limits on y are 0 to ∞ as x varies over (μ, ∞). Thus the variance is given by

$$\frac{\sqrt{2}\,2\sigma^2}{\sqrt{2\pi}} \int_{\mu}^{\infty} \frac{1}{\sqrt{2}} \left(\frac{x - \mu}{\sigma} \right) \exp\left[-\frac{1}{2}\left(\frac{x - \mu}{\sigma} \right)^2 \right] \left(\frac{x - \mu}{\sigma} \right) \frac{dx}{\sigma}$$

$$= \frac{2\sigma^2}{\sqrt{\pi}} \int_{0}^{\infty} y^{3/2 - 1} e^{-y}\, dy$$

$$= \frac{2\sigma^2}{\sqrt{\pi}} \Gamma\left(\frac{3}{2} \right) = \sigma^2 \cdot \frac{2}{\sqrt{\pi}} \cdot \frac{1}{2} \Gamma\left(\frac{1}{2} \right) = \sigma^2.$$

The linearity of expectation (Eq. 4.13) can be applied to provide an alternative formula for computing σ^2, using the expansion $(x - \mu)^2 = x^2 - 2\mu x + \mu^2$. Thus

$$E[(X - \mu)^2] = E(X^2 - 2\mu X + \mu^2) = E(X^2) - 2\mu E(X) + \mu^2$$
$$= E(X^2) - 2\mu^2 + \mu^2 = E(X^2) - \mu^2 = V(X). \qquad (4.14)$$

This formula, sometimes called the computational formula for variance, allows us to compute the variance by calculating the first and second moments, which is often easier than calculating μ_2 directly.

EXAMPLE 4.26   Suppose X is a Bernoulli random variable. Then $\mu = p$, so that

$$V(X) = E[(X - p)^2] = (1 - p)^2 p + (0 - p)^2 q$$
$$= q^2 p + p^2 q = pq(q + p) = pq.$$

Alternatively, we could compute $E(X^2) = \Sigma_x x^2 p_X(x) = 1^2 \cdot p + 0^2 \cdot q = p$, so by Eq. (4.14), $\sigma^2 = p - p^2 = p(1 - p) = pq$. The function $f(p) = p - p^2$ has a unique maximum at $p = \frac{1}{2}$, so $\sigma^2 \le \frac{1}{4}$ for all choices of p in $(0, 1)$.

Aside from the moment of inertia of a physical system, it is somewhat difficult to interpret the numerical value of $V(X)$. If X and Y are two random variables having the same mean μ, and if $V(X) > V(Y)$, then, speaking intuitively, in repeated experiments, the values of X would be more widely dispersed about μ than would those of Y. Figure 4.6 illustrates the effect of changing σ^2 for several normal density functions. Sometimes it is more convenient to use the positive square root of $V(X)$, which will then be in

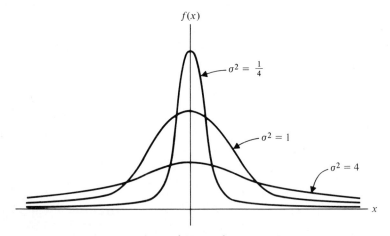

FIGURE 4.6   *Several Normal Density Functions*

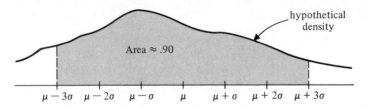

FIGURE 4.7   *The Probability Within 3σ units of μ*

the same units of measure as the values of X, as an equivalent measure of variability. This quantity, denoted σ, is called the *standard deviation* of X. When it is important to show the dependence on X, the standard deviation is denoted σ_X.

The standard deviation of a random variable measures the amount of variability in a rather interesting manner. Since we are discussing the dispersion of values of X in its range, it is natural to consider the distance of a given point x from the mean of X, given by $|x - \mu|$. If you were to consider certain events measured in σ-units along the x-axis for various distributions (see Fig. 4.7), you might discover that the amount of probability within 3σ units of μ was never less than about .90; within 4σ units that figure is more like .94. In other words, it is not very likely that, when the experiment is performed, the outcome will fall beyond three or four standard deviations in either direction from μ. Indeed, if X is $N(0, 1)$, then since $\sigma = 1$, the probability area included within 3σ units of the mean of 0 is given by $2\Phi(3) - 1 = .997$. Practically all the area of a normal density is included within 4σ units of the mean $[2\Phi(4) - 1 \approx 1.0]$.

The reason we can be definitive about how much area is included within so many σ units of μ is due to a result by the Russian, Chebyshev. We now state a simple version of Chebyshev's result.

Theorem 4.2   —*The Chebyshev Inequality*   Let X be a random variable having mean μ and variance σ^2. Then, for any $\delta > 0$, $P(|X - \mu| \geq \delta) \leq \sigma^2/\delta^2$.

PROOF   Suppose X is continuous with density function f, and let $\delta > 0$ be specified. Define A to be the set of real numbers

$$A = \{x : |x - \mu| \geq \delta\} = \{x : (x - \mu)^2 \geq \delta^2\}.$$

Then $\overline{A} = \{x : (x - \mu)^2 < \delta^2\}$, and

$$\sigma^2 = \int_{-\infty}^{\infty} (x - \mu)^2 f(x)\, dx = \int_A (x - \mu)^2 f(x)\, dx + \int_{\overline{A}} (x - \mu)^2 f(x)\, dx.$$

Now if $x \in A$, $(x - \mu)^2 \geq \delta^2$, so

$$\int_A (x - \mu)^2 f(x) \, dx \geq \delta^2 \int_A f(x) \, dx = \delta^2 P(X \in A) = \delta^2 P(|X - \mu| \geq \delta),$$

while if $x \in \bar{A}$, $(x - \mu)^2 f(x) \geq 0$, in which case $\int_{\bar{A}} (x - \mu)^2 f(x) \, dx \geq 0$. Hence, $\sigma^2 \geq \delta^2 P(|X - \mu| \geq \delta)$, and since $\delta^2 > 0$, $P(|X - \mu| \geq \delta) \leq \sigma^2 / \delta^2$, as required. The proof for the discrete case is similar (see Exercise 4.46).

___

If $\delta = k\sigma$, then Eq. (4.15) may be written as

$$P(|X - \mu| \geq k\sigma) \leq \frac{1}{k^2}, \tag{4.16}$$

which demonstrates that the amount of probability beyond 3σ units in either direction of μ cannot exceed $\frac{1}{9} \approx .11$ for *any* probability distribution. [Compare this result with the exact value for the $N(0, 1)$ distribution, .003.] The Chebyschev bound is .0625 for $\delta = 4\sigma$ units and .05 for $\delta = 5\sigma$. Equivalent ways of writing the Chebyshev inequality are, for any $\delta > 0$ and any positive integer k,

$$P(|X - \mu| < \delta) \geq 1 - \frac{\sigma^2}{\delta^2} \quad \text{and} \quad P(|X - \mu| < k\sigma) \geq 1 - \frac{1}{k^2}. \tag{4.17}$$

These express lower bounds on the probability included within a certain distance of the mean.

The surprising part of the Chebyshev inequality is the fact that the bounds apply to any probability distribution having a mean and variance. Because of that, you would not expect the bound to be very sharp in specific cases, as you have already seen when X is $N(0, 1)$. To see how bad it can be, suppose X is Bernoulli with $p = \frac{1}{2}$. Then $\sigma = \frac{1}{2} = \mu$ and $P(|X - \mu| \geq 3\sigma) = P(|X - .5| > 1.5)$. But the values of X are either 0 or 1, so, in fact, $|x - .5|$ can never exceed 1.5; hence, $P(|X - \mu| \geq 3\sigma) = 0$, while the Chebyshev bound is .11. On the other hand, an example can be constructed (Exercise 4.47) where the bound is actually achieved, which demonstrates that it is useless to search for a bound that will in general improve on the Chebyshev bound.

The linearity of expectation can be further exploited to give a general result for the variance of $Y = aX + b$.

EXAMPLE 4.27 Suppose X is a random variable having mean μ_X and variance σ_X^2. It has already been established that $\mu_Y = a\mu_X + b$, so $\mu_Y^2 = a^2\mu_X^2 + 2ab\mu_X + b^2$. Since $Y^2 = a^2X^2 + 2abX + b^2$, then $E(Y^2) = a^2 E(X^2) + 2abE(X) + b^2$. Substracting yields

$$\sigma_Y^2 = a^2 \sigma_X^2 \quad \text{and} \quad \sigma_Y = |a|\sigma_X. \tag{4.18}$$

Adding a constant to each of the possible values of X does not affect the variance. As a particular application (taking $a = 1$ and $b = -\mu_X$), the random variables X and $X - \mu_X$ have the same variance. If $a = 1/\sigma_X$ and $b = -\mu_X/\sigma_X$, the random variable $Z = (X - \mu_X)/\sigma_X$ has mean zero and variance one. This result is quite independent of the particular distribution of X, and for this reason, Z is sometimes called a *standardized random variable*. As an example, if X is $N(\mu, \sigma^2)$, than $Z = (X - \mu)/\sigma$ is standard normal, $N(0, 1)$.

The Cauchy density lacks a variance as well as a mean. That the second moment μ_2' does not exist can be observed readily by noting that $x^2/(1 + x^2)$ can be written $1 - [1/(1 + x^2)]$ so that, if $E(X^2)$ existed,

$$E(X^2) = \frac{1}{\pi} \int_{-\infty}^{\infty} \frac{x^2}{1 + x^2}\, dx$$

$$= \int_{-\infty}^{\infty} \frac{1}{\pi}\, dx - \int_{-\infty}^{\infty} \frac{1}{\pi(1 + x^2)}\, dx = \int_{-\infty}^{\infty} \frac{1}{\pi}\, dx - 1.$$

But the last integral written cannot be finite. In general, it may be shown that, if $E(X^n)$ exists, then $E(X^k)$ exists for all $k < n$. Thus the existence of the second moment guarantees the existence of the first moment. In the Cauchy case, and in the St. Petersburg paradox example, the nonexistence of $E(X)$ allows one to infer $E(X^2)$ does not exist either.

When factorials are involved in a probability distribution (such as the binomial and Poisson), it is often convenient to calculate $E[X(X - 1)]$, perhaps using RUS, and then use the fact that $X(X - 1) = X^2 - X$ and the linearity of E for finding $E(X^2)$ and, ultimately, σ^2. The reason is that $x(x - 1)$ divides $x!$ to leave $(x - 2)!$, which can be very convenient. This moment is a special case of a general type of moment.

DEFINITION 4.5   The *nth factorial moment* of a random variable X is, for each positive integer n,

$$E[X(X - 1)(X - 2) \cdots (X - n + 1)].$$

EXAMPLE 4.28   Let X be Poisson with parameter λ. Then

$$E[X(X - 1)] = \sum_{x=0}^{\infty} x(x - 1)e^{-\lambda}\frac{\lambda^x}{x!} = e^{-\lambda}\sum_{x=2}^{\infty}\frac{\lambda^x}{(x - 2)!},$$

since the terms for $x = 0$ and $x = 1$ vanish. Now, make the change of variable $y = x - 2$, so $x = y + 2$ and

$$E[X(X - 1)] = e^{-\lambda} \sum_{y=0}^{\infty} \frac{\lambda^{y+2}}{y!} = \lambda^2 \sum_{y=0}^{\infty} e^{-\lambda} \frac{\lambda^y}{y!} = \lambda^2,$$

since the last series written is the sum of all the Poisson mass values. Now, since $\lambda^2 = E[X(X - 1)] = E(X^2) - E(X) = E(X^2) - \lambda$, it follows that $E(X^2) = \lambda^2 + \lambda$ and, hence, $\sigma^2 = \lambda^2 + \lambda - \mu^2 = \lambda$. For Poisson distributions, coincidentally, the mean and variance are the same.

In general, if the second factorial moment has been computed, σ^2 may be recovered (as in our example) from the formula

$$\sigma_X^2 = E[X(X - 1)] + \mu_X - \mu_X^2. \tag{4.19}$$

All our attention has focused on first and second moments. While higher-order moments are used in some applications, we confine our attention to the first two. There are other characteristics of a probability distribution that also enter applications. Among these are *percentiles* of a distribution. The intuitive idea is that, for a given percentage P or proportion $.01P$, there is a point ξ that divides the probability mass into two parts, with $P\%$ below that point and $(1 - P)\%$ above. (See Fig. 4.8.) Basically, this idea requires that $F_X(\xi) = .01P$. But the situation is not so simple when the CDF is not strictly increasing, leading to a general definition that is somewhat awkward.

DEFINITION 4.6   Let X be a random variable and P be any percentage, with $p = .01P$ the corresponding proportion. The *Pth percentile* or *pth quantile* of X (or its distribution) is defined as any number ξ_p satisfying

$$P(X < \xi_p) \le p$$

and

$$P(X > \xi_p) \le 1 - p. \tag{4.20}$$

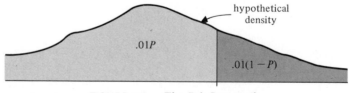

FIGURE 4.8   *The Pth Percentile*

EXAMPLE 4.29 Let X have range $\{1, 2, 3, 4\}$ with mass function $p_X(1) = .1$, $p_X(2) = .2$, $p_X(3) = .3$, and $p_X(4) = .4$. Let $P = 65$, for example, and find $\xi_{.65}$.

SOLUTION A little trial and error will reveal that 4 is the only real number that will do, since $P(X < 4) = .60 < .65$ and $P(X > 4) = 0 < .35$. No value less than 4 will do for $\xi_{.65}$ since then $P(X > \xi_p) \geq p_X(4) = .40 > .35$. As another example, if $P = 60$, then any number between 3 and 4 is a .60 quantile (including 3 but not 4), as you may verify.

When F_X is strictly increasing over R_X, ξ_p is uniquely defined by

$$F_X(\xi_p) = p \qquad \text{or} \qquad \xi_p = F_X^{-1}(p). \qquad (4.21)$$

EXAMPLE 4.30 Suppose X is exponential with parameter λ. Since $F_X(x) = 1 - e^{-\lambda x}$ for $x > 0$, the equation $1 - e^{-\lambda x} = p$ leads to the unique solution $\xi_p = \ln(1 - p)/-\lambda$ for any specified value of p between 0 and 1.

Certain quantiles have special names and applications. When $p = .5$, ξ_p is called the *median*, sometimes denoted Md. The median is occasionally used in place of μ as a measure of the center of a distribution. The *first quartile* is $\xi_{.25}$, and the *third quartile* is $\xi_{.75}$. The difference $\xi_{.75} - \xi_{.25}$ is called the *interquartile range* and is sometimes taken as a measure of variability since it measures the length of span required to enclose the middle 50% of the distribution.

A *mode* (Mo) of a distribution (or of its associated random variable X) is defined to be a point at which the maximum of the mass function or density function is attained, provided that such a maximum exists. Thus Mo is the mode of X if, for all x, $p_X(\text{Mo}) \geq p_X(x)$, when X is discrete, or $f_X(\text{Mo}) \geq f_X(x)$, when X is continuous. The mode is also sometimes used as a measure of the center of a distribution. As a measure of the center, the mode represents a "typical value" in the sense of being a value of X "most likely" to occur when the experiment is performed. In the discrete case, this result is strictly true, whereas in the continuous case, Mo is typically a point in whose neighborhood one is most likely to observe the value of X. This idea is sometimes extended to call any relative maximum of p_X (or f_X) a mode. Thus a density whose graph has two humps might be called a "bimodal" distribution, for example.

EXERCISES

THEORY

4.45 Show that Eq. (4.12) is valid when g is differentiable and strictly decreasing. *Hint:* Apply Eq. (4.10) and observe that in this case $|dx/dy| = -dx/dy$, but the limits of integration are also reversed by the change of variable $x = g^{-1}(y)$.

4.46 Prove Chebyshev's inequality when X is assumed discrete with mean μ and variance σ^2.

4.47 Let k be a fixed positive integer greater than 1. Suppose X has range $\{-k, 0, k\}$ with mass function $p_X(0) = (k^2 - 1)/k^2$, while $p_X(\pm k) = 1/2k^2$. Show that $\mu_X = 0$, $\sigma_X^2 = 1$, and $P(|X - \mu_X| \geq k\sigma) = 1/k^2$. Thus the Chebyshev bound cannot, in general, be improved upon.

4.48 Differentiate both sides of the identity $1/(1 - q)^2 = \sum_{x=1}^{\infty} xq^{x-1}$ with respect to q (as derived in Example 4.3 of Section 4.1) to find an expression for $E[X(X - 1)]$ when X is geometric with parameter p. Use this result and Eq. (4.19) to verify that $V(X) = q/p^2$.

4.49 Use techniques similar to those in Exercise 4.48 and the fact that $1/(1 - q)^r = \sum_{x=0}^{\infty} (-1)^x \binom{-r}{x} q^x$ to verify that, if X is negative binomial with parameters p and r, then $E(X) = rq/p$ and $V(X) = rq/p^2$.

4.50 Suppose X is the number of trials up to and including the rth success in independent Bernoulli trials, while Y is the number of successes. Clearly, $X = Y + r$. Use the fact that Y is negative binomial to show that $E(X) = r/p$ and $V(X) = rq/p^2$. *Note:* When $r = 1$, X is geometric.

4.51 Use Eq. (4.19) to show that, if X is binomial with parameters n and p, then $V(X) = npq$. Also show that, if Y is hypergeometric with parameters N, M, and n, then, with $p = M/N$, $V(Y) = npq(N - n)/(N - 1)$.

4.52 Show that, if X is degenerate, then $V(X) = 0$.

4.53 Suppose the range of a discrete random variable X has at least two points. Show that $V(X) > 0$ and, hence, infer that a discrete random variable with zero variance is degenerate. Can a continuous random variable have zero variance?

4.54 Show that $E(-X) = -E(X)$ and $V(-X) = V(X)$.

4.55 Show that, if X is uniform over (a, b), then $V(X) = (b - a)^2/12$.

4.56 Show that, if $X = I_A$, the indicator function of A, then $V(X) = P(A)P(\overline{A})$.

4.57 If X is exponential with parameter λ, show that $\sigma_X = 1/\lambda$.

4.58 Show that, for any real constant c, $E[(X - c)^2] = V(X) + (\mu - c)^2$. Hence, infer that $c = \mu_X$ minimizes the second moment of X about c. This result explains why σ^2 is called the moment of inertia.

4.59 If X is continuous with a density symmetric about c, show that c is the median of X. Is c also a mode of X? The mean of X?

4.60 Show that the expression $E(|X - c|)$ is minimized by $c = \text{Md}$, the median of X. *Hint:* First show that

$$E(|X - c|) = E(|X - \text{Md}|) + 2 \int_{\text{Md}}^{c} (c - x)f_X(x)\, dx.$$

The value of $E(|X - \text{Md}|)$ is sometimes called the *mean absolute deviation* (MAD) of X.

4.61 Suppose f_X is an even function $[f_X(-x) = f_X(x)$ for all $x \in R_X]$. Show that $E(X^k) = 0$ for odd integer k.

4.62 Show that, if X is geometric (p), then, for $0 < \alpha < 1$, the 100αth percentile may be taken as that integer ξ_α satisfying

$$\frac{\log(1 - \alpha)}{\log q} \le \xi_\alpha < 1 + \frac{\log(1 - \alpha)}{\log q}.$$

4.63 Show that, if p is a mass function symmetric about some point c, and if p has a mean, then (a) $c = \mu$ and (b) $c = $ Md.

4.64 Show that, if F is a CDF continuous at $\frac{1}{2}$, then Md $= F^{-1}(\frac{1}{2})$.

APPLICATION

4.65 Let X be an exponential random variable with $\lambda = 1$. Let g be defined by $g(x) = e^{(1/2)x}$. Find $E[g(X)]$. Repeat the exercise for $g(x) = e^{(1/10)x}$.

4.66 Suppose R is discrete with mass function

$$p_R(r) = \begin{cases} \frac{1}{6} & \text{if } r = -2, 2, \\ \frac{1}{3} & \text{if } r = -1, 1, \\ 0 & \text{otherwise.} \end{cases}$$

Evaluate (a) $E(R)$, (b) $E(R^3)$, (c) σ_R^2, and (d) $\xi_{.4}$.

4.67 The time (in days) between earthquakes of magnitude 4 or more is an exponential random variable Z; it is known that $P(Z > 14) = .632$.
 a. What are the mean and variance of Z?
 b. What is $P(Z \le 10)$?
 c. What is the median interarrival time?

4.68 Let T be continuous with density $f_T(t) = 3t^2, 0 < t < 1$.
 a. Find the standard form of T, and find its distribution.
 b. Calculate the expected value of $Y = \sqrt{T}$ in two ways.

4.69 A game is played by allowing three shots at a target with a rifle. A prize of $10 is given for three hits, $5 for two hits, $1 for a single hit, and nothing otherwise. The probability of scoring a hit is .8 for an individual. How much should he or she be willing to pay to play the game?

4.70 Suppose V is discrete with mass function

$$p_V(v) = \begin{cases} \frac{1}{4} & \text{if } v = 5, 10, 15, \\ \frac{1}{8} & \text{if } v = 20, 25, \\ 0 & \text{otherwise.} \end{cases}$$

Evaluate (a) $E(V/5)$; (b) $E(V - 15)^3$; (c) $E(5/V)$; (d) the median of V; (e) the interquartile, range of V.

4.71 Suppose Y is continuous with density function $f_Y(y) = \frac{2}{57}(y + y^2)$, $1 \le y \le 4$.
 a. Find μ_Y and σ_Y^2.
 b. Find an equation whose solution is ξ_p.

4.72 Let T be continuous with density $f_T(t) = \frac{1}{2}(t + 1), -1 < t < 1$.
 a. Find the standard form for T.
 b. Determine $E[1/(T + 1)]$ in two ways.

4.73 A manufacturer of television sets guarantees the sets for one year and assumes any costs of repair. Past records make it reasonable to assume that the number of times a randomly selected set will need repairs in a year is a random variable R with probability distribution given by

$$p_R(r) = \begin{cases} \frac{1}{2} \text{ if } r = 0, \\ \frac{1}{3} \text{ if } r = 1, \\ \frac{1}{9} \text{ if } r = 2, \\ \frac{1}{18} \text{ if } r = 3, \\ 0 \text{ otherwise.} \end{cases}$$

The cost of the first repair is $6, the cost of the second is $9, and the cost of the third is $18. What is the expected cost of the guarantee to the manufacturer, per set? What is the median cost?

4.74 Suppose Z is a discrete random variable with probability function

$$P_Z(z) = \begin{cases} \frac{1}{6} \text{ if } z = 6, 7, 8, 10, \\ \frac{1}{3} \text{ if } z = 9, \\ 0 \text{ otherwise.} \end{cases}$$

Evaluate (a) μ_Z, (b) σ_Z^2, (c) $E[(Z - 8)^2]$, and (d) $\xi_{.75}$.

4.75 Let X be continuous with CDF

$$F_X(t) = \begin{cases} 0 \text{ if } t < -1, \\ (t + 1)^2 \text{ if } -1 \leq t < 0, \\ 1 \text{ if } 0 \leq t. \end{cases}$$

a. Find μ_X by using the results of Exercise 4.10 in Section 4.1.
b. Determine f_X and compute μ_X and σ_X^2 from it.
c. What is the interquartile range for X?

4.76 The 90th percentile of a normal random variable is 116; its 40th percentile is 110.
a. What are the values of μ and σ^2?
b. Evaluate $P(110 < X < 116)$.
c. Evaluate $P(X \leq 115)$.

4.77 X is an exponential random variable with median equal to 1.
a. What is the 90th percentile of X?
b. What is the mean value of X?

4.78 U is a uniform random variable with mean 2 and 10th percentile equal to .2.
a. What is the density of U?
b. Determine the standard deviation for U.

4.79 A normal random variable Y has $\mu = 5$ and $\sigma = 10$.
a. Find the value a such that $P(|Y - 5| < a) = .9$.
b. Find the value b such that $P(Y > b) = .9$.

4.80 A standard mathematical aptitude test for sixth grade students is used nationwide; the scores made on the test are well described by a normal distribution (meaning that, if one score is selected at random, it can be assumed to be the outcome on a normal random variable X). In a large sample of students, it was found that 80% of the scores were less than or equal to 584 and 40% were less than or equal to 475. As a model, suppose these numbers are true percentiles for X.

    a. What are the mean and standard deviation of X?

    b. What is the 10th percentile?

4.81 A binomial random variable X has mean 8 and standard deviation 2.

    a. What are the values of n and p?

    b. Evaluate $P(X = 5)$.

    c. Calculate $P(|X - \mu_X| > \sigma_X)$. *Hint:* $|x - \mu_X| \leq \sigma_X$ if and only if $\mu_X - \sigma_X \leq x \leq \mu_X + \sigma_X$, so that $P(|X - \mu_X| > \sigma_X) = 1 - P(\mu_X - \sigma_X \leq X \leq \mu_X + \sigma_X)$.

4.82 Assume that fatal automobile accidents occur along a çertain segment of a freeway at a rate of one every month.

    a. What is the probability that there will be at least two such accidents in a six-month period?

    b. Evaluate the probability that there are no such accidents in a six-month period.

    c. Calculate $P(|X - \mu_X| > 1)$, where X is the number of accidents in a month.

    d. What would be the median number of fatal accidents per year?

4.83 A door-to-door salesman makes an average of two sales per day. Let X denote the number of sales he makes in a week (five days).

    a. Calculate μ_X and σ_X.

    b. What is the probability that he will make more than ten sales in a week?

    c. Calculate $P(|X - \mu_X| > \sigma_X)$ and compare the answer with the Chebyshev bound.

4.84 Phone calls arrive at the switchboard of a large corporation at the rate of one every 2 minutes. Let X denote the number to arrive in 10 minutes.

    a. What is the probability that the value of X will be less than 5?

    b. Calculate the mean and variance of X.

    c. Determine $P(|X - \mu_X| > \sigma_X)$.

4.85 A fair die is rolled until three dots are showing faceup. Let Z be the number of rolls required.

    a. What is the expected value of Z?

    b. Calculate the probability that the value of Z will be even.

    c. Determine $P(|Z - \mu_Z| > 1)$.

4.86 In the context of Chebyshev's inequality, show that, for any $\epsilon > 0$, $P(|X - \mu_X| > \epsilon) \leq \sigma_X^2/\epsilon^2$. *Hint:* If $a < b$ (or $a > b$), then surely $a \leq b$ (or $a \geq b$).

4.87 Suppose X is uniform on the interval $(0, 10)$.

    a. Evaluate $P(|X - 5| < 1.2)$.

    b. Find the Chebyshev bound in part (a).

    c. Find a formula for ξ_p.

4.88 A professional baseball player has probability .3 of making a hit each time he is at bat. Assume his attempts at getting a hit are independent and that he is at bat 400 times in one year. Find a bound on the probability that he gets between 110 and 130 hits this year.

4.89 Orders for a replacement part arrive at a supply depot at the rate of one per day and are modeled as arrivals in a Poisson process.

    a. Find a bound on the probability that between 5 and 25 orders arrive in 15 days.

    b. What is the exact probability in part (a)?

4.90 Assume that slippage along a fault in the earth's crust each year is a random variable X. From a long history of observations, it is known that the mean is 2 inches and the standard deviation is 1 inch, but the distribution of X is unknown. What can be said about the probability that slippage within the year will be at most 4 inches?

4.91 Patients with a communicable disease arrive at a hospital during an epidemic at the rate of five per day, like events in a Poisson process.
a. For a two-day period, what is the probability that at most 20 patients arrive?
b. What is the Chebyshev bound on the probability in part (a)?

4.92 A random variable X has unknown mean and variance.
a. How small should σ be so that $P(|X - \mu| \leq 1) \geq .9$?
b. Suppose it is known that X is normal. How small, then, should σ be?

4.4 MOMENT-GENERATING FUNCTIONS

In this section, yet another method of characterizing a probability distribution is presented. When this method can be used, it provides one of the most powerful tools a probabilist has at his or her disposal, called *moment-generating functions*. As a by-product, moment-generating functions are sometimes useful also for computing means, standard deviations, and other characteristics of distributions. Mathematically, these generating functions are *transforms* of densities and mass functions.

EXAMPLE 4.31 Let X be exponential with $\lambda = 1$, and for a real value of t, let $g(x) = e^{tx}$. The expected value of $g(X)$ is

$$E[g(X)] = \int_0^\infty e^{tx} e^{-x}\, dx = \int_0^\infty e^{(t-1)x}\, dx = \frac{e^{(t-1)x}}{t-1}\bigg|_0^\infty. \tag{4.22}$$

Now this last expression is finite only if the exponent coefficient $t - 1$ is negative, that is, if $t < 1$. Then we obtain $E[g(X)] = 1/(1 - t)$ if $t < 1$. But this expression defines a function $M(t)$ of the real variable t, $M(t) = 1/(1 - t)$, $t < 1$. The function M has some very special properties associated with the probability distribution of X. For all values of t, e^{tx} can be expanded in the power series $e^{tx} = \sum_{k=0}^\infty (tx)^k/k!$; substitute this expression in Eq. (4.22) and interchange the summation and integration, which gives

$$M(t) = \int_0^\infty \sum_{k=0}^\infty \frac{t^k x^k}{k!} e^{-x}\, dx = \sum_{k=0}^\infty \frac{t^k}{k!} \int_0^\infty x^k e^{-x}\, dx.$$

But $\int_0^\infty x^k e^{-x}\, dx$ is the kth moment of X, so

$$M(t) = \sum_{k=0}^\infty E(X^k) \frac{t^k}{k!}, \tag{4.23}$$

which shows that the coefficient of $t^k/k!$ in the series expansion of M is the kth moment of X. We found $M(t) = 1/(1 - t)$, and this expression has the series expansion

$$M(t) = \frac{1}{1 - t} = 1 + t + t^2 + \cdots = \sum_{k=0}^{\infty} t^k = \sum_{k=0}^{\infty} k! \frac{t^k}{k!}. \qquad (4.24)$$

Equating coefficients in Eqs. (4.23) and (4.24), we have $E(X^k) = k!$ for an exponential (1) random variable, a result that can easily be verified directly. Because of this relationship between M and the probability distribution of X, we say that M *generates* the moments of X; and we call M the moment-generating function (or mgf) of X.

The expected value of e^{tX} does not exist for some probability distributions, even for t in a neighborhood of the origin [i.e., for all t in some open interval $(-t_0, t_0)$, where $t_0 > 0$]. Fortunately, this function does exist for most of the probability distributions encountered in practice.

DEFINITION 4.7   Let X be a random variable. If $E(e^{tX})$ exists for all t in some open interval $(-t_0, t_0)$, then the function M_X defined by $M_X(t) = E(e^{tX})$ is called the *moment-generating function of X.*

When M_X exists as required in Definition 4.7, all the derivatives of M_X exist at the origin; and the coefficient of $t^k/k!$ in the series expansion of M_X is the kth derivative $M_X^{(k)}$ of M_X, evaluated at 0. On the other hand, interchanging differentiation and integration in the expression $M_X(t) = \int_{-\infty}^{\infty} e^{tx} f_X(x)\, dx$, we obtain*

$$M_X'(t) = \int_{-\infty}^{\infty} xe^{tx} f_X(x)\, dx,$$

$$M_X''(t) = \int_{-\infty}^{\infty} x^2 e^{tx} f_X(x)\, dx,$$

$$M_X^{(3)}(t) = \int_{-\infty}^{\infty} x^3 e^{tx} f_X(x)\, dx,$$

$$\vdots$$

$$M_X^{(k)}(t) = \int_{-\infty}^{\infty} x^k e^{tx} f_X(x)\, dx.$$

When $t = 0$ is substituted in these expressions, the only term affected is

---

* For the discrete case, replace $\int$ by Σ.

e^{tx}, which becomes 1 for any x. Thus $M_X(0) = 1$, $M'_X(0) = E(X)$, $M''_X(0) = E(X^2)$, . . . , $M_X^{(k)}(0) = E(X^k)$.

Theorem 4.3   If X has mgf M_X, then $M_X^{(k)}(0) = E(X^k)$, $k = 1, 2, 3, \ldots$.

It follows that, if we have a formula (in t) for M_X, we may obtain the moments of X by first differentiating the expression and then substituting 0 for t. This procedure may be computationally simpler than integrating (or summing, in the discrete case) to find the moments directly.

EXAMPLE 4.32   Let X be degenerate at a. Then $M_X(t) = E(e^{tX}) = e^{ta}p_X(a) = e^{ta}$, for all real t. Differentiating, we obtain $M_X^{(k)}(t) = a^k e^{ta}$ and $E(X^k) = M_X^{(k)}(0) = a^k$.

EXAMPLE 4.33   If Y denotes the outcome in a Bernoulli trial with $p = \frac{1}{2}$, then $M_Y(t) = e^t p_Y(1) + e^0 p_Y(0) = (\frac{1}{2})e^t + \frac{1}{2}$ for all real t. Upon differentiating, we have $M_X^{(k)}(t) = (\frac{1}{2})e^t$, from which we obtain $E(Y^k) = \frac{1}{2}$ for all k.

EXAMPLE 4.34   If X has a Poisson mass function with $\lambda = 1$, then $p_X(x) = e^{-1}/x!$, $x = 0, 1, 2, \ldots$, and

$$M_X(t) = \sum_{x=0}^{\infty} \frac{e^{tx}e^{-1}}{x!} = e^{-1}\sum_{x=0}^{\infty}\frac{(e^t)^x}{x!} = e^{-1}e^{e^t} = e^{(e^t - 1)}.$$

Accordingly, $M'_X(t) = e^t e^{(e^t - 1)}$ and $E(X) = 1$;

also, $M''_X(t) = e^{2t}e^{(e^t - 1)} + e^t e^{(e^t - 1)}$,

from which $E(X^2) = M''_X(0) = 2$. Finding higher moments of X is tedious but feasible with this approach.

Of far more importance than assisting in the calculation of moments is the fact that the moment-generating function, when it exists, is unique. Unfortunately, the proof of this fact is beyond the scope of this book. Nevertheless, we will appeal to this result freely and state it here as a theorem without proof.

Theorem 4.4   Let X and Y be random variables having mgf's. Then X and Y are *identically distributed* if and only if $M_X(t) = M_Y(t)$ for all t in some neighborhood of 0.

Moment-generating functions thus provide another way of character-izing the probability distributions, in addition to CDFs and density or mass functions. For if M_X and M_Y agree over some neighborhood of 0, then the theorem assures us that X and Y have the same mass function or density function, as the case may be, and the same CDF in any case. As remarked before, this result does not mean necessarily that $X = Y$ as functions, but it does mean that, for every set B, $P(X \in B) = P(Y \in B)$. If X and Y are identically distributed, they will also have the same moments (a weaker statement). Since M_X characterizes X, the term "probability distribution" is now taken to include the mgf of X. For example, we may state, "X is distributed $M(t) = 1/(1 - t)$," meaning X is exponential (1) (i.e., exponential with parameter $\lambda = 1$).

Ordinarily, finding the moment-generating function for a function $g(X)$ of X will not be particularly simple, depending, for example, on our ability to integrate the expression $e^{tg(x)}$ with respect to the density f_X. One useful case, where g is a linear function, is straightforward, however.

EXAMPLE 4.35   Suppose M_X is known and it is desired to find M_Y, where $Y = g(X) = aX + b$. Then, for the continuous case,

$$M_Y(t) = E(e^{tY}) = E[e^{t(aX+b)}] = \int_{-\infty}^{\infty} e^{atx+bt} f_X(x)\, dx$$

$$= e^{bt} \int_{-\infty}^{\infty} e^{atx} f_X(x)\, dx = e^{bt} E(e^{atX}).$$

But $E(e^{atX})$ is just the value of M_X evaluated at (at) instead of t. Hence,

$$M_{aX+b}(t) = e^{bt} M_X(at). \qquad (4.25)$$

This result also holds in the discrete case. In particular, if $a = 0$, then $M_Y(t) = e^{bt}$, which implies that Y is degenerate at b from the uniqueness theorem, a fact previously established. Also, for $b = 0$ and $a = -1$, it follows that $M_{-X}(t) = M_X(-t)$. This result implies that, if X has a density symmetric about 0 (so that X and $-X$ are identically distributed), then the mgf is also symmetric about 0.

We have already made a case for computing factorial moments in certain types of probability distributions. There is a generating function that may be used to generate such moments.

> **DEFINITION 4.8**   If $E(t^X)$ exists for each t in some neigh-borhood of 1, then the function defined by $N_X(t) = E(t^X)$ is called the *factorial moment-generating function* for X.

To see how this function generates moments, we need only compute successive derivatives. Although valid for continuous X, N_X is most useful for discrete cases, so our discussion assumes X has mass function p_X, in which case

$$N_X(t) = \sum_{x \in R_X} t^x p_X(x). \tag{4.26}$$

Differentiating with respect to t, we obtain

$$N_X'(t) = \sum_{x \in R_X} x t^{x-1} p_X(x),$$

$$N_X''(t) = \sum_{x \in R_X} x(x-1) t^{x-2} p_X(x),$$

$$\vdots$$

$$N_X^{(k)}(t) = \sum_{x \in R_X} x(x-1) \cdots (x-k+1) t^{x-k} p_X(x).$$

Substituting $t = 1$, we have $N_X^{(k)}(1) = E[X(X-1) \cdots (X-k+1)]$, the kth factorial moment. Since $z = e^{\log z}$ for any $z > 0$, the factorial moment-generating function is related to the moment-generating function by means of

$$N_X(t) = E(e^{X \log t}] = M_X(\log t). \tag{4.27}$$

Thus if M_X is known, N_X may be computed by substituting $\log t$ for t throughout the formula that defines M_X. Conversely, if $N_X(t)$ is given, substituting e^t throughout for t gives the mgf,

$$M_X(t) = N_X(e^t). \tag{4.28}$$

EXAMPLE 4.36   If X is Poisson with $\lambda = 1$, $M_X(t) = e^{(e^t - 1)}$, as computed in Example 4.34. Then $N_X(t) = e^{t-1}$ and $N_X^{(k)}(t) = e^{t-1}$ for all k. Thus all factorial moments are equal to $N_X^{(k)}(1) = 1$ for the Poisson distribution with parameter $\lambda = 1$. (See Exercise 4.93 for a more general case.)

A special case arises when $R_X = \{0, 1, 2, 3, \ldots\}$, as in the Poisson and negative binomial distributions. Such random variables are called *integer-valued*. Letting $p_k = p_X(k)$ and using the dummy variable z in place of t in Eq. (4.26), we may write

$$N_X(z) = \sum_{k=0}^{\infty} p_k z^k.$$

This factorial moment-generating function is called the *probability-generating function* of the sequence $\{p_0, p_1, p_2, \ldots\}$ (or of X). Observe that $N_X(z)$ generates probabilities rather than moments, in the sense that $p_k = N_X^{(k)}(0)/k!$ (see Exercise 4.94). If it should happen that $R_X = \{0, 1,$

2, . . . , n} for some n, then

$$N_X(z) = \sum_{k=0}^{n} p_k z^k. \tag{4.29}$$

EXAMPLE 4.37 If X is geometric, then the moment-generating function of X is easy to compute, being related to the geometric series again. We need only be careful about the values of t for which convergence is assured. By definition,

$$M_X(t) = p \sum_{x=1}^{\infty} e^{tx} q^{x-1} = \frac{p}{q} \sum_{x=1}^{\infty} (qe^t)^x = \frac{pqe^t}{q} \sum_{x=1}^{\infty} (qe^t)^{x-1}.$$

Since the series $\sum_{n=1}^{\infty} r^{n-1}$ converges if $|r| < 1$, the last series converges to $1/(1 - qe^t)$, provided that $|qe^t| < 1$, which is equivalent to $t < -\log q$, a positive number for fixed p since $0 < q < 1$. Since the origin is included in the interval $(-\infty, -\log q)$, it follows that

$$M_X(t) = \frac{pe^t}{1 - qe^t} \qquad \text{for } t < -\log q.$$

By differentiating M_X and doing some simplification, we obtain

$$M'(t) = \frac{pe^t}{(1 - qe^t)^2} \quad \text{and} \quad M''(t) = \frac{pe^t + pqe^{2t}}{(1 - qe^t)^3}.$$

Substituting $t = 0$, we have $E(X) = 1/p$, $E(X^2) = (1 + q)/p^2$, and $V(X) = q/p^2$. Using Eq. (4.27), we obtain $N_X(t) = pt/(1 - qt)$ if $0 < t < 1/q$. Since X is integer-valued, the probability-generating function of the geometric distribution is given by

$$N_X(z) = \frac{pz}{1 - qz}, \qquad 0 < z < \frac{1}{q}.$$

EXAMPLE 4.38 —*The Negative Binomial Probability-Generating Function*   If an experiment consists of independent, repeated Bernoulli trials, and Y denotes the number of failures until the rth success is observed, then Y has a negative binomial distribution with mass function

$$p_Y(y) = \binom{y + r - 1}{y} p^r q^y, \qquad y = 0, 1, 2, \ldots.$$

The probability-generating function of Y is easily obtained by using the binomial series expansion of $(a - b)^{-r}$ as follows:

$$N_Y(z) = p^r \sum_{y=0}^{\infty} \binom{y + r - 1}{y} (qz)^y = p^r(1 - qz)^{-r} = \frac{p^r}{(1 - qz)^r}.$$

From this result, it is a simple matter to compute the moments of a random variable Y having a negative binomial distribution. They are $E(Y) = rq/p$, $E[Y(Y - 1)] = r(r + 1)q^2/p^2$, and $V(Y) = rq/p^2$, as you may have already seen in Exercise 4.49 of Section 4.3.

EXAMPLE 4.39 —*The Normal mgf*   First, assume Z is $N(0, 1)$. The moment-generating function of Z is computed as follows:

$$M(t) = \frac{1}{\sqrt{2\pi}} \int_{-\infty}^{\infty} e^{tz} e^{-(1/2)z^2} \, dz = \frac{e^{(1/2)t^2}}{\sqrt{2\pi}} \int_{-\infty}^{\infty} e^{-(1/2)(z-t)^2} \, dz.$$

In order to obtain the last expression, we used the identity $-\frac{1}{2}z^2 + tz = -\frac{1}{2}(z - t)^2 + \frac{1}{2}t^2$.

Now let $u = z - t$ (so $du = dz$). Then

$$M(t) = e^{(1/2)t^2} \int_{-\infty}^{\infty} \frac{1}{\sqrt{2\pi}} e^{-(1/2)u^2} \, du = e^{(1/2)t^2}.$$

Now let $X = \sigma Z + \mu$, so X is normally distributed with mean $E(\sigma Z + \mu)$ $= \mu$ and variance $V(\sigma Z + \mu) = \sigma^2$; that is, X is $N(\mu, \sigma^2)$. (See Exercise 4.95.) But by Eq. (4.25), the mgf of X is

$$M_X(t) = e^{\mu t} M_Z(\sigma t) = e^{\mu t} e^{(1/2)(\sigma t)^2} = e^{\mu t + (1/2)\sigma^2 t^2}.$$

This expression is the mgf of an $N(\mu, \sigma^2)$ distribution. It is easily verified that $M_X'(0) = \mu$ and $M_X''(0) = \sigma^2 + \mu^2$, so $E(X) = \mu$ and $V(X) = \sigma^2$, as before.

EXAMPLE 4.40 —*Log Normal Distributions*   A distribution family related to the normal family is the *log normal*. A nonnegative random variable X has the log normal distribution with parameters μ and σ^2 if the logarithm of X is $N(\mu, \sigma^2)$. Let $Y = \log X$, so Y is $N(\mu, \sigma^2)$, $X = e^Y$, and

$$E(X^k) = E(e^{kY}) = M_Y(k) = e^{\mu k + (1/2)\sigma^2 k^2}, \qquad k = 1, 2, 3, \ldots,$$

expressing the moments of X explicitly. In particular, if Y is $N(0, 1)$, then $E(X^k) = e^{(1/2)k^2}$. The distribution function for X can be expressed in terms of the $N(0, 1)$ CDF as follows: For $x > 0$,

$$F_X(x) = P(e^Y \leq x) = P(Y \leq \log x) = F_Y(\log x) = \Phi\left(\frac{\log x - \mu}{\sigma}\right).$$

In particular, the median of X, Md_X, is the solution to

$$\Phi\left(\frac{\log \mathrm{Md} - \mu}{\sigma}\right) = \frac{1}{2}.$$

so that $Md_X = e^\mu$. The log normal is encountered in a variety of applications, providing models for incomes and classroom sizes. In Exercise 4.96, you are asked to discover relationships between moments of X and Y.

EXAMPLE 4.41 —*Gamma Distributions*   The gamma function, introduced in Chapter 3, is defined by $\Gamma(r) = \int_0^\infty x^{r-1} e^{-x}\, dx$ for each $r > 0$. The function $f(x) = x^{r-1} e^{-x}/\Gamma(r)$, $x > 0$, thus defines a density function for each such choice of the parameter r. Expanding slightly to introduce an additional "scale parameter" $\lambda > 0$, we can easily verify that

$$f(x) = \frac{\lambda^r x^{r-1} e^{-\lambda x}}{\Gamma(r)}, \qquad x > 0, \tag{4.30}$$

defines a family of density functions, the *gamma distributions*, for each choice of the two parameters $r > 0$ and $\lambda > 0$. The moment-generating function of a gamma distribution can be computed as follows:

$$M_X(t) = K \int_0^\infty x^{r-1} e^{-(\lambda - t)x}\, dx,$$

where $K = \lambda^r/\Gamma(r)$ is the normalizing constant. Letting $z = (\lambda - t)x$, we obtain, for $t < \lambda$,

$$M_X(t) = \frac{K}{\lambda^r[1 - (1/\lambda)t]^r} \int_0^\infty z^{r-1} e^{-z}\, dz = \frac{K\Gamma(r)}{\lambda^r[1 - (1/\lambda)t]^r} = \frac{1}{[1 - (1/\lambda)t]^r}.$$

From this expression, the first two moments of X are $E(X) = r/\lambda$ and $V(X) = r/\lambda^2$. If X is gamma with $r = 1$, then X is exponential (λ). Thus the exponential family is a subfamily of the gamma family.

EXAMPLE 4.42 —*Beta Distribution*   The beta function is defined as the function

$$B(r, \lambda) = \int_0^1 x^{r-1}(1 - x)^{\lambda - 1}\, dx \tag{4.31}$$

for $r > 0$ and $\lambda > 0$. The beta function is related to the gamma function through the identity

$$B(r, \lambda) = \frac{\Gamma(r)\Gamma(\lambda)}{\Gamma(r + \lambda)}. \tag{4.32}$$

If the integrand in Eq. (4.31) is divided by the expression in Eq. (4.32), the resulting function,

$$f(x) = \frac{\Gamma(r + \lambda)}{\Gamma(r)\Gamma(\lambda)} x^{r-1}(1 - x)^{\lambda - 1}, \qquad 0 < x < 1, \tag{4.33}$$

is a density function. It is a member of the *beta family,* a family indexed by two positive parameters, r and λ. For $r = 1$ and $\lambda = 1$, since $\Gamma(2) = 1!$ $= 1$ and $\Gamma(1) = 0! = 1$,

$$f_X(x) = 1, \qquad 0 < x < 1,$$

a uniform density function. For other choices of parameters, the beta densities exhibit a wide variety of shapes. Some typical graphs are displayed in Fig. 4.9.

The moment-generating function for a beta distribution exists but is not particularly tractable. On the other hand, the moments are easy to compute directly because of the functional form of the density. Thus if X is beta (r, λ) and k is any positive integer, then

$$E(X^k) = \frac{\Gamma(r + \lambda)}{\Gamma(r)\Gamma(\lambda)} \int_0^1 x^{k+r-1}(1 - x)^{\lambda-1}\, dx$$

$$= \frac{\Gamma(r + \lambda)}{\Gamma(r)\Gamma(\lambda)} \cdot \frac{\Gamma(k + r)\Gamma(\lambda)}{\Gamma(k + r + \lambda)} = \frac{\Gamma(r + \lambda)\Gamma(k + r)}{\Gamma(r)\Gamma(k + r + \lambda)}.$$

In particular,

$$E(X) = \frac{\Gamma(r + \lambda)\Gamma(r + 1)}{\Gamma(r)\Gamma(r + \lambda + 1)} = \frac{r}{r + \lambda} \quad \text{and} \quad V(X) = \frac{r\lambda}{(r + \lambda)^2(r + \lambda + 1)}.$$

The beta CDF is also not very tractable, but it is important enough to have been calculated numerically and tabulated rather extensively as the "incomplete beta function."

There is a useful relationship between the beta and the binomial. (See Exercise 4.108.) That relationship is the following: If X is beta with $r = k$

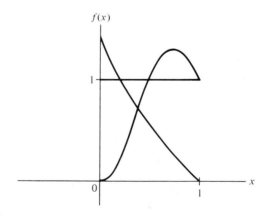

FIGURE 4.9   *Typical Beta Densities*

and $\lambda = n - k + 1$, then

$$F_X(p) = \sum_{x=k}^{n} \binom{n}{k} p^x (1 - p)^{n-x}, \qquad 0 < p < 1. \qquad (4.34)$$

So, for these special integer parameters, the value of the beta CDF can be found by evaluating the tail probabilities of the appropriate binomial mass function as though p were one of its parameters. For example, suppose we wanted $F_X(.2)$, where X is beta with parameters $r = 2$ and $\lambda = 6$. Then we can imagine $k = r = 2$ and $6 = n - k + 1 = n - 2 + 1 = n - 1$, which implies $n = 7$. Now we evaluate

$$\sum_{x=2}^{7} \binom{7}{x} (.2)^x (.8)^{7-x} = .4233$$

(from a binomial table or a calculator routine).

EXERCISES

THEORY

4.93 Suppose X is Poisson with arbitrary parameter $\lambda > 0$.
    a. Verify $M_X(t) = \exp[\lambda(e^t - 1)]$.
    b. Use part (a) to compute $V(X)$ by taking derivatives of M_X.
    c. Find $N_X(t)$.
    d. Use part (c) to find $V(X)$.
    e. Find the kth factorial moment of X.

4.94 Differentiate $N_X(z) = \sum_{k=0}^{\infty} p_k z^k$ formally with respect to z, k times, to verify that $p_k = N_X^{(k)}(0)/k!$ for $k = 0, 1, 2, \ldots$.

4.95 Suppose Z is $N(0, 1)$. Use an argument not involving mgf's to verify that $X = \sigma Z + \mu$ is $N(\mu, \sigma^2)$.

4.96 Relationships between normal and log normal moments are examined. Suppose X is $N(\mu_X, \sigma_X^2)$, so $Y = e^X$ has a log normal distribution with mean μ_Y and variance σ_Y^2.
    a. Show that $\mu_Y = \exp(\mu_X + \frac{1}{2}\sigma_X^2)$; $\sigma_Y^2 = \exp(2\mu_X + 2\sigma_X^2) - \exp(2\mu_X + \sigma_X^2)$.
    b. Show that $\mu_X = 2 \log \mu_Y - \frac{1}{2} \log(\sigma_Y^2 + \mu_Y^2)$; $\sigma_X^2 = \log(\sigma_Y^2 + \mu_Y^2) - 2\mu \log \mu_Y$.

4.97 a. Find the general exponential (λ) mgf.
    b. Use the results of part (a) to get the mean and variance of the exponential (λ) distribution.
    c. Find the 100αth percentile of this distribution.

4.98 Show that, if X has an even density, (a) X and $-X$ are identically distributed and (b) M_X is symmetric about 0.

4.99 Show that, if X is uniform over (a, b), then $M_X(t) = (e^{bt} - e^{at})/(b - a)t$ for

$t \neq 0$ and $M_X(0) = 1$. Use this function to verify that $E(X^k) = (b^{k+1} - a^{k+1})/(b - a)(k + 1)$ for $k = 1, 2$.

4.100   For X uniform over $(0, 1)$, use Eq. (4.25) to show that $Y = cX + d$ is uniform over $(d, c + d)$.

4.101   Use an mgf argument to show that, if X has mean μ and variance σ^2, then the standardization of X, $(X - \mu)/\sigma$, has mean zero and variance one.

4.102   a. If X is binomial, show that $M_X(t) = (pe^t + q)^n$. *Hint:* Use the binomial theorem.
    b. Use part (a) to find $E(X)$, $V(X)$, $E(X^3)$.

4.103   Using mgf's, show that, if X is binomial with parameters n and p, then $Y = n - X$ is also binomial with parameters n and q.

4.104   Find the binomial factorial generating function by (a) using the definition and (b) using M_X of Exercise 4.102 evaluated at log z.

4.105   If X is uniform over $(0, 1)$, show that $Y = 1 - X$ has this same probability distribution.

4.106   Prove the probability integral transformation for the case where continuous X has a moment-generating function. *Hint:* In $E(e^{tF_X(X)}) = \int_{-\infty}^{\infty} e^{tF_X(x)} f_X(x) \, dx$ observe that $f_X(x) = F_X'(x)$; use Theorem 4.4.

4.107   Prove that, if X is continuous with distribution function F, then the random variable $1 - F(X)$ is uniform over $(0, 1)$.

4.108   Let X be beta with parameters $r = k$, a positive integer, and $\lambda = n - k + 1$ with positive integer $n \geq k$. The CDF for X evaluated at $0 < p < 1$ is given by

$$F_X(p) = \frac{\Gamma(n + 1)}{\Gamma(k)\Gamma(n - k + 1)} \int_0^p t^{k-1}(1 - t)^{n-k} \, dt.$$

Let $u = (1 - t)^{n-k}$ and $dv = t^{n-1} \, dt$, and integrate repeatedly by parts to verify Eq. (4.34).

APPLICATION

4.109   If it happens that the value of an mgf can be written explicitly for each t as $M(t) = ae^t + be^{3t} + ce^{10t} + d$, what must be the range of the associated random variable?

4.110   Let X have the mass function m defined by $m(x) = (\frac{1}{2})^x$, $x = 1, 2, 3, \ldots$. Determine the moment-generating function for X.

4.111   Suppose a moment-generating function M is given by the expression $M(t) = e^{2t+3t^2}$.
    a. For a random variable X having this mgf, find $E(X)$ and $V(X)$.
    b. Find $P(X < 2.5)$.

4.112   Suppose M is defined by $M(t) = (1 - t)^{-m}$, where $m > 0$. For X with M as an mgf, find $E(X)$ and $V(X)$. Describe the distribution of X.

4.113   Suppose M is defined by $M(t) = \frac{1}{3} + (\frac{2}{3}) \sum_{k=0}^{\infty} t^k/k!$. What probability mass function is necessarily associated with this moment-generating function?

4.114   Three people are in a store. Let X be the random number of those who will make a purchase before they leave. All that is known is the probability-generating function, $N_X(z) = \frac{1}{10}(4 + 3z + 2z^2 + z^3)$. What unique mass function for X does this function represent?

4.115   The mgf for a random variable X is asserted to be $M_X(t) = 1/(1 - 2t)^3$ provided $t < \frac{1}{2}$. If this is so, compute μ_X and σ_X^2.

4.116   Five rounds are fired at a target. The number of hits scored is a random variable H with probability function as given in the accompanying table. Let Y be the number of *misses* fired from the same five rounds. Determine the mgf for Y. Deduce the probability function from this mfg.

h	0	1	2	3	4	5
$p_H(h)$	.17	.36	.31	.13	.02	.01

4.117   The random variable V has probability function

$$p_V(v) = \begin{cases} \binom{4}{v} (.2)^v (.8)^{4-v} & \text{if } v = 0, 1, 2, 3, 4, \\ 0 \text{ otherwise.} \end{cases}$$

   a. Compute the mgf for V.
   b. Compute the probability-generating function for $4 - V$.

4.118   Suppose a random variable U has the following probability generating function: $N_U(r) = \frac{1}{8} + \frac{1}{2}r + \frac{1}{4}r^2 + \frac{1}{8}r^3$.
   a. Find the distribution of U.
   b. Find $E(U)$ and $V(U)$.
   c. Find $P[U \le E(U)]$.

4.119   A random variable X has mgf given by $M_X(t) = e^{4t + 6t^2}$. Find the mean and variance of X. What is the distribution of X?

4.120   Suppose the density function for X is given by $f(x) = \frac{1}{2}e^{-|x|}$. Find the mgf for X and verify that $\mu_X = 0$, the point of symmetry. *Hint:* In integrating $E(e^{tX})$, recall that $|x| = -x$ if $x < 0$, so that the range of integration must be divided into two parts.

4.121   A loaded die is rolled one time, with outcome X. The factorial mgf for X is asserted to be $N_X(t) = \frac{1}{2}t + \frac{1}{4}t^2 + \frac{1}{4}t^3$. For all practical purposes, what faces are missing on the die?

4.122   A basketball player shoots from the free throw line until she makes a basket. It is supposed that the probability of her making a basket on any throw is .8.
   a. Derive the mgf for X, the number of tosses needed.
   b. What is the expected number of such tosses?

4.123   The time of arrival of a commercial airliner at its destination is a random variable Y with density function $f_Y(y) = \frac{1}{10}$ for $0 < y < 10$. Derive the mgf for Y and use this expression to find the mgf for $X = Y + 5$. Identify the distribution of X.

4.124   Four missiles are shot at the same target. The number of hits X is a random variable with the probability function shown in the accompanying table.

x	0	1	2	3	4
$p(x)$	.05	.10	.20	.40	.25

a. Derive the mgf for X and find its factorial moment-generating function.
b. Compute the first three moments for X.

4.125 A random variable Z is asserted to have mgf $M_Z(t) = (1 - t)^{-3}$ for $t < 1$.
a. Find μ_Z and σ_X^2. Guess an expression for $E(X^k)$.
b. Find the general expression for $E(X^k)$.

4.126 Let X be the random time until the next earthquake of magnitude 4 or more on the Richter scale somewhere in the world. For $F_X(x) = 1 - e^{-(1/2)x}$, $x > 0$, determine the mgf for X, and find μ_X and σ_X^2.

4.127 The number of orders received per week for a specific part is a random variable Y with probability-generating function $N_Y(t) = e^{2(t-1)}$ valid for all t.
a. Derive the probability mass function for Y.
b. Find an expression for the mgf for Y and compute μ_Y and σ_Y^2.

4.128 The number of TV sets sold per day is a random variable X with factorial moment-generating function $N_X(t) = (.9 + .1t)^3$.
a. Determine the probability function for X.
b. Find μ_X and σ_X^2.

4.129 Suppose the mgf for a random variable X is given as $M_X(t) = e^{t^2}$.
a. Find the mean and variance of X.
b. Determine the mgf for the standard form Z of X, and verify that $\mu_Z = 0$ and $\sigma_Z^2 = 1$.

4.130 The random variable Z has probability function

$$p_Z(z) = \begin{cases} \frac{1}{4} \text{ if } z = 0, 2, \\ \frac{1}{2} \text{ if } z = 1, \\ 0 \text{ otherwise.} \end{cases}$$

a. Find the mgf for Z.
b. Determine the probability-generating function for Z.
c. Calculate the first two moments of Z.

4.131 The mgf for a random variable X is asserted to be $M_X(t) = t/(1 - t)$ for $t < 1$. Tell why this cannot be so.

4.132 Suppose that X is uniform on $(0, 1)$. Find the probability distribution of $Y = X^2$.

4.133 Suppose that X is exponential (λ). Find the density function of $Y = e^{-X}$.

4.134 Under what circumstances will the mean and median of the exponential density coincide?

4.135 Find the mean of a general gamma distribution directly, without using the moment-generating function.

4.136 Suppose the moment-generating function of a random variable X is given by the formula $(1 - 2t)^{-4}$. Determine the distribution family of X, and find its mean and variance.

4.137   Suppose X is uniform on $(0, 1)$. Identify the probability distribution of the composition $Y = -2 \log X$.

4.5   SUMMARY

If X is a discrete random variable with mass function $p(x)$, the mean (expected value, first moment) of X is $\Sigma_x \, xp(x)$ and is denoted μ_X or $E(X)$. If X is continuous with density $f(x)$, the mean is given by $\mu_X = \int_{-\infty}^{\infty} xf(x) \, dx$. The second central moment, $E(x - \mu_X)$, is the variance of X, denoted σ_X^2 or $V(X)$. The square root of the variance is the standard deviation σ_X. The expectation of a function g of X can be calculated (when it exists) by the rule of the unconscious statistician, $E(g(X)) = \int_{-\infty}^{\infty} g(x)f(x) \, dx$ [or $\Sigma_x \, g(x)p(x)$ if X is discrete].

The moment-generating function of X is $M_X(t) = E(e^{tX})$, provided this expectation exists for all t in some neighborhood of 0. The moment-generating function for a distribution characterizes that distribution; two random variables having the same mgf are identically distributed. Moments of X can be obtained by differentiation of $M_X(t)$; $E(X^k) = M_X^{(k)}(0)$. If $Y = aX + b$, then $M_Y(t) = e^{bt}M_X(at)$. The factorial moment-generating function for X is $N_X(t) = E(t^X)$, provided this expectation exists for all t in some neighborhood of 1. The identities $M_X(t) = N_X(e^t)$ and $N_X(t) = M_X(\log t)$ hold. If X is integer-valued with mass function p_k, the probability-generating function of X is $N_X(z) = \Sigma_k \, p_k z^k$.

If Y is a function g of X and the distribution of X is known, the distribution of Y may be found by using several possible approaches:

1. If X is discrete, so is $g(X)$, and the mass function of Y at the point y in its range is given by $p_Y(y) = \Sigma_{g^{-1}(\{y\})} \, p_X(x)$.

2. The CDF of Y is given by $F_Y(y) = P[X \in g^{-1}((-\infty, y])]$, which may be a sum or integral, depending on the distribution of X. If X is continuous, Y may be discrete or continuous. If Y is continuous, the density of Y may be obtained by taking the derivative of the expression for F_Y.

3. If X is continuous and g is strictly monotone (either increasing or decreasing), then the density of Y is $f_Y(y) = f_X(g^{-1}(y))|\psi'(y)|$, $y \in R_Y$, where $\psi(y)$ denotes $g^{-1}(y)$.

4. The moment-generating function of Y may be found by taking $E(e^{tg(X)})$.

If X has a continuous distribution, the random variable $Y = F_X(X)$ is uniform over $(0, 1)$. This result is called the probability integral transformation and is useful both in statistical theory and in stimulation applications.

The nth moment of X is $\mu_n' = E(X^n)$. The nth central moment of X is $\mu_n = E[(X - \mu)^n]$. The first moment is the mean, μ. The second central moment is the variance, σ^2. The nth factorial moment of X is $E[X(X - 1) \cdots (X - n + 1)]$. The Pth percentile, or the $p = .01P$ quantile, of X is any point ξ_p satisfying $P(X < \xi_p) \le p$ and $P(X > \xi_p) \le 1 - p$. If F_X is strictly increasing, $\xi_p = F^{-1}(p)$. The median of X is $\xi_{.5}$, and the first and third quartiles are $\xi_{.25}$ and $\xi_{.75}$, respectively. The value $\xi_{.75} - \xi_{.25}$ is the inter-quartile range of X. The mode of X is the point (if any) at which the density (or mass function) of X is at a maximum. The mean, median, and mode are measures of location of a distribution. The variance, standard deviation, and interquartile range are measures of dispersion of a distribution.

For any random variable having a mean and variance, the Chebyshev inequality asserts that, for any $\delta > 0$, $P(|X - \mu| \ge \delta) \le \sigma^2/\delta^2$. This inequality gives a bound on how dispersed a distribution can be about its mean, in terms of σ^2. This bound is sloppy for some particular distributions, such as normal distributions, but it is sharp for other distributions.

A random variable Y has a log normal distribution with parameters μ and σ^2 if log Y is $N(\mu, \sigma^2)$; the CDF of Y is related to the standard normal CDF by $F_Y(y) = \Phi[(\log y - \mu)/\sigma]$. The beta function, $B(r, \lambda) = \Gamma(r)\Gamma(\lambda)/\Gamma(r + \lambda)$ for $r > 0, \lambda > 0$, is the normalizing constant for the beta distribution with those parameters; the beta density is $f(x) = x^{r-1}(1 - x)^{\lambda - 1}/B(r, \lambda)$, $0 < x < 1$. Members of the beta family assume a wide variety of shapes over the interval $(0, 1)$, including the uniform distribution over $(0, 1)$. The gamma distribution has density $f(x) = \lambda^r x^{r-1} e^{-\lambda x}/\Gamma(r)$, $x > 0$. The exponential distribution is a member of the two-parameter gamma family.

Chapter 5

JOINT DISTRIBUTIONS

5.1 INTRODUCTION

We are now ready to consider models that involve making several measurements each time the underlying experiment is performed. These models enable us to study connections between various attributes of the outcome of an experiment. For example, we might be interested in possible relationships between the air temperature, barometric pressure, and wind velocity at a given location and the weather conditions at the same place 24 hours later. We might view the experiment as making the necessary measurements, together with the weather outcome (say inches of precipitation) a day later. Such experiments may be modeled by using the notions of *jointly distributed random variables*.

Mathematically, a joint distribution is a change in dimensionality from random variables with values in the one-dimensional set of real numbers to *n*-tuples of random variables, with values in a space of *n* dimensions. We will first establish the properties of probability models associated with ordered pairs of random variables, including in our consideration methods of describing associated probability functions such as joint CDFs and joint densities or joint mass functions. We will then consider how to form and use new probability distributions associated with a joint distribution, including the notions of conditional distributions and marginal distributions. Finally, we will show how these notions can easily be extended from the two-dimensional case to more general *n*-dimensional models.

5.2 TWO-DIMENSIONAL RANDOM PAIRS

So far, the outcomes of most of our experimental situations have been conveniently summarized by means of a single random variable. But it is clear that, in some situations, such a model may not be desirable. In several of our examples, such as those involving compound experiments, we found it convenient to use ordered n-tuples for the outcomes. If we are studying the relationship between two traits in some human population—for example, age and blood pressure or art appreciation and tennis ability—it is not clear how these *two* traits can be summarized by a *single* number. Even if we use single random variables to model each trait, it seems reasonable that the information provided by the outcome of the experiment (selecting an individual and measuring the two traits for this individual) should be kept together, since they both relate to the same individual. In other words, if age and blood pressure are both measured on the same person, then this information should be preserved into two separate pieces of information. One way to accomplish this separation is to report each observation as an ordered pair. This procedure, in turn, amounts to defining a random pair with values in the two-dimensional plane.

Like a random variable, a random pair may be viewed as a function mapping outcome o of the underlying experiment to a numerical pair, or point in the plane. Thus if a person is selected in the underlying experiment, the outcome on the random pair might be (35 years, 130 mmHg); if $o \in S$ is the underlying outcome, $(x, y) = (X(o), Y(o))$ is the outcome on the random pair (X, Y), which represents (age, blood pressure).

We wish to make predictions, in the form of probabilities, about outcomes that will be observed on a random pair. The events in this case are subsets of the plane, which may be thought of as two-dimensional regions. The probabilities of two-dimensional events may be specified in a manner similar to that used in the one-dimensional case. We will find it convenient to consider separately discrete and continuous random pairs, with probabilities specified by mass functions and density functions, respectively.

If the components X and Y of a random pair (X, Y) are discrete random variables, we say the pair is *jointly discrete*. This term means that there is a countable set of points in the plane such that the outcome on (X, Y) will be in this set with probability 1. The probability of occurrence of any one of these points is the *probability mass* at that point. Thus in the joint discrete case, there is defined a mass function with positive values at individual points in the plane, and these values sum to 1. We summarize this result in Definition 5.1.

EXAMPLE 5.1   Consider an experiment of testing two manufactured devices selected at random from a large shipment lot. Let X be a Bernoulli (p) random variable representing that the first item is defective $(X = 0)$ or not defective $(X = 1)$; similarly, let Y denote the quality of the second item. Then (X, Y) is discrete, and the joint mass function for (X, Y) is positive only at the four points in the set $R = \{(0, 1), (0, 1), (1, 0), (1, 1)\}$. Indeed, since this situation involves sampling from a large lot, a reasonable model would be to define $p(x, y)$ as follows: $p(0, 0) = (1 - p)(1 - p); p(0, 1) = (1 - p) \cdot p; p(1, 0) = p \cdot (1 - p); p(1, 1) = p \cdot p$.

DEFINITION 5.1   If the components X and Y of a random pair (X, Y) are discrete random variables, the pair is said to be *jointly discrete*. There is, then, a joint mass function $p(x, y)$ for the pair, with the following properties:

1. $p(x, y) \geq 0$ for all (x, y) in the plane.
2. $\Sigma_R \, p(x, y) = 1$, where R is the (countable) range of the pair (X, Y).
3. For any two-dimensional event A, the probability that A occurs is given by $P[(X, Y) \in A] = \Sigma_{A \cap R} \, p(x, y)$. As a special case [with $A = \{(x, y)\}$], $P[(X, Y) = (x, y)]$ $= P(X = x, Y = y) = p(x, y)$.

EXAMPLE 5.2   Suppose the underlying experiment consists of reading the binary contents of two two-bit registers, a low-order and a high-order register. Let ℓ denote the contents of the low-order register and h denote those of the high-order register, and let the underlying experimental sample point o be an ordered pair (ℓ, h). There are 16 such points in S, as depicted in Fig. 5.1. Consider the random variables X and Y defined to be, respectively, the decimal equivalents of the low-order register and the maximum of the two registers. Now, when X and Y, as defined, are considered jointly, each does something to o and produces a *pair* of values (x, y). (It is merely coincidental that o is also two-dimensional in this particular example.) Thus if the underlying outcome is $o = (11, 10)$, then $x = X(o) = 3$ and $y = Y(o)$ $= 3$. The final outcome is the pair $(3, 3)$. Let us summarize our discussion thus far in terms of the present example. As before, X and Y are real-valued functions with common domain S. Just as a random variable is a function mapping S into the real line, a pair of random variables, a *random pair*, is

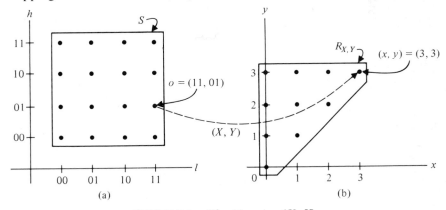

FIGURE 5.1   *The Mapping (X, Y)*

a function mapping S into the plane. Just as a random variable, as a function, has a range, so too a random pair has a range. Indeed, for the example, the range of (X, Y), say $R_{X,Y}$, is the collection of points displayed in Fig. 5.1. In Fig. 5.1, you can see how (X, Y) maps each point o into a point in $R_{X,Y}$. From the rule given, you see, for example, that $(X, Y)(o) = (3, 3)$ for $o = (11, 00)$ or $o = (11, 01)$ or $o = (11, 10)$ or $o = (11, 11)$. Consequently, $p(3, 3) = P[(X, Y) = (3, 3)] = P(\{11, 00), (11, 01), (11, 10), (11, 11)\}) = \frac{4}{16}$ if the register bits are random (have equal likelihoods).

For many situations involving discrete random pairs, we will find it convenient to summarize the joint mass function values in tabular form. In order to associate the table entries with corresponding points in the plane, we list these tables so that the table headers closely resemble the coordinate axes. Thus for Example 5.1, which concerned sampling two devices from a large lot having reliability $p = .8$ (say), the joint mass function is conveniently listed as follows:

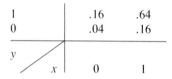

We can easily associate this table with one of the graphs of the mass function $p(x, y)$ shown in Fig. 5.2.

We may define the joint cumulative distribution function (CDF) for a random pair (X, Y) in a manner analogous to that for random variables. Thus we define $F(x, y)$ over the two-dimensional plane such that, at each point (x, y),

$$F(x, y) = P(X \le x, Y \le x) = P[(X, Y) \in A],$$

where A is the lower quarter plane with corner at (x, y), as shown in Fig. 5.3. It may be seen that such a function of two variables has monotonicity and continuity properties similar to those for one-dimensional CDFs. Since we will be making little direct use of joint CDFs, we won't concern ourselves further with these properties.

EXAMPLE 5.3   Suppose X and Y are jointly distributed with the bivariate mass function given by

$$P_{(X,Y)}(0, 1) = .3, \qquad P_{(X,Y)}(4, 10) = .2,$$

$$P_{(X,Y)}(-2, 7) = .1, \qquad P_{(X,Y)}(6, 6) = .4.$$

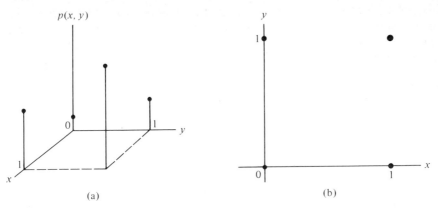

FIGURE 5.2   *Graphs of p(x, y)*

The random pair (X, Y) is, of course, discrete (since the masses sum to 1), and it has a discrete CDF, say F. Some values of this joint CDF are $F(0, 8) = .4$, $F(8, 8) = .8$, $F(7, 1) = .3$, and $F(6, 10) = 1$.

Now, we consider the case in which X and Y are *jointly continuous*. Such a model is appropriate in most instances where there are no points in the plane with positive probability; that is, there are no discrete masses in the distribution of (X, Y).

EXAMPLE 5.4   Consider an experiment that consists of throwing a dart into a circular target. (Imagine the thrower is allowed new tries until the dart lands in the target.) Let the outcome be the coordinates of the impact point of the dart. Then there is a continuous distribution of the unit of probability over the disk representing the target. In this case, there is no point, at least in theory, such that the dart can be predicted to hit that specific point with positive probability.

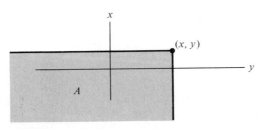

FIGURE 5.3   *Plot of the Event $[X \leq x, Y \leq y]$*

We could use a joint CDF to define the allocation of probabilities to two-dimensional continuous distributions such as that described in the dart example. It is almost always more convenient to work with the associated joint density function, however.

DEFINITION 5.2   Suppose the random pair (X, Y) has CDF $F(x, y)$, and suppose there is a nonnegative function $f(x, y)$ such that, for each point (x, y) in the plane,

$$F(x, y) = \int_{-\infty}^{x} \int_{-\infty}^{y} f(x, y) \, dy \, dx. \tag{5.1}$$

Then the nonnegative function $f(x, y)$ is called the *joint density* of the random pair (X, Y), and the random pair is said to be *jointly continuous* or to have a *continuous distribution*.

EXAMPLE 5.5   For the dart-throwing experiment described in Example 5.4, let the target be represented by the unit disk $\{(x \ y) : x^2 + y^2 \leq 1\}$. If the dart impact is "random"—for example, generated in an electronic game—the density function might be taken as uniform over the disk. Since $F(\infty, \infty) = 1$, the integral of the joint density must be 1, and it follows that

$$f(x, y) = \begin{cases} \dfrac{1}{\pi} & \text{if } x^2 + y^2 \leq 1, \\[2mm] 0 & \text{if } x^2 + y^2 > 1. \end{cases}$$

It can be seen from Eq. (5.1) that, if $F(x, y)$ is a known CDF for a random pair (X, Y) with continuous distribution, then the joint density can be obtained by differentiation:

$$f(x, y) = \frac{\partial^2 F(x, y)}{\partial x \, \partial y} .$$

It can also be seen that any nonnegative function $f(x, y)$ for which $\int_{-\infty}^{\infty} \int_{-\infty}^{\infty} f(x, y) \, dy \, dx = 1$ can serve as a density function. A plot of a hypothetical density surface is shown in Fig. 5.4. One of our goals is to consider how such functions should be selected for models of various common phenomena. Once we have defined such a function of two variables, the probabilities of two-dimensional events are determined by integration.

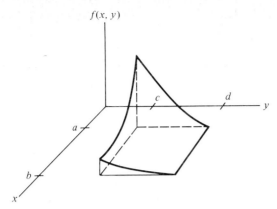

FIGURE 5.4   *Hypothetical Joint Density $f(x, y)$*
for a Random Pair (X, Y) with
Range $R_{(X,Y)} = (a, b) \times (c, d)$

Thus if A is a two-dimensional event, the probability that A occurs is given by

$$P[(X, Y) \in A] = \int \int_A f(x, y) \, dy \, dx.$$

You can see that Eq. (5.1) is a special case of this expression, where A is a quarter plane as shown in Fig. 5.3.

In contrast to the discrete case, here the density function values are *not* probabilities. Indeed, you will encounter many examples of density functions that have values exceeding 1 at various points. By the law of the mean for integrals, the product $f(x, y) \, \Delta x \, \Delta y$ can be viewed as approximating a probability, as follows:

$$f(x, y) \, \Delta x \, \Delta y \approx P(x \le X \le x + \Delta x, y \le Y \le y + \Delta y), \qquad (5.2)$$

where Δx and Δy are positive variables. This approximation becomes exact as Δx and Δy tend to 0, provided the density is continuous at (x, y). If you interpret the right-hand expression in Eq. (5.2) as a double integral, you can see that this expression amounts to approximating the volume under the surface defined by $z = f(x, y)$, which is above the base rectangle with corner (x, y) and dimensions Δx and Δy. The approximating volume is the area of this base times an approximate height, which is constant rather than varying according to $f(x, y)$. It should seem intuitively clear to you that the volume under the surface and the volume of the box become the same (i.e., their ratio tends to 1) as Δx and Δy tend to 0.

EXAMPLE 5.6 Suppose it is known that (X, Y) is continuous and $P(x \le X \le x + \Delta x, y \le Y \le y + \Delta y)$ is proportional to $x \, \Delta x \, \Delta y$ for small Δx and Δy and for $0 \le x \le 1$ and $0 \le y \le 1$. Then dividing Eq. (5.2) by $\Delta x \, \Delta y$ and then taking a limit as Δx and Δy go to 0, we have $f(x, y) \propto x$,

$0 \le x \le 1$, $0 \le y \le 1$. The proportionality constant is such that $\int_0^1 \int_0^1 f(x, y)\, dy\, dx = 1$, so we have, finally, $f(x, y) = 2x$ for $0 \le x \le 1$ and $0 \le y \le 1$. (The density is 0 outside the unit square.)

We will often be interested in using knowledge of the distributions of the individual random variables X and Y to determine the joint distribution of the random pair (X, Y). This determination can be done if certain additional conditions are met. It is also of interest to find the distributions of the individual random variables from a knowledge of the joint distribution of the random pair. For reasons you will see shortly, the individual one-dimensional probability functions are called *marginal probability functions*. A distribution associated with the joint probability model is called a *joint distribution*, while those associated with the individual spaces are called *marginal distributions*. A distribution associated with a single random variable is called a *univariate distribution*; a joint distribution associated with two random variables is called a *bivariate distribution*; and (you'll see later) similarly for *trivariate*, and so on. In general, the term *multivariate* is used to distinguish such distributions from univariate distributions.

EXAMPLE 5.7   Let $f(x, y) = 1$ for $0 \le x \le 1$, $0 \le y \le 1$, and assume f is 0 elsewhere in the plane. Is f a density?

SOLUTION   To verify that f is a bivariate density function, we must check two conditions:

1. f is nonnegative.
2. $\int_{-\infty}^{\infty} \int_{-\infty}^{\infty} f(x, y)\, dx\, dy = 1$.

The first condition is obviously satisfied; the second follows easily from the fact that, since the integrand is zero outside the square $[0, 1] \times [0, 1]$, the integral in condition 2 is equal to $\int_0^1 \int_0^1 1 \, dx\, dy$. This density is the two-dimensional analogue of the uniform distribution over $(0, 1)$ and is sometimes called a *bivariate uniform density*. If (X_1, X_2) has this density, then, for example,

$$P(-10 < X_1 < \tfrac{1}{2}, \tfrac{1}{2} < X_2 \le \tfrac{3}{4}) = \int_0^{1/2} \int_{1/2}^{3/4} dx_1\, dx_2 = \tfrac{1}{2} \cdot \tfrac{1}{4} = \tfrac{1}{8}.$$

EXAMPLE 5.8   —*A Bivariate Exponential Density Function*   Suppose the miles to failure of tires on two randomly selected automobiles is a random pair (X_1, X_2) having density

$$f(x_1, x_2) = e^{-x_1 - x_2}, \qquad x_1 > 0, x_2 > 0,$$

in units of 10,000 miles. As usual, the absence of specification to the contrary means we agree that f is zero if $x_1 \le 0$ or $x_2 \le 0$. It is easy to verify that

f is a two-dimensional density function. If (X_1, X_2) has this density, then

$$F(x_1, x_2) = \int_0^{x_2} \int_0^{x_1} e^{-t_1-t_2} \, dt_1 \, dt_2$$

$$= \int_0^{x_2} e^{-t_2} \, dt_2 \int_0^{x_1} e^{-t_1} \, dt_1$$

$$= (1 - e^{-x_2})(1 - e^{-x_1}), \qquad x_1 > 0, x_2 > 0.$$

Thus the event "Both of the automobiles experience tire failure within 15,000 miles" has probability $F(1.5, 1.5) \approx .60$. Again, our convention is that F is zero outside the positive quadrant. Here, we might indicate the probability model associated with (X_1, X_2) by specifying either f or F and stating "(X_1, X_2) is distributed f" or "(X_1, X_2) is distributed F."

EXERCISES

THEORY

5.1   a. Interpret $[X_1 \leq x_1]$ as a two-dimensional event.
      b. Show that $[X_1 \leq x_1, X_2 \leq x_2] = \cap_{j=1}^2 [X_j \leq x_j]$.

5.2   Properties of joint CDF functions are explored here. Suppose (X_1, X_2) is distributed by the CDF F.
      a. Let $F_a(x_2) = F(a, x_2)$ denote the section of $F(x_1, x_2)$ at $x_1 = a$. (The *section* of F at a is the function of x_2 that results when $x_1 = a$ is held fixed.)
         i. Verify that $0 \leq F_a(x_2) \leq 1$ for all x_2.
         ii. Verify that F_a is monotone nondecreasing.
         Hint: If $x_2' \leq x_2''$, then $(-\infty, x_2'] \subset (-\infty, x_2'']$.
      b. Use the results of Exercise 5.1 and the intuitive notion that $[X_2 \leq x_2]$ becomes empty as $x_2 \to -\infty$ to argue that $\lim_{x_2 \to -\infty} F_a(x_2) = 0$.
      c. Give an intuitive argument like the one in part (b) to verify that $\lim_{x_2 \to \infty} F_a(x_2) = P(X_1 \leq a) = F_{X_1}(a)$.
      d. Use parts (b) and (c) to argue that

$$\lim_{x_1 \to \infty} \lim_{x_2 \to \infty} F(x_1, x_2) = 1, \qquad \lim_{x_1 \to -\infty} F(x_1, x_2) = 0,$$

$$\lim_{x_1 \to \infty} F(x_1, x_2) = F_{X_2}(x_2).$$

      e. Show that F is right continuous in each variable.

5.3   Suppose B_1 is a lower quarter plane with corner (a, b) (i.e., $B_1 = \{(x, y) : x \leq a, y \leq b\}$), and similarly for B_2 with corner (c, d), where $a < c$ and $b < d$. Show that the set $B_2 - B_1$ may be written as the union of three mutually disjoint sets A_1, A_2, A_3, where

$$A_1 = \{(x, y) : x \leq a, b < y \leq d\},$$

$$A_2 = \{(x, y) : a < x < c, b < y \leq d\},$$

$$A_3 = \{(x, y) : a < x \leq c, y \leq b\}.$$

5.4   Suppose (X, Y) is continuous with known CDF F. Using Eq. (5.1), verify that the joint density is given by $\partial^2 F/\partial x\, \partial y$.

5.5   If (X, Y) is continuous, show that the probability of any one-dimensional "curve" in the plane, such as the event $[X = Y]$, has probability 0.

APPLICATION

5.6   Suppose a box contains two red, one white, and three green balls, and suppose k balls are drawn from the box in succession. Define the random variables X_i, $i = 1, 2$, as follows: The value x_i of X_i is given by

$$x_i = \begin{cases} 1 \text{ if the } i\text{th ball drawn is red,} \\ 2 \text{ if the } i\text{th ball drawn is white,} \\ 3 \text{ if the } i\text{th ball drawn is green.} \end{cases}$$

a. Describe the joint CDF of (X_1, X_2) if the balls are drawn with replacement.
b. Describe the CDF of X_2 if the balls are drawn with replacement.
c. Describe the joint CDF of (X_1, X_2) at $(2, 1)$ if the balls are drawn without replacement.
d. Same as part (b), but the balls are drawn without replacement.

5.7   Suppose

$$F_{(X_1,X_2)}(x_1, x_2) = \begin{cases} 0 \text{ if } x_1 < 0 \text{ or } x_2 < 0, \\ \frac{1}{4}x_1x_2 \text{ if } 0 \le x_1 \le 2 \text{ and } 0 \le x_2 \le 2, \\ \frac{1}{2}x_1 \text{ if } 0 \le x_1 \le 2 \text{ and } x_2 > 2, \\ \frac{1}{2}x_2 \text{ if } 0 \le x_2 \le 2 \text{ and } x_1 > 2, \\ 1 \text{ if } x_1 > 2 \text{ and } x_2 > 2. \end{cases}$$

a. Verify that $F_{(X_1,X_2)}$ satisfies the conditions discussed in Exercise 5.2.
b. Find $P(X_1 < 1, X_2 < 1)$.   c. Find $P(X_1 < 5, X_2 < .5)$.
d. Find $P(X_1 > X_2)$.   e. Find $P(X_1 > \frac{3}{2})$.
f. Find the probability that X_1 is less than $\frac{1}{2}$ or X_2 is greater than $\frac{3}{8}$.

5.8   a. Graph the joint probability mass function defined by $p(x_1, x_2) = \frac{1}{3}$ if $(x_1, x_2) = (0, 0)$ or $(x_1, x_2) = (0, 1)$, while $p(x_1, x_2) = \frac{1}{6}$ if $(x_1, x_2) = (1, 1)$ or $(x_1, x_2) = (\frac{1}{2}, \frac{1}{2})$.
b. Verify that p is a (bivariate) mass function.
c. If (X_1, X_2) is distributed p, find $P(X_1 \le \frac{1}{2})$.
d. Continuing part (c), find the CDF of X_1.
e. Find the joint CDF of (X_1, X_2).

5.9   Suppose you are taking five courses currently and believe that your probabilities of getting grades of A, B, and C in any course are, respectively, .20, .75, .05. Compute the probability that you will get all B's this quarter. Specify your model. Let Z denote your lowest grade this quarter. Find the distribution of Z. (Let A have value 4.0, B have value 3.0, etc.)

5.10   Determine whether the following are bivariate density functions:
a. $f(x, y) = x^2 + \frac{1}{2}xy;\ 0 \le x \le 1,\ 0 \le y \le 2$.
b. $f(x, y) = 2(x + y - 2xy);\ 0 \le x \le 1,\ 0 \le y \le 1$.

5.11 A discrete random pair has the mass function partially listed in the accompanying table.

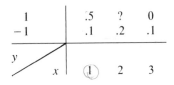

a. Complete the table defining the mass function.
b. Find $F(2, -1)$.
    c. Find $P(X = 2, Y \leq 5)$.
d. Find $P(X \leq Y)$.
    e. Find $P(X > 1)$.

5.12 Suppose (X, Y) has a continuous distribution, and the density function is a positive constant over the region above the curve $y = x^2$ and below the line $y = x$ (assume the density is zero elsewhere).
a. Find a mathematical expression for $f(x, y)$.
b. Find $P(X < .5, Y < .5)$.
c. Find $P(X < Y - \frac{1}{2})$.
    d. Find $P(X < .8)$.

5.13 Suppose (X, Y) has a discrete distribution, and the mass function is positive only on the set $R = \{(x, y) : 0 < x < y < 4, x \text{ and } y \text{ integers}\}$. Suppose that at each $(x, y) \in R$, $p(x, y)$ is proportional to $x \cdot y$.
a. Find $p(x, y)$.
    b. Find $P(X < Y - \frac{1}{2})$.
c. Find $P(X < 4, Y \leq 2)$.
    d. Find $P(X \geq 2)$.

5.14 For the tire life model in Example 5.8, what are the probabilities of the following events, in terms of F?
a. $[X_1 < 1.5, X_2 < 1.5]$.
b. At least one of the cars does not experience tire failure within 15,000 miles.
c. Car 1 experiences tire failure within 5,000 miles and car 2 experiences tire failure within one million miles.

5.15 Relate the event in Exercise 5.14(c) to an event in the marginal model for X_1 alone having very nearly the same probability.

5.3 JOINT AND MARGINAL DISTRIBUTIONS

We have discussed briefly the connection between multivariate distributions and the corresponding marginal distributions. We will now investigate the methods of finding marginal distributions when the joint distribution is given, in two important cases: when the random variables are jointly discrete, and when they are jointly continuous.

Let us consider first a discrete bivariate distribution. Suppose (X, Y) has a mass function $p(x, y)$ that is positive only for (x, y) in a finite set R.

We have defined $p(x, y)$ to be a probability, $P(X = x, Y = y)$, for each (x, y) in R. Now, suppose we wish to compute $P(X = x)$, without regard to the outcome on Y. This computation may be accomplished for any given argument x as follows: Find the set of points in R with first member equal to the given x, say $\{(x, y_1), (x, y_2), \ldots, (x, y_n)\}$. [If there are no such points, $P(X = x)$ must be zero.] Then

$P(X = x)$

$= P(X = x, Y \in \{y_1, y_2, \ldots, y_n\})$

$= P[(X = x, Y = y_1), \text{ or } (X = x, Y = y_2), \ldots, \text{ or } (X = x, Y = y_n)]$

$= P(X = x, Y = y_1) + P(X = x, Y = y_2) + \cdots + P(X = x, Y = y_n),$

since the events in the second line of the equation are mutually exclusive. But the probabilities in the third line are, by definition, given by the mass function values. Thus we have

$$P(X = x) = \sum_{j=1}^{n} p(x, y_j).$$

That is, the marginal mass function of X is found by summing over y in the joint mass function of (X, Y), for each fixed x. In a similar manner, the marginal mass function for Y, $p_Y(y)$, is given by $p_Y(Y) = P(Y = y) = \sum_x p(x, y)$, where we use the notation convention $\sum_x$ to mean summing over each x-value such that (x, y) is a point in R. We also use the convention that if there are no such x-values, the sum above is "empty," which is defined to be zero.

In cases in which the range of (X, Y), $R_{(X,Y)}$, is finite, it is possible to list the bivariate mass function in a table such as Table 5.1. In such cases, we can take $R_{(X,Y)}$ to be a product set (so the table is a rectangular layout), without loss of generality, simply by adding points with zero mass as required to fill out the rectangle, as we have shown in Table 5.1.

It is easy to see that the marginals of X and Y are obtained from the joint simply by adding the values in individual rows or columns of the joint table. These row and column totals can be listed in the margins of the table of joint mass values—hence the term "marginal" mass function.

TABLE 5.1   *A Hypothetical Bivariate Mass Function*

y		x	0	1
1			$\frac{1}{4}$	0
0			$\frac{1}{2}$	$\frac{1}{4}$

EXAMPLE 5.9 Suppose the bivariate mass function for the random pair (X, Y) has values as shown in the accompanying table. The marginal mass functions p_X and p_Y are as shown.

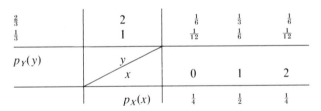

$p_Y(y)$	y \ x	0	1	2
$\frac{2}{3}$	2	$\frac{1}{6}$	$\frac{1}{3}$	$\frac{1}{6}$
$\frac{1}{3}$	1	$\frac{1}{12}$	$\frac{1}{6}$	$\frac{1}{12}$
	$p_X(x)$	$\frac{1}{4}$	$\frac{1}{2}$	$\frac{1}{4}$

As a check, we see that $\Sigma_{i,j}\, p_{X,Y}(i, j) = \Sigma_i\, p_X(i) = \Sigma_j\, p_Y(j) = 1$, where $\Sigma_{i,j}$ denotes a summation over all i and j, that is, $\Sigma_{i=0}^2\, \Sigma_{j=1}^2$. You might have noticed that, in *this* example, for each i and j, $p_{(X,Y)}(i, j) = p_X(i) \cdot p_Y(j)$.

If (X, Y) is jointly continuous, with bivariate density $f(x, y)$, the marginal distributions of X and Y are continuous. To find the marginal density $f_X(x)$ of X, for example, we replace summation of the discrete mass function with integration of the density function:

$$f_X(x) = \int_{-\infty}^{\infty} f(x, y)\, dy. \tag{5.3}$$

This expression may be interpreted as follows: For each fixed x, $f_X(x)\, \Delta x$ is approximately the probability that (X, Y) falls in the infinite strip parallel to the y-axis, with base $(x, x + \Delta x]$:

$$f_X(x)\, \Delta x \approx P(x < X \le x + \Delta x, \, -\infty < Y < \infty) \approx \Delta x \int_{-\infty}^{\infty} f(x, y)\, dy.$$

Now, divide both sides by Δx and take a limit as Δx goes to zero to obtain Eq. (5.3). If $f(x, y)$ is zero outside a specified region such as the unit square, the integral in Eq. (5.3) can be given as an expression such as $\int_0^1 f(x, y)\, dy$. Of course, similar comments pertain to finding the marginal density of Y by integrating the joint with respect to x.

Graphically, integrating the joint density to find a marginal can be interpreted as a projection of volumes under the joint density surface to areas under the marginal density. For example, consider a joint density $f(x, y)$ of a random pair (X, Y) and the marginal density $f_Y(y)$. The probability of the event $[y_0 < Y < y_0 + \Delta y]$, when considered as the two-dimensional event $(-\infty, \infty) \times (y_0, y_0 + \Delta y]$, is the volume of a "slice" of the joint density, parallel to the x-axis (see Fig. 5.5). This volume is approximately the area of a face of the slice times Δy:

$$P(y_0 < Y < y_0 + \Delta y) \approx \left[\int_{-\infty}^{\infty} f(x, y_0)\, dx\right] \cdot \Delta y = f_Y(y_0) \cdot \Delta y.$$

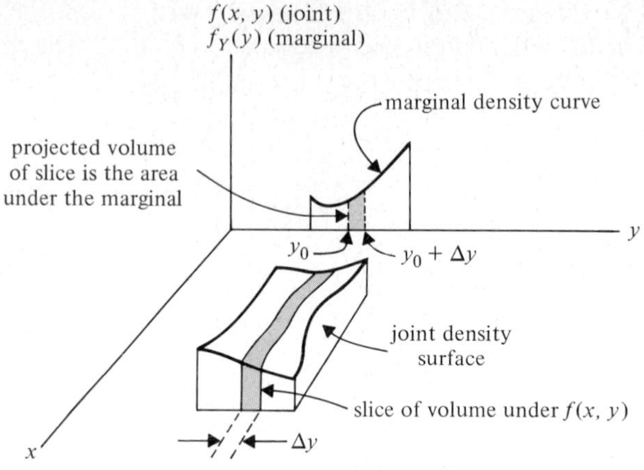

FIGURE 5.5   *Projection of Volume by Integration*

Viewed as a one-dimensional event, the probability of $[y < Y < y + \Delta y]$ is approximately the marginal density height at y times Δy, which agrees with the approximate volume of the slice found above. In Fig. 5.5, this area is shown on the two-dimensional plane containing the y-axis and the vertical axis, on which f_Y is plotted.

EXAMPLE 5.10   Suppose (X, Y) has a density that is uniform over the unit square; $f(x, y) = 1, 0 < x < 1, 0 < y < 1$. The marginal density of X is, for $0 < x < 1$,

$$f_X(x) = \int_0^1 f(x, y)\, dy = 1, \qquad 0 < x < 1.$$

It follows that the univariate distribution of X is uniform over $(0, 1)$; similarly, the marginal distribution of Y is uniform over $(0, 1)$.

It is instructive to consider Eq. (5.3) from another point of view. If X_1 and X_2 are jointly continuous with joint cumulative distribution function F and joint density function f, we have observed, in Exercise 5.2, that the cumulative distribution function for X_1 is given by $F_{X_1}(x_1) = \lim_{x_2 \to \infty} F(x_1, x_2)$. On the other hand,

$$F(x_1, x_2) = \int_{-\infty}^{x_1} \int_{-\infty}^{x_2} f(t_1, t_2)\, dt_2\, dt_1,$$

and we have $$F_{X_1}(x_1) = \int_{-\infty}^{x_1} \int_{-\infty}^{\infty} f(t_1, t_2)\, dt_2\, dt_1.$$

Differentiating with respect to x_1 (using Leibniz's rule), we find the density function for X_1 to be given by

$$f_{X_1}(x_1) = \frac{dF_{X_1}(x_1)}{dx_1} = \int_{-\infty}^{\infty} f_X(x_1, t_2) \, dt_2,$$

which is, of course, Eq. (5.3). Similarly, $f_{X_2}(x_2) = \int_{-\infty}^{\infty} f(t_1, x_2) \, dt_1$.

EXAMPLE 5.11   A bivariate density that is of theoretical interest, and one we will refer to several times in later sections, is defined as follows: Let $c = \Phi^{-1}(\frac{3}{4})$, where Φ is the standard normal CDF (so that $c \approx .6745$ from a table of the normal CDF). Let $A_1 = \{(x, y) : x < -c, y > 0\}$, $A_2 = \{(x, y) : x > c, y > 0\}$, and $A_3 = \{(x, y) : -c < x < c, y < 0\}$, as in Fig. 5.6. Define the function g by

$$g(x, y) = \begin{cases} \dfrac{1}{\pi} e^{-(1/2)(x^2 + y^2)} & \text{if } (x, y) \in A_1 \cup A_2 \cup A_3, \\ 0 & \text{otherwise.} \end{cases}$$

We now verify that g is a density. Certainly, g is nonnegative. Note that

$$\int\int_{A_1} g(x, y) \, dx \, dy = \frac{1}{\pi} \int_0^\infty \int_{-\infty}^{-c} e^{-(1/2)x^2} e^{-(1/2)y^2} \, dx \, dy$$

$$= 2 \int_0^\infty \frac{1}{\sqrt{2\pi}} e^{-(1/2)y^2} \, dy \int_0^{-c} \frac{1}{\sqrt{2\pi}} e^{-(1/2)x^2} \, dx$$

$$= 2 \cdot \tfrac{1}{2} \cdot \tfrac{1}{4} = \tfrac{1}{4}.$$

Similarly, $\int \int_{A_2} g(x, y) \, dx \, dy = \frac{1}{4}$, and $\int \int_{A_3} g(x, y) \, dx \, dy = \frac{1}{2}$. Since g is

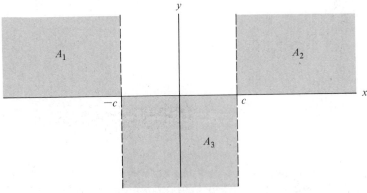

FIGURE 5.6   *The Regions A_1, A_2, and A_3*

zero outside $\cup_{i=1}^{3} A_i$, and the sets A_1, A_2, A_3 are disjoint, we have

$$\int_{-\infty}^{\infty}\int_{-\infty}^{\infty} g(x, y)\, dx\, dy = \int_{\underset{i=1}{\overset{3}{\cup} A_i}} g(x, y)\, dx\, dy = \sum_{i=1}^{3}\int\int_{A_i} g(x, y)\, dx\, dy = 1.$$

Thus g is a two-dimensional density function for a random pair (X, Y) whose range is $R_{(X,Y)} = A_1 \cup A_2 \cup A_3$.

Now, let us find the marginal distributions of X and Y. To do so, we must consider three separate regions for the corresponding variable. Note, for example, that if $-\infty < x < -c$, then

$$f_X(x) = \frac{1}{\pi} e^{-(1/2)x^2} \int_0^{\infty} e^{-(1/2)y^2}\, dy = \frac{1}{\pi} e^{-(1/2)x^2} \sqrt{\frac{\pi}{2}} = \frac{1}{\sqrt{2\pi}} e^{-(1/2)x^2}.$$

If $-c \le x \le c$, then

$$f_X(x) = \frac{1}{\pi} e^{-(1/2)x^2} \int_{-\infty}^0 e^{-(1/2)y^2}\, dy$$

$$= \frac{1}{\pi} e^{-(1/2)x^2} \int_0^{\infty} e^{-(1/2)y^2}\, dy = \frac{1}{\sqrt{2\pi}} e^{-(1/2)x^2}.$$

If $c < x < \infty$, then

$$f_X(x) = \frac{1}{\pi} e^{-(1/2)x^2} \int_0^{\infty} e^{-(1/2)y^2}\, dy = \frac{1}{\sqrt{2\pi}} e^{-(1/2)x^2}.$$

Thus in all cases, $f_X(x) = (1/\sqrt{2\pi})e^{-(1/2)x^2}$, so we conclude that X is distributed $N(0, 1)$. In a similar manner, we can show Y is also standard normal. A class of bivariate distributions also having univariate normal marginals is introduced in Exercise 5.22.

EXAMPLE 5.12   Suppose in the bivariate exponential example, Example 5.8, in which $f(x_1, x_2) = e^{-x_1 - x_2}$, $x_1 > 0$, $x_2 > 0$, it is desired to find the marginal density of X_2 for fixed positive values of x_2. By integrating with respect to x_1, we obtain

$$f_{X_2}(x_2) = \int_0^{\infty} e^{-x_1 - x_2}\, dx_1 = e^{-x_2}, \qquad x_2 > 0.$$

Thus in this case, X_2 is distributed exponential (1), so the miles to tire failure on the second car is exponentially distributed with mean 10,000 miles.

Random pairs need not have both components discrete or both continuous. Applications often arise in which some components are discrete, some continuous, and some mixed. The evaluation of probabilities and the derivation of marginal distributions in many such cases can be carried out by formally using summation over discrete variables and integration over continuous variables, as appropriate. We illustrate this procedure with an example in a reliability setting.

EXAMPLE 5.13 *—A Beta-Binomial Distribution*   Microprocessors are manufactured in large batches in which chips are sliced from a large silicone crystal. Individual chips sampled from within a given batch have a common reliability r, where the underlying model is that a given processor either works or does not, so it is associated with a Bernoulli random variable X with parameter r. Now, suppose the reliability varies over manufactured batches in such a way that selection of a batch at random corresponds to determining a specific value r on a random variable R. Assume R is distributed beta (v, λ). As we will see when we consider conditional distributions later, it follows that the joint distribution of X and R is thus related to

$$f(x, r) = r^x(1 - r)^{1-x} \frac{1}{B(v, \lambda)} r^{v-1}(1 - r)^{\lambda-1}, \qquad 0 < r < 1, x = 0, 1.$$

This expression is neither a mass function nor a density function, since the variable x varies only over the discrete values (0 and 1), whereas the variable r varies over the interval (0, 1). Nevertheless, by formal summation and integration of f, we find the following:

$$\sum_{x=0}^{1} \int_0^1 f(x, r) \, dr = 1$$

and f is nonnegative;

$$P(X = 1, R < .5) = \int_0^{.5} f(1, r) \, dr;$$

$$f_R(r) = \sum_{x=0}^{1} f(x, r) = \frac{1}{B(v, \lambda)} r^{v-1}(1 - r)^{\lambda-1} \qquad 0 < r < 1;$$

$$f_X(x) = \int_0^1 f(x, r) \, dr = \frac{1}{B(v, \lambda)} \int_0^1 r^{v-1+x}(1 - r)^{\lambda-x} \, dr$$

$$= \frac{B(v + x, \lambda - x + 1)}{B(v, \lambda)} \int_0^1 \frac{1}{B(v + x, \lambda - x + 1)} r^{(v+x)-1} \cdot$$

$$(1 - r)^{(\lambda-x+1)-1} \, dr = \frac{B(v + x, \lambda - x + 1)}{B(v, \lambda)};$$

$$f_X(x) = \begin{cases} \dfrac{\lambda}{\lambda + v} & \text{if } x = 0, \\[4mm] \dfrac{v}{\lambda + v} & \text{if } x = 1. \end{cases}$$

Thus from the expressions for $f_R(r)$ and $f_X(x)$, we see that R is marginally beta (v, λ) (as specified in the model) and X is marginally Bernoulli $[v/(\lambda + v)]$. The latter may be interpreted as follows: A microprocessor selected at random from a random lot has reliability $v/(\lambda + v)$.

EXERCISES

THEORY

5.16 a. Let R be any two-dimensional event such that $P[(X_1, X_2) \in R] = 1$, where R is not necessarily the entire two-dimensional plane. Show that for any two-dimensional event B, $P[(X_1, X_2) \in B] = P[(X_1, X_2) \in B \cap R]$.
   b. Let R_i be any event such that $P(X_i \in R_i) = 1$, $i = 1, 2$. Show that R above is not necessarily the same as $R_1 \times R_2$.
   c. Show that, in any case, $R \subset R_1 \times R_2$.

5.17 a. Show that, if f_i is a density function for a random variable X_i with range R_i for $i = 1, 2$, then the function defined by $f(x_1, x_2) = f_1(x_1)f_2(x_2)$ is a two-dimensional density function for a random pair (X_1, X_2) with range $R = R_1 \times R_2$.
   b. Derive a result similar to that in part (a) for mass functions.

5.18 Suppose (X, Y) is distributed p[i.e., (X, Y) has mass function $p(x, y)$]. Show that $P(X \in A)$ can be obtained either as $\Sigma_{x \in A}\, p_X(x)$ or as $\Sigma_{y \in R_Y}\, \Sigma_{x \in A}\, p(x, y)$.

5.19 a. If F and G are CDFs, show that $H = F \cdot G$ is a CDF.
   b. Compare the result in part (a) with the bivariate exponential model cited in Example 5.12.

5.20 Let f_1 and f_2 be two density functions with corresponding cumulative distribution functions F_1 and F_2. For each $0 \le \lambda \le 1$, define f_λ by

$$f_\lambda(x_1, x_2) = f_1(x_1)f_2(x_2)\{1 + \lambda[2F_1(x_1) - 1][2F_2(x_2) - 1]\}.$$

Show that f_λ is a two-dimensional density function for each λ and that the marginal density functions are f_1 and f_2. *Hint:* Note that

$$\int_{-\infty}^{\infty} f_i(x)F_i(x)\, dx = \left.\frac{F_i^2(x)}{2}\right|_{-\infty}^{\infty} = \frac{1}{2}.$$

5.21 Use the fact given in the hint of Exercise 5.20 to verify that, if X is distributed uniformly over $(0, 1)$, then $E(X) = \frac{1}{2}$.

5.22 The general bivariate normal density function is explored here. Suppose that (X, Y) has density

$$f(x, y) = \frac{1}{2\pi\sigma_1\sigma_2 \sqrt{1 - \rho^2}} \exp\left\{-\frac{1}{2(1 - \rho^2)}\left[\left(\frac{x - \mu_1}{\sigma_1}\right)^2 - 2\left(\frac{x - \mu_1}{\sigma_1}\right)\left(\frac{y - \mu_2}{\sigma_2}\right) + \left(\frac{y - \mu_2}{\sigma_2}\right)^2\right]\right\},$$

where $\sigma_1, \sigma_2 > 0$, and μ_1, μ_2, and $\rho \in (-1, 1)$ are five real parameters.

a. Verify that f is a density function. *Hint:* Complete the square in the exponent.

b. Find the marginal densities of X and Y.

c. Show that if $\rho = 0$, then $f_{(X,Y)} = f_X \cdot f_Y$.

5.23   In the beta-binomial situation described in Example 5.13, suppose Y denotes the number of nondefective chips in a sample of size n taken from a common randomly selected batch. Find the marginal distribution of Y.

APPLICATION

5.24   Suppose X and Y are jointly discrete random variables with a mass function partially listed as in the accompanying table.

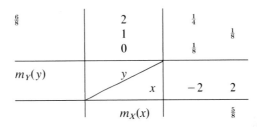

a. Complete the table for $m_{(X,Y)}$, $m_X(x)$, and $m_Y(y)$.

b. Find F_Y and $E(Y)$.

5.25   Define a bivariate mass function such that (a) both marginals are geometric mass functions and (b) one marginal is binomial and the other is Poisson. *Hint:* Use the result of Exercise 5.17.

5.26   Suppose $f(x, y)$ is as described in each part below. Find the constant k and sketch a graph of the joint density surface; find and sketch graphs of the marginals.

a. $f(x, y) = kxy;\ x \geq 0,\ y \geq 0,\ x^2 + y^2 \leq 1$.

b. $f(x, y) = k(x^2 + y^2);\ x^2 + y^2 \leq 1$.

c. $f(x, y) = k(x^2 y + 1);\ x \geq 0,\ y \geq 0,\ y \leq 2 - x/2$.

5.27   In Example 5.12, verify that X_1 and X_2 are identically distributed, that is, that X_1 and X_2 have the same marginal distributions.

5.28   Suppose (X, Y) has the particular bivariate normal distribution described in Exercise 5.22, where $\mu_1 = \mu_2 = \rho = 0$ and $\sigma_1 = \sigma_2 = 1$. Show that X and Y are each (marginally) distributed in $N(0, 1)$.

5.29   a. Verify that X_2 in Example 5.7 of the preceding section has a marginal uniform distribution over $(0, 1)$.

b. Find an explicit expression for the joint CDF corresponding to the joint density of (X_1, X_2).

5.30   In Exercise 5.28, a selection of parameters in the joint normal density of Exercise 5.22 is given such that X and Y are identically $N(0, 1)$ distributed. Compare this result with that of Example 5.11. *Note:* This exercise shows that different joint distributions can result in identical marginals. (See also Exercises 5.20 and 5.31.)

5.31 Suppose a joint mass function is specified by the accompanying table. Show that this distribution leads to exactly the same marginal mass functions as those specified by Table 5.1.

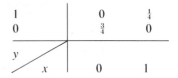

b. Find $E(X)$ and $E(Y)$. *Hint:* Use the marginal Bernoulli distributions.

5.32 Suppose that the components of (X_1, X_2) are jointly distributed with joint mass function $p(x_1, x_2) = \frac{1}{4}$ if $(x_1, x_2) \in \{(1, 0), (0, 1), (0, 0), (1, 1)\}$. Show that each X_i is Bernoulli-distributed with $p = .5$.

5.33 In the binary register experiment (Example 5.2 of the preceding section), with the equal-likelihood model for the registers, let A be the event that the iow-order register has 01, B the event that the high-order register has 11, and C the event that both registers have the same contents. Let $X_1 = I_A$, $X_2 = I_B$, and $X_3 = I_C$ (indicator functions).
a. Find the joint mass function for (X_1, X_2) and for (X_2, X_3).
b. Find the marginal mass function of X_1 and of X_3.

5.34 Suppose $f(x_1, x_2) = k$ for $x_1^2 + x_2^2 \leq 10$ (and is zero elsewhere).
a. Determine k so that f is a two-dimensional density function.
b. If (X_1, X_2) has this density, find $P(X_1 < X_2)$. *Hint:* Integrate f over the two-dimensional event $\{(x_1, x_2) : x_1 < x_2\}$.
c. Find f_{X_1}.
d. Find $E(X_1)$.

5.35 a. For Exercise 5.34, find $P(X_1 > 2X_2)$ and $P(X_1^2 + X_2^2 \leq 1)$.
b. Find $P(X_1^2 < 5)$.

5.36 Verify that the accompanying table defines a bone fide joint mass function, and compute the marginals.

y \ x	3	4	5
2	$\frac{1}{8}$	$\frac{1}{16}$	$\frac{3}{8}$
1	$\frac{1}{16}$	$\frac{1}{16}$	$\frac{1}{8}$
0	$\frac{1}{16}$	$\frac{1}{16}$	$\frac{1}{16}$

5.37 Suppose $f_{X,Y}(x, y) = c$ over the region $0 < x < \frac{1}{2}, 0 < y < x$. Determine c so that $f_{X,Y}$ is a bivariate density function, and find formulas for the marginals.

5.38   a. Determine $F_{X,Y}(4, 2)$ for the distribution of Exercise 5.36.

   b. For the density of Exercise 5.37, calculate $P(0 < X < \frac{1}{4}, Y > \frac{1}{4})$.

5.4   INDEPENDENCE OF RANDOM VARIABLES

We have seen how the components of a random pair (X_1, X_2) may them-selves be viewed as random variables X_1 and X_2. We have seen how to find the distributions of X_1 and X_2, given the distribution of (X_1, X_2). We also mentioned that, in some cases, several univariate random variables and their corresponding marginal distributions might be given directly, where the random variables may be considered as having common domain and, hence, being jointly distributed. Such a situation gives rise to the problem of de-termining the joint distribution, given the marginals. In the present section, we consider this problem in a framework that includes the important notion of independence of random variables; in the next section, we will investigate the use of conditional distributions in solving the problem. In both sections, we will continue to concentrate on the two-dimensional case. Later, we indicate how the definitions and results for higher-dimensional cases can be given easily by analogy.

Recall that events A and B are defined to be independent provided $P(A \cap B) = P(A) \cdot P(B)$. We wish to extend this idea to random variables, so the joint distribution of independent random variables can be formed simply as the product of marginals. Suppose A and B are one-dimensional events, so $A \times B$ is a two-dimensional event in the plane, and suppose

$$P[(X, Y) \in A \times B] = P(X \in A, Y \in B) = P(X \in A) \cdot P(Y \in B).$$

Then the events $[X \in A]$ and $[X \in B]$ are independent. For the purposes outlined above, we must have this factorization property for *any* choices of A and B; if this property is so, we say X and Y are independent.

Definition 5.3   The random variables X and Y are said to be *independent* provided

$$P(X \in A, Y \in B) = P(X \in A) \cdot P(Y \in B) \quad (5.4)$$

for any choice of (one-dimensional) events A and B. If X and Y are not independent, they are said to be *dependent*.

EXAMPLE 5.14   Suppose (X, Y) has joint mass function $p(x, y)$ described in the accompanying below. Then

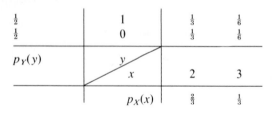

$$\tfrac{1}{3} = P\{(X, Y) \in (-1, 2] \times (-1, .5)]\}$$

$$= \frac{2}{3} \cdot \frac{1}{2} = P\{X \in (-1, 2]\} \cdot P[Y \in (-1, .5)];$$

$$\tfrac{1}{6} = P\{(X, Y) \in (2.5, 5) \times [1, 4)\}$$

$$= \frac{1}{3} \cdot \frac{1}{2} = P[X \in (2.5, 5)] \cdot P\{Y \in [1, 4)\};$$

$$\tfrac{1}{2} = P\{(X, Y) \in [1, 4] \times (-.5, .5)\}$$

$$= 1 \cdot \frac{1}{2} = P(X \in [1, 4]) \cdot P[Y \in (-.5, .5)].$$

Indeed, in this discrete case, it is easy to verify that, for any simple events $\{a\}$ and $\{b\}$,

$$P(X = a, Y = b) = P[(X, Y) \in \{a\} \times \{b\}] = P(X = a) \cdot P(Y = b);$$

that is, $p_{(X, Y)}(a, b) = p_X(a) \cdot p_Y(b)$. From this result, it follows that, for *any* events A and B, $P[(X, Y) \in A \times B] = P(X \in A) \cdot P(Y \in B)$, so X and Y are independent random variables.

This example suggests a very useful fact: If X and Y are jointly discrete, then X and Y are independent provided the joint mass function factors into a product of marginals. We state this fact and its converse formally as follows:

> **Theorem 5.1**   If the discrete random variables X and Y are jointly distributed with bivariate mass function $p_{(X, Y)}$, and if the marginal mass functions of X and Y are p_X and P_Y, respectively, then X and Y are independent if and only if the following condition holds: For any (x, y),
>
> $$p_{(X, Y)}(x, y) = p_X(x) \cdot p_Y(y). \qquad (5.5)$$

PROOF   Assume that Eq. (5.5) is satisfied. We wish to show that Eq. (5.4) is satisfied. If A or B is empty, then Eq. (5.4) is vacuously satisfied. Suppose, then, that $A \cap R_X = \{a_1, a_2, \ldots, a_n\}$ and that $B \cap R_Y = \{b_1, b_2, \ldots, b_m\}$.*
Now,

$$P[(X, Y) \in A \times B] = \sum_{(x,y) \in A \times B} p_{(X,Y)}(x, y) = \sum_{j=1}^{n} \sum_{i=1}^{m} p_{(X,Y)}(a_j, b_i)$$

$$= \sum_{j=1}^{n} \sum_{i=1}^{m} [p_X(a_j) \cdot p_Y(b_i)] = \sum_{j=1}^{n} p_X(a_j) \cdot \sum_{i=1}^{m} p_Y(b_i)$$

$$= \left[\sum_{j=1}^{n} p_X(a_j) \right] \left[\sum_{i=1}^{m} p_Y(b_i) \right] = P(X \in A) \cdot P(Y \in B).$$

Now, suppose Eq. (5.4) is satisfied. Then taking the particular choice of A and B as the simple events $\{x\}$ and $\{y\}$, respectively, we obtain

$$p_{(X,Y)}(x, y) = P[(X, Y) \in \{x\} \times \{y\}] = P(X \in \{x\}) \cdot P(Y \in \{y\})$$

$$= p_X(x) p_Y(y).$$

In practice, the independence of certain random variables is often *assumed* as part of the model. This assumption enables us to form the joint mass function from a knowledge of the individual marginal mass functions.

EXAMPLE 5.15   Students arriving for enrollment at a certain college campus are given a psychological profile exam. One of the composite scores of the exam is the "emotional stability category." Imagine that an experiment consists of selecting a student record at random and extracting the emotional stability category and gender of the student. Past experience shows that about one-half of the students are females and that the emotional stability category has an equal likelihood over the set $\{A, B, C, D, F\}$. Let X denote the gender and Y the numerical equivalent of the emotional category. What is the joint distribution of (X, Y)?

SOLUTION   The random pair (X, Y) will, upon performance of the experiment, map to an image in the range $R_{(X,Y)} = \{(0, 1), (0, 2), \ldots, (0, 5), (1, 1), \ldots, (1, 5)\}$. Let us assume that the marginal distributions of X and Y are each equal-likelihood models, so that X is distributed as a Bernoulli (.5) and Y is distributed as a discrete uniform on the set $\{1, \ldots, 5\}$. The problem of finding the joint distribution of (X, Y) is difficult, even in this simple

---

* The notation can be modified in an obvious manner if $A \cap R_X$ or $B \cap R_Y$ is denumerably infinite.

example, unless it is assumed, in addition, that X and Y are *independent*. In that case, the joint mass function of X and Y is simply the product of the uniform marginals:

$$p_{(X,Y)}(i,j) = p_X(i) \cdot p_Y(j) = \tfrac{1}{2} \cdot \tfrac{1}{5} = \tfrac{1}{10}, \qquad i = 0, 1, \qquad j = 1, 2, \ldots, 5.$$

Thus each of the simple two-dimensional events in the range $R_{(X,Y)}$ has probability $\tfrac{1}{10}$, so this is a bivariate discrete uniform distribution. Intuitively, the assumption of independence of X and Y may seem justified by a belief that, in this case, the outcome on one of the random variables does not carry information concerning the accompanying outcome on the other. (We will return to this type of conditional argument in Section 5.6.) It might be of interest to perform statistical analyses on a sample of student record data to investigate the tenability of the independence assumption in an actual application.

An alternative characterization of the independence of two jointly distributed random variables can be given in terms of factorization of the joint CDF into a product of marginal CDFs. It should be noted that the following theorem does not require assumptions concerning whether the random variables are discrete, continuous, or mixed.

> **Theorem 5.2**   Suppose (X, Y) is distributed $F_{(X,Y)}$. Then X and Y are independent if and only if, for any (x, y),
> $$F_{(X,Y)}(x, y) = F_X(x) \cdot F_Y(y).$$

PROOF   Suppose X and Y are independent. Then the events $[X \le x]$ and $[Y \le y]$ are independent, so

$$F_{(X,Y)}(x, y) = P(X \le x, Y \le y) = P(X \le x) \cdot P(Y \le y) = F_X(x) \cdot F_Y(y).$$

To prove rigorously that the factorization condition implies independence (especially if X and Y are not discrete) is more difficult. To make it plausible, consider the following argument for the special case $A = (a, c]$, $B = (b, d]$: In this case (see Fig. 5.7),

$$\begin{aligned}
P[(X, Y) \in A \times B] &= P(a < X \le c, b < Y \le d) \\
&= F_{(X,Y)}(c, d) - F_{(X,Y)}(a, d) \\
&\quad - F_{(X,Y)}(c, b) + F_{(X,Y)}(a, b).
\end{aligned}$$

(See Exercise 5.42.) Now, if $F_{(X,Y)}(x, y) = F_X(x) \cdot F_y(y)$ for any (x, y), the

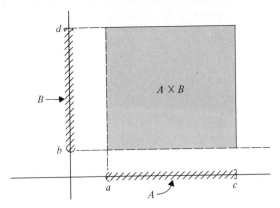

FIGURE 5.7   *Graph of $(a, c] \times (b, d]$*

above equation may be written as

$P[(X, Y) \in A \times B] =$

$$F_X(c)F_Y(d) - F_X(a)F_Y(d) - F_X(c)F_Y(b) + F_X(a)F_Y(b)$$

$$= [F_X(c) - F_X(a)][F_Y(d) - F_Y(b)]$$

$$= P(X \in A) \cdot P(Y \in B).$$

For continuous distributions, we can use the result of Theorem 5.2 to derive a condition similar to that for the mass function factorization.

> **Corollary 5.1**   If (X, Y) is jointly continuous with CDF $F_{(X,Y)}$, then X and Y are independent if and only if $f_{(X,Y)}(x, y) = f_X(x)f_Y(y)$ for each (x, y).

PROOF   If X and Y are independent, then

$$f_{(X,Y)}(x, y) = \frac{\partial^2 F_{(X,Y)}(x, y)}{\partial x \, \partial y} = \frac{\partial}{\partial x}\left[F_X(x)\right]\frac{\partial}{\partial y}\left[F_Y(y)\right] = f_X(x)f_Y(y).$$

On the other hand, if $f_{(X,Y)}(x, y) = f_X(x)f_Y(y)$, then

$$F_{(X,Y)}(x, y) = \int_{-\infty}^{x} \int_{-\infty}^{y} f_{(X,Y)}(u, v) \, du \, dv$$

$$= \int_{-\infty}^{x} f_X(u) \, du \int_{-\infty}^{y} f_Y(v) \, dv = F_X(x) \cdot F_Y(y),$$

which implies the independence of X and Y by Theorem 5.2.

EXAMPLE 5.16 Suppose X_1 is distributed $N(\mu_1, \sigma_1^2)$ and X_2 is distributed $N(\mu_2, \sigma_2^2)$, and X_1 and X_2 are independent. Then

$$f_{(X_1, X_2)}(x_1, x_2) = \frac{1}{2\pi\sigma_1\sigma_2} \exp -\frac{1}{2}\left[\left(\frac{x_1 - \mu_1}{\sigma_1}\right)^2 + \left(\frac{x_2 - \mu_2}{\sigma_2}\right)^2\right],$$

and

$$F_{(X_1, X_2)}(x_1, x_2) = \Phi\left(\frac{x_1 - \mu_1}{\sigma_1}\right)\Phi\left(\frac{x_2 - \mu_2}{\sigma_2}\right),$$

where Φ is the standard normal CDF. *Note:* This is a bivariate normal distribution with parameter $\rho = 0$ (see Exercise 5.22 in Section 5.3).

The following theorem is a useful result that can be employed to establish the independence of continuous random variables by inspection of the joint density function.

> **Theorem 5.3** Let (X_1, X_2) be jointly continuous with density function f, and suppose $R = R_{X_1} \times R_{X_2}$, where $R = \{(x_1, x_2) : f_X(x_1, x_2) > 0\}$ and similarly for the ranges R_{X_1} and R_{X_2} of X_1 and X_2. If there exist functions f_1 and f_2 such that f_1 does not depend on x_2 and f_2 does not depend on x_1, and $f(x_1, x_2) = f_1(x_1)f_2(x_2)$ for all $(x_1, x_2) \in R$, then X_1 and X_2 are independent.

PROOF Define $f_1(x_1) = 0$ if $x_1 \notin R_{X_1}$, and similarly for f_2, so that each f_i is defined on $(-\infty, \infty)$. Then there exist constants c_1 and c_2 such that $c_i = \int_{-\infty}^{\infty} f_i(x)\, dx$, $i = 1, 2$. Since f_1 does not depend on x_2,

$$f_{X_1}(x_1) = \int_{-\infty}^{\infty} f_1(x_1)f_2(x_2)\, dx_2 = f_1(x_1)\int_{-\infty}^{\infty} f_2(x_2)\, dx_2 = c_2 f_1(x_1).$$

Similarly, $f_{X_2}(x_2) = c_1 f_2(x_2)$, and

$$c_1 c_2 = \int_{-\infty}^{\infty} f_1(x_1)\, dx_1 \int_{-\infty}^{\infty} f_2(x_2)\, dx_2$$

$$= \int_{-\infty}^{\infty}\int_{-\infty}^{\infty} f_1(x_1)f_2(x_2)\, dx_1\, dx_2 = \int_{-\infty}^{\infty}\int_{-\infty}^{\infty} f_X(x_1, x_2)\, dx_1\, dx_2 = 1.$$

Finally,

$$f(x_1, x_2) = f_1(x_1)f_2(x_2) = \left[\frac{f_{X_1}(x_1)}{c_1}\right]\left[\frac{f_{X_2}(x_2)}{c_2}\right] = f_{X_1}(x_1)f_{X_2}(x_2),$$

so that X_1 and X_2 are independent by Corollary 5.1.

Remark: If the joint density f factors into a function of x_1 alone times a function of x_2 alone, the latter functions are, except for normalizing constants, the marginals of X_1 and X_2.

EXAMPLE 5.17   Suppose $f_{(X_1,X_2)}(x_1, x_2) = \frac{1}{6}x_1 x_2^2$, $0 < x_1 < 1$, $0 < x_2 < 1$. Then by Theorem 5.3, we can immediately assert that X_1 and X_2 are independent and that the marginal density function for X_1 is x_1, except possibly for a multiplicative constant. Indeed, it is easy to verify that $f_{X_1}(x_1) = \frac{1}{2}x_1$, $0 < x_1 < 1$, so that necessarily $f_{X_2}(x_2) = \frac{1}{3}x_2^2$, $0 < x_2 < 1$.

EXAMPLE 5.18   Care must be exercised in establishing that f_1 depends only on x_1 and similarly for f_2. Suppose, in the preceding example, we had $f_{(X_1,X_2)}(x_1, x_2) = kx_1 x_2^2$, $x_1 \geq 0$, $x_2 \geq 0$, $x_1 + x_2 \leq 1$. It does not follow that $f_{(X_1,X_2)}(x_1, x_2) = (kx_1)(x_2^2)$ factors into a product $f_1(x_1) \cdot f_2(x_2)$, where f_1 depends only on x_1, since, for example, the domain of f_1 depends on x_2. That is, we must take $f_1(x_1) = kx_1$ for $0 \leq x_1 \leq 1 - x_2$. It is easy to find the marginals f_{X_1} and f_{X_2} in this case and to verify that their product is *not* $f_{(X_1,X_2)}$. Thus X_1 and X_2 are dependent.

It is sometimes easy to establish that jointly distributed random variables X_1 and X_2 are dependent by finding a point (x_1, x_2) such that $f_1(x_1) > 0$ and $f_2(x_2) > 0$, yet $f(x_1, x_2) = 0$. This technique amounts to showing that the range R of (X_1, X_2) is not the product set $R_{X_1} \times R_{X_2}$. On the other hand, the mere fact that $R = R_{X_1} \times R_{X_2}$ does not guarantee that the random variables are independent, as the following example shows.

EXAMPLE 5.19   Let

$$f(x_1, x_2) = \begin{cases} \dfrac{1}{2x_1} & \text{if } 0 \leq x_2 < x_1 \leq 1, \\[2mm] 1 & \text{if } 0 \leq x_1 \leq x_2 \leq 1. \end{cases}$$

Then the range R of (X_1, X_2) is $[0, 1] \times [0, 1]$, f is indeed a bivariate density, and the marginals are given by $f_{X_1}(x_1) = \frac{3}{2} - x_1$, $0 \leq x_1 \leq 1$, and $f_{X_2}(x_2) = x_2 - \log \sqrt{x_2}$, $0 \leq x_2 \leq 1$. The random variables X_1 and X_2 are not independent, and yet $R = R_{X_1} \times R_{X_2}$.

The following theorem is extremely useful, and it shows the assumption of independence of random variables has far-reaching consequences.

> **Theorem 5.4**   If X and Y are independent, then, for any functions g and h, the random variables $g(X)$ and $h(Y)$ are independent.

PROOF We need only show that, for any events A and B,

$$P[g(X) \in A, h(Y) \in B] = P[g(X) \in A] \cdot P[h(Y) \in B].$$

But the event $[g(X) \in A]$ occurs if and only if the outcome on X is in the set $\{x : g(x) \in A\}$, which we denote by $g^{-1}(A)$. Similarly, $[h(Y) \in B]$ is equivalent to $[Y \in h^{-1}(B)]$. Thus

$$P[g(X) \in A, h(Y) \in B] = P[X \in g^{-1}(a), Y \in h^{-1}(B)]$$
$$= P[X \in g^{-1}(A)] \cdot P[Y \in h^{-1}(B)],$$

by the assumed independence of X and Y. Now, we reverse the process on each of the factors above; the event $[X \in g^{-1}(A)]$ is, by definition, the event $[X \in \{x : g(x) \in A\}]$, which occurs if and only if $g(X)$ falls in A. Thus $P[X \in g^{-1}(A)] = P[g(X) \in A]$, and similarly for the second factor.

EXAMPLE 5.20   Suppose X and Y are independent with marginal $N(0, 1)$ distributions. Then the joint density function of the vector (X^2, Y^2) is the product of the marginal $\chi^2_{(1)}$ density functions of X^2 and Y^2, since X^2 and Y^2 are themselves independent. Similarly, $\text{Cos}\,|X|$ and $3Y^3 - 1$ are independent.

EXAMPLE 5.21   Suppose X is Bernoulli with parameter .5. Let $Y = 2X - 1$ and $Z = 1 - 2X$. It is easy to verify that Y and Z are dependent, yet Y^2 and Z^2 are independent (see Exercise 5.48).

EXERCISES

THEORY

5.39 Suppose (X, Y) has CDF $F_{(X, Y)}$ and X and Y are discrete random variables. Prove that, if $F_{(X, Y)}(x, y) = F_X(x) \cdot F_Y(y)$ for all $(x, y) \in R_{(X, Y)}$, then X and Y are independent. That is, prove Theorem 5.2 rigorously for the discrete case.

5.40 Show that, if U and V are jointly distributed and V is degenerate at a, then U and V are independent.

5.41 Show that the converse of Theorem 5.3 is true.

5.42 Suppose $A = (a, c]$ and $B = (b, d]$, so the two-dimensional event $A \times B$ is as shown in Fig. 5.7. Assume the random pair (X, Y) has CDF F. Show that

$$P[(X, Y) \in A \times B] = F(c, d) - F(a, d) - F(c, b) + F(a, b).$$

5.43 Let A and B be events contained in a given sample space. Show that the events A and B are independent if and only if the random variables I_A and I_B are independent. I_A is the indicator function

$$I_A(o) = \begin{cases} 1 \text{ if } o \in A, \\ 0 \text{ if } o \notin A. \end{cases}$$

APPLICATION

5.44 An experiment consists of tossing a fair coin and then tossing the coin again if the first toss was zero; otherwise, tossing a fair die. The outcome Y is the second number tossed.
a. List R_Y for this experiment.
b. List the mass function for Y.
c. Find $E(Y)$.

5.45 Suppose that, in Exercise 5.44, we let X denote the outcome on the first toss of the coin.
a. Are X and Y independent?   b. Find $P(Y = 1|X = 1)$.
c. Find $P(X = 1|Y = 1)$.   d. Find $p(x, y)$.

5.46 Suppose (X_1, X_2) has the mass function shown in the accompanying table. Determine whether X_1 and X_2 are independent.

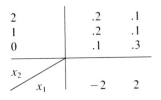

5.47 In Exercise 5.46, determine whether X_1^2 and X_2^2 are independent.

5.48 Suppose X is Bernoulli with parameter .5, $Y = 2X - 1$, and $Z = 1 - 2X$.
a. Find the joint mass function of Y and Z.
b. Verify that X and Z are identically distributed.
c. Verify that Y and Z are dependent.
d. Show that Y^2 and Z^2 are independent.

5.49 The traffic counts taken on each of two different freeways are assumed to obey a Poisson (λ) probability law. Determine the joint mass function.

5.50 For Examples 5.17 and 5.18, verify that $f_{(X_1, X_2)}$ is a density and that the marginals are as specified.

5.51 Suppose $f_{(X,Y)}(x, y) = (1/2\pi)e^{-(1/2)(x^2 + y^2)}$; that is, (X, Y) is bivariate normal with $\mu_X = \mu_Y = 0$, $\sigma_X^2 = \sigma_Y^2 = 1$, and $\rho = 0$.
a. Show that X and Y are identically distributed and independent.
b. Find $P(X^2 + Y^2 < 1)$. *Hint:* Use polar coordinates.
c. Find $P(X^2 < 1)$, after verifying that X^2 is distributed $\chi_{(1)}^2$.

5.52 a. A piece of radio equipment has an exponentially distributed life with a mean time to failure of $\frac{1}{2}$; that is, L is distributed exponential (2). Find the time t_0 such that, with probability .9, the equipment is still operating at t_0.

b. Assume that five pieces of equipment have independent identically distributed lifetimes L_i, $i = 1, 2, \ldots, 5$, where L_i is distributed exponential (2). Find the time t_0 such that, with probability .9, all five pieces of equipment are operating at time t_0.

c. What is the probability that at least four pieces of equipment are still operating at time .8?

5.53   Discuss whether random variables representing the quantities in the following situations are independent:

a. Height and weight of an individual chosen at random.

b. Age and weight of a penny selected at random.

c. The price of gasoline and temperature at a certain location and time.

d. Your blood pressure and shoe size, five minutes from now.

5.54   Two points are chosen at random, independently, on a circle. What is the probability that the shortest arc length between the points is less than the radius of the circle?

5.5   JOINT MOMENTS

In many situations, model specification involves moments, such as means, variances, and covariances. We are now in a position to discuss these concepts in the setting of bivariate distributions and to relate them to our earlier considerations of expected values of random variables. We will continue to limit our remarks to the bivariate case; extensions to n-dimensional distributions will be discussed in Section 5.7.

Suppose the random pair (X, Y) has a continuous distribution, with known density $f_{X,Y}(x, y)$, and suppose it is desired to find the expected value of some function g of X and Y, such as $X^2 + Y^2$. In general, such a function of X and Y is itself a random variable, which we denote by Z. Now $Z = g(X, Y)$ has a univariate distribution, which, depending on g, may or may not be continuous. Let us assume for now that Z has a continuous distribution, with density function f_Z. If we can find f_Z, then the expected value of $g(X, Y)$ is, by definition, given by $E[g(X, Y)] = E(Z) = \int_{-\infty}^{\infty} z f_Z(z)\, dz$. We will show how to find f_Z, using the known joint density $f_{X,Y}$ and the function g, later, in Chapter 6. For now, a form of the rule of the unconscious statistician (RUS) is invoked to compute the mean for Z directly from the joint density of (X, Y). Recall, in the univariate case, that RUS states that $E[g(X)] = \int_{-\infty}^{\infty} g(x) f_X(x)\, dx$; the bivariate version similarly states that

$$\int_{-\infty}^{\infty} z f_Z(z)\, dx = E[g(X, Y)] = \int_{-\infty}^{\infty} \int_{-\infty}^{\infty} g(x, y) f_{X,Y}(x, y)\, dy\, dx.$$

This rule is extremely useful, since it allows us to find moments of $g(X, Y)$ without having to worry about first finding f_Z.

EXAMPLE 5.22 Suppose (X, Y) is discrete, and the values of the mass function $p_{(X,Y)}$ are given in the accompanying table.

y		
1	$\frac{1}{3}$	$\frac{1}{2}$
0	$\frac{1}{6}$	0
	1	2
	x	

Suppose $g(X, Y) = X + Y$. Then (using RUS)

$$E(X + Y) = \sum_{(x,y)\in R_{(X,Y)}} (x + y)p_{(X,Y)}(x, y) = \sum_{x=1}^{2}\sum_{y=0}^{1} (x + y)p_{(X,Y)}(x, y)$$

$$= (1 + 0)\cdot\tfrac{1}{6} + (1 + 1)\cdot\tfrac{1}{3} + (2 + 0)\cdot 0 + (2 + 1)\cdot\tfrac{1}{2} = \tfrac{7}{3}.$$

On the other hand, it is easy to find the mass function for the random variable $Z = X + Y$ in this simple discrete example. For example, the only way $[X + Y = 3]$ can occur is if $[X = 2, Y = 1]$ occurs, so $P(Z = 3) = P(X + Y = 3) = P(X = 2, Y = 1) = \tfrac{1}{2}$. Using a similar argument for the other values in R_Z gives the results in the accompanying table. In this case, using the definition of $E(Z)$, we have $E(Z) = (1)\cdot\tfrac{1}{6} + (2)\cdot\tfrac{1}{3} + (3)\cdot\tfrac{1}{2} = \tfrac{7}{3}$, as before.

z	1	2	3
$p_Z(z)$	$\frac{1}{6}$	$\frac{1}{3}$	$\frac{1}{2}$

As a further example, let $h(X, Y) = X$, so (using RUS)

$$E[h(X, Y)] = E(X) = \sum_{(x,y)\in R_{(X,Y)}} xp_{(X,Y)}(x, y)$$

$$= (1)\cdot\tfrac{1}{6} + (1)\cdot\tfrac{1}{3} + (2)\cdot 0 + (2)\cdot\tfrac{1}{2} = \tfrac{3}{2},$$

whereas, by using the definition of $E(X)$ with the marginal p_X, we have $E(X) = (1)\cdot\tfrac{1}{2} + (2)\cdot\tfrac{1}{2} = \tfrac{3}{2}$.

You have seen that expectation is a linear operator, in the sense that $E(aX + b) = aE(X) + b$. The multivariate version of this result can be seen as follows: Suppose (X, Y) is jointly discrete with joint mass function $p_{X,Y}(x, y)$. Then by RUS, the mean of $(X + Y)$ can be calculated as follows:

$$E(X + Y) = \sum_{(x,y)\in R_{(X,Y)}} (x + y)p_{(X,Y)}(x, y)$$

$$= \sum_{(x,y)} xp_{(X,Y)}(x, y) + \sum_{(x,y)} yp_{(X,Y)}(x, y)$$

$$= \sum_{x} x \sum_{y} p_{(X,Y)}(x, y) + \sum_{y} y \sum_{x} p_{(X,Y)}(x, y)$$

$$= \sum_{x} xp_X(x) + \sum_{y} yp_Y(y),$$

since the inner sums in the preceding line give the marginal mass functions of X and Y, respectively. But the latter expression is $E(X) + E(Y)$. A similar argument can be used for cases in which (X, Y) is not jointly discrete. Also, since expectation with respect to a univariate distribution has been seen to be a linear operator, it follows that $E(aX + bY + c) = aE(X) + bE(Y) + c$. We summarize as follows:

Theorem 5.5   Suppose X and Y are jointly distributed and have means μ_X and μ_Y, respectively. Then for any constants a, b, and c,

$$E(aX + bY + c) = a\mu_X + b\mu_Y + c.$$

Since sums are of great importance in applications, there is interest also in $V(X + Y)$. It would be natural to wonder how $V(X + Y)$ is associated with the individual variances $V(X)$ and $V(Y)$. We will see that, in general, $V(X + Y)$ is the sum of the individual variances plus a term involving a "cross product variance," called the covariance of X and Y. Suppose $E(X) = \mu_X$ and $E(Y) = \mu_Y$ while $V(X) = \sigma_X^2$ and $V(Y) = \sigma_Y^2$. Then by definition,

$$V(X + Y) = E\{[X + Y - E(X + Y)]^2\} = E\{[X + Y - (\mu_X + \mu_Y)]^2\}$$

$$= E\{[(X - \mu_X) + (Y - \mu_Y)]^2\}$$

$$= E[(X - \mu_X)^2 + (Y - \mu_Y)^2 + 2(X - \mu_X)(Y - \mu_Y)]$$

$$= E[(X - \mu_X)^2] + E[(Y - \mu_Y)^2] + 2E[(X - \mu_X)(Y - \mu_Y)],$$

where Theorem 5.5 is used in the last step, allowing the expected value of the sum to be taken term by term. The first two expected values are σ_X^2 and σ_Y^2, respectively. The third term arises often, so we give it a name; $E[(X - \mu_X)(Y - \mu_Y)]$ is defined to be the *covariance* of X and Y, denoted $\text{Cov}(X, Y)$. In the exercises, you are asked to track through this argument with the function $aX + bY + c$; we summarize this result as follows:

Theorem 5.6   Suppose X and Y are jointly distributed with variances σ_X^2 and σ_Y^2, respectively. Then $V(aX + bY + c) = a^2\sigma_X^2 + b^2\sigma_Y^2 + 2ab\,\text{Cov}(X, Y)$.

Remark: Note that the additive constant c does not affect the variance of the (univariate) random variable $aX + bY$, as we have seen before. Note also that $\text{Cov}(aX, bY) = ab\,\text{Cov}(X, Y)$.

EXAMPLE 5.23   Suppose (X, Y) is jointly continuous, with density $f(x, y)$ $= cx$, $0 < x < y < 1$. Compute the variance of $Z = X + Y$.

SOLUTION   First, we evaluate c:

$$1 = \int_0^1 \int_x^1 f(x, y) \, dy \, dx = c \cdot \int_0^1 x(1 - x) \, dx,$$

so $c = 1/B(2, 2) = 6$. Note that

$$E(X) = \int_0^1 \int_x^1 x \cdot f(x, y) \, dy \, dx = 6 \int_0^1 x^2(1 - x) \, dx = 6 \cdot B(3, 2) = \frac{1}{2};$$

$$E(Y) = \int_0^1 \int_x^1 y \cdot f(x, y) \, dy \, dx = \frac{6}{2} \int_0^1 (1 - x^2)x \, dx = \frac{3}{4};$$

$$V(X) = \int_0^1 \int_x^1 x^2 f(x, y) \, dy \, dx - \frac{1}{4} = 6 \int_0^1 x^3(1 - x) \, dx - \frac{1}{4} = \frac{1}{20};$$

$$V(Y) = \int_0^1 \int_x^1 y^2 f(x, y) \, dy \, dx - \frac{9}{16} = \frac{6}{3} \int_0^1 (1 - x^3)x \, dx - \frac{9}{16} = \frac{3}{80}.$$

Now, consider $Z = X + Y$. By Theorem 5.5, $E(Z) = \frac{1}{2} + \frac{3}{4}$, and by Theorem 5.6, $V(Z) = \frac{1}{20} + \frac{3}{80} + 2 \, \text{Cov}(X, Y)$. Finally, by definition,

$$\text{Cov}(X, Y) = \int_0^1 \int_x^1 (x - \tfrac{1}{2})(y - \tfrac{3}{4}) f(x, y) \, dy \, dx$$

$$= 6 \int_0^1 x(x - \tfrac{1}{2})(-\tfrac{1}{4} + \tfrac{3}{4}x - \tfrac{1}{2}x^2) \, dx = \tfrac{1}{40}.$$

So $V(Z) = \frac{11}{80}$.

The computation of $\text{Cov}(X, Y)$ in the preceding example is somewhat tedious. This computation can be simplified by using a computational formula similar to that used for variance, $\sigma_Z^2 = E(Z^2) - E^2(Z)$.

> **Theorem 5.7** $\text{Cov}(X, Y) = E(XY) - E(X) \cdot E(Y)$.

PROOF

$$E[(X - \mu_X)(Y - \mu_Y)] = E(XY - \mu_X Y - \mu_Y X + \mu_X \mu_Y)$$

$$= E(XY) - \mu_X E(Y) - \mu_Y E(X) + \mu_X \mu_Y$$

$$= E(XY) - \mu_X \mu_Y.$$

EXAMPLE 5.24   In the preceding example, we could have computed

$$\text{Cov}(X, Y) = \int_0^1 \int_x^1 xyf(x, y)\, dy\, dx - (\tfrac{1}{2}) \cdot (\tfrac{3}{4})$$

$$= \tfrac{6}{2} \int_0^1 x^2(1 - x^2)\, dx - \tfrac{3}{8} = \tfrac{6}{15} - \tfrac{3}{8} = \tfrac{1}{40},$$

as before.

You may have been surprised to see that $V(X) + V(Y)$ is not, in general, the variance of $X + Y$. As will be discussed in more detail later, the correction term, involving the covariance of X and Y, is necessary when X and Y are dependent. We now state a sequence of results leading to the fact that, when X and Y are independent, $V(X + Y)$ is just $V(X) + V(Y)$.

> **Theorem 5.8**   If X and Y are independent, then $E(X \cdot Y) = E(X) \cdot E(Y)$.

PROOF   Suppose X and Y are jointly continuous with density function f. By RUS, $E(XY) = \int_{-\infty}^{\infty} \int_{-\infty}^{\infty} xyf(x, y)\, dx\, dy$. Since X and Y are independent, $f(x, y) = f_X(x) \cdot f_Y(y)$, so

$$E(XY) = \int_{-\infty}^{\infty} xf_X(x) \int_{-\infty}^{\infty} yf_Y(y)\, dy\, dx = \int_{-\infty}^{\infty} xf_X(x)\mu_Y\, dx = \mu_X \cdot \mu_Y.$$

The case for X and Y not jointly continuous is similar.

From Theorem 5.8, there are two immediate applications, which you are asked to consider in the exercises.

> **Corollary 5.2**   If X and Y are independent, then $\text{Cov}(X, Y) = 0$.
> **Corollary 5.3**   If X and Y are independent, then $V(X + Y) = V(X) + V(Y)$.

EXAMPLE 5.25   Suppose an experiment consists of firing at a target in the xy-plane, and suppose X and Y denote the deviation (left-right) and range (along the gun-target line) miss distances, respectively. Assume that the marginal distributions of X and Y are normal—that is, X is $N(\mu_X, \sigma_X^2)$ and Y is $N(\mu_Y, \sigma_Y^2)$—and assume that X and Y are independent. Find the mean and variance of the squared radial miss distance $R^2 = X^2 + Y^2$.

SOLUTION   By Theorem 5.5, $E(R^2) = E(X^2 + Y^2) = E(X^2) + E(Y^2)$. Since $\sigma_X^2 = E(X^2) - \mu_X^2$, it follows that $E(X^2) = \sigma_X^2 + \mu_X^2$, and similarly for $E(Y^2)$. Thus $E(R^2) = \sigma_X^2 + \sigma_Y^2 + \mu_X^2 + \mu_Y^2$. Since X and Y are components of miss, or "error," R^2 is the "squared error" of the projectile, and $E(R^2)$ is its "mean squared error" (MSE). This quantity, or its square root, the "root mean square" (RMS), is sometimes used as a measure of the overall quality of the gun system. These measures can only be small if the gun has good *accuracy* (μ_X and μ_Y near zero) and good *precision* (σ_X^2 and σ_Y^2 near zero). When the MSE of the gun system is small, it can be anticipated that a projectile fired from the gun will hit near the target.

By Corollary 5.3, the variance of R^2 is given by $V(X^2) + V(Y^2)$. To compute $V(X^2)$, we use the fact that $[(X - \mu_X)/\sigma_X]^2$ is $\chi_{(1)}^2$, so

$$V\left(\frac{X^2 - 2\mu_X X + \mu_X^2}{\sigma_X^2}\right) = 2.$$

It follows that

$$V(X^2) = 2\sigma_X^4 - 4\mu_X^2 V(X) + 4\mu_X \operatorname{Cov}(X, X^2)$$

$$= 2\sigma_X^4 - 4\mu_X^2\sigma_X^2 + 4\mu_X[\mu_X^* - \mu_X(\sigma_X^2 + \mu_X^2)],$$

where $\mu_X^* = E(X^3)$ is the third central moment of an $N(\mu_X, \sigma_X^2)$ random variable. By differentiation of the $N(\mu_X, \sigma_X^2)$ moment-generating function, it is easily seen that $\mu_X^* = \mu_X^3 + 3\mu_X\sigma_X^2$. Using a similar expression for $V(Y^2)$, we obtain

$$V(R^2) = 2\sigma_X^4 - 4\mu_X^2\sigma_X^2 + 4\mu_X(\mu_X^3 + 3\mu_X\sigma_X^2 - \mu_X\sigma_X^2 - \mu_X^3) + V(Y^2)$$

$$= 2(\sigma_X^4 + \sigma_Y^4 + 2\mu_X^2\sigma_X^2 + 2\mu_Y^2\sigma_Y^2).$$

EXAMPLE 5.26   —*Dependent Random Variables with Zero Covariance* The fact that $\operatorname{Cov}(X, Y) = 0$ does *not* imply independence may be shown as follows: Suppose (X_1, X_2) has joint density g, where the bivariate density g is as described in Example 5.11 of Section 5.3, $g(x, y) = (1/\pi) \exp[-(\frac{1}{2})(x^2 + y^2)]$ for $(x, y) \in A_1 \cup A_2 \cup A_3$ (see Fig. 5.6). We have observed that X_1 and X_2 are then marginally identically distributed $N(0, 1)$, so it follows that $\operatorname{Cov}(X_1, X_2) = E(X_1 \cdot X_2)$. But

$$E(X_1, X_2) = \frac{1}{\pi}\left[\int_0^\infty ye^{-(1/2)y^2}\,dy \int_{-\infty}^{-c} xe^{-(1/2)x^2}\,dx\right.$$

$$+ \int_{-\infty}^0 ye^{-(1/2)y^2}\,dy \int_{-c}^c xe^{-(1/2)x^2}\,dx$$

$$\left. + \int_0^\infty ye^{-(1/2)y^2}\,dy \int_c^\infty xe^{-(1/2)x^2}\,dx\right]$$

$$= \frac{1}{\pi}[-e^{-(1/2)c^2} + e^{-(1/2)c^2}] = 0.$$

Thus $\text{Cov}(X_1, X_2) = 0$. But X_1 and X_2 are not independent, since $A_1 \cup A_2 \cup A_3$ is not a product set. (See Exercise 5.62 for a discrete example.)

In spite of the fact that $\text{Cov}(X_1, X_2) = 0$ does not imply the independence of X_1 and X_2, $\text{Cov}(X_1, X_2)$ is often used by the statistical practitioner as a measure of the "dependence" of X_1 and X_2. Part of this misuse has arisen from a perhaps unfortunate use of the term "dependence," since there is already an inherent concept of dependence between X_1 and X_2 as functions. Thus we sometimes say that X_2 *depends* on X_1 if X_2 can be written as a composition of some function h with X_1, or else depends on X_1 through its range; otherwise, X_1 and X_2 might be (incorrectly) called independent. The idea is that if $X_2 = h(X_1)$, then, given any $o \in S$, the value of X_2 at o is determined, once the value of X_1 at o is given. We might refer to such dependence as *functional dependence* to contrast it with *statistical* or *stochastic independence*. But $\text{Cov}(X_1, X_2)$ is not even a good measure of functional dependence. If X_2 is a function of X_1, we might expect X_1 and X_2 to be statistically dependent. For example, for any nondegenerate X_1, if $X_2 = X_1$, then

$$\text{Cov}(X_1, X_2) = E(X_1^2) - [E(X_1)]^2 = \sigma_{X_1}^2 > 0,$$

and so X_1 and X_2 could not be independent. But even when X_2 is a function of X_1, the covariance may be zero. (See Exercise 5.62 for an example of this result.) Thus covariance is not in itself a good measure of either functional or statistical independence. There is a restricted sense in which a comparison can be made, involving linear functions. It is useful to define the dimensionless quantity $\rho(X, Y) = \text{Cov}(X, Y)/\sqrt{\sigma_X^2 \sigma_Y^2}$, called the *correlation* of X and Y. The quantity ρ is dimensionless because the units in which X and Y are measured occur in both the numerator and denominator of the expression above. Thus, for example, the same value of ρ would result with X measured in miles and Y in pounds as with X in feet and Y in tons. This result is not true of $\text{Cov}(X, Y)$. If $\rho = 0$, it is said that X and Y are *uncorrelated*, which means that $\text{Cov}(x, y) = 0$. Thus independent random variables are uncorrelated, but not conversely (see Exercises 5.60 and 5.62). In Chapter 6 it is shown that $\rho(X, Y)$ is always between -1 and 1; and if $|\rho| = 1$, then X and Y are linearly related.

EXERCISES

THEORY

5.55   Verify Corollaries 5.2 and 5.3.

5.56   Generalize Theorem 5.8 as follows: Suppose X and Y are independent. Then $E[g(X) \cdot h(Y)] = E[g(X)]E[h(Y)]$.

5.57 Suppose (X, Y) has the general bivariate normal density function (see Exercise 5.22 of Section 5.3), with parameters σ_X^2, σ_Y^2, μ_X, μ_Y, and ρ. Show that $\text{Cov}(X, Y) = \rho \sqrt{\sigma_X^2 \sigma_Y^2}$, so $\rho = \text{Cov}(X, Y)/\sqrt{V(X)V(Y)}$ is the correlation of X and Y.

5.58 Show that $\text{Cov}(X, X) = V(X)$.

5.59 Suppose X_1, X_2, and X_3 are random variables with means μ_i and variances σ_i^2, $i = 1, 2, 3$. Find $V(X_1 + X_2 + X_3)$, using Theorem 5.6. *Hint:* First, consider the variance of $(X_1 + X_2) + X_3$, and use the fact that $\text{Cov}(X_1 + X_2, X_3) = \text{Cov}(X_1, X_3) + \text{Cov}(X_2, X_3)$.

5.60 Show that, if X and Y are independent, then the correlation of X and Y, $\rho(X, Y) = \text{Cov}(X, Y)/\sqrt{V(X)V(Y)}$, must be zero.

APPLICATION

5.61 Suppose X and Y have the bivariate mass function listed in the accompanying table.
   a. Find $E(X)$ and $E(Y)$.   b. Verify $E(XY) \neq E(X)E(Y)$.
   c. Compute $\text{Cov}(X, Y)$.   d. Find $V(2X - 3Y + 5)$.   e. Find $\rho(X, Y)$.

y	-2	2
2	.2	.1
1	.2	.1
0	.1	.3
	x	

5.62 Find two discrete uncorrelated random variables that are dependent. *Hint:* Note that $\rho = 0$ if and only if $E(XY) = E(X)E(Y)$. Suppose X is distributed uniform on $(\{-1, 0, 1\})$, so $E(X) = 0$, and take $Y = X^2$. Show that X and Y are dependent but $E(XY) = 0$.

5.63 Show that $\text{Cov}(X, aY + b) = a\,\text{Cov}(X, Y)$.

5.64 Suppose X, Y, and Z are independent, each with a marginal uniform distribution over $(0, 1)$.
   a. Find $P(X + Z < \frac{1}{3})$.   b. Find $E(X^2 Y)$.
   c. Find $E(e^X)$.   d. Find $P(X^2 + Y^2 + Z^2 < \frac{1}{2})$.
   e. Find the probability that at least k of the random variables are less than α, $\alpha \in (0, 1)$, $k = 0, 1, 2, 3$. *Hint:* Use the binomial tables.

5.65 Show that $\text{Cov}(X, Y)$ may be positive or negative. *Hint:* Find examples of (nonindependent) random variables for which each occurs.

5.66 Suppose X has mean μ_X and variance σ_X^2, and Y has mean μ_Y and variance σ_Y^2. Assume $\rho(X, Y) = \rho$.
   a. Find $\text{Cov}(aX, bY)$.   b. Find $\text{Cov}(X + Y, X - Y)$.   c. Find $E[(X + Y)^2]$.

5.67 A box contains five good and two defective items. Two items are drawn, without replacement, from the box and tested. Let X_1 and X_2 denote the Bernoulli random variables representing the outcomes on the two tests.
   a. Find $\text{Cov}(X_1, X_2)$.   b. Find $V(X_1 - X_2)$.

5.68 The random pair (X, Y) has a uniform density over the region $0 < x < y < 1$.
   a. Find $\text{Cov}(X, Y)$.   b. Find $\rho(X, Y)$.   c. Find $V(X + Y)$.

5.69 A stick one unit in length is broken at a random point. Let S and L be the lengths of the shortest and longest parts, respectively.
   a. Find $\text{Cov}(S, L)$.   b. Find $\rho(S, L)$.   c. Find $V(S + L)$.

5.70 Two points are chosen at random on the portion in the first quadrant of a unit circle with center at the origin. Let U and V denote the distances along the arc from the x-axis to the points. Find $\rho(U, V)$.

5.6   CONDITIONAL DISTRIBUTIONS AND CONDITIONAL MOMENTS

We are now in a position to extend the notions of conditional probabilities to situations involving joint distributions. Let us review briefly what was accomplished with the earlier work on conditional probabilities. Recall that, if A and B are two events with B fixed such that $P(B) \neq 0$, then $P(A|B) = P(A \cap B)/P(B)$ defines a probability function. As a function of A, this conditional probability function could be written $P(\cdot|B)$. It satisfies all the Kolmogorov axioms for a probability, and, hence it also enjoys all the properties of probability functions in general. This conditional probability is useful in applications since $P(A|B)$ may be interpreted as a revised prediction of the likelihood that A would occur in the presence of the additional information that B also occurs. And if that information does not change the likelihood of A occurring, A and B are independent events.

Let us begin with the simplest case, where (X_1, X_2) is discrete. Let R_1 and R_2 be the respective ranges of X_1 and X_2, and suppose $b \in R_2$ so that $P(X_2 = b) > 0$. Then for each x_1, we can evaluate the conditional probability of the event $[X_1 = x_1]$ given the event $[X_2 = b]$. By definition, this probability is

$$P(X_1 = x_1 | X_2 = b) = \frac{P(X_1 = x_1, X_2 = b)}{P(X_2 = b)}.$$

In terms of probability mass functions, this result may be expressed as

$$P(X_1 = x_1 | X_2 = b) = \frac{p_{X_1,X_2}(x_1 b)}{p_{X_2}(b)}. \tag{5.6}$$

Now, with b held fixed, Eq. (5.6) defines a function of x_1, which we temporarily denote as p_b.

Let us verify that, for each b in R_2, p_b is a mass function. First, observe that $p_b(x_1)$ is nonnegative. Next, note that

$$\sum_{x_1 \in R_1} p_b(x_1) = \frac{1}{p_{X_2}(b)} \sum_{x_1 \in R_1} p_{X_1 X_2}(x_1, b) = \frac{p_{X_2}(b)}{p_{X_2}(b)} = 1.$$

Moreover, if $x_1 \notin R_1$, then $(x_1, b) \notin R_{X_2, X_2}$ and $p_{X_1, X_2}(x_1, b) = 0$. Thus p_b is a probability mass function whose range is a subset of R_1 and that depends on b, R_1, and the joint mass function $p_{X_1 X_2}$. We define p_b to be the *conditional probability mass function for X_1 given $X_2 = b$*. It is traditional to write $P_b(x_1)$ as $p(x_1 | X_2 = b)$, as $p_{X_1 | X_2}(x_1 | b)$, or, simply, as $P(x_1 | b)$. This expression may be read, "the conditional probability that the value of X_1 is x_1 given that the value of X_2 is b." We observe that there are as many conditional probability mass functions for X_1 as there are elements in R_2 (possible values of b).

Similarly, for each $a \in R_1$, the formula

$$p_{X_2 | X_1}(x_2 | a) = p(x_2 | X_1 = a) = P(X_2 = x_2 | X_1 = a) = \frac{p_{X_1, X_2}(a, x_2)}{p_{X_1}(a)} \quad (5.7)$$

defines a probability mass function indexed by a, the *conditional probability mass function for X_2 given $X_1 = a$*. We sometimes write $p(x_2 | a)$ in Eq. (5.7) when it is clear from the context which variable is associated with a. There are as many conditional mass functions defined by Eq. (5.7) as there are elements in R_1.

EXAMPLE 5.27   Consider again the experiment of reading the binary contents of two two-bit registers (Example 5.2 of Section 5.2), with the underlying 16-element sample space $S = \{(00, 00), (00, 01), \ldots, (11, 10), (11, 11)\}$. Again, let X denote the decimal outcome in the low-order register and Y denote the larger decimal outcome of the two registers. The joint mass function, together with the two marginals of X and Y, are given in the accompanying table.

$p_Y(y)$	y		x	0	1	2	3
$\frac{7}{16}$	3			$\frac{1}{16}$	$\frac{1}{16}$	$\frac{1}{16}$	$\frac{4}{16}$
$\frac{5}{16}$	2			$\frac{1}{16}$	$\frac{1}{16}$	$\frac{3}{16}$	0
$\frac{3}{16}$	1			$\frac{1}{16}$	$\frac{2}{16}$	0	0
$\frac{1}{16}$	0			$\frac{1}{16}$	0	0	0
			$p_X(x)$	$\frac{4}{16}$	$\frac{4}{16}$	$\frac{4}{16}$	$\frac{4}{16}$

Now consider the two events $[Y = 1]$ and $[X = 0]$. From the marginal distributions, $P(Y = 1) = \frac{3}{16}$, while $P(X = 0) = \frac{4}{16}$. From the definition of conditional probability,

$$P(Y = 1 | X = 0) = \frac{P(X = 0, Y = 1)}{P(X = 0)} = \frac{1}{16} \div \frac{4}{16} = \frac{1}{4}.$$

Interestingly, the event $[Y = 1]$ is more likely to occur in the presence of the information that $[X = 0]$ occurs than without it. What about the other possible values of Y? It is easy to verify that $P(Y = 0|X = 0) = \frac{1}{4}$, $P(Y = 2|X = 0) = \frac{1}{4}$, and $P(Y = 3|X = 0) = \frac{1}{4}$, so that the *conditional probabilities* given the event $[X = 0]$ follow a discrete uniform model, in contrast to the *unconditional probabilities* (as reflected by the marginal probabilities). The collection of conditional probabilities just computed (all with respect to the same conditioning event, $[X = 0]$) defines a probability mass function on the set $R_Y = \{0, 1, 2, 3\}$. It can be used to compute conditional probabilities *of events*, just as $p_Y(y)$ is used to compute unconditional probabilities. For example, if A is the event "largest register content is less than 2," that is, $A = [Y = 0] \cup [Y = 1]$, then $P(A) = p_Y(0) + p_Y(1) = \frac{1}{2}$, while

$$P(A|X = 0) = P(Y = 0|X = 0) + P(Y = 1|X = 0)$$
$$= p_{Y|X}(0|0) + p_{Y|X}(1|0) = \tfrac{1}{2}.$$

There are four conditional mass functions for Y in Example 5.27, one for each possible fixed value of X in $\{0, 1, 2, 3\}$. Mechanically, the values in each of these conditional mass functions are found by dividing the quantities in the joint mass function table column corresponding to the given x by X's marginal mass function at that given value. In a similar fashion, dividing each row entry corresponding to a given value y of Y by the marginal entry for y generates the conditional mass function of X given that particular y. For example, for $y = 1$,

$$p(x|1) = \begin{cases} \frac{1}{3} & \text{if } x = 0, \\ \frac{2}{3} & \text{if } x = 1, \\ 0 & \text{otherwise.} \end{cases}$$

All these mass functions can be displayed in a table of all conditional and marginal distributions, as shown below:

y	$p(y\|3)$	$p(y\|2)$	$p(y\|1)$	$p(y\|0)$	$p_Y(y)$	y	$x=0$	$x=1$	$x=2$	$x=3$
1	$\frac{1}{4}$	$\frac{1}{4}$	$\frac{1}{4}$	$\frac{7}{16}$		3	$\frac{1}{16}$	$\frac{1}{16}$	$\frac{1}{16}$	$\frac{4}{16}$
0	$\frac{3}{4}$	$\frac{1}{4}$	$\frac{1}{4}$	$\frac{5}{16}$		2	$\frac{1}{16}$	$\frac{1}{16}$	$\frac{3}{16}$	0
0	0	$\frac{1}{2}$	$\frac{1}{4}$	$\frac{3}{16}$		1	$\frac{1}{16}$	$\frac{2}{16}$	0	0
0	0	0	$\frac{1}{4}$	$\frac{1}{16}$		0	$\frac{1}{16}$	0	0	0

x	$x=0$	$x=1$	$x=2$	$x=3$
$p_X(x)$	$\frac{4}{16}$	$\frac{4}{16}$	$\frac{4}{16}$	$\frac{4}{16}$
$p(x\|0)$	1	0	0	0
$p(x\|1)$	$\frac{1}{3}$	$\frac{2}{3}$	0	0
$p(x\|2)$	$\frac{1}{5}$	$\frac{1}{5}$	$\frac{3}{5}$	0
$p(x\|3)$	$\frac{1}{7}$	$\frac{1}{7}$	$\frac{1}{7}$	$\frac{4}{7}$

EXAMPLE 5.28 Suppose (X, Y) and (U, V) are discrete random pairs with mass functions as given in the accompanying tables.

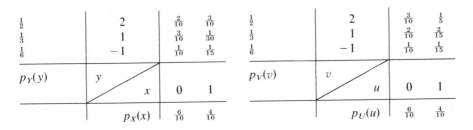

$p_Y(y)$	y	x	0	1
$\frac{1}{2}$	2		$\frac{2}{10}$	$\frac{3}{10}$
$\frac{1}{3}$	1		$\frac{3}{10}$	$\frac{1}{30}$
$\frac{1}{6}$	-1		$\frac{1}{10}$	$\frac{1}{15}$
		$p_X(x)$	$\frac{6}{10}$	$\frac{4}{10}$

$p_V(v)$	v	u	0	1
$\frac{1}{2}$	2		$\frac{3}{10}$	$\frac{1}{5}$
$\frac{1}{3}$	1		$\frac{2}{10}$	$\frac{2}{15}$
$\frac{1}{6}$	-1		$\frac{1}{10}$	$\frac{1}{15}$
		$p_U(u)$	$\frac{6}{10}$	$\frac{4}{10}$

Then some of the possible conditional mass functions are as shown next:

x	$p_{X|Y}(x|1)$	y	$p_{Y|X}(y|0)$	u	$p_{U|V}(u|-1)$	v	$p_{V|U}(v|0)$
0	$\frac{9}{10}$	-1	$\frac{1}{6}$	0	$\frac{6}{10}$	-1	$\frac{1}{6}$
1	$\frac{1}{10}$	1	$\frac{1}{2}$	1	$\frac{4}{10}$	1	$\frac{1}{3}$
		2	$\frac{1}{3}$			2	$\frac{1}{2}$

While X and U, and Y and V, are identically distributed, the conditional distributions of Y given $X = 0$ and V given $U = 0$ are not the same. Also, the conditional mass functions $p_{U|V}$ and $p_{V|U}$ are the same as the marginal mass functions p_U and p_V.

The fact that $p_{U|V}$ is the same as p_U in Example 5.28 is true because, in this case, U and V are independent. You can verify this independence by showing that the joint mass values are, in each case, equal to the product of the corresponding marginal mass values. Suppose, in general, that Z and W are independent with joint mass function $p(z, w)$. Then for any w for which $p_W(w) > 0$,

$$p_{Z|W}(z|w) = \frac{p(z, w)}{p_W(w)} = \frac{p_Z(z) \cdot p_W(w)}{p_W(w)} = p_Z(z),$$

so if the random variables are independent, the conditional mass functions are the same as the marginal. In Exercise 5.72, you are asked to show that the converse is also true.

EXAMPLE 5.29 Suppose X is binomial (n, p) and Y is binomial (m, p) where X and Y are independent. The conditional distribution of X, given

$X + Y = z$, is hypergeometric. This useful connection between the binomial and hypergeometric distributions is established as follows:

$$p_{X|X+Y}(x|z) = P(X = x|X + Y = z) = \frac{P(X = x, X + Y = z)}{P(X + Y = z)}$$

$$= \frac{P(X = x, Y = z - x)}{P(X + Y = z)} = \frac{p_X(x) \cdot p_Y(z - x)}{p_{X+Y}(z)}.$$

Since X and Y are independent, $X + Y$ is binomial $(n + m, p)$, as we will show formally in Chapter 6. This result may be argued informally as follows: If $X + Y$ is the number of successes in $n + m$ independent Bernoulli trials, with constant probability p of success on each trial, then $X + Y$ is binomial $(n + m, p)$. But this result seems reasonable if X is the number of successes in n independent trials and Y is the number of successes in m additional independent trials. It then follows that $p_{X|X+Y}(x|z)$ is given by

$$\frac{\binom{n}{x}p^x q^{n-x}\binom{m}{z-x}p^{z-x}q^{m-z+x}}{\binom{n+m}{z}p^z q^{n+m-z}} = \frac{\binom{n}{x}\binom{m}{z-x}}{\binom{n+m}{z}},$$

which is the hypergeometric mass function.

If X_1 and X_2 are jointly continuous, then the preceding analysis does not apply directly since the values of density functions are not probabilities. We can, however, proceed analogously, as follows: Let (X_1, X_2) have joint density f, and suppose $f_{X_1}(x_1) > 0$, that is, $x_1 \in R_{X_1}$. We can define a function $f_{x_1}(x_2)$, indexed by x_1, by means of the equation

$$f_{x_1}(x_2) = \frac{f(x_1, x_2)}{f_{X_1}(x_1)}.$$

It is easy to see that f_{x_1} is a nonnegative function, and

$$\int_{-\infty}^{\infty} f_{x_1}(x_2)\, dx_2 = \frac{1}{f_{X_1}(x_1)} \int_{-\infty}^{\infty} f(x_1, x_2)\, dx_2 = \frac{f_{X_1}(x_1)}{f_{X_1}(x_1)} = 1.$$

Thus f_{X_1} is a density function, which we will shortly define to be the conditional density of X_2 given $X_1 = x_1$. Since the probability of the conditioning event is zero, let us first discuss an interpretation of this function.

Suppose f is positive and continuous in a neighborhood of (x_1, x_2). The conditional probability that X_2 falls in the interval $(x_2, x_2 + \Delta x_2)$, given X_1 is in $[x_1, x_1 + \Delta x_1)$, is, for small positive Δx_1 and Δx_2, given approximately by

$$P(x_2 < X_2 < x_2 + \Delta x_2 | x_1 \leq X_1 < x_1 + \Delta x_1)$$

$$= \frac{P(x_1 \leq X_1 < x_1 + \Delta x_1, x_2 < X_2 < x_2 + \Delta x_2)}{P(x_1 \leq X_1 < x_1 + \Delta x_1)}$$

$$\approx \frac{f(x_1, x_2) \, \Delta x_1 \, \Delta x_2}{f_{X_1}(x_1) \, \Delta x_1} = f_{x_1}(x_2) \, \Delta x_2.$$

Thus letting Δx_1 go to zero, we see that $f_{x_1}(x_2)$ can be interpreted as the conditional density function of X_2, given $[X_1 = x_1]$. Since this conditioning event has probability zero, the direct definition of conditional probability could not be used, as in the discrete case. Because of the above interpretation of $f_{x_1}(x_2)$, and by analogy with the discrete case, $f_{x_1}(x_2)$ is traditionally written $f_{X_2|X_1}(x_2|x_1)$, or $f(x_2|X_1 = x_1)$, or simply $f(x_2|x_1)$, and is called the *conditional density of X_2 given $X_1 = x_1$*.

Definition 5.4   Let X_1 and X_2 be jointly continuous with joint density $f(x_1, x_2)$. The functions defined, for fixed $x_1 \in R_{X_1}$, by

$$f(x_2|x_1) = \frac{f(x_1, x_2)}{f_{X_1}(x_1)}$$

and, for fixed $x_2 \in R_{X_2}$, by

$$f(x_1|x_2) = \frac{f(x_1, x_2)}{f_{X_2}(x_2)}$$

are, respectively, the *conditional density of X_2 given $X_1 = x_1$ and the conditional density of X_1 given $X_2 = x_2$*.

EXAMPLE 5.30   Suppose an experiment consists of drawing a number X_1 at random from $(0, 1)$ and then, given the value x_1 of X_1, drawing a number X_2 at random from $(0, x_1)$. Then X_1 is distributed uniformly over $(0, 1)$, and it is reasonable, from the description, to suppose that $f(x_2|x_1) = 1/x_1$ for $0 < x_2 \leq x_1$. That is, the conditional density function of X_2 given x_1 is uniform over $(0, x_1)$. Accordingly, from Definition 5.4,

$$f(x_1, x_2) = f(x_2|x_1)f_{X_1}(x_1) = \frac{1}{x_1} \qquad \text{for } 0 < x_2 \leq x_1 \leq 1.$$

It then follows that the marginal density of X_2 is $f_{X_2}(x_2) = -\log x_2$ for $0 < x_2 < 1$. Further, it follows that

$$f(x_1|x_2) = \frac{-1}{x_1 \log x_2} \qquad \text{for } 0 < x_2 < x_1 < 1$$

is the conditional density of X_1 given x_2.

As before, we observe that there are as many conditional density functions as there are points in R_{X_1} plus those in R_{X_2}. Suppose, once again, that $x_1 \in R_{X_1}$. Since the conditional density of X_2 given x_1 is a true density function, it possesses a mean value that usually depends on x_1. This mean value is traditionally denoted by $E(X_2|x_1)$ [or by $E(X_2|X_1 = x_1)$ when needed for clarity] and is called the *conditional expected value of X_2 given x_1 (or given $X_1 = x_1$)*. More generally, for any function ϕ defined on R_{X_2}, we define the *conditional expectation of the random variable* $\phi(X_2)$, *given* x_1, by

$$E[\phi(X_2)|x_1] = \int_{-\infty}^{\infty} \phi(x_2) f_{X_2|X_1}(x_2|x_1)\, dx_2. \tag{5.8}$$

Note that since Eq. (5.8) is valid for each fixed $x_1 \in R_{X_1}$, we can interpret it as a function x_1 with domain R_{X_1}. We will later examine such functions.

EXAMPLE 5.31   For the double-uniform-drawing situation discussed in Example 5.30,

$$E(X_2|x_1) = \int_0^{x_1} x_2 \cdot \frac{1}{x_1}\, dx_2 = \frac{x_1}{2},$$

and, similarly, $$E(X_2^2|x_1) = \int_0^{x_1} \frac{x_2^2}{x_1}\, dx_2 = \frac{x_1^2}{3},$$

for each $x_1 \in (0, 1) = R_{X_1}$.

EXAMPLE 5.32   If X and Y are binomial random variables as in Example 5.29, and we let $Z = X + Y$, then

$$E(X|z) = \sum_x x \cdot \frac{\binom{n}{x}\binom{m}{z-x}}{\binom{n+m}{z}} = \frac{nz}{n+m},$$

the mean of the hypergeometric distribution involved.

As a special instance, we may define, for fixed x_1, $\phi(x_2) = [x_2 - E(X_2|x_2)]^2$. In that case, with $E(X_2|x_1)$ constant (for fixed x_1), it is quite reasonable to call

$$E[\phi(X_2)|x_1] = E\{[X_2 - E(X_2|x_1)]^2|x_1\}$$

the *conditional variance of X_2 given x_1*. This variance is sometimes denoted by $V(X_2|x_1)$. It is the variance of the conditional distribution (density or mass function) of X_2 given x_1.

EXAMPLE 5.33   Again, for the double-uniform-drawing situation of Example 5.30,

$$V(X_2|x_1) = \int_0^{x_1} \left(x_2 - \frac{x_1}{2} \right)^2 \frac{1}{x_1} \, dx_2 = \frac{x_1^2}{12}.$$

Alternatively, we may compute

$$E(X_2^2|x_1) - E^2(x_2|x_1) = \frac{x_1^2}{3} - \frac{x_1^2}{4} = \frac{x_1^2}{12}$$

(see Example 5.31 and Exercise 5.74).

The following two theorems concerning conditional expectation will be useful in the work that follows. One is proved for the continuous case and the other for the discrete case. You are asked to supply proofs for the remaining cases in Exercise 5.75. We have observed that $E(Y|x)$ may be thought of as a function of x. Then $E(Y|X)$ is the composition of this function with the (function) X and, hence, is a random variable. The following theorem, concerning the mean of this random variable, is extremely useful.

Theorem 5.9   Suppose (X, Y) has density $f_{(X,Y)}$, and let $E(Y|x)$ be viewed as a function $\phi(x)$. Then $\phi(X) = E(Y|X)$ is a random variable with mean $E(Y)$. That is, $E[E(Y|X)] = E(Y)$.

PROOF   By definition,

$$E[E(Y|X)] = \int_{R_X} E(Y|x) f_X(x) \, dx$$

$$= \int_{R_X} \left[\int_{-\infty}^{\infty} y \cdot f_{Y|X}(y|x) \, dy \right] f_X(x) \, dx$$

$$= \int_{R_X} \int_{-\infty}^{\infty} y \, \frac{f_{(X,Y)}(x, y)}{f_X(x)} f_X(x) \, dy \, dx$$

$$= \int_{-\infty}^{\infty} \int_{-\infty}^{\infty} y \cdot f_{(X,Y)}(x, y) \, dy \, dx = E(Y).$$

EXAMPLE 5.34   Consider once again the double-uniform-drawing experiment of Example 5.30. In Example 5.31, we found $E(X_2|x_1) = x_1/2$, so $E[E(X_2|X_1)] = E(X_1/2) = \frac{1}{4}$. As a check, we saw that the marginal density of X_2 is $f_{X_2}(x_2) = -\log x_2$ for $0 < x_2 < 1$, so

$$E(X_2) = \int_0^1 x_2(-\log x_2)\, dx_2 = \tfrac{1}{4},$$

as before.

We have remarked that the problem of finding the joint distribution of random variables having known properties is usually a difficult problem, unless the random variables happen to be independent. The problem can also be solved easily if the marginal distribution of one of the random variables and the conditional distribution of the other are known. For example, if X has known density $f_X(x)$ and the conditional density of Y given x, $f_{Y|X}(y|x) = f_{X,Y}(x, y)/f_X(x)$, is known, then it follows that the joint density is the product of the marginal and conditional,

$$f_{X,Y}(x, y) = f_X(x) \cdot f_{Y|X}(y|x). \tag{5.9}$$

EXAMPLE 5.35   Recall the example in which microprocessor chips selected from a given manufactured batch have fixed reliability r [so $P(X = 1) = r$, where X is a Bernoulli random variable representing a defective or non-defective chip when $X = 0$ or 1, respectively]. The reliability varies over batches, and the reliability value for a randomly selected batch is modeled as an outcome on a random variable R having a beta distribution. The joint distribution of (X, R) was given (somewhat awkwardly, since X is discrete and R is continuous) as a product of a Bernoulli mass function times the beta density of R. This example is an application of Eq. (5.9). The Bernoulli distribution is the conditional distribution of X, given the reliability $R = r$. Recall that the marginal distribution of X turned out also to be Bernoulli but with a success parameter involving the parameters of the distribution of R.

In applications such as the one in the preceding example, the distribution of reliability over batches is sometimes called a *prior distribution*; the conditional distribution of R given the outcome on X is called the *posterior distribution* of reliability. This terminology is consistent with that used in connection with Bayes's formula, as we will now demonstrate. Suppose it is desired to find the (posterior) distribution of reliability in the selected batch of chips, given that one chip from the batch has been tested and found to be nondefective ($X = 1$). For small Δr,

$$f_{R|X}(r|1)\, \Delta r \approx P(r < R \leq R + \Delta r | X = 1) = \frac{P(r < R \leq r + \Delta r, X = 1)}{P(X = 1)}$$

$$= \frac{P(X = 1 | r < R \leq r + \Delta r) \cdot P(r < R \leq r + \Delta r)}{P(X = 1)}.$$

The numerator in this expression is approximately $r^1(1 - r)^0 \cdot f_R(r)\Delta r$. The denominator can be found as follows: Consider a partition of the interval $[0, 1]$ into small intervals of width Δr; let the ith subinterval be denoted by $(r_i, r_{i+1}]$. Now

$$P(X = 1) = \sum_i P(X = 1|r_i < R \leq r_{i+1}) \cdot P(r_i < R \leq r_{i+1})$$

$$\approx \sum_i r_i \cdot f_R(r_i) \, \Delta r.$$

We thus have

$$f_{R|X}(r|1) \, \Delta r \approx \frac{r \cdot f_R(r) \, \Delta r}{\sum_i r_i f_R(r_i) \, \Delta r}.$$

Dividing both sides by Δr and then taking a limit as Δr tends to zero yields

$$f_{R|X}(r|1) = \frac{r \cdot f_R(r)}{\int_0^1 r f_R(r) \, dr}, \qquad 0 < r < 1.$$

If the prior distribution of R is beta (v, λ), the posterior distribution of the reliability of a batch, given that a chip from the batch is tested and found to be nondefective, is beta $(v + 1, \lambda)$. Thus testing an item and finding it to be nondefective has moved the mean reliability upward from $v/(v + \lambda)$ to $(v + 1)/(v + 1 + \lambda)$; had the test uncovered a defective item, the mean reliability would have decreased from $v/(v + \lambda)$ to $v/(v + \lambda + 1)$.

This type of reasoning, involving Bayes's formula, is often used to update a model, given some information such as the outcome on X. It can be carried out also when both random variables are continuous or both are discrete. (See Exercise 5.71.)

EXAMPLE 5.36 Suppose a unit amount of rocket propellant, if ignited at time x, moves the rocket a random amount Y (measured in some suitable units). But suppose x itself is the observed value of a random variable X, which is uniform over the interval $(0, 10)$. Suppose the state of the art concerning rocket propulsion makes it reasonable to assume $P(Y \leq y|X = x) = 1 - e^{-xy}$, an exponential model. It is desired to find $P(Y > 1)$. Using the expectation of the conditional probability of $[Y > 1]$ given X, we can calculate immediately

$$P(Y > 1) = \int_{-\infty}^{\infty} P(Y > 1|X = x)f_X(x) \, dx = \frac{1}{10} \int_0^{10} e^{-x} \, dx$$

$$= -\frac{1}{10} \left[e^{-x} \right]_0^{10} = \frac{1 - e^{-10}}{10} = .10.$$

We have evaluated $P(Y > 1)$ without explicitly finding the distribution of Y. If the latter is desired, we can proceed as follows:

$$F_Y(y) = E[F(y|X)] = \frac{1}{10} \int_0^{10} (1 - e^{-xy}) \, dx = \frac{1}{10} \left[x + \frac{1}{y} (e^{-xy}) \right]_0^{10}$$

$$= 1 + \frac{1}{10y} e^{-10y} - \frac{1}{10y}.$$

Differentiation yields the marginal density for Y.

In the preceding example, the probability of an event $[Y > 1]$ was found by taking the expectation of the conditional probability $P(Y > 1|X)$. There are also useful applications for conditional distributions of random variables given an event A.

EXAMPLE 5.37 Suppose X is $N(0, 1)$ and $E = (0, \infty)$. Then the conditional CDF of X, given E, could be written as $\Phi(x|E)$ or $\Phi(x|X > 0)$, and it is given by

$$\Phi(x|X > 0) = \frac{P(X \le x, X > 0)}{P(X > 0)} = \begin{cases} 0 \text{ if } x \le 0, \\ 2\Phi(x) \text{ if } x > 0. \end{cases}$$

The corresponding conditional density is given by

$$f_X(x|X > 0) = \frac{\sqrt{2}}{\sqrt{\pi}} e^{-(1/2)x^2} \qquad \text{for } x > 0.$$

EXAMPLE 5.38 —*Exponential Distributions are Memoryless*   Suppose Y is distributed exponential (λ) and $E = (a, \infty)$. Then the conditional CDF of Y, given $Y > a$, is obtained as follows:
If $y > a$, then

$$F_Y(y|E) = P(Y \le y|Y > a) = P(a < Y \le y|Y > a) = \frac{P(a < Y \le y)}{P(Y > a)}$$

$$= \frac{1 - e^{-\lambda y} - (1 - e^{-\lambda a})}{e^{-\lambda a}} = \frac{e^{-\lambda a}[1 - e^{-\lambda(y - a)}]}{e^{-\lambda a}}.$$

Thus

$$F(y|Y > a) = 1 - e^{-\lambda(y - a)} \qquad \text{for } y > a. \tag{5.10}$$

This expression is just an exponential (λ) CDF translated a units to the right. The interpretation of this fact is quite interesting. Suppose a light bulb has an exponentially distributed lifetime. If we begin observing the bulb while it is in operation at some time after it has been put into service, the distribution of the remaining lifetime of the bulb is the same as the original lifetime distribution! {Here it is assumed that observation begins at some time a and that time a is taken to be a new origin [replace $y - a$ by x in Eq. (5.10)] for the conditional experiment.} Thus in assuming an exponential model, we assume that the device does not wear or experience fatigue. Its lifetime distribution (distribution of time to failure) remains the same from any new time at which we find the device still operating. For this reason, the exponential distributions are sometimes called *memoryless*; devices having such life distributions do not ''remember'' (take into account) the length of time they have operated, as long as they are still operating. The converse of the implication in this example is also true, so that the exponential is the only memoryless continuous distribution family.

We end this section with two examples that show how the preceding results may be useful in finding probabilities and expected values, especially in contexts that involve independent repeated trials.

EXAMPLE 5.39 —*Mean of a Geometric Distribution Using Difference Equations*   Suppose X denotes the number of independent attempts at a task that is required to observe the first success (say 1). Under these circumstances, it may be reasonable to assume that X is geometric (p), where p is the probability of a 1 on each attempt. It follows, as previously seen, that $E(X) = 1/p$.

However, consider the following argument for establishing this result: Let A denote the event ''a 1 occurs on the first attempt.'' Then $E(X|A) = 1$. On the other hand, if the first attempt results in 0, it is as if we were beginning the experiment over again, so that the expected number of tosses required (in addition to the first) is $E(X)$. Thus we have $E(X|\overline{A}) = 1 + E(X)$. Since $E(X) = E(X|A) \cdot P(A) + E(X|\overline{A}) \cdot P(\overline{A})$, we have

$$E(X) = 1 \cdot P(A) + [1 + E(X)] \cdot P(\overline{A}) = p + [1 + E(X)]q.$$

Thus $E(X)(1 - q) = p + q = 1$; whence $E(X) = 1/(1 - q) = 1/p$.

EXAMPLE 5.40   A taxi repeatedly circles past a certain point, using one of two routes. Route 1 requires 4 minutes to complete and is used with probability .6, while route 2 takes 7 minutes and is used with probability .4. Assume that the taxi has probability .1 of getting a fare at the point each time it passes and that this is the only source of fares for the taxi. Given that the empty taxi has just passed the point, what is the expected length of time required before the taxi gets a fare?

SOLUTION   We condition on the routes taken, as follows: Let T denote the length of the time required to get a fare, let I denote the event "route 1 taken," and let F denote the event "fare obtained." If route 1 is taken on the first loop, the taxi either gets a fare (with probability .1) or it does not, requiring 4 minutes. Given that a fare is obtained on the first return, the expected time required is 4: $E(T|I \cap F) = 4$. If no one is waiting when the taxi first returns, then $E(T|I \cap \bar{F}) = 4 + E(T)$. Similarly, $E(T|\bar{I} \cap F) = 7$ and $E(T|\bar{I} \cap \bar{F}) = 7 + E(T)$. Thus

$$E(T) = [E(T|I \cap F) \cdot P(I \cap F)] + [E(T|I \cap \bar{F}) \cdot P(I \cap \bar{F})]$$
$$+ [E(T|\bar{I} \cap F) \cdot P(\bar{I} \cap F)] + [E(T|\bar{I} \cap \bar{F}) \cdot P(\bar{I} \cap \bar{F})].$$

If the events I and F are assumed to be independent, this expression reduces to

$$E(T) = 4(.6)(.1) + [4 + E(T)](.6)(.9) + 7(.4)(.1)$$
$$+ [7 + E(T)](.4)(.9)$$
$$= 5.2 + .9E(T),$$

or $E(T) = 52$ minutes.

EXERCISES

THEORY

5.71   Suppose X has density f_X, and the conditional density of Y given $X = x$ is $g(y|x)$.
   a. Find the joint distribution of (X, Y).
   b. Find the marginal distribution of Y.
   c. Find the conditional distribution of X given $Y = y$.
   d. Interpret the expression in part (c) in terms of Bayes's formula, where f_X is the prior, and the conditional in part (c) is the posterior distribution of X, given $Y = y$.

5.72   Show that, if (X, Y) has mass function $p_{(X,Y)}$ and range $R_X \times R_Y$, and for each $j \in R_Y$, $p_{X|Y}(x|j) = p_X(x)$ for all $x \in R_X$, then X and Y are independent. *Note:* This exercise gives yet another characterization of independence, inasmuch as the above statement is also true for the continuous case, and the converse is stated in the paragraph following Example 5.28.

5.73   Suppose (X, Y) induces the general bivariate normal density function (see Exercise 5.22 of Section 5.3). Show that

$$f_{X|Y}(x|y) = \frac{1}{\sqrt{2\pi}\,\sigma_1 \sqrt{1 - \rho^2}}$$
$$\exp\left[-\frac{1}{2(1 - \rho^2)\sigma_1^2} \left\{ x - \left[\mu_1 + \rho\left(\frac{\sigma_1}{\sigma_2}\right)(y - \mu_2) \right] \right\}^2 \right].$$

Hence, the conditional distribution of X given y is

$$N\left(\mu_1 + \rho\left(\frac{\sigma_1}{\sigma_2}\right)(y - \mu_2), (1 - \rho^2)\sigma_1^2\right),$$

whereas the marginal distribution of X is $N(\mu_1, \sigma_1^2)$.

5.74   a. Show that $V(Y|x) = E\{[Y - E(Y|x)]^2|x\}$ may be calculated by $V(Y|x)$
$= E(Y^2|x) - E^2(Y|x)$.
b. Show that $V(Y)$ may be calculated by $V(Y) = E[V(Y|X)] + V[E(Y|X)]$.

5.75   State and prove a theorem analogous to Theorem 5.9 for the discrete case.

5.76   Suppose (X, Y) has mass function $p_{(X,Y)}$, i is in R_X, and $A = \{j_1, j_2, \ldots, j_n\} \subset R_Y$.
a. Show that $P(X = i|Y \in A) = \sum_{k=1}^n p_{(X,Y)}(i, j_k)/\sum_{k=1}^n p_Y(j_k)$.
b. Show that

$$P(Y \in A|X = i) = \frac{\sum_{k=1}^n p_{(X,Y)}(i, j_k)}{p_X(i)} = \sum_{y \in A} p_{Y|X}(y|i).$$

5.77   Show that $E[g(X)|x] = g(x)$.

5.78   Suppose X and Y are jointly discrete.
a. Show that $P(X = i, Y = k|Y = j) = 0$ if $j \neq k$.
b. Show that $P(X = i, Y = j|Y = j) = p_{X|Y}(i|j)$.

5.79   Suppose Y_k has a negative binomial distribution with parameters k and p. Find the mean of Y_k by using a difference equation. *Hint:* Let X denote the outcome of the first trial, and let $p_k = E(Y_k)$. Argue that

$$p_k = E(Y_k|X = 1)P(X = 1) + E(Y_k|X = 0)P(X = 0)$$

$$= p_{k-1}p + (1 + p_k)q.$$

Note that $p_1 = E(X - 1)$, where X is distributed geometric (p).

5.80   Suppose X is a continuous random variable with density f. Show that $E(X|X > a) = \int_a^\infty xf(x)\, dx/\int_a^\infty f(x)\, dx$.

5.81   Show that the geometric is a family of discrete memoryless distributions; that is, if X is geometric (p), then $P(X > n + m|X > n) = P(X > m)$.

5.82   a. Show that $E(a - X|X \leq a)F(a) \geq a - \mu$.
b. Show that $E[(a - X)^2|X \leq a]F(a) \leq \sigma^2 + (a - \mu)^2$. *Hint:* $E(a - X) = E(a - X|X \leq a)P(X \leq a) + E(a - X|X > a)P(X > a)$.

5.83   This exercise concerns mixtures of distributions. Suppose a compound experiment consists of first selecting, by the outcome on a Bernoulli (p) random variable Z, which population to sample: population "zero" associated with the continuous random variable X or population "one" associated with a continuous random variable Y. Let T be the outcome on the compound experiment. Assume X, Y, and Z are independent.
a. Show that $f_T(t) = p \cdot f_X(t) + (1 - p)f_Y(t)$, and interpret this expression as an expectation of the conditional distribution of T given Z. *Note:* T is called a *mixture* of X and Y.

b. Find similar expressions for F_T and the moment-generating function of T.

c. Generalize these results to cover a random choice among k populations.

APPLICATION

5.84 In the microprocessor chip example, suppose 10 items are sampled from the selected batch and it is found that 8 are nondefective and 2 are defective. Suppose the prior distribution of reliability is beta $(4, 7)$. Find the posterior mean reliability, given 8 successes in 10 tests.

5.85 Suppose X denotes the outcome of one fair die and Y denotes the outcome of another. It is known that the total number of spots showing on a particular trial was 6. Find the (conditional) probabilities that (a) $X = 3$; (b) $X \le 3$; (c) $X < Y$.

5.86 For Example 5.11 of Section 5.3, verify that if $x_1 < -c$ or $x_1 > c$, $f(x_2|x_1) = 2\phi(x_2)$ for $x_2 > 0$, while if $-c < x_1 < c$, $f(x_2|x_1) = 2\phi(x_2)$ for $x_2 < 0$, where ϕ is the $N(0, 1)$ density.

5.87 a. For Example 5.30, compute the conditional mean of X_2 given any x_1 between 0 and 1. Repeat for X_1 given x_2.

b. Compute $P(\frac{1}{2} \le X_1 \le \frac{2}{3}|x_2 = \frac{1}{2})$.

c. Find $V(X_1|x_2)$.

5.88 a. For Example 5.28, verify that each conditional mass function associated with U and V is the same as the corresponding marginal.

b. Compute $V(X|Y = i)$ for $i = -1, 1, 2$.

c. Using the results of part (b), calculate $\sum_{i \in R_Y} V(X|Y = i)p_Y(i) = E[V(X|Y)]$ and compare with $V(X)$.

5.89 Suppose $f_{(X,Y)}(x, y) = 2$, $x \ge 0$, $y \ge 0$, $x + y \le 1$.

a. Find $f_{X|Y}(x|y)$.   b. Find $E(X|y)$.

5.90 Suppose (X_1, X_2) has density $f(x_1, x_2) = 1/\pi$, $x_1^2 + x_2^2 \le 1$.

a. Find $f_{X_1|X_2}(x_1|x_2)$.   b. Find $E(X_1 + X_2|x_2)$.   c. Find $V(X_2|X_1 = -.5)$.

5.91 Suppose (X, Y) has a general bivariate normal distribution with parameters $\mu_X, \mu_Y, \sigma_X^2, \sigma_Y^2, \rho$. *Hint:* See Exercise 5.73.

a. Find $E(X|y)$.   b. Find $V(X|y)$.

5.92 Suppose X is $N(0, 1)$, and the conditional distribution of Y, given $X = x$, is $N(\rho x, 1 - \rho^2)$. Find the conditional distribution of X given $Y = y$.

5.93 Suppose (X, Y) induces the joint mass function listed in the accompanying table.

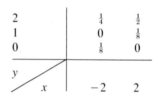

2		$\frac{1}{4}$	$\frac{1}{2}$
1		0	$\frac{1}{8}$
0		$\frac{1}{8}$	0
y	x	-2	2

a. Show that X and Y are not independent.

    b. Find $p_{X|Y}(x|2)$.   c. Find $E(X|Y = 2)$.   d. Find $F(x|Y < 2)$.
    e. Find $E(X|Y < 2)$.   f. Find $V(Y|X = -2)$.

5.94   The number of accidents occurring in a factory in a week is a random variable with mean μ and variance σ^2. The numbers of individuals injured in single accidents are independently distributed, each with mean ν and variance τ^2. Find the mean and variance of the number of individuals injured in a week, assuming that the number of individuals injured per accident is independent of the number of accidents.

5.95   a. For what values of x is the argument of Example 5.29 valid?
    b. What are the possible values of the parameters and variable in the hypergeometric mass function of Example 5.29?

5.96   Suppose the number of particles from a radioactive source arriving at a counter during a 1-second period is Poisson (4) distributed. Assume, in addition, that each particle that arrives at the counter is actually counted only with probability p, independent of other arrivals or counts. Find the distribution of the number of particles counted in 1-second period. *Hint:* The conditional distribution of the number of counts, given n arrivals, is binomial (n, p).

5.97   Show that, in Example 5.30, the conditional distributions are not the same as the marginals, and conclude that the random variables involved are therefore dependent.

5.98   Display all the conditional density functions associated with the joint density function in the accompanying table. Are X and Y independent? Identically distributed?

1	.40	.01	.01
0	.01	.20	.05
-1	.01	.01	.30

    y

         x -1 0 1

5.99   Independent and dependent repeated trials are considered here. Suppose an urn contains three balls, numbered 1, 2, and 3, and a sample of size 2 is drawn from the urn. Let X denote the result of the first draw and Y the result of the second.
    a. Find the joint mass function of X and Y when sampling is done *with* replacement. Also, find the marginals and a conditional.
    b. Repeat part (a) but for sampling done *without* replacement.
    c. The marginal distributions are the same in the two cases in parts (a) and (b), whereas the conditionals are not. Explain.
    d. In the with-replacement case, the marginals and conditionals are the same; whereas in the nonreplacement case, they are not. Explain.
    e. Generalize the results in parts (c) and (d) to the case of sampling m items from an urn containing n items ($m \leq n$).

5.100   A ring holds n keys; keys are selected at random, one at a time, and are tried in a lock. Exactly one of the keys fits the lock. Find the expected number of tries until the correct key is found, (a) if selection is with replacement; (b) if selection is without replacement. (c) Examine the difference in parts (a) and (b) as n gets large.

5.101 Suppose n addressed letters are randomly inserted in envelopes with the same set of addresses.
   a. Find the expected number of letters that are placed in envelopes with the correct corresponding envelope.
   b. Discuss the solution to part (a) as n varies.

5.102 Suppose X and Y have joint density $f_{X,Y}(x, y) = 2x/(y \ln 2)$, $0 < x < 1$, $1 < y < 2$. Find expressions for $f_{X|Y}(x|y)$ and $f_{Y|X}(y|x)$. Are X and Y independent?

5.103 Display a two-way table so that the random variables X and Y of Exercise 5.98 will be independent. Calculate $P(X^2 = 1, Y^2 = 1)$.

5.104 For the distribution of Exercise 5.98, let $A = [Y \geq 0]$. Compute $P(A)$ in two ways.

5.105 Suppose $f_{X,Y}(x, y) = 6(1 - x - y)$, $x > 0$, $y > 0$, $x + y \leq 1$.
   a. Find expressions for $f_{Y|X}(y|x)$ and $f_{X|Y}(x|y)$.
   b. Are X and Y independent? Identically distributed?
   c. Construct a joint density function having the same marginal distributions as those above but such that the random variables X and Y are independent. Calculate $P(X^2 < .25, Y^2 < .25)$.
   d. Calculate $P(Y < .5, X > .5)$.
   e. Calculate $E(Y|X = .5)$; $V(Y|X = .5)$.

5.106 For each of the following distributions, find (i) $E(X)$, (ii) $E(X|X < a)$, and (iii) $V(X|X < a)$.
   a. $f_X(x) = 1 - |1 - x|$, $0 < x < 2$, $a = .5$.

   b. $p_X(x) = \begin{cases} \frac{1}{3} \text{ for } x = 1, \\ \frac{2}{3} \text{ for } x = 5, \end{cases}$ $a = 4$.

   c. $F_X(x) = \begin{cases} 0 \text{ for } x < 0, \\ x^2 \text{ for } 0 \leq x \leq 1, \\ 1 \text{ for } x > 1, \end{cases}$ $a = .2$.

5.107 A coin is tossed n times. Show that the probability that an even number of heads is tossed is $[1 + (q - p)^n]/2$, where p is the probability of tossing heads on a given toss.

5.108 a. Two players roll a fair die alternately. Whoever rolls a 6 first wins. Find the probability that the first player wins.
   b. Extend part (a) to a game in which the probability of winning on a given trial is p.

5.109 a. The thief of Baghdad has been placed in a dungeon with three doors. One of the doors leads into a tunnel that returns him to the dungeon after one day's travel through the tunnel. Another door leads to a similar tunnel whose traversal requires three days rather than one day. The third door leads to freedom. Assume that the thief is equally likely to choose each door each time he makes a choice. (The thief has a poor memory!) Find the expected number of days the thief will be imprisoned, from the moment he first chooses a door to the moment he chooses the door leading to freedom. (Assume that the thief immediately chooses a door each time he enters the dungeon.)

b. Is it possible to design two tunnel lengths so that the thief has an expected imprisonment of 90 days?

5.7 EXTENSIONS TO HIGHER DIMENSIONS*

The considerations of random pairs and their associated distributions and properties extend easily to random n-tuples. It is often the case in applications that outcomes are observed on several (2, 3, 100, or more) random variables upon a single performance of the underlying experiment. For example, collecting data, such as making repeated measurements of line voltage, can be viewed as observing an outcome on a single experiment, where the outcome (the set of data) is an n-tuple. We will discuss such samples in detail in Chapter 7. The only difficulty encountered in extending to higher dimensions is the inability to provide simple graphical displays and interpretations. This is certainly not a serious problem, but it does mean we must proceed on purely mathematical grounds. It also suggests that the notational burden will escalate somewhat, and you must be prepared to pay that price. Fortunately, the concepts generalize in a simple way, and we will be able to make extensions by analogy with what we have seen for the bivariate case. Consequently, the discussion that follows is in the form of a summary and review of the notation and terminology we will use later.

The random variables $X_1, X_2, \ldots, X_n$ are *jointly distributed* provided that their values are determined once the underlying experiment is performed. That is, $X_1, X_2, \ldots, X_n$ have common domain S so that once $o \in S$ is observed, the numbers $X_1(o), X_2(o), \ldots, X_n(o)$ are determined. In this case, it is useful to view the n-tuple $(X_1, X_2, \ldots, X_n)$ as an object, a *random n-tuple*, having a joint distribution. This joint distribution may be discrete, continuous, or mixed.

The random n-tuple is *discrete* if there is a countable set R of points in n-dimensional space such that the outcome on the n-tuple will be a point in R with probability 1. Then the probabilities that the outcome on the random n-tuple will be a given point in R are the joint mass function values at that point. The joint mass function values are nonnegative and sum to 1. A simple tabular format cannot be used to specify the joint mass function, and other descriptive schemes must be used.

---

* This section may be omitted on a first reading.

For a discrete random n-tuple $(X_1, X_2, \ldots, X_n)$, there are n univariate marginal mass functions. In addition, there are $\binom{n}{k}$ k-dimensional joint marginal mass functions for any k of the component random variables. These are found by summing the joint mass function over all the possible values of the remaining variables.

EXAMPLE 5.41 Suppose X_1, X_2, and X_3 are jointly discrete with the mass function

$$p_{(X_1, X_2, X_3)}(x_1, x_2, x_3) = \begin{cases} \frac{1}{8} \text{ if } (x_1, x_2, x_3) = (1, 0, 0), \\ \frac{1}{16} \text{ if } (x_1, x_2, x_3) = (0, 1, 0), \\ \frac{1}{16} \text{ if } (x_1, x_2, x_3) = (0, 0, 1), \\ \frac{1}{4} \text{ if } (x_1, x_2, x_3) = (1, 1, 0), \\ \frac{1}{2} \text{ if } (x_1, x_2, x_3) = (0, 1, 1). \end{cases}$$

The components of X are each marginally Bernoulli-distributed with respective parameters $\frac{3}{8}$, $\frac{13}{16}$, $\frac{9}{16}$. In addition, as an example of a bivariate marginal,

$$p_{(X_1, X_2)}(x_1, x_2) = \begin{cases} \frac{1}{8} \text{ if } (x_1, x_2) = (1, 0), \\ \frac{9}{16} \text{ if } (x_1, x_2) = (0, 1), \\ \frac{1}{16} \text{ if } (x_1, x_2) = (0, 0), \\ \frac{1}{4} \text{ if } (x_1, x_2) = (1, 1). \end{cases}$$

The marginal mass function of X_1 can also be obtained from this bivariate marginal, and it agrees with that obtained from $p_{(X_1, X_2, X_3)}$.

A similar situation is encountered with more general multivariate distributions. Thus if $X_1, X_2, \ldots, X_n$ are jointly distributed random variables with joint mass function $p_{(X_1, X_2, \ldots, X_n)}$, then the marginal mass function of X_2, X_3, and X_5, for example, is obtained by summing over the remaining arguments:

$$p_{(X_2, X_3, X_5)}(x_2, x_3, x_5) = \sum_{x_1, x_4, x_6, \ldots, x_n} p_{(X_1, X_2, \ldots, X_n)}(x_1, x_2, \ldots, x_n).$$

In general, the term "marginal" is used in connection with *any* distribution (not necessarily univariate) that gives the distribution of some, but not all, of several jointly distributed random variables.

We now come to the second major category of the joint distributions that we wish to consider: joint continuous distributions. We will say that $X_1, X_2, \ldots, X_n$ are *jointly continuous* if the joint CDF can be described in terms of a joint density function in a way analogous to that for univariate continuous CDF functions. A real-valued function f of n variables is called an (*n-dimensional*) *density function* if $f(x_1, x_2, \ldots, x_n) \geq 0$ for all $x_1, x_2, \ldots, x_n$ and

$$\int_{-\infty}^{\infty} \int_{-\infty}^{\infty} \cdots \int_{-\infty}^{\infty} f(t_1, t_2, \ldots, t_n) \, dt_1 \cdots dt_n = 1. \tag{5.11}$$

For technical reasons, it is also required that the cumulative distribution function defined by

$$F(x_1, x_2, \ldots, x_n)$$

$$= \int_{-\infty}^{x_1} \int_{-\infty}^{x_2} \cdots \int_{-\infty}^{x_n} f(t_1, t_2, \ldots, t_n) \, dt_1 \, dt_2 \cdots dt_n \tag{5.12}$$

be differentiable (except possibly at countably many points), and that

$$\frac{\partial^n F(x_1, x_2, \ldots, x_n)}{\partial x_1 \, \partial x_2 \cdots \partial x_n} = f(x_1, x_2, \ldots, x_n). \tag{5.13}$$

The latter condition will be satisfied by the density functions considered in this book.

Now, let $(X_1, X_2, \ldots, X_n)$ be a random n-tuple with joint distribution given by the CDF F. If $(X_1, X_2, \ldots, X_n)$ is continuous (so its components $X_1, X_2, \ldots, X_n$ are jointly continuous), there exists a density function f of n variables such that

$$P[(X_1, X_2, \ldots, X_n) \in B]$$

$$= \int \cdots \int_B f(t_1, t_2, \ldots, t_n) \, dt_1 \, dt_2 \cdots dt_n \tag{5.14}$$

for any n-dimensional event B. The function f is then denoted $f_{(X_1, \ldots, X_n)}$ and is the joint density function for $(X_1, X_2, \ldots, X_n)$. In case $(X_1, X_2, \ldots, X_n)$ is continuous, the probability that it will assume any specified value is zero. In addition, the probability that $(X_1, \ldots, X_n)$ will assume a value in any k-dimensional subset of n-dimensional space, where $k < n$, is also zero. Thus, for example, in the plane, the probability is zero that a continuous random pair will assume a value at any specified point or on a line in the plane, since the double integral of the density over such a set is zero.

If $(X_1, X_2, \ldots, X_n)$ is continuous with density f, then there are n univariate marginal densities,

$$f_{X_k}(x_k) = \int_{-\infty}^{\infty} \int_{-\infty}^{\infty} \cdots \int_{-\infty}^{\infty} f(x_1, \ldots, x_{k-1}, x_k, x_{k+1}, \ldots, x_n) \times$$

$$dx_1 \cdots dx_{k-1} dx_{k+1} \cdots dx_n, \qquad k = 1, 2, 3, \ldots, n.$$

In addition, there are $\binom{n}{k}$ k-dimensional marginal joint density functions for k components of $(X_1, X_2, \ldots, X_n)$, found by holding those k coordinates fixed and integrating the joint density with respect to the remaining variables.

The moments of functions of the components of a random n-tuple can be found by using RUS, as in the bivariate case. For example, if $(X_1, X_2, \ldots, X_n)$ is jointly continuous with density f, the mean of X_i is given by

$$E(X_i) = \int_{-\infty}^{\infty} \cdots \int_{-\infty}^{\infty} x_i f(x_1, x_2, \ldots, x_n) \, dx_1 \, dx_2 \cdots dx_n.$$

The mean of a function of some (or all) of the X's, say $\phi(X_1, \ldots, X_n)$, is given by

$$E[\phi(X_1, X_2, \ldots, X_n)]$$

$$= \int_{-\infty}^{\infty} \cdots \int_{-\infty}^{\infty} \phi(x_1, x_2, \ldots, x_n) f(x_1, x_2, \ldots, x_n) \, dx_1 \, dx_2 \cdots dx_n.$$

Similar results hold in the discrete case, with multiple summation with respect to the mass function replacing multiple integration with respect to the density function.

EXAMPLE 5.42　For the trivariate distribution defined in Example 5.41, we have

$$E(X_1) = \sum_{x_1, x_2, x_3} x_1 \cdot p(x_1, x_2, x_3)$$

$$= (1)(\tfrac{1}{8}) + (0)(\tfrac{1}{16}) + (0)(\tfrac{1}{16}) + (1)(\tfrac{1}{4}) + (0)(\tfrac{1}{2}) = \tfrac{3}{8}$$

and

$$E(X_1^2 - X_2) = (1^2 - 0)(\tfrac{1}{8}) + (0^2 - 1)(\tfrac{1}{16}) + (0^2 - 0)(\tfrac{1}{16})$$

$$+ (1^2 - 1)(\tfrac{1}{4}) + (0^2 - 1)(\tfrac{1}{2})$$

$$= \tfrac{7}{16}.$$

We define the components of a random n-tuple $(X_1, X_2, \ldots, X_n)$ to be *independent* provided that, for any one-dimensional events $A_1, A_2, \ldots, A_n$,

$$P[(X_1, X_2, \ldots, X_n) \in A_1 \times A_2 \times \cdots \times A_n]$$

$$= P(X_1 \in A_1) P(X_2 \in A_2) \cdots P(X_n \in A_n).$$

Otherwise, the random variables are said to be *dependent*. This condition is characterized by factorization of the joint CDF F into a product of marginals $F(x_1, x_2, \ldots, x_n) = F_{X_1}(x_1) F_{X_2}(x_2) \cdots F_{X_n}(x_n)$ for all $(x_1, x_2, \ldots, x_n)$. If $(X_1, X_2, \ldots, X_n)$ is continuous or discrete, the independence of its components may also be characterized by factorization of the joint density function or joint mass function into a product of marginals. It follows that if $X_1, X_2, \ldots, X_n$ are independent, then so are the random variables

$g_1(X_1), g_2(X_2), \ldots, g_n(X_n)$. Indeed, any function of some of the X_i's is independent of any function of the remaining X_i's. For example, if $X, Y, U,$ and V are independent, then so are the random variables $\sin X$, $Y^2 + 2U,$ and $|V|$.

EXAMPLE 5.43   If $(X_1, X_2, \ldots, X_n)$ is continuously distributed with joint density $f_{(X_1,X_2,\ldots,X_n)}$, where

$$f_{(X_1,X_2,\ldots,X_n)}(x_1, x_2, \ldots, x_n)$$

$$= \exp\left(-\sum_{i=1}^{n} x_i\right), \qquad x_i > 0, i = 1, 2, \ldots, n,$$

then the marginal density functions are exponential (1), so that $f_{X_i}(x_i) = e^{-x_i}$ for $x_i > 0$. Since $f_{(X_1,X_2,\ldots,X_n)} = f_{X_1} \cdot f_{X_2} \cdots f_{X_n}$, it follows that $X_1, X_2, \ldots, X_n$ are independent. Thus, for example, $(X_1 - X_2)$ and X_3^2 are independent.

EXAMPLE 5.44   —*Random Variables That Are Pairwise Independent but Not Independent*   Suppose X denotes the outcome of a fair coin on the first toss and Y denotes the outcome of the second (independent) trial. Define the random variable Z to have value 1 if the two tosses result in outcomes that are alike and let Z have value zero otherwise. Now, X and Y are independent by assumption; X and Z have the joint mass function $p_{(X,Z)}$ shown in the accompanying table.

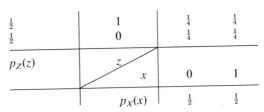

It is readily verified that $p_{(X,Z)}(x, z) = p_X(x)p_Z(z)$, with a similar result for Y and Z. It follows that the components of each of the pairs $(X, Y),$ $(X, Z),$ and (Y, Z) are independent. *But the random variables $X, Y,$ and Z are not independent,* since, for example,

$$0 = p_{(X,Y,Z)}(1, 1, 0) \neq p_X(1)p_Y(1)p_Z(0) = \tfrac{1}{8}.$$

The various possible conditional distributions can be defined in a manner analogous to that for the bivariate case. To avoid unnecessary detail and repetition, we suppose that all denominators in the following discussion are positive.

Definition 5.5   Let $(X_1, X_2, \ldots, X_n)$ be a random n-tuple, and suppose that for $k < n$, $\alpha_1, \alpha_2, \ldots, \alpha_k$ are k distinct subscripts, while $\beta_1, \beta_2, \ldots, \beta_{n-k}$ are the remaining subscripts among $1, 2, \ldots, n$.

1. If $(X_1, X_2, \ldots, X_n)$ has joint density f, then the *conditional density function* of $(X_{\alpha_1}, X_{\alpha_2}, \ldots, X_{\alpha_k})$, given the respective values $x_{\beta_1}, x_{\beta_2}, \ldots, x_{\beta_{n-k}}$ of $(X_{\beta_1}, X_{\beta_2}, \ldots, X_{\beta_{n-k}})$, is defined by

$$f(x_{\alpha_1}, x_{\alpha_2}, \ldots, x_{\alpha_k} | x_{\beta_1}, x_{\beta_2}, \ldots, x_{\beta_{n-k}})$$

$$= \frac{f(x_1, x_2, \ldots, x_n)}{f_*(x_{\beta_1}, x_{\beta_2}, \ldots, x_{\beta_{n-k}})}.$$

2. If $(X_1, X_2, \ldots, X_n)$ is discrete with joint mass function p, then the *conditional mass function* of $(X_{\alpha_1}, X_{\alpha_2}, \ldots, X_{\alpha_k})$, given the value $(x_{\beta_1}, x_{\beta_2}, \ldots, x_{\beta_{n-k}})$, is defined by

$$p(x_{\alpha_1}, x_{\alpha_2}, \ldots, x_{\alpha_k} | x_{\beta_1}, x_{\beta_2}, \ldots, x_{\beta_{n-k}})$$

$$= \frac{p(x_1, x_2, \ldots, x_n)}{p_* x_{\beta_1}, x_{\beta_2}, \ldots, x_{\beta_{n-k}})}.$$

Here, f_* and p_* denote the marginal distributions of $X_{\beta_1}, X_{\beta_2}, \ldots, X_{\beta_{n-k}}$. The conditional mass function above is the value of the conditional probability

$$p(x_{\alpha_1}, x_{\alpha_2}, \ldots, x_{\alpha_k} | x_{\beta_1}, x_{\beta_2}, \ldots, x_{\beta_{n-k}})$$

$$= P(X_{\alpha_1} = x_{\alpha_1}, X_{\alpha_2} = x_{\alpha_2}, \ldots, X_{\alpha_k} = x_{\alpha_k} | X_{\beta_1}$$

$$= x_{\beta_1}, X_{\beta_2} = x_{\beta_2}, \ldots, X_{\beta_{n-k}} = x_{\beta_{n-k}}).$$

EXAMPLE 5.45   Suppose (X, Y, Z) has mass function

$$p_{(X,Y,Z)}(x, y, z) = \begin{cases} \frac{1}{4} \text{ if } (x, y, z) = (0, 0, 0), \\ \frac{1}{8} \text{ if } (x, y, z) = (1, 0, 0), \\ \frac{1}{16} \text{ if } (x, y, z) = (0, 0, 1), \\ \frac{3}{8} \text{ if } (x, y, z) = (0, 1, 1), \\ \frac{3}{16} \text{ if } (x, y, z) = (1, 1, 0). \end{cases}$$

Then

$$p_{X,Z|Y}(x, z|0) = \frac{p_{(X,Y,Z)}(x, 0, z)}{p_Y(0)} = \begin{cases} \frac{4}{7} \text{ if } (x, z) = (0, 0), \\ \frac{2}{7} \text{ if } (x, z) = (1, 0), \\ \frac{1}{7} \text{ if } (x, z) = (0, 1). \end{cases}$$

Also.

$$p_{Y|X,Z}(y|0, 1) = \begin{cases} \frac{1}{7} \text{ if } y = 0, \\ \frac{6}{7} \text{ if } y = 1. \end{cases}$$

As seen in the preceding example, Definition 5.5 may be used to find the univariate conditional density functions

$$f(x_i|x_1, \ldots, x_{x-1}, x_{i+1}, \ldots, x_n) = \frac{f(x_1, x_2, \ldots, x_n)}{f_*(x_1, \ldots, x_{i-1}, x_{i+1}, \ldots, x_n)},$$

where f_* is the marginal density of $(X_1, \ldots, X_{i-1}, X_{i+1}, \ldots, X_n)$. In turn, the means of these one-dimensional conditional distributions may be computed as in the bivariate case. The *conditional expected value* of the random variable $\phi(X_i)$ given $x_1, \ldots, x_{i-1}, x_{i+1}, \ldots, x_n$ is defined to be

$$E[\phi(X_i)|x_1, \ldots, x_{i-1}, x_{i+1}, \ldots, x_n]$$

$$= \int_{-\infty}^{\infty} \phi(x_i) f(x_i|x_1, \ldots, x_{i-1}, x_{i+1}, \ldots, x_n) \, dx_i.$$

In particular, for $\phi(x_i) = x_i$, we obtain the *conditional mean* of X_i given $x_1, \ldots, x_{i-1}, x_{i+1}, \ldots, x_n$. With other choices for ϕ, we may obtain higher conditional moments. The corresponding formulas for the discrete case are similar.

EXAMPLE 5.46   For the preceding example, we have

$$E[(1 + X)^{Z+1}|Y = 0] = 1(\tfrac{4}{7}) + 2(\tfrac{2}{7}) + 1^2(\tfrac{1}{7}) = \tfrac{9}{7}$$

and $E(Y|X = 0, Z = 1) = \tfrac{6}{7}$ and $V(Y|X = 0, Z = 1) = \tfrac{6}{49}$. The latter is also immediate from the observation that the conditional distribution of Y, given $X = 0$ and $Z = 1$, is binomial $(1, \tfrac{6}{7})$.

There are many other conditional densities. Thus the joint marginal distribution $f_{(X_{\alpha_1}, X_{\alpha_2}, \ldots, X_{\alpha_k})}$ of the random variables $X_{\alpha_1}, X_{\alpha_2}, \ldots, X_{\alpha_k}$ is itself a k-dimensional joint density function. As such, the conditional density of $m < k$ of these random variables, given values of the remaining $k - m$, can be defined by applying Definition 5.5. For example,

$$f(x_2|x_1) = \frac{f_{(X_1,X_2)}(x_1, x_2)}{f_{X_1}(x_1)}$$

$$= \frac{\int_{-\infty}^{\infty} \cdots \int_{-\infty}^{\infty} f(x_1, x_2, x_3, \ldots, x_n)\, dx_3 \cdots dx_n}{\int_{-\infty}^{\infty} \cdots \int_{-\infty}^{\infty} f(x_1, x_2, \ldots, x_n)\, dx_2 \cdots dx_n}.$$

Similarly,

$$f_3(x_3|x_1, x_2) = \frac{f_{(X_1, X_2, X_3)}(x_1, x_2, x_3)}{f_{(X_1, X_2)}(x_1, x_2)} = \frac{f_{(X_1, X_2, X_3)}(x_1, x_2, x_3)}{f_2(x_2|x_1)f_{X_1}(x_1)}.$$

In fact, if we proceed recursively in this manner, we can write

$$f(x_1, x_2, \ldots, x_n)$$

$$= f_1(x_1)f_2(x_2|x_1)f_3(x_3|x_1, x_2) \cdots f_n(x_n|x_1, x_2, \ldots, x_{n-1}) \quad (5.15)$$

with an identical formula for the discrete case. This is a fact that is often useful in computing a joint distribution when the components are not independent.

It is possible to exploit conditioning to find means and variances for multivariate situations, just as for the bivariate case. The following example illustrates this technique for an $(n + 1)$-dimensional application.

EXAMPLE 5.47 —*A Random Sum of Random Variables*   Suppose N, X_1, X_2, X_3, $\ldots$, X_n are independent, N is binomial (n, p), and $Z = \Sigma_{i=1}^{N} X_i$. In addition, assume $E(X_i) = \mu$ and $V(X_i) = \sigma^2$, $i = 1, 2, \ldots, n$. Then the mean and variance of Z may be calculated as follows:

$$E(Z) = E[E(Z|N)] = E\left[E\left(\sum_{i=1}^{N} X_i|N\right)\right] = E\left[\sum_{i=1}^{N} E(X_i|N)\right] = E(N\mu)$$

$$= E(N)E(X_i) = np\mu.$$

Also,

$$E(Z^2) = E[E(Z^2|N)] = E\left[E\left(\sum_{i=1}^{N} X_i^2 + \sum_{\substack{i,j \\ i \neq j}} X_i X_j|N\right)\right]$$

$$= E\left[\sum_{i=1}^{N} E(X_i^2|N) + \sum_{\substack{i,j \\ i \neq j}} E(X_i X_j|N)\right]$$

$$= E[NE(X_i^2)] + E\{N(N-1)[E(X_i)]^2\}$$

$$= E(N)[E(X_i^2) - E^2(X_i)] + E(N^2)E^2(X_i).$$

Thus

$$V(Z) = E(Z^2) - E^2(Z) = E(N)V(X_i) + E(N^2)E^2(X_i) - E^2(N)E^2(X_i)$$
$$= E(N)V(X_i) + V(N)E^2(X_i) = np(\sigma^2 + q\mu^2).$$

EXERCISES

THEORY

5.110   Show that if $X_1, X_2, \ldots, X_n$ are jointly continuous, then

$$f(x_1, x_2, \ldots, x_n) = f_{X_1}(x_1)f_{X_2}(x_2) \cdots f_{X_n}(x_n)$$

if and only if $F(x_1, x_2, \ldots, x_n) = F_{X_1}(x_1)F_{X_2}(x_2) \cdots F_{X_n}(x_n)$.

5.111   If $X = (X_1, X_2, \ldots, X_n)$ and $Y = (Y_1, Y_2, \ldots, Y_m)$ are jointly distributed (i.e., if $X_1, X_2, \ldots, X_n, Y_1, Y_2, \ldots, Y_m$ are jointly distributed), X and Y are defined to be independent provided that

$$f_{(X_1, X_2, \ldots, X_n, Y_1, Y_2, \ldots, Y_m)}(x_1, x_2, \ldots, x_n, y_1, y_2, \ldots, y_m)$$

$$= f_{(X_1, X_2, \ldots, X_n)}(x_1, x_2, \ldots, x_n)f_{(Y_1, Y_2, \ldots, Y_m)}(y_1, y_2, \ldots, y_m).$$

Suppose $Z_1, Z_2, \ldots, Z_k$ are independent and that $\{\alpha_1, \alpha_2, \ldots, \alpha_t\}$ and $\{\beta_1, \beta_2, \ldots, \beta_s\}$ are disjoint subsets of $\{1, 2, \ldots, k\}$. Argue that $Z_\alpha = (Z_{\alpha_1}, Z_{\alpha_2}, \ldots, Z_{\alpha_t})$ and $Z_\beta = (Z_{\beta_1}, Z_{\beta_2}, \ldots, Z_{\beta_s})$ are independent. (Note that $t + s$ is not necessarily as large as k.)

5.112   In Example 5.43, show that $f_{(X_1, X_2, X_3)}$ may be obtained as follows: First, obtain $F_{(X_1, X_2, \ldots, X_n)}$; then obtain $F_{(X_1, X_2, X_3)}(x_1, x_2, x_3)$ by taking

$$\lim_{x_4 \to \infty} \cdots \lim_{x_n \to \infty} F_{(X_1, X_2, \ldots, X_n)}(x_1, x_2, \ldots, x_n);$$

then find the joint density as $\partial^3 F_{(X_1, X_2, X_3)}(x_1, x_2, x_3)/\partial x_1 \, \partial x_2 \, \partial x_3$.

5.113   Find $V(Z)$ in Example 5.47 by using $V(Z) = E[V(Z|N)] + V[E(Z|N)]$.

APPLICATION

5.114   For Example 5.44, show that, while X and Y are each independent of Z, the random variables $X + Y$ and Z are dependent.

5.115   Suppose X_1, X_2, X_3, X_4 have joint density $f(x_1, x_2, x_3, x_4) = 4x_1x_2$ for $0 < x_i < 1$, $i = 1, 2, 3, 4$.
   a. Show that X_1, X_2, X_3, X_4 are independent.
   b. Find $\text{Cov}(X_1, X_4)$.   c. Find $f_{X_2|X_1}(x_2|x_1)$.

5.116   Suppose $X_1, X_2, \ldots, X_n$ are jointly continuous and independent, each distributed with marginal density f.
   a. What is the probability that two or more x_i's have the same value ("ties")?
   b. Find $P(X_1 < X_2 < \cdots < X_n)$.
   c. Find the probability that the largest value among the X's is X_1.
   d. If the X's represent annual rainfalls at a given location, find the distribution

of the number of years until the first year's rainfall (X_1) is exceeded for the first time.

5.117   If $(X_1, X_2, \ldots, X_n)$ has density f, and g is a function of n variables, then by RUS,

$$E[g(X_1, X_2, \ldots, X_n)]$$
$$= \int \cdots \int g(x_1, x_2, \ldots, x_n) f(x_1, x_2, \ldots, x_n) \, dx_1 \, dx_2 \cdots dx_n.$$

Suppose X_1 is uniform on $(0, 1)$, X_2 is uniform on $(1, 2)$, and X_3 is exponential (1), and that these random variables are independent.
a. Find $V(X_1 + X_2)$.   b. Find $E(X_1 X_2 X_3)$.   c. Find $E(X_1 | X_3)$.
d. Find $P(X_1 < X_3)$.   e. Find $\rho(X_3, X_2)$.

5.118   Suppose (X_1, X_2, X_3, X_4) is a random 4-tuple with independent discrete components.
a. Suppose it is specified that X_1 and X_3 are Bernoulli $(\frac{1}{3})$, X_2 is Poisson (1), and X_4 is geometric $(\frac{1}{2})$. Determine the nature of the range of R of (X_1, X_2, X_3, X_4), and specify the joint mass function p.
b. Find the marginal mass function of (X_2, X_4).

5.119   a. For Example 5.41, find $p_{X_1|X_2}(x_1|1)$.
b. Find $E(X_1 | X_2 = 1)$.   c. Find $V(X_1 | X_2 = 1)$.
d. Find $E(X_2 - X_1 + X_3^2)$.   e. Find $\text{Cov}(X_1, X_2)$.

5.120   An experiment has two phases. In phase I, an observation is made on a random variable, with the distribution given in the accompanying table. The performance of phase II depends on what happened in phase I. If X is less than μ_X, a fair coin is tossed ($Y = 0$ if tails; $Y = 1$ if heads); otherwise, a die is tossed (Y has equal likelihood of being 1, 2, $\ldots$, 6).

x	-2	-1	0	3
$p(x)$	.1	.4	.2	.3

a. Find $E(Y)$.   b. Find $V(X)$.   c. Find $P(X > 0)$.
d. Find $P(X > 0 | Y > 0)$.   e. Find $E(Y | X = -1)$.
f. List the joint mass function of X and Y.

5.121   Two smoke detectors are installed side by side in my attic. One, brand A, gives a false alarm (in a period of a month) with probability .013. The other, brand B, gives a false alarm (in a period of a month) with probability .004. Last month, I heard a false alarm.
a. What is the probability that it was the brand A detector?
b. What is the probability that, in the next 12 months, I will encounter at least one month in which at least one false alarm occurs?

5.8   SUMMARY

A random pair (X, Y) consists of two jointly distributed random variables; when $o \in S$ is the outcome in the underlying sample space, the real number pair $(X(o), Y(o)) = (x, y)$ is the outcome on (X, Y).

The distribution of (X, Y) may be jointly discrete, jointly continuous, or mixed. The distribution of a jointly discrete random pair is determined by either a joint mass function $p(x, y)$ or a joint CDF, $F(x, y)$. The distribution of a jointly continuous random pair is determined by a joint density $f(x, y)$ or a joint CDF, $F(x, y)$.

The probability of occurrence of a two-dimensional event B in the plane is found by summing (integrating) the mass function (density) over B when (X, Y) is jointly discrete (continuous). A marginal mass function (density) is found by summing (integrating) the joint distribution with respect to the unwanted variable.

If X and Y are independent, the joint mass function (density) factors into the product of marginal mass functions (densities), and conversely. The covariance of X and Y, $\text{Cov}(X, Y)$ is defined to be $E[(X - \mu_X)(Y - \mu_Y)]$, and the correlation of X and Y is $\rho(X, Y) = \text{Cov}(X, Y)/\sqrt{\sigma_X^2 \sigma_Y^2}$. If X and Y are independent, then they are uncorrelated ($\rho = 0$), but the converse is not true. If X and Y are independent, then $g(X)$ and $h(Y)$ are also independent.

$$E(X + Y) = E(X) + E(Y); \quad V(X + Y) = V(X) + V(Y) + 2\,\text{Cov}(X, Y).$$

If X and Y are jointly discrete, the conditional mass function for X, given $Y = y$, where $y \in R_Y$, is given by $p_1(x|y) = p(x, y)/p_Y(y)$. Similarly, in the continuous case, the conditional density of X, given $Y = y$, is $f_1(x|y) = f(x, y)/f_Y(y)$. The mean and variance of the conditional distribution of X given $Y = y$ are denoted $E(X|y)$ and $V(X|y)$, respectively, and called the conditional mean of X given $Y = y$ and the conditional variance of X given $Y = y$.

The following identities hold:

$$E(X) = E[E(X|Y)], \quad V(X) = E[V(X|Y)] + V[E(X|Y)].$$

The exponential distribution and the geometric distribution are memoryless.

Concepts for random pairs extend to n-dimensional distributions.

Chapter 6

ANALYTICAL METHODS FOR JOINT DISTRIBUTIONS

6.1 INTRODUCTION

In applications involving jointly distributed random variables, various functions of the random variables and their distributions are considered. For example, imagine a system composed of components connected in series. A model may have been developed for the joint distribution of the lifetimes of the individual components, and it is desired to find the distribution of the length of life of the entire system. The latter is a function of the individual life lengths of the components; indeed, the system life in this case is the minimum of the individual component lives. The question we now consider is, "How do we find the distribution of $g(X_1, X_2)$, given the joint distribution of X_1 and X_2?" As we will see, there are several possible answers to this question.

In Chapter 4, two methods of finding the distribution of $g(X)$, given the distribution of X, were discussed. One of the approaches involved the CDF of $g(X)$; the other involved using the density or mass function of X directly. The latter approach will be used in extending to functions $g(X_1, X_2)$ of random pairs. These methods are easily extended to functions of random n-tuples, although we do not pursue this case formally. (Limited use is made of the more general case in Chapter 7 in connection with functions of random samples.) The use of joint moment-generating functions is also considered. Applications of these methods to several important models involving joint distributions are discussed.

234

6.2   FUNCTIONS OF RANDOM PAIRS

It is relatively easy to visualize how the function g maps probabilities from the joint distribution of (X_1, X_2) to the distribution of $g(X_1, X_2)$ when (X_1, X_2) is discrete. In Fig. 6.1, a hypothetical situation is shown in which the joint mass function of the discrete random pair (X_1, X_2) is positive at four points in the x_1x_2-plane, and where the function g maps these two-dimensional points to three images in the range R_Y of the random variable $Y = g(X_1, X_2)$. In this figure, you can see how the mass function for Y is obtained, at each point y in R_Y, by summing the joint masses of (X_1, X_2) over the (x_1, x_2) points that g maps to y. This function is essentially a bookkeeping task; symbolically, for each $y \in R_Y$,

$$p_Y(y) = P[(X_1, X_2) \in A_y] = \sum_{(x_1, x_2) \in A_y} p_{X_1, X_2}(x_1, x_2), \qquad (6.1)$$

where A_y is the set of two-dimensional points that g maps to y, or $A_y = \{(x_1, x_2) : g(x_1, x_2) = y\}$.

EXAMPLE 6.1   A system is composed of two (dependent) components in parallel, and its lifetime L is measured in terms of the number of use cycles to failure of the system. Suppose the components have undergone extensive testing, and the cycles to failure of the components X and Y have the joint distribution shown in the accompanying table.

4	.01	.1	.1	.1
3	.05	.05	.01	.01
2	.1	.05	.01	.01
1	.1	.2	.1	0
y				
x	1	2	3	4

The system life is a function of the component lives, $L = \max\{X, Y\}$. The mass function for L is positive on $R_L = \{1, 2, 3, 4\}$ and is determined as follows:

$$p_L(1) = P(L = 1) = P(X = 1 \text{ and } Y = 1) = p_{X,Y}(1, 1) = .1;$$

$$p_L(2) = P(L = 2) = P[(X, Y) \in \{(1, 2), (2, 1), (2, 2)\}]$$

$$= p_{X,Y}(1, 2) + p_{X,Y}(2, 1) + p_{X,Y}(2, 2) = .35;$$

$$p_L(3) = P[(X, Y) \in \{(1, 3), (2, 3), (3, 3), (3, 2), (3, 1)\}] = .22;$$

$$p_L(4) = \sum_{i=1}^{4} p(i, 4) + \sum_{j=1}^{3} p(4, j) = .33.$$

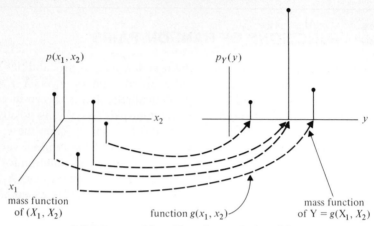

FIGURE 6.1   *Mass Function p_Y Induced by g*

EXAMPLE 6.2   Consider the random variables X_1 and X_2 denoting the respective outcomes on the toss of two fair dice. Let $g(x_1, x_2) = x_1 + x_2$ and let Y denote the composition $g(X_1, X_2) = X_1 + X_2$. Then the range of Y is $R_Y = \{2, 3, \ldots, 12\}$, and the set A_y is the set of (x_1, x_2) combinations such that $x_1 + x_2 = y$. Then by Eq. (6.1), for each $y \in R_Y$,

$$p_Y(y) = \sum_{\substack{x_1 \ x_2 \\ x_1+x_2=y}} \frac{1}{36}.$$

Thus, for example, $P(Y = 2) = \frac{1}{36}$ and $P(Y = 6) = \frac{5}{36}$. A graph of this mass function was considered before in Fig. 3.3.

EXAMPLE 6.3   Suppose X_1 and X_2 are independent and identically distributed binomial (n, p) random variables, and suppose $Y = X_1 + X_2$. Then Y is discrete with range $R_Y = \{0, 1, 2, \ldots, 2n\}$. For any $y \in R_Y$,

$$p_Y(y) = \sum_{\{(x_1,x_2): x_1+x_2=y\}} p_{(X_1,X_2)}(x_1, x_2) = \sum_{x_2=0}^{y} p_{(X_1,X_2)}(y - x_2, x_2)$$

$$= \sum_{x_2=0}^{y} \binom{n}{y - x_2} p^{y-x_2} q^{n-y+x_2} \binom{n}{x_2} p^{x_2} q^{n-x_2}$$

$$= p^y q^{2n-y} \sum_{x_2=0}^{y} \binom{n}{y - x_2} \binom{n}{x_2} = \binom{2n}{y} p^y q^{2n-y}.$$

Thus Y is binomial $(2n, p)$. This result is plausible, since, if we consider the number X_1 of successes in n independent, repeated Bernoulli (p) trials plus the number X_2 of successes in another n independent, repeated Bernoulli (p) trials (independent of the first n trials), the result is the number $X_1 + X_2$ of successes in $2n$ independent, repeated Bernoulli (p) trials. This result rests heavily on the independence of X_1 and X_2 and the commonality of the parameter p in both marginal distributions. A generalization is given in Exercise 6.1. Note that $p_Y(y)$ can be found either as $\sum_{x_2} p(y - x_2, x_2)$, as done here, or as $\sum_{x_1} p(x_1, y - x_1)$.

A second case of great interest is that in which (X_1, X_2) is continuous with joint density f and $Y = g(X_1, X_2)$ is also continuous. In this case, we may attempt to derive the distribution of Y by using a CDF method. Thus for a given $y \in R_Y$, let $B_y = \{(x_1, x_2) : g(x_1, x_2) \leq y\}$. Then

$$F_Y(y) = P(Y \leq y) = P[(X_1, X_2) \in B_y] = \int \int_{B_y} f(x_1, x_2) \, dx_1 \, dx_2.$$

How much further the formula can be carried depends on the tractability of the integral of f over the set B_y. Evaluating such a multiple integral depends on proper limits of integration, which, in turn, depend critically on the functional form of g. If this step can be accomplished, the result may be differentiated to obtain the density function of Y. Since no completely general result can be established, let us consider some special cases. Other cases may be found in the exercises at the end of the section.

EXAMPLE 6.4 Suppose a system is composed of two components in parallel, and the lifetimes of these components are independent with common marginal CDF F_X. Let Y denote the system life, $Y = \max\{X_1, X_2\}$. Then B_y is a square in the first quadrant, $B_y = \{(x_1, x_2) : 0 \leq x_1 \leq y, 0 \leq x_2 \leq y\}$, and

$$F_Y(y) = P(Y \leq y) = P(X_1 \leq y, X_2 \leq y)$$

$$= \int_0^y \int_0^y f_{X_1, X_2}(x_1, x_2) \, dx_1 \, dx_2.$$

Because of the assumed independence of X_1 and X_2, this integral can be evaluated as $F_X^2(y)$. Thus the density of system life is $f_Y(y) = 2F_X(y)f_X(y)$, $y > 0$. As a special case, if X_i is uniform over $(0, 1)$, we find Y is beta $(2, 1)$.

For a somewhat general formula, suppose (X_1, X_2) is distributed with density f, and $g(x_1, x_2) = x_1/x_2$ for $x_2 \neq 0$. It is desired to derive the density of $Y = X_1/X_2$. For any y, $B_Y = \{(x_1, x_2) : x_1/x_2 \leq y\}$ is a set of points in two-dimensional space determined by a line, as shown in Fig. 6.2. The appropriate limits of integration are easily determined. Thus

$$F_Y(y) = \int_{-\infty}^0 \int_{yx_2}^\infty f(x_1, x_2) \, dx_1 \, dx_2 + \int_0^\infty \int_{-\infty}^{yx_2} f(x_1, x_2) \, dx_1 \, dx_2.$$

Differentiating with respect to the inner integral, assuming that interchange

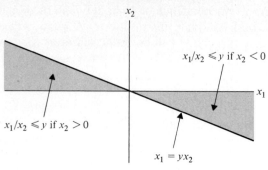

FIGURE 6.2   *Region B_y*

with the outer integral is valid, yields

$$f_Y(y) = - \int_{-\infty}^{0} x_2 f(yx_2, x_2) \, dx_2 + \int_{0}^{\infty} x_2 f(yx_2, x_2) \, dx_2$$

$$= \int_{-\infty}^{\infty} |x_2| f(yx_2, x_2) \, dx_2. \qquad (6.2)$$

EXAMPLE 6.5   Suppose X_1 and X_2 are independent, and each is $N(0, 1)$ so that

$$f(x_1, x_2) = \frac{1}{2\pi} e^{-(1/2)(x_1^2 + x_2^2)},$$

and let $Y = X_1/X_2$. Substituting in Eq. (6.2) gives

$$f_Y(y) = - \frac{1}{2\pi} \int_{-\infty}^{0} x_2 e^{-(1/2)x_2^2(y^2 + 1)} \, dx_2 + \frac{1}{2\pi} \int_{0}^{\infty} x_2 e^{-(1/2)x_2^2(y^2 + 1)} \, dx_2$$

$$= \left[\frac{1}{2\pi} \cdot \frac{e^{-(1/2)x_2^2(y^2 + 1)}}{y^2 + 1} \right]_{-\infty}^{0} - \left[\frac{1}{2\pi} \cdot \frac{e^{-(1/2)(x_2^2(y^2 + 1))}}{y^2 + 1} \right]_{0}^{\infty}$$

$$= \frac{1}{2\pi} \cdot \frac{1}{y^2 + 1} + \frac{1}{2\pi} \cdot \frac{1}{y^2 + 1} = \frac{1}{\pi(1 + y^2)}.$$

This expression is readily recognized to be the Cauchy density. In words, a quotient of two independent $N(0, 1)$ random variables is a Cauchy random variable.

_____

    As an application of this result, suppose a dart is thrown at a target with center at the origin, and suppose the vertical and horizontal components of miss are independent standard normal variates. If a line is drawn through the point of impact and the target center, the line has a slope that is Cauchy-distributed. Because the joint density $f(x_1, x_2)$ has circular symmetry (contours on the density surface are circles), the angle Θ between the x-axis and the line described above is uniform over $(0, \pi)$. It follows that, if Θ is uniform $(0, \pi)$, then $Z = \tan \Theta$ is Cauchy-distributed.

EXAMPLE 6.6   As another illustration of this technique, where finding the limits of integration in rectangular coordinates is not as easy, let $g(x_1, x_2) = x_1^2 + x_2^2$. It is desired to find the density function for the composition $Y = X_1^2 + X_2^2$.

SOLUTION   In this case, we know that $f_Y(y) = 0$ if $y < 0$. For $y > 0$, $B_Y = \{(x_1, x_2) : x_1^2 + x_2^2 \leq y\}$ is a disc in the plane, so the limits of integration in

$$F_Y(y) = \int \int_{B_Y} f(x_1, x_2) \, dx_1 \, dx_2$$

may be expressed in rectangular coordinates. The integration is easier if we convert to polar coordinates, in which case

$$F_Y(y) = \int_0^{2\pi} \int_0^{\sqrt{y}} f(r \cos \theta, r \sin \theta) r \, dr \, d\theta, \qquad y > 0.$$

Differentiating with respect to y (using Leibniz's rule), for $y > 0$, yields

$$f_Y(y) = \frac{1}{2\sqrt{y}} \int_0^{2\pi} f(\sqrt{y} \cos \theta, \sqrt{y} \sin \theta) \sqrt{y} \, d\theta$$

$$= \frac{1}{2} \int_0^{2\pi} f(\sqrt{y} \cos \theta, \sqrt{y} \sin \theta) \, d\theta.$$

This expression is a general formula for f_Y in terms of f_{X_1, X_2}. As a case in point, suppose again that X_1 and X_2 are independent, each $N(0, 1)$. Then

$$f_Y(y) = \frac{1}{4\pi} \int_0^{2\pi} e^{-(y/2)} \, d\theta = \tfrac{1}{2} e^{-(y/2)}.$$

This expression is recognized to be a $\chi_{(2)}^2$ [or exponential $(\tfrac{1}{2})$] density function.

A general formula for the distribution of the sum $Y = X_1 + X_2$ of two jointly distributed random variables is found as follows. For each y, $B_Y = \{(x_1, x_2) : x_1 + x_2 \leq y\}$ is the set of points on and to the left of a line in the plane. Consequently,

$$F_Y(y) = \int_{-\infty}^{\infty} \int_{-\infty}^{y - x_2} f(x_1, x_2) \, dx_1 \, dx_2 = \int_{-\infty}^{\infty} \int_{-\infty}^{y - x_1} f(x_1, x_2) \, dx_2 \, dx_1,$$

and differentiating yields

$$f_Y(y) = \int_{-\infty}^{\infty} f(y - x_2, x_2) \, dx_2 = \int_{-\infty}^{\infty} f(x_1, y - x_1) \, dx_1. \qquad (6.3)$$

If X_1 and X_2 are independent, Eq. (6.3) may be written as

$$f_Y(y) = \int_{-\infty}^{\infty} f_{X_1}(y - x_2) f_{X_2}(x_2) \, dx_2$$

$$= \int_{-\infty}^{\infty} f_{X_1}(x_1) f_{X_2}(y - x_1) \, dx_1. \qquad (6.4)$$

These integrals are called *convolution integrals,* and the resulting function (f_Y, in our case) is said to be the convolution of f_{X_1} and f_{X_2}, written as $f_Y = f_{X_1} * f_{X_2}$. A similar treatment can be given in the discrete case, as was seen in Example 6.3, where it was shown that a convolution of certain binomial mass functions is again a binomial mass function. You are asked to investigate some properties of convolutions in Exercise 6.2. When R_{X_1, X_2} is a bounded set, added caution must be taken on setting the limits of integration.

EXAMPLE 6.7   Suppose X_1 and X_2 are independent, each uniform over $(0, 1)$, and let $Y = X_1 + X_2$. In this case, we know that $f(x_1, x_2)$ is zero outside the unit square. We will have to take account of this fact in substituting into Eq. (6.3). The range of g (restricted to $[0, 1] \times [0, 1]$ for this case) is the interval $[0, 2]$ (see Fig. 6.3). Consequently, for $0 \le y \le 1$, x_2 varies from 0 to y, so $f_Y(y) = \int_0^y dx_2 = y$. On the other hand, if $1 \le y \le 2$, then x_2 varies from $(y - 1)$ to 1, so $f_Y(y) = \int_{y-1}^{1} dx_2 = 2 - y$. Hence, Y has the *triangular density* shown in Fig. 6.4.

EXAMPLE 6.8   Let us view the problem of Example 6.6 from the vantage point of convolution integrals. Recall that, if X_1 is $N(0, 1)$, X_1^2 is $\chi^2_{(1)}$. Since the independence of X_1 and X_2 implies the independence of X_1^2 and X_2^2, one can also find the density of $Y = X_1^2 + X_2^2$ by the convolution of two $\chi^2_{(1)}$

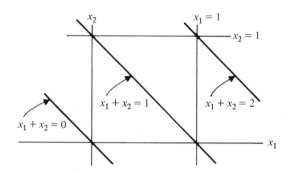

FIGURE 6.3   *Range of $x_1 + x_2$: $[0, 2]$*

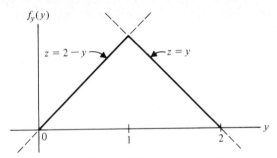

FIGURE 6.4   *Triangular Density*

density functions. For $y > 0$,

$$f_Y(y) = \int_{-\infty}^{\infty} f_{X_1}(x_1) f_{X_2}(y - x_1)\, dx_1$$

$$= \int_0^y \frac{1}{\sqrt{2\pi}} x_1^{-(1/2)} e^{-(1/2)x_1} \cdot \frac{1}{\sqrt{2\pi}} (y - x_1)^{-(1/2)} e^{-(1/2)(y - x_1)}\, dx_1$$

$$= \tfrac{1}{2} e^{-(1/2)y} \int_0^y \frac{1}{\pi\sqrt{x_1(y - x_1)}}\, dx_1$$

$$= \tfrac{1}{2} e^{-(1/2)y} \cos^{-1}\left[\frac{(y/x_1) - x_1}{y/2} \right] \Big|_0^y = \tfrac{1}{2} e^{-(1/2)y},$$

the $\chi^2_{(2)}$ density, as was found before.

It is possible, of course, that $g(X_1, X_2)$ is discrete even though (X_1, X_2) is continuous. In such a case, we examine the range R_Y of Y and, for each $y \in R_Y$, form the set of points A_y that g maps to y. The mass function for Y at the argument y is then given by

$$p_Y(y) = \int \int_{A_y} f(x_1, x_2)\, dx_1\, dx_2.$$

Again, this process is essentially a bookkeeping task, with a major effort being the formation of $A_y = \{(x_1, x_2) : g(x_1, x_2) = y\}$ for each given y. The ease of carrying out the integration in closed form depends on how complicated the sets A_y turn out to be.

EXAMPLE 6.9   Suppose (X_1, X_2) is uniform over the unit square $[0, 1] \times [0, 1]$, and suppose

$$Y = \begin{cases} 0 \text{ if } X_1 > 2X_2, \\ 1 \text{ if } X_1 < 2X_2. \end{cases}$$

Then Y is discrete [in fact, it is Bernoulli (p)], and the mass at 0 is

$$p(Y = 0) = p_Y(0) = \int_0^{1/2} \int_{2x_2}^1 dx_1\, dx_2 = \tfrac{1}{4}.$$

Similarly,

$$p = p_Y(1) = \int_0^1 \int_{x_{1/2}}^1 dx_2\, dx_1 = \tfrac{3}{4}.$$

EXERCISES

THEORY

6.1 a. Show that $\sum_{i=0}^{j} \binom{n}{i}\binom{n}{j-i} = \binom{2n}{j}$. *Hint:* Note that

$$(1 + x)^{2n} = \sum_{j=0}^{2n} \binom{2n}{j} x^j = (1 + x)^n(1 + x)^n$$

$$= \left[\sum_{i=0}^{n} \binom{n}{i} x^i \right] \left[\sum_{k=0}^{n} \binom{n}{k} x^k \right].$$

Now find the coefficient of x^j in each of these expressions.

b. Show that if X is binomial (n, p), Y is binomial (m, p), and X and Y are independent, then $X + Y$ is binomial $(n + m, p)$.

6.2 Properties of convolutions are explored. Assume that f_1, f_2, and f_3 are density functions, and let $h * g$ denote the convolution $\int_{-\infty}^{\infty} h(x)g(y - x)\, dx$. *Note:* We can view * as an operation over a suitable class of density functions.
a. Show that $f_1 * f_2 = f_2 * f_1$.
b. Show that $f_1 * (f_2 * f_3) = (f_1 * f_2) * f_3$.
c. $f_1 * f_2$ is a density function.

6.3 a. Derive the density for $Y = X_1 X_2$ by using methods similar to those used to establish Eq. (6.2). *Hint:* For fixed y, the equation $y = x_1 x_2$ graphs as a hyperbola in the $x_1 x_2$-plane. Consider both of the cases $y > 0$ and $y < 0$.
b. Find the density of $X_1 X_2$ when X_1 and X_2 are independent, each uniform over $(0, 1)$.

6.4 Suppose X is uniform over $(0, 1)$ and the outcome x is thought to be written in a number system with base m, so that $x = \sum_{i=1}^{\infty} (x_i/m^i)$, where x_i is the ith "digit" in the "decimal" expansion of x. Suppose, however, the number is, in fact, written with a number system with base n, $1 < n < m$. Show that the expected error (the difference between what the experimenter *thinks* the number is and what it *actually* is) in "reading" x is $(m - n)/2(m - 1)$.

6.5 Show that, if (X_1, X_2) is discrete, then for any function g of two variables, $Y = g(X_1, X_2)$ is a discrete random variable.

APPLICATION

6.6 Two trains arrive at a station independently at random times in the interval $(0, t)$. Find the distribution of the time interval between the arrivals of the trains. *Hint:* If X and Y denote the arrival times, the interarrival time is $Z = |X - Y|$.

6.7   Suppose a system is composed of six components in parallel, each with lifetime distributed exponential (λ). Suppose the components are independent.
    a. Find the distribution of system lifetime.
    b. Verify, from part (a), that the mean system life is about 2.5 times the mean component life. Discuss the value of redundancy in this parallel design.

6.8   Find the density of X_1/X_2 when X_1 and X_2 are independent, each exponential (1).

6.9   This exercise shows that the sum of independent Poisson random variables is Poisson. Show that, if X_1 is Poisson (μ) and X_2 is Poisson (λ), with X_1 and X_2 independent, then $X_1 + X_2$ is Poisson ($\mu + \lambda$).

6.10   Generalize Example 6.7 to determine the density of $Y = X_1 + X_2$ when X_1 and X_2 are independent, X_1 is uniform (a, b), and X_2 is uniform (c, d). *Hint:* Assume that $b - a > d - c$ and consider three cases: $a + c \le y \le a + d$, $a + d \le y \le b + c$, and $b + c \le y \le b + d$.

6.11   Let X_1, X_2, X_3, X_4 have joint density $f_X(x_1, x_2, x_3, x_4) = 1$ if $0 < x_i < 1$, $i = 1, 2, 3, 4$.
    a. Find the density function for $Y = X_1 + X_2 + X_3$.
    b. Find the distribution of $(X_1 + X_2)/X_2$.

6.12   Suppose that X_1 and X_2 are independent, where X_i is Poisson (λ). Show that the conditional distribution of X_1, given $X_1 + X_2 = n$, is binomial. *Hint:* See Exercise 6.9.

6.13   Suppose X and Y are jointly discrete with mass function listed in the accompanying table.

y \ x	-1	0	1	2
3	.2	.1	0	.1
2	.1	0	.1	0
1	.1	.2	0	.1

    a. Find the distribution (list the mass function) of $U = X^2 - 2Y$.
    b. Find the joint distribution of U and Y.
    c. Find the conditional distribution of Y given $U = -2$.
    d. Find the joint distribution of U and $V = Y + X$.

6.14   Suppose X has density f_X, and $Y = 1/X$.
    a. Find a general formula for f_Y.
    b. Suppose, in particular, that X is uniform (0, 1). Find F_Y and determine whether $E(Y)$ exists. Compare $E(Y)$ with $1/E(X)$.
    c. Repeat part (b) where X is exponential (2).

6.15   Suppose the components of horizontal and vertical miss distance in shooting at a target are independent and each is standard normal. Find the distribution of the radial miss distance. Compute its expected value.

6.16   a. Find the distribution of $X_1 + X_2$, where X_1 and X_2 have the joint density $f(x_1, x_2) = 1/x_1, 0 < x_2 \le x_1 < 1$.
    b. Find the distribution of X_1X_2, and use it to calculate $E(X_1X_2)$. Compare the latter value with $E(X_1)E(X_2)$.

6.17 Consider each of the joint distributions shown below.
a. Compute $E(X|Y = y)$ for each y, and verify that $E(X) = E[E(X|Y)]$.
b. Display $V(X|Y)$ as a random variable, and calculate $E[V(X|Y)]$. Compare the result with $V(X)$.
c. Calculate $\text{Cov}(X, Y)$.
d. Find the distribution of $X + Y$.

Distribution 1:

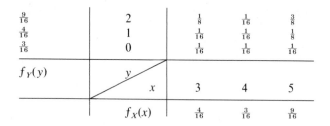

$f_Y(y)$	y		x	3	4	5
$\frac{9}{16}$	2			$\frac{1}{8}$	$\frac{1}{16}$	$\frac{3}{8}$
$\frac{4}{16}$	1			$\frac{1}{16}$	$\frac{1}{16}$	$\frac{1}{8}$
$\frac{3}{16}$	0			$\frac{1}{16}$	$\frac{1}{16}$	$\frac{1}{16}$
			$f_X(x)$	$\frac{4}{16}$	$\frac{3}{16}$	$\frac{9}{16}$

Distribution 2:

$$f_{X,Y}(x, y) = 6(1 - x - y), \qquad x > 0, y > 0, x + y < 1$$

$$f_x(x) = 3(1 - x)^2, \qquad 0 < x < 1$$

$$f_Y(y) = 3(1 - y)^2, \qquad 0 < y < 1$$

$$f_{X|Y}(x|y) = \frac{6(1 - x - y)}{3(1 - y)^2}, \qquad 0 < y < 1, 0 < x < 1 - y$$

6.18 Suppose $X_1, X_2, \ldots, X_n$ are independent and identically distributed, and say X_i has density f. Let $X_{(i)}$ denote the ith largest of the X_i, $i = 1, 2, \ldots, n$. Thus $X_{(1)}$ denotes the smallest outcome among the n outcomes on the X_i's, whereas $X_{(n)} = \max\{X_1, X_2, \ldots, X_n\}$.
a. Find the marginal density functions of $X_{(1)}$ and $X_{(n)}$ directly by forming the CDFs involved and then differentiating. *Hint:*

$$1 - F_{X_{(1)}}(x) = P(X_1 > x, X_2 > X, \ldots, X_n > x) = [1 - F(x)]^n.$$

b. Suppose X_i is uniform $(0, 1)$. Show that $X_{(1)}$ is beta-distributed with mean $1/(n + 1)$. Show that $X_{(n)}$ is beta-distributed with mean $n/(n + 1)$.
c. Give an intuitive interpretation of the results in part (b) in terms of drawing n points at random in the interval $(0, 1)$.

6.3   THE CHANGE-OF-VARIABLE TECHNIQUE*

The CDF method of finding the distribution of $g(X_1, X_2)$ has obvious limitations related to the form of the set B_y over which we must attempt to integrate f_{X_1, X_2}. This difficulty increases when we consider joint distribu-

---

* This section is optional.

tions of dimension higher than that considered in Section 6.2. In this section, we will continue our consideration of finding the distribution of $g(X_1, X_2)$, assuming we know the joint density of (X_1, X_2). We will discuss an alternative to the CDF method, which deals directly with the densities involved. This alternative procedure, which is sometimes easier to employ and which allows a wider variety of applications, is a consequence of the technique of changing variables in integration. The discussion below will be easier to understand if you recall the following result for the univariate case: Suppose X has density f_X and g is a monotone function with $g'(x)$ either positive for all x or negative for all x. Now, if $Y = g(X)$, we have, if g is increasing,

$$F_Y(y) = P(Y \le y) = P[g(X) \le y] = P[X \le g^{-1}(y)] = F_X(g^{-1}(y)).$$

In g is decreasing, we have

$$F_Y(y) = P[X \ge g^{-1}(y)] = 1 - F_X(g^{-1}(y)).$$

Thus in either case,

$$f_Y(y) = f_X(g^{-1}(y)) \left| \frac{d}{dy} g^{-1}(y) \right| \tag{6.5}$$

or

$$f_Y(y) = \frac{f_X(g^{-1}(y))}{|g'(g^{-1}(y))|}. \tag{6.6}$$

We will now consider an extension of these ideas to two-dimensional spaces. Suppose (X_1, X_2) is continuous with density f_{X_1, X_2} and range R_{X_1, X_2}. Let g_1 and g_2 be functions of x_1 and x_2, so that y_1 and y_2 are defined by the system of equations

$$y_1 = g_1(x_1, x_2), \tag{6.7}$$

$$y_2 = g_2(x_1, x_2).$$

It is desired to find the joint density of the random pair (Y_1, Y_2) defined by the compositions $Y_1 = g_1(X_1, X_2)$ and $Y_2 = g_2(X_1, X_2)$. Suppose (g_1, g_2) is one to one, and let (ψ_1, ψ_2) be the inverse mapping defined by the system of equations

$$x_1 = \psi_1(y_1, y_2), \tag{6.8}$$

$$x_2 = \psi_2(y_1, y_2).$$

In practice, ψ_1 and ψ_2 are found by solving the simultaneous equations in (6.7) for x_1 and x_2. Since Eqs. (6.7) are satisfied if and only if Eqs. (6.8) are satisfied, there is a correspondence between two-dimensional events for (Y_1, Y_2) and for (X_1, X_2). For example, if C is an event in the $y_1 y_2$-plane, then the corresponding event B in the $x_1 x_2$-plane is defined by $B = \{(x_1, x_2) : [g_1(x_1, x_2), g_2(x_1, x_2)] \in C\}$. Then the event $[(Y_1, Y_2) \in C]$ is equivalent to the event $[(X_1, X_2) \in B]$, so

$$P[(Y_1, Y_2) \in C] = P[(X_1, X_2) \in B] = \int \int_B f(x_1, x_2) \, dx_1 \, dx_2. \quad (6.9)$$

The value of the latter integral can be determined, at least in theory, since $f(x_1, x_2)$ is assumed to be known.

EXAMPLE 6.10 Suppose (X_1, X_2) is uniform over the unit square $[0, 1] \times [0, 1]$, and let (Y_1, Y_2) be defined by

$$Y_1 = X_1 - X_2,$$

$$Y_2 = 3X_2.$$

What is the probability of the event $[Y_1 < Y_2]$?

SOLUTION   From the equations defining (y_1, y_2) in terms of (x_1, x_2), it is easy to see that $y_1 < y_2$ if and only if $x_1 - x_2 < 3x_2$, that is, if and only if $x_1 < 4x_2$. Thus $B = \{(x_1, x_2) : 0 \le x_1 < 4x_2 \le 1\}$, so

$$P(Y_1 < Y_2) = \int \int_B f_{X_1, X_2}(x_1, x_2) \, dx_1 \, dx_2 = \int_0^{1/4} \int_0^{4x_2} dx_1 \, dx_2 = \tfrac{2}{16}.$$

At this point, we wish to make a change of variables in the integral in Eq. (6.9). To do so, we must impose some additional, though not seriously restrictive, assumptions. With regard to the inverse mappings ψ_1 and ψ_2, it is assumed that the four partial derivatives with respect to y_1 and y_2 exist and are continuous. Corresponding to the term $(d/dy)g^{-1}(y)$ found in Eq. (6.5) for the univariate case, the *Jacobian, J,* is defined as the determinant consisting of the derivatives of these inverse mappings. Thus

$$J = \begin{vmatrix} \dfrac{\partial \psi_1}{\partial y_1} & \dfrac{\partial \psi_1}{\partial y_2} \\[2mm] \dfrac{\partial \psi_2}{\partial y_1} & \dfrac{\partial \psi_2}{\partial y_2} \end{vmatrix} = \dfrac{\partial \psi_1}{\partial y_1} \cdot \dfrac{\partial \psi_2}{\partial y_2} - \dfrac{\partial \psi_1}{\partial y_2} \cdot \dfrac{\partial \psi_2}{\partial y_1}. \quad (6.10)$$

This expression defines a formula in y_1 and y_2, so that $J = J(y_1, y_2)$. Assuming the Jacobian does not vanish for all points in C (which will always be satisfied in our applications), the double integral in Eq. (6.9) may be evaluated by

$$P[(Y_1, Y_2) \in C] = \int \int_C f_{X_1, X_2}(\psi_1(y_1, y_2), \psi_2(y_1, y_2)) |J| \, dy_1 \, dy_2, \quad (6.11)$$

an integral with respect to y_1 and y_2* Since this result holds for every event

---

* It should be observed that $|J|$ is the absolute value of J, which is, in turn, a determinant.

C, in particular, then,

$$F_{Y_1, Y_2}(y_1, y_2) = \int_{-\infty}^{y_1} \int_{-\infty}^{y_2} f_{X_1, X_2}(\psi_1(u, v), \psi_2(u, v))|J(u, v)|\, du\, dv. \quad (6.12)$$

Differentiating this result, we find the joint density of Y_1 and Y_2 to be

$$f_{Y_1, Y_2}(y_1, y_2) = f_{X_1, X_2}(\psi_1(y_1, y_2), \psi_2(y_1, y_2))|J(y_1, y_2)|. \quad (6.13)$$

[Compare this result with that for the univariate case, Eq. (6.5).] Furthermore, the individual density functions for Y_1 and Y_2 may now be found as marginal density functions.

EXAMPLE 6.11   Consider the linear transformation determined by the equations

$$y_1 = g_1(x_1, x_2) = x_1 + x_2,$$

$$y_2 = g_2(x_1, x_2) = x_1 - x_2.$$

Then the inverses are given by

$$x_1 = \psi_1(y_1, y_2) = \tfrac{1}{2}(y_1 + y_2),$$

$$x_2 = \psi_2(y_1, y_2) = \tfrac{1}{2}(y_1 - y_2).$$

The Jacobian of this transformation is therefore

$$J = \begin{vmatrix} \tfrac{1}{2} & \tfrac{1}{2} \\ \tfrac{1}{2} & -\tfrac{1}{2} \end{vmatrix} = -\tfrac{1}{4} - \tfrac{1}{4} = -\tfrac{1}{2}.$$

Consequently, the joint density of the random variables $Y_1 = X_1 + X_2$ and $Y_2 = X_1 - X_2$ is given by

$$f_{Y_1, Y_2}(y_1, y_2) = \tfrac{1}{2} f_{X_1, X_2}\left(\frac{y_1 + y_2}{2}, \frac{y_1 + y_2}{2} \right).$$

From this expression, the marginal density function of y_1 is

$$f_{Y_1}(y_1) = \int_{-\infty}^{\infty} \tfrac{1}{2} f_{X_1, X_2}\left(\frac{y_1 + y_2}{2}, \frac{y_1 - y_2}{2} \right) dy_2,$$

or letting $z = (y_1 + y_2)/2$, we have

$$f_{Y_1}(y_1) = \int_{-\infty}^{\infty} f_{X_1, X_2}(z, y_1 - z)\, dz,$$

a formula that agrees with the convolution integral found earlier. In addition, we are now in a position to give a general formula for the density of the difference of two random variables by computing the marginal density for Y_2. Thus

$$f_Y(y_2) = \int_{-\infty}^{\infty} \tfrac{1}{2} f_{X_1,X_2} \left(\frac{y_1 + y_2}{2}, \frac{y_1 - y_2}{2} \right) dy_1,$$

or letting $z = (y_1 - y_2)/2$, we have

$$f_{Y_2}(y_2) = \int_{-\infty}^{\infty} f_{X_1,X_2}(z + y_2, z)\, dz.$$

As usual, if R_{X_1,X_2} is bounded, care must be exercised in setting the limits of integration, because f_{X_1,X_2} is zero outside R_{X_1,X_2}. Thus f_{Y_1,Y_2} is positive only for values of (y_1, y_2) in the range of (g_1, g_2).

EXAMPLE 6.12   Let $f_{X_1,X_2}(x_1, x_2) = 1$ on the unit square $[0, 1] \times [0, 1]$, and let Y_1 and Y_2 be the sum and difference as in Example 6.11. Then by Eq. (6.13), $f_{Y_1,Y_2}(y_1, y_2) = \tfrac{1}{2}$ for $(y_1, y_2) \in R_{Y_1,Y_2}$. In the determination of the marginal densities, the limits of integration depend on the values of the variables involved. Thus, for example,

$$f_{Y_1}(y_1) = \begin{cases} \displaystyle\int_{-y_1}^{y_1} \tfrac{1}{2}\, dy_2 \text{ if } 0 \le y_1 \le 1 \\[2mm] \displaystyle\int_{y_1-2}^{2-y_1} \tfrac{1}{2}\, dy_2 \text{ if } 1 \le y_1 \le 2 \end{cases} = \begin{cases} y_1 \text{ if } 0 \le y_1 \le 1, \\[2mm] 2 - y_1 \text{ if } 1 \le y_1 \le 2, \end{cases}$$

which agrees with Fig. 6.3. In addition,

$$f_{Y_2}(y_2) = \begin{cases} \displaystyle\int_{y_2}^{2-y_2} \tfrac{1}{2}\, dy_1 \text{ if } 0 \le y_2 \le 1 \\[2mm] \displaystyle\int_{-y_2}^{y_2+2} \tfrac{1}{2}\, dy_1 \text{ if } -1 \le y_2 \le 0 \end{cases} = \begin{cases} 1 - y_2 \text{ if } 0 \le y_2 \le 1, \\[2mm] 1 + y_2 \text{ if } -1 \le y_2 \le 0. \end{cases}$$

Thus the density of $X_1 - X_2$ is also triangular, a translation of the density graphed in Fig. 6.4.

In part, the success of this change-of-variables method depends on our ability to determine R_{Y_1,Y_2} explicitly, so the marginal density functions can be computed.

EXAMPLE 6.13   Let X_1 and X_2 be independent and each distributed exponential (1), so $f_{X_1,X_2}(x_1, x_2) = e^{-x_1-x_2}$, $x_1 > 0$, $x_2 > 0$. Let (Y_1, Y_2) be defined by $y_1 = g_1(x_1, x_2) = x_1 + x_2$ and $y_2 = g_2(x_1, x_2) = x_1/(x_1 + x_2)$. The transformation (g_1, g_2) is a one-to-one transformation, with inverse defined by $x_1 = \psi_1(y_1, y_2) = y_1 y_2$ and $x_2 = \psi_2(y_1, y_2) = y_1 - y_1 y_2$.

Since y_1 and y_2 are both nonnegative, y_1 can be as large as either x_1 or x_2—hence, arbitrarily large; but $y_2 = x_1/(x_1 + x_2) < 1$, since $x_2 > 0$ implies $x_1 < x_1 + x_2$. Hence, the joint density for $Y_1 = X_1 + X_2$ and Y_2

$= X_1/(X_1 + X_2)$ will be zero outside the strip $(0, \infty) \times (0, 1)$. In this case,

$$J = \begin{vmatrix} y_2 & y_1 \\ 1 - y_2 & -y_1 \end{vmatrix} = -y_1 y_2 - y_1 + y_1 y_2 = -y_1 .$$

Thus by Eq. (6.13),

$$f_{Y_1, Y_2}(y_1, y_2) = y_1 f_{X_1, X_2}(y_1 y_2, y_1 - y_1 y_2)$$

$$= y_1 e^{-y_1}, \qquad y_1 > 0, 0 < y_2 < 1.$$

This function does not depend on y_2, and R_{Y_1, Y_2} is a product set; so as a bonus for this case, we have discovered that Y_1 and Y_2 are *independent*. In addition, the marginal density functions can be computed by integration, as follows:

$$f_{Y_1}(y_1) = \int_0^1 f_{Y_1, Y_2}(y_1, y_2) \, dy_2 = y_1 e^{-y_1} \int_0^1 dy_2$$

$$= y_1 e^{-y_1}, \qquad y_1 > 0,$$

$$f_{Y_2}(y_2) = \int_0^\infty y_1 e^{-y_1} \, dy_1 = 1, \qquad 0 < y_2 < 1.$$

Thus Y_1 is gamma $(2, 1)$, and Y_2 is uniform over $(0, 1)$.

All these results extend immediately for $n > 2$, because the corresponding result on a change of variables in multiple integrals is valid for $n > 2$. We will not consider these formally here, but we state merely that the procedure is analogous to the bivariate case we have been discussing. You should have no difficulty understanding derivations that employ the change-of-variable technique for higher-dimensional situations. In many applications of this technique, we may actually desire to find the density function of only one function of X_1 and X_2, say $Y_1 = g_1(X_1, X_2)$. But the technique we have described requires that there be as many equations as unknowns. This requirement may be met by choosing a second function g_2 arbitrarily to "build up" the required dimension, keeping the mapping one to one. Typically, we might arbitrarily take $g_2(x_1, x_2) = x_2$, for example. Let us illustrate this technique with an example involving a higher-dimension situation. (You should feel free to track through this example taking $n = 2$ and performing a single integration of y_2 over the limits 0 to y_1.)

EXAMPLE 6.14   —*A Sum of Independent, Identically Distributed, Exponential Random Variables is Gamma-Distributed*   Suppose X_1, X_2, $\ldots$, X_n are independent, each exponential (1), and it is desired to compute the density function of the random variable $Y_1 = X_1 + X_2 + \cdots + X_n$. This computation may be done by taking $y_1 = x_1 + x_2 + \cdots + x_n$ and $y_j = x_j$ for $j = 2, 3, \ldots, n$. In this case, we can easily solve for the inverses, obtaining the system

$$x_1 = y_1 - y_2 - \cdots - y_n,$$

$$x_2 = y_2,$$

$$\vdots$$

$$x_n = y_n.$$

The transformation is one to one and maps the region $x_1 > 0$, $x_2 > 0, \ldots, x_n > 0$ onto the region $y_1 > 0$, $y_2 > 0, \ldots, y_n > 0$, $y_2 + y_3 + \cdots + y_n < y_1$, the latter restriction stemming from the fact that x_1 must be positive. It follows that

$$J = \begin{vmatrix} 1 & -1 & \cdots & -1 \\ 0 & 1 & \cdots & 0 \\ \vdots & \vdots & \ddots & \vdots \\ 0 & 0 & \cdots & 1 \end{vmatrix} = 1,$$

the product of the diagonal elements. Since

$$f_{X_1, X_2, \ldots, X_n}(x_1, x_2, \ldots, x_n) = \exp\left(-\sum_{i=1}^{n} x_i \right),$$

it follows that $f_{Y_1, Y_2, \ldots, Y_n}(y_1, y_2, \ldots, y_n) = e^{-y_1}$ for $y_i > 0$ and $y_2 + \cdots + y_n < y_1$. Although the formula does not depend on $y_2, \ldots, y_n$, the random variables $Y_1, Y_2, \ldots, Y_n$ are dependent because $R_{Y_1, Y_2, \ldots, Y_n}$ is not a product set.

To complete the task of finding the distribution of Y_1, we must integrate the joint density with respect to $y_2, \ldots, y_n$; this procedure requires establishing the proper limits of integration. For a given $y_1 > 0$, no variable can exceed y_1. If y_n is an arbitrary choice between 0 and y_1, then y_{n-1} can only vary between 0 and $y_1 - y_n$. Similarly, y_{n-2} varies between 0 and $y_1 - y_{n-1} - y_n$, and so on. Hence, for $y_1 > 0$,

$$f_{Y_1}(y_1) = \int_0^{y_1} \int_0^{y_1 - y_n} \int_0^{y_1 - y_{n-1} - y_n} \cdots \int_0^{y_1 - y_3 - \cdots - y_n}$$

$$\times e^{-y_1} \, dy_2 \cdots dy_{n-2} \, dy_{n-1} \, dy_n$$

$$= e^{-y_1} \int_0^{y_1} \int_0^{y_1 - y_n} \int_0^{y_1 - y_{n-1} - y_n} \cdots \int_0^{y_1 - y_4 - \cdots - y_n}$$

$$\times (y_1 - y_3 - \cdots - y_n) \, dy_3 \cdots dy_{n-2} \, dy_{n-1} \, dy_n$$

$$\vdots$$

$$= e^{-y_1} \int_0^{y_1} \frac{(y_1 - y_n)^{n-2}}{(n-2)!} \, dy_n$$

$$= -e^{-y_1} \frac{(y_1 - y_n)^{n-1}}{(n-1)!} \bigg|_0^{y_1} = \frac{y_1^{n-1}}{(n-1)!} e^{-y_1}, \qquad y_1 > 0.$$

It thus follows that Y_1 is gamma $(n, 1)$.

EXERCISES

THEORY

6.19   Suppose X_1 and X_2 are independent and jointly continuous. Let $Y_1 = g_1(X_1)$ and $Y_2 = g_2(X_2)$. Use the joint density of (Y_1, Y_2) to show that Y_1 and Y_2 are independent.

6.20   This exercise shows that a sum of independent chi-square random variables is chi-square. Suppose that X_1 and X_2 are independent, X_1 is distributed $\chi^2_{(n_1)}$, and X_2 is distributed $\chi^2_{(n_2)}$.
   a. Show that $Y_1 = X_1 + X_2$ and $Y_2 = X_1/X_2$ are independent and that Y_1 is distributed $\chi^2_{(n_1 + n_2)}$.
   b. Find the density of $(n_2/n_1)Y_2 = Z$. *Note:* This distribution is a member of the family of distributions called F-distributions, and we may state that Z is distributed $F_{(n_1, n_2)}$.
   c. Use part (a) to show that, if $X_1, X_2, \ldots, X_n$ are independent $N(0, 1)$ random variables, then $X_1^2 + X_2^2 + \cdots + X_n^2$ is distributed $\chi^2_{(n)}$.

6.21   a. Use the transformation method (with Jacobian) to find the density function of the random variable $|X_1 - X_2|$.
   b. Find the density of $|X_1 - X_2|$ where X_1 and X_2 are independent, each uniform over $(0, 1)$.

6.22   Suppose (X_1, X_2) has density g as described in Example 5.11 of Section 5.3. Show that, even though X_1 and X_2 are normal, $X_1 + X_2$ is not normally distributed.

6.23   The change of variable to polar coordinates in calculus can be viewed as a mapping defined by the system of equations $x_1 = y_1 \cos y_2$ and $x_2 = y_1 \sin y_2$, defined for $\{(y_1, y_2) : 0 \le y_1, 0 \le y_2 < 2\pi\}$. Establish whether or not this transformation is one to one, and specify the domain and range. Find the associated Jacobian.

6.24   In a manner similar to that of Exercise 6.23, a change of variable to cylindrical coordinates is a three-dimensional transformation defined by the system of equations $x_1 = y_1 \cos y_2$, $x_2 = y_1 \sin y_2$, and $x_3 = y_3$, for $y_1 \ge 0$, $0 \le y_2 < 2\pi$, and $-\infty < y_3 < \infty$. Examine this transformation for the inverse, domain, and range, and find the Jacobian.

6.25* Linear transformations are explored here. If

$$y_1 = a_{11}x_1 + a_{12}x_2,$$

$$y_2 = a_{21}x_1 + a_{22}x_2,$$

we may write

$$\begin{pmatrix} Y_1 \\ Y_2 \end{pmatrix} = \begin{pmatrix} a_{11} & a_{12} \\ a_{21} & a_{22} \end{pmatrix} \begin{pmatrix} X_1 \\ X_2 \end{pmatrix} = A \begin{pmatrix} X_1 \\ X_2 \end{pmatrix}$$

where

$$A = \begin{pmatrix} a_{11} & a_{12} \\ a_{21} & a_{22} \end{pmatrix}.$$

---

* Omit if you do not have a background in matrix algebra.

Now, if A is nonsingular,

$$\begin{pmatrix} X_1 \\ X_2 \end{pmatrix} = A^{-1} \begin{pmatrix} Y_1 \\ Y_2 \end{pmatrix},$$

where A^{-1} is given by

$$A^{-1} = \begin{pmatrix} a_{11}^* & a_{12}^* \\ a_{21}^* & a_{22}^* \end{pmatrix} = \frac{1}{a_{11}a_{22} - a_{12}a_{21}} \begin{pmatrix} a_{22} & -a_{12} \\ -a_{21} & a_{11} \end{pmatrix},$$

so

$$x_1 = a_{11}^* y_1 + a_{12}^* y_2,$$

$$x_2 = a_{21}^* y_1 + a_{22}^* y_2.$$

a. Find the Jacobian of this transformation.
b. Assume that X_1 and X_2 are independent and $N(0, 1)$ distributed. Find the joint density of Y_1 and Y_2.
c. Find the marginal distribution of Y_1, concluding that a linear function of two independent, normally distributed random variables is normally distributed.
d. Show that $J = 1/|A|$. *Note:* In general, we may replace the factor $|J|$ in the expression for f_{Y_1, Y_2} by a factor $1/|K|$, where

$$K = \begin{vmatrix} \dfrac{\partial g_1}{\partial x_1} & \dfrac{\partial g_1}{\partial x_2} \\[2mm] \dfrac{\partial g_2}{\partial x_1} & \dfrac{\partial g_2}{\partial x_2} \end{vmatrix},$$

provided that the latter determinant is nonzero.

APPLICATION

6.26 a. Let X_1 and X_2 be continuous with joint density f_{X_1, X_2}. Find the joint density of $Y_1 = X_1 X_2$ and $Y_2 = X_1/X_2$.
   b. Compute the marginal densities to verify those found in Section 6.2.
   c. Repeat parts (a) and (b) for the special case in which $f_X(x_1, x_2) = 2$, $x_1 > 0$, $x_2 > 0$, $x_1 + x_2 < 1$.

6.27 A point (X_1, X_2) is selected at random in the disc with unit radius [i.e., $f(x_1, x_2) = 1/\pi$, $x_1^2 + x_2^2 \le 1$]. Find the joint density of the distance Y_1 from the origin to the point and the positive angle Y_2 between the positive abscissa and the point. Are Y_1 and Y_2 independent?

6.28 Suppose X_1 and X_2 are independent, each $N(0, 1)$. Show that $X_1^2 + X_2^2$ and X_1/X_2 are independent.

6.29 Discuss the relationships between the univariate and bivariate transformation-of-variable results in Eqs. (6.5) and (6.13).

6.30 Let X_1 and X_2 be independent, each exponential (1). Prove that $Y_1 = X_1$ and $Y_2 = X_1/X_2$ are independent. Compute the marginal densities and compare them with those found in Example 6.13.

6.31 Suppose $X_1, X_2, \ldots, X_n$ are independent, each distributed uniformly over $(0, 1)$.

a. Find the smallest value of n such that $P(\max\{X_1, X_2, \ldots, X_n\} \geq .95) \geq .90$.

b. Find $P(.1 < \min\{X_1, X_2, \ldots, X_{10}\}, \max\{X_1, X_2, \ldots, X_{10}\} < .9)$.

c. Find $P(\max\{X_1, X_2, \ldots, X_{10}\} - \min\{X_1, X_2, \ldots, X_{10}\} > .9)$.

6.32 Suppose that X_1 is gamma $(\alpha_1, 1)$ and X_2 is gamma $(\alpha_2, 1)$, where X_1 and X_2 are independent. Show that $X_1/(X_1 + X_2)$ is beta (α_1, α_2) and that $X_1 + X_2$ is gamma $(\alpha_1 + \alpha_2, 1)$.

6.33 Suppose (X_1, X_2) has a joint mass function as listed in the accompanying table.

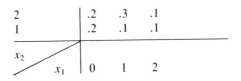

a. Find the joint distribution of $Y_1 = X_1/X_2$ and $Y_2 = X_1 + X_2$.

b. Are Y_1 and Y_2 independent?

c. Find the marginal distribution of Y_1.

6.34 Suppose (X_1, X_2) is uniform over the unit square, and

$$Y_1 = \begin{cases} 0 \text{ if } X_1 < X_2, \\ 1 \text{ if } X_1 - .5 < X_2 < X_1, \\ 2 \text{ otherwise,} \end{cases}$$

while

$$Y_2 = \begin{cases} -1 \text{ if } X_2 < 1 - X_1, \\ 1 \text{ otherwise.} \end{cases}$$

a. Find the joint distribution of (Y_1, Y_2).

b. Find the marginal distribution of Y_1.

c. Determine whether Y_1 and Y_2 are dependent.

6.4   THE LINEAR CORRELATION COEFFICIENT AND JOINT MOMENT-GENERATING FUNCTIONS*

> **Definition 6.1**   Let X_1 and X_2 be jointly distributed with means μ_1 and μ_2, respectively. Then $E(X_1^{\alpha_1} X_2^{\alpha_2})$ is called a *joint moment* of order $\alpha_1 + \alpha_2$, denoted by $\mu'_{\alpha_1, \alpha_2}$. The quantity $\mu_{\alpha_1, \alpha_2} = E[(X_1 - \mu_1)^{\alpha_1}(X_2 - \mu_2)^{\alpha_2}]$ is called a *joint central moment* of order $\alpha_1 + \alpha_2$.

---

* This section is optional.

Thus, for example, $E(X_2) = \mu'_{0,1}$ and $V(X_1) = \mu_{2,0}$. As defined previously, the joint central moment $\mu_{1,1} = E[(X_1 - \mu_1)(X_2 - \mu_2)]$ is the *covariance* of X_1 and X_2, usually denoted as $\text{Cov}(X_1, X_2)$. One of the difficulties in interpreting $\text{Cov}(X_1, X_2)$ stems from the fact that its values can be arbitrarily large or small. As mentioned before, a related measure is the *correlation coefficient*,

$$\rho(X_1, X_2) = \frac{\text{Cov}(X_1, X_2)}{\sigma_{X_1}\sigma_{X_2}}. \tag{6.14}$$

Of course, if $\sigma_{X_1} = 0$ or $\sigma_{X_2} = 0$, then Eq. (6.14) is meaningless. But in that case, X_1 or X_2 is degenerate, and, hence, X_1 and X_2 are independent. When this is the case, it makes sense to simply *define* $\rho(X_1, X_2) = 0$, since $\text{Cov}(X_1, X_2) = 0$. With this agreement, $\rho(X_1, X_2) = 0$ if and only if $\text{Cov}(X_1, X_2) = 0$. It is sometimes said that ρ is a *dimensionless quantity* since it will not involve the units in which X_1 and X_2 are measured. Furthermore, ρ always lies in the interval $[-1, 1]$, as we now show.

Theorem 6.1   Let X_1 and X_2 be jointly distributed random variables. Then $-1 \le \rho(X_1, X_2) \le 1$.

PROOF If X_1 or X_2 is degenerate, the theorem holds by the convention that $\rho = 0$. Otherwise, let $U = X_1 - \mu_{X_1}$ and $W = X_2 - \mu_{X_2}$. Now, for any ξ, $E[(\xi U + W)^2] \ge 0$ since the term in the integrand (or summand, as the case may be) is nonnegative. On the other hand, expanding $(\xi U + W)^2$ and using the linearity of the expectation operator yields

$$E[(\xi U + W)^2] = \xi^2 E(U^2) + 2\xi E(UW) + E(W^2).$$

As a quadratic equation in ξ, $\xi^2 E(U^2) + 2\xi E(UW) + E(W^2) \ge 0$ if and only if

$$4[\text{Cov}(X_1, X_2)]^2 \le 4\sigma_{X_1}^2\sigma_{X_2}^2, \tag{6.15}$$

that is, if and only if $|\text{Cov}(X_1, X_2)| \le \sigma_{X_1}\sigma_{X_2}$. Since $\sigma_{X_1} > 0$ and $\sigma_{X_2} > 0$, this result implies that $|\rho| \le 1$.

The previous comments concerning independence and zero covariance may be extended to correlation. Thus if X_1 and X_2 are independent, then $\rho(X_1, X_2) = 0$ (X_1 and X_2 are uncorrelated), but not conversely. The danger of interpreting the covariance as a measure of dependence applies also to correlation. We have seen examples that show that a cause-and-effect relationship between X_1 and X_2 (as indicated by a general functional relationship) cannot be safely inferred simply from the fact that $\rho \neq 0$. Concluding that X_1 (whatever it measures) *causes* X_2 (whatever it represents) simply because $\rho(X_1, X_2) > 0$ is dangerous, at best. In fact, from the symmetry of the definition, $\rho(X_1, X_2) = \rho(X_2, X_1)$; so the conclusion that X_2 causes X_1 would then be just as valid.

There is, however, a weak sense in which a degree of relationship between X_1 and X_2 is measured by $\rho(X_1, X_2)$. Suppose X_1 and X_2 are linearly related; that is, $X_2 = \alpha X_1 + \beta$. Let $\mu_i = E(X_i)$, $\sigma_i^2 = V(X_i)$, $i = 1, 2$, and let $\rho = \rho(X_1, X_2)$. Now, $E(X_2) = \alpha\mu_1 + \beta$ and $\sigma_2^2 = \alpha^2\sigma_1^2$. Also,

$$E(X_1 X_2) = E(\alpha X_1^2 + \beta X_1) = \alpha E(X_1^2) + \beta E(X_1) = \alpha\sigma_1^2 + \alpha\mu_1^2 + \beta\mu_1$$

and $$\text{Cov}(X_1, X_2) = \alpha\sigma_1^2 + \alpha\mu_1^2 + \beta\mu_1 - \alpha\mu_1^2 - \beta\mu_1 = \alpha\sigma_1^2.$$

Hence, $$\rho = \frac{\alpha\sigma_1^2}{\sqrt{\alpha^2\sigma_1^4}} = \frac{\alpha}{|\alpha|} = \begin{cases} 1 \text{ if } a > 0, \\ -1 \text{ if } a < 0. \end{cases}$$

Thus if X_1 and X_2 are linearly related, then they possess one of the most extreme possible correlations, ± 1.

The converse is not true but is nearly so. Suppose $\rho = 1$ or $\rho = -1$, so that $\rho^2 = 1$. Using the notation of Theorem 6.1, we see that $\rho^2 = 1$ implies that the discriminant of the quadratic equation vanishes; hence, the equation has a single root, say $\xi = \xi_0$. But this expression is equivalent to $E[(\xi_0 U + W)^2] = 0$. Recalling the definition of U and W, we have $E(\xi_0 U + W) = \xi_0 E(U) + E(W) = 0$. Hence, $V(\xi_0 U + W) = 0$, so that $\xi_0 U + W = \xi_0 X_1 + X_2 - \xi_0 \mu_{X_1} + \mu_{X_2}$ is degenerate at 0. Letting $\alpha = -\xi_0$ and $\beta = \xi_0 \mu_{X_1} + \mu_{X_2}$, we may assert that there are constants α and β such that $P(X_2 = \alpha X_1 + \beta) = 1$, or, in words, "$X_2 = \alpha X_1 + \beta$ *with probability one*." This result means that the two-dimensional distribution of X_1 and X_2 is concentrated on a line in the $x_1 x_2$-plane. In the continuous case, this distribution is degenerate, since a line is a one-dimensional subspace. In the discrete case, it merely means that all the positive entries in the table for the joint mass function lie on one diagonal of the table. Thus while ρ may not be a good measure of a general functional relationship between X_1 and X_2, it is, in the above restricted sense, a good measure of the *linear* relationship between the two random variables. For this reason, ρ is sometimes called the *linear correlation coefficient*.

General formulas for the variance of a sum $X_1 + X_2 + \cdots + X_n$ of random variables, or, more generally, of a linear combination $Y = \sum_{i=1}^{n} \alpha_i X_i + b$, are useful in statistical applications. These formulas are simplified when the random variables are uncorrelated. Since

$$E\left(\sum_{i=1}^{n} \alpha_i X_i + b\right) = \sum_{i=1}^{n} \alpha_i E(X_i) + b,$$

it is convenient to let $Y_i = \alpha_i[X_i - E(X_i)]$, $i = 1, 2, \ldots, n$, so that

$E(Y_i^2) = \alpha_i^2 V(X_i)$ and $E(Y_i Y_j) = \alpha_i \alpha_j \operatorname{Cov}(X_i, X_j)$. Then

$$Y - E(Y) = \sum_{i=1}^{n} \alpha_i X_i + b - \left[\sum_{i=1}^{n} \alpha_i E(X_i) + b\right]$$

$$= \sum_{i=1}^{n} \alpha_i [X_i - E(X_i)] = \sum_{i=1}^{n} Y_i.$$

Consequently,

$$V(Y) = E\left[\left(\sum_{i=1}^{n} Y_i\right)^2\right] = E\left(\sum_{i=1}^{n} Y_i^2 + \sum_{\substack{i=1 \\ i \neq j}}^{n} \sum_{j=1}^{n} Y_i Y_j\right)$$

$$= \sum_{i=1}^{n} E(Y_i^2) + \sum_{\substack{i=1 \\ i \neq j}}^{n} \sum_{j=1}^{n} E(Y_i Y_j),$$

or, in terms of the original linear function of the X_i's,

$$V\left(\sum_{i=1}^{n} \alpha_i X_i + b\right) = \sum_{i=1}^{n} \alpha_i^2 V(X_i) + \sum_{\substack{i=1 \\ i \neq j}}^{n} \sum_{j=1}^{n} \alpha_i \alpha_j \operatorname{Cov}(X_i, X_j). \quad (6.16)$$

Letting $\rho_{ij} = \rho(X_i, X_j)$ and $\sigma_i^2 = V(X_i)$, so that $\operatorname{Cov}(X_i, X_j) = \rho_{ij}\sigma_i\sigma_j$, we may write Eq. (6.16) as

$$V\left(\sum_{i=1}^{n} \alpha_i X_i + b\right) = \sum_{i=1}^{n} \alpha_i^2 \sigma_i^2 + \sum_{\substack{i=1 \\ i \neq j}}^{n} \sum_{i=1}^{n} \alpha_i \alpha_j \sigma_i \sigma_j \rho_{ij}$$

$$= \sum_{i=1}^{n} \alpha_i^2 \sigma_i^2 + 2 \sum_{\substack{i=1 \\ i < j}}^{n} \sum_{j=1}^{n} \alpha_i \alpha_j \sigma_i \sigma_j \rho_{ij}. \quad (6.17)$$

Equation (6.17) may be used as a general formula for computing the variance of any linear combination of $X_1, X_2, \ldots, X_n$. The random variables are said to be *pairwise uncorrelated* if $\rho(X_i, X_j) = 0$ whenever $i \neq j$. Naturally, if $X_1, X_2, \ldots, X_n$ are independent, they are pairwise uncorrelated. If the random variables are pairwise uncorrelated, then

$$V\left(\sum_{i=1}^{n} \alpha_i X_i\right) = \sum_{i=1}^{n} \alpha_i^2 V(X_i).$$

Let us now turn to moment-generating functions. Suppose $Y = g(X_1, X_2)$, where the distribution of (X_1, X_2) is known. By definition, the moment-generating function of Y is $M_Y(t) = E(e^{tY})$; this function can be evaluated by summing $\exp[tg(x_1, x_2)]$ with respect to the joint mass function of (X_1, X_2) (or integrating with respect to f_{X_1, X_2} if (X_1, X_2) is continuous). Since moment-generating functions are unique, if a function of t is obtained that is recognized to be a known moment-generating function for some distribution, then the distribution of Y must be that distribution. This feature provides a valuable technique for finding the distribution of a function of several random variables, in addition to the CDF method and transformation-of-variables technique studied earlier.

EXAMPLE 6.15   It has been claimed that a binomial distribution provides a reasonable model for the number of successes in n independent Bernoulli trials with probability p of success on each trial. Let us now prove that if $X_1, X_2, \ldots, X_n$ are independent and identically Bernoulli (p) distributed, then $\Sigma_{i=1}^{n} X_i$ is binomial (n, p). [Note that $\Sigma_{i=1}^{n} X_i$ is precisely the number of successes (ones) in the n trials $X_1, X_2, \ldots, X_n$.] Now, under the above assumptions,

$$p_{X_1, X_2, \ldots, X_n}(x_1, x_2, \ldots, x_n) = \prod_{i=1}^{n} p^{x_i}(1 - p)^{1 - x_i}$$

$$= p^{x_1 + x_2 + \cdots + x_n}(1 - p)^{n - (x_1 + x_2 + \cdots + x_n)},$$

where each x_i is 0 or 1. Let $Y = \Sigma_{i=1}^{n} X_i$. Then

$$M_Y(t) = \sum_{y=0}^{n} e^{ty} p_Y(y)$$

$$= \sum_{x_1=0}^{1} \sum_{x_2=0}^{1} \cdots \sum_{x_n=0}^{1} (pe^t)^{x_1 + x_2 + \cdots + x_n}(1 - p)^{n - (x_1 + x_2 + \cdots + x_n)}$$

$$= \sum_{x_1=0}^{1} (pe^t)^{x_1}(1 - p)^{1 - x_1} \sum_{x_2=0}^{1} (pe^t)^{x_1}(1 - p)^{1 - x_1} \cdots$$

$$\times \sum_{x_n=0}^{1} (pe^t)^{x_n}(1 - p)^{1 - x_n}$$

$$= (q + pe^t)(q + pe^t) \cdots (q + pe^t) = (q + pe^t)^n.$$

But the latter expression is recognized to be the moment-generating function for a binomial (n, p) distribution. In words, "The sum of n independent, identical, Bernoulli random variables is a binomial random variable."

In the preceding example, the final form of the moment-generating function came about by multiplying individual moment-generating functions for the X_i's. This result suggests the following theorem concerning sums of independent random variables.

Theorem 6.2   Suppose $X_1, X_2, \ldots, X_n$ are independent and X_i has moment-generating function M_i, $i = 1, 2, \ldots, n$. Let $Y = \alpha_1 X_1 + \alpha_2 X_2 + \cdots + \alpha_n X_n$, where the α_i's are real numbers. Then the moment-generating function of Y is

$$M_Y(t) = M_1(\alpha_1 t) \cdot M_2(\alpha_2 t) \cdots M_n(\alpha_n t).$$

PROOF   For given $\alpha_1, \alpha_2, \ldots, \alpha_n$,

$$e^{tY} = \exp\left[t\left(\sum_{i=1}^{n} \alpha_i X_i \right) \right] = e^{t\alpha_1 X_1} e^{t\alpha_2 X_2} \cdots e^{t\alpha_n X_n}$$

is a product of mutually independent random variables. Hence,

$$M_Y(y) = E(e^{tY}) = E(e^{t\alpha_1 X_1} e^{t\alpha_2 X_2} \cdots e^{t\alpha_n X_n})$$
$$= E(e^{t\alpha_1 X_1})E(e^{t\alpha_2 X_2}) \cdots E(e^{t\alpha_n X_n}) = M_1(\alpha_1 t)M_2(\alpha_2 t) \cdots M_n(\alpha_n t).$$

If $X_1, X_2, \ldots, X_n$ are independent, and if each X_i has the same probability distribution and, hence, the same moment-generating function M (so we may state that the X_i's are identically distributed M), then the moment-generating function of $Y = X_1 + X_2 + \cdots + X_n$ is simply M^n; that is, $M_Y(t) = [M(t)]^n$ and $X_1 + X_2 + \cdots + X_n$ is distributed M^n.

A family of probability distributions has the *reproductive property* if, for independent members $X_1, X_2, \ldots, X_n$ of the family, the distribution of $\sum_{i=1}^{n} X_i$ also belongs to the family. Among such families are certain subfamilies of the normal, Poisson, and binomial distributions and certain members of the gamma family. Indeed, in the case of the gamma, the chi-square subfamily has this property, as is seen in the following example.

EXAMPLE 6.16 —*Reproductive Property of the Chi-square Family*
Suppose $X_1, X_2, \ldots, X_n$ are independent and X_i is $\chi^2_{(n_i)}$, $i = 1, 2, \ldots, n$, and let $Y = \sum_{i=1}^{n} X_i$. Then by Theorem 6.2,

$$M_Y(t) = \prod_{i=1}^{n} M_{X_i}(t) = \prod_{i=1}^{n} (1 - 2t)^{(n_i/2)} = (1 - 2t)^{(m/2)},$$

where $m = \sum_{i=1}^{n} n_i$. But this expression is a chi-square moment-generating function, so Y is a chi-square random variable with $\sum_{i=1}^{n} n_i$ degrees of freedom.

We now consider multivariate, joint moment-generating functions, which play a role analogous to moment-generating functions in the univariate case. These multivariate, joint moment-generating functions may be employed to assist in the computation of joint moments. A more practical use, however, is to exploit their uniqueness in determining multivariate distributions of random n-tuples. In what follows, the bivariate case is emphasized; extensions to higher dimensions are straightforward.

Definition 6.2   The *joint moment-generating function* of a random pair (X_1, X_2) is the function $M_{X_1,X_2}(t_1, t_2)$ defined by

$$M_{X_1,X_2}(t_1, t_2) = E(e^{t_1 X_1 + t_2 X_2}),$$

provided this expected value exists for all (t_1, t_2) in a neighborhood of $(0, 0)$.

EXAMPLE 6.17   Suppose (X_1, X_2) has a joint mass function as listed in the accompanying table. Then

x_2		-1	1
4		.1	.3
2		.4	.2
		x_1	

$$M_{X_1}(t_1, t_2) = .4e^{-t_1 + 2t_2} + .1e^{-t_1 + 4t_2} + .2e^{t_1 + 2t_2} + .3e^{t_1 + 4t_2}.$$

When the joint moment-generating function exists, it can be shown that

$$E(X_i) = \frac{\partial M_{X_1,X_2}}{\partial t_i}\bigg|_{t_1=0, t_2=0}, \qquad i = 1, 2,$$

$$E(X_1 \cdot X_2) = \frac{\partial^2 M_{X_1,X_2}}{\partial t_1 \, \partial t_2}\bigg|_{t_1=0, t_2=0}, \qquad (6.18)$$

$$E(X_i^2) = \frac{\partial^2 M_{X_1,X_2}}{\partial t_i^2}\bigg|_{t_1=0, t_2=0}, \qquad i = 1, 2.$$

EXAMPLE 6.18   For the random pair of Example 6.17,

$$E(X_1) = [.4e^{-t_1 + 2t_2}(-1) + .1e^{-t_1 + 4t_2}(-1)$$
$$+ .2e^{t_1 + 2t_2}(1) + .3e^{t_1 + t_2}(1)] \Big|_{t_1 = 0, t_2 = 0}$$
$$= .4(-1) + .1(-1) + .2(1) + .3(1),$$

which is the expression for $E(X_1)$ by using RUS directly with the bivariate mass function.

It is also easy to identify the marginal moment-generating functions (and, hence, in principle, the marginal *distributions*) when the joint moment-generating function is given. Thus

$$M_{X_1}(t_1) = M_{X_1, X_2}(t_1, t_2) \Big|_{t_2 = 0},$$

and similarly for M_{X_2}.

EXAMPLE 6.19   For the joint distribution of Example 6.17,

$$M_{X_1}(t_1) = [.4e^{-t_1 + 2t_2} + .1e^{-t_1 + 4t_2} + .2e^{t_1 + 2t_2} + .3e^{t_1 + t_2}]_{t_2 = 0}$$
$$= .4e^{-t_1} + .1e^{-t_1} + .2e^{t_1} + .3e^{t_1} = .5e^{-t_1} + .5e^{t_1},$$

which is the moment-generating function of a random variable with mass function assigning $P(X_1 = -1) = P(X_1 = 1) = .5$, the marginal mass function of X_1.

We end this section with yet another characterization of independence of jointly distributed random variables.

> **Theorem 6.3**   Suppose (X_1, X_2) has joint moment-generating function M_{X_1, X_2}. Then X_1 and X_2 are independent if and only if
>
> $$M_{X_1, X_2}(t_1, t_2) = M_{X_1}(t_1) \cdot M_{X_2}(t_2),$$
>
> where, for $i = 1, 2$, $M_{X_i}(t_i)$ is a function of t_i alone.

PROOF   Suppose X_1 and X_2 are independent. Then so are the random variables $e^{t_1 X_1}$ and $e^{t_2 X_2}$, so

$$M_{X_1, X_2}(t_1, t_2) = E(e^{t_1 X_1} e^{t_2 X_2}) = E(e^{t_1 X_1}) \cdot E(e^{t_2 X_2})$$
$$= M_{X_1}(t_1) \cdot M_{X_2}(t_2).$$

Conversely, if $M_{X_1,X_2}(t_1, t_2) = M_{X_1}(t_1) \cdot M_{X_2}(t_2)$ and (X_1, X_2) is jointly continuous, then

$$M_{X_1,X_2}(t_1, t_2) = \int_{-\infty}^{\infty} e^{t_1 x_1} f_{X_1}(x_1)\, dx_1 \cdot \int_{-\infty}^{\infty} e^{t_2 x_2} f_{X_2}(x_2)\, dx_2$$

$$= \int_{-\infty}^{\infty} \int_{-\infty}^{\infty} e^{t_1 x_1 + t_2 x_2} f_{X_1}(x_1) \cdot f_{X_2}(x_2)\, dx_1\, dx_2.$$

But

$$M_{X_1,X_2}(t_1, t_2) = \int_{-\infty}^{\infty} \int_{-\infty}^{\infty} e^{t_1 x_1 + t_2 x_2} f_{X_1,X_2}(x_1, x_2)\, dx_1\, dx_2.$$

Since M_{X_1,X_2} is unique, the two joint density functions above that give rise to M_{X_1,X_2} must be the same, that is, $f_{X_1,X_2}(x_1, x_2) = f_{X_1}(x_1)f_{X_2}(x_2)$, which, in turn, implies X_1 and X_2 are independent.

EXAMPLE 6.20   Suppose (X_1, X_2) has joint density $f_{X_1,X_2}(x_1, x_2) = 2$, $0 < x_1 < x_2 < 1$. The joint moment-generating function of (X_1, X_2) is given by

$$M_{X_1,X_2}(t_1, t_2) = \int_0^1 \int_0^{x_2} e^{t_1 x_1 + t_2 x_2} \cdot 2\, dx_1\, dx_2$$

$$= 2 \int_0^1 e^{t_2 x_2} \cdot \frac{1}{t}(e^{t_1 x_2} - 1)\, dx_2$$

$$= \frac{2}{t_1}\left[\frac{1}{t_1 + t_2}(e^{t_1 + t_2} - 1) - \frac{1}{t_2}(e^{t_2} - 1) \right].$$

Since M_{X_1,X_2} does not factor into a function of t_1 alone times a function of t_2 alone, X_1 and X_2 are not independent, as could readily be seen by inspecting the joint density. We could, in principle, use this joint moment-generating function to find joint moments of X_1 and X_2. However, differentiation of M_{X_1,X_2} with respect to the t_i's is tedious, so it is easier to deal directly with the joint density. For example, just to find $E(X_1)$ by using M_{X_1,X_2}, we must first find $\partial M_{X_1,X_2}/\partial t_1$, a difficult task because of the presence of t_1 in the denominators. The point is, moment-generating functions are usually most useful for theoretical purposes, such as showing that a family is reproductive or examining independence. Finding moments is often best accomplished by other means.

EXERCISES

THEORY

6.35   Discuss a definition for a joint probability-generating function, and give an example.

6.36   Show that the normal family is reproductive; that is, if X_i is $N(\mu_i, \sigma_i^2)$, $i = 1$,

$2, \ldots, n$, and the X_i's are independent, then $\sum_{i=1}^{n} X_i$ is $N(\sum_{i=1}^{n} \mu_i, \sum_{i=1}^{n} \sigma_i^2)$.

6.37 Show that the subfamily of the binomial family having p fixed is reproductive; that is, if X_i is binomial (n_i, p), $i = 1, 2, \ldots, n$, and $X_1, X_2, \ldots, X_n$ are independent, then $\sum_{i=1}^{n} X_i$ is binomial $(\sum_{i=1}^{n} n_i, p)$.

6.38 Show that the subfamily of the gamma family having λ fixed is reproductive; that is, if X_i is gamma (r_i, λ), $i = 1, 2, \ldots, n$, and $X_1, X_2, \ldots, X_n$ are independent, then $\sum_{i=1}^{n} X_i$ is gamma $(\sum_{i=1}^{n} r_i, \lambda)$.

6.39 Suppose X_i is $N(\mu_i, \sigma_i^2)$ for $i = 1, 2, \ldots, n$ and $X_1, X_2, \ldots, X_n$ are independent. Show that $\sum_{i=1}^{n} [(X_i - \mu_i)/\sigma_i]^2$ is $\chi^2_{(n)}$.

6.40 Use RUS to verify that $E(\sum_{i=1}^{n} \alpha_i X_i + b) = \sum_{i=1}^{n} \alpha_i E(X_i) + b$.

6.41 Suppose $g_1(x_1, x_2) = t_1(x_1)$, a function of x_1 alone, and $g_2(x_1, x_2) = t_2(x_2)$. Use a formal expression for the joint distribution of $g_1(X_1, X_2)$ and $g_2(X_1, X_2)$ to show that

$$E[g_1(X_1, X_2)] = \int_{-\infty}^{\infty} t_1(x) f_{X_1}(x) \, dx,$$

for the continuous case, and that

$$E[g_2(X_1, X_2)] = \sum_{y \in R_{X_2}} t_2(y) p_{X_2}(y)$$

if (X_1, X_2) is discrete.

6.42 Assuming Theorem 6.3 is true, use a moment-generating function argument to show that, if X_1 and X_2 are independent, then $E(X_1 \cdot X_2) = E(X_1) \cdot E(X_2)$.

6.43 a. Show that $\rho(X_1 - \mu_{X_1}, X_2 - \mu_{X_2}) = \rho(X_1, X)$. More generally, let

$$Z_1 = \frac{X_1 - \mu_{X_1}}{\sigma_{X_1}} \quad \text{and} \quad Z_2 = \frac{X_2 - \mu_{X_2}}{\sigma_{X_2}}$$

be the standardized variables. Show that $\rho(Z_1, Z_2) = \rho(X_1, X_2)$.
b. Relate Cov (Z_1, Z_2) to $\rho(X_1, X_2)$.

6.44 Show that, for any real numbers $\alpha_1, \alpha_2, \alpha_3, \alpha_4$,

$$\text{Cov}(\alpha_1 X_1 + \alpha_3, \alpha_2 X_2 + \alpha_4) = \alpha_1 \alpha_2 \text{ Cov}(X_1, X_2).$$

APPLICATION

6.45 Show that the Poisson family is reproductive; that is, if X_i is Poisson (λ_i), $i = 1, 2, \ldots, n$, and $X_1, X_2, \ldots, X_n$ are independent, then $\sum_{i=1}^{n} X_i$ is Poisson $(\sum_{i=1}^{n} \lambda_i)$.

6.46 Find an expression for $\rho(X_1 + X_2, X_1 - X_2)$ for two jointly distributed random variables. Write the special case when X_1 and X_2 are independent.

6.47 Show that, if X_1 and X_2 are independent, then $E(X_2|X_1)$ is a degenerate random variable.

6.48 Suppose the components of $X = (X_1, X_2, X_3)$ are jointly distributed by the joint mass function $p(x_1, x_2, x_3) = \frac{1}{4}$ if $(x_1, x_2, x_3) \in \{(1, 0, 0), (0, 1, 0), (0, 0, 1), (1, 1, 1)\}$.
a. Compute $\rho(X_1 + X_2, X_3)$.
b. Compute the conditional covariance of X_1 and X_2, given $X_3 = 1$.

6.49 Suppose the joint density of X_1 and X_2 is $f(x_1, x_2) = 1/x_1, 0 < x_2 \leq x_1 \leq 1$.
a. Find $\rho(X_1, X_2)$.   b. Find $M_{X_1,X_2}(t_1, t_2)$.

6.50 Let A and B be two events, and let I_A and I_B be indicator functions [$I_A(o)$ $= 1$ if $o \in A$, $I_A(o) = 0$ if $o \notin A$].
a. Find an expression for $\rho(I_A, I_B)$.
b. Find the joint moment-generating function of I_A and I_B.

6.51 As part of her duties, an operator monitors a meter in a nuclear power plant control room. If the meter indicates a value not within the specified tolerance of the desired value, the operator makes an adjustment in the system to bring it back to the nominal (desired) value. Examination of chart records indicates that the operator completes required corrections within t minutes ($0 \leq t \leq 1$) of the occurrence of the need for correction with probability t^3. It is known that the correction time T is composed of reaction time R and action time A (i.e., $T = R + A$), and that action time is uniformly distributed within any given correction period.
a. Find the means and variances of T, R, and A.
b. Find $\rho(A, T)$.
c. Find $E(A|T = t) = r(t)$.
d. Calculate $V[r(T)]$ and $E[r(T)]$, and compare $Er(T)$ with $E(A)$. Explain your result.
e. Find $V(A|T = t) = u(t)$.
f. Find $E[u(T)]$, and compare the sum $V[r(T)] + E[u(T)]$ with $V(A)$. [Use results from parts (a) and (d).] Explain your result.

6.52 A machine produces an item whose measurable characteristic is assumed to be the value of an $N(\mu, \sigma^2)$ random variable. If the characteristic lies between $\mu - 2\sigma$ and $\mu + 2\sigma$, the item is judged satisfactory; below $\mu - 2\sigma$, the item is discarded; above $\mu + 2\sigma$, the item is sent back for reworking.
a. Describe the experiment of inspecting an item in terms of a Bernoulli model.
b. What is the expected number of items discarded in a day's inspection of 2000 items?
c. Suppose the reworked items have the measurable characteristic distributed as $N(\mu_r, \sigma_r^2)$, and if this characteristic is not between $\mu - 2\sigma$ and $\mu + 2\sigma$, the item is discarded. What are the mean and variance of the number of items ultimately judged satisfactory in a day's run of 2000 items?

6.53 Suppose (U_1, U_2, U_3) is a three-dimensional Bernoulli random triple whose distribution is identified with the parameters .1, .2, and .7.
a. Find the distribution of the number of independent trials required to first observe $(1, 0, 0)$.
b. Find the distribution of the number of independent trials required to first observe an outcome precisely like a previous outcome.
c. Find the mean and variance in part (b).

6.54 Suppose X_1, X_2, and X_3 are jointly discrete with the mass function

$$p_{(X_1,X_2,X_3)}(x_1, x_2, x_3) = \begin{cases} \frac{1}{8} & \text{if } x = (1, 0, 0), \\ \frac{1}{16} & \text{if } x = (0, 1, 0), \\ \frac{1}{16} & \text{if } x = (0, 0, 1), \\ \frac{1}{4} & \text{if } x = (1, 1, 0), \\ \frac{1}{2} & \text{if } x = (0, 1, 1). \end{cases}$$

a. Find $\rho(X_1, X_2)$, $\rho(X_1, X_3)$, and $\rho(X_2, X_3)$.
b. Compute $V(2X_1 + X_2 - X_3)$ (i) by finding the distribution of $2X_1 + X_2 - X_3$ and (ii) by using Eq. (6.16).
c. Find the joint moment-generating function of X_1, X_2, and X_3, and use it to find $E(X_1)$.

6.55 Suppose (X, Y) is distributed M, where

$$M(t_1, t_2) = \tfrac{1}{2}e^{t_1 + t_2} + \tfrac{1}{8}e^{t_1 + 2t_2} + \tfrac{1}{8}e^{2t_1 + t_2} + \tfrac{1}{4}e^{2t_1 + 2t_2}.$$

a. Find $E(X)$, $E(Y)$, $E(XY)$, and $Cov(X, Y)$.
b. Determine, from part (a), whether X and Y are independent.
c. Deduce the joint mass function of X and Y.

6.5 THE BIVARIATE NORMAL DISTRIBUTION*

In the preceding chapters, we considered three essentially different methods of developing an appropriate joint model (probability distribution for the components of a random pair). When an experiment is such that a random pair is appropriate, the nature of the experiment will usually dictate which of these methods should be employed. Let us briefly review these methods of developing joint distributions.

First, it may be natural to specify the marginal distributions of the components. Then if independence of these components can be assumed, the joint distribution is determined by multiplying the marginals. This method is by far the most common one of determining the multivariate distributions used in most applied problems.

A second method involves the use of conditional distributions and is exemplified by the microprocessor reliability example in which it is natural to specify the conditional distribution of the status of a chip selected from a given batch, as well as the marginal distribution of reliability over batches. The joint distribution is then the product of the conditional and marginal. This approach is very important, and it, too, is commonly used.

A third method of completing a model when a random pair is involved is to specify the joint distribution directly, specifying the appropriate density function (mass function, CDF, or moment-generating function) as a function of two variables. Of course, the marginal distributions are then automatically determined, and it may or may not turn out that the components are independent.

In practice, many applications that use this third approach involve only

---

* This section is optional.

two types of distributions, one for the discrete case and one for the continuous case. The most common discrete example is the multinomial distribution, which will be considered in detail in the next section. The most
common continuous case employs a two-dimensional version of the normal
distribution. You may have already gained an appreciation of certain aspects
of bivariate normal distributions through earlier exercises. Since bivariate
normal distributions are extremely important in applications, it is appropriate
to review and extend those considerations. We use an approach that exhibits
the bivariate normal density as a product of a marginal and a conditional
univariate normal density.

Suppose X, Y, and W are random variables that are related by the
equation

$$Y = \alpha + \beta X + W,$$

so Y is a linear function of X plus random "noise." Assume that X and W
are independent, that each is normally distributed, and that the noise W has
mean zero and variance σ^2. Here, we might think of X as a random stimulus
and Y as a measured response, where W is the measurement error.

Theorem 6.4   Under the above conditions, Y is normally
distributed, $E(Y|x) = \alpha + \beta x$, and $V(Y|x) = V(W) = \sigma^2$.

PROOF   Since X and W are independent and normal, so are $\alpha + \beta X$ and W,
and the sum $\alpha + \beta X + W$ is normal by the reproductive property of the
normal family. By the linearity of expectation,

$$E(\alpha + \beta X + W|X = x) = E(\alpha + \beta X|X = x) + E(W|X = x)$$

$$= \alpha + \beta x + E(W) = \alpha + \beta x$$

since X and W are independent. Similarly, since $\alpha + \beta X$ and W are independent,

$$V(\alpha + \beta X + W|X = x) = V(\alpha + \beta X|X = x) + V(W|X = x) = 0 + \sigma^2.$$

Now, consider the joint distribution of X and Y. The joint density of
X and Y is the product of two normal densities, the marginal $N(\mu_X, \sigma_X^2)$
density of X and the conditional $N(\alpha + \beta x, \sigma^2)$ density of Y given x. It is
useful to parametrize the joint distribution in terms of moments of X and Y,
$\mu_X, \sigma_X^2, \mu_Y, \sigma_Y^2$, and $\rho(X, Y) = \rho$, eliminating α, β, and σ^2.

Theorem 6.5   Under the above assumptions,

$$\beta = \frac{\sigma_Y}{\sigma_X} \cdot \rho, \tag{6.19}$$

$$\alpha = \mu_Y - \frac{\sigma_Y}{\sigma_X} \rho \cdot \mu_X, \tag{6.20}$$

and

$$\sigma^2 = \sigma_Y^2 (1 - \rho^2). \tag{6.21}$$

PROOF   First, we calculate $\rho(X, Y) = \text{Cov}(X, Y)/\sigma_X\sigma_Y$; the numerator is obtained as follows:

$$\begin{aligned}
\text{Cov}(X, Y) &= E(X \cdot Y) - \mu_X\mu_Y = E[E(X \cdot Y|X)] - \mu_X\mu_Y \\
&= E[X \cdot E(Y|X)] - \mu_X\mu_Y = E[X(\alpha + \beta X)] - \mu_X\mu_Y \\
&= \alpha\mu_X + \beta E(X^2) - \mu_X\mu_Y = \alpha\mu_X + \beta(\sigma_X^2 + \mu_X^2) - \mu_X\mu_Y,
\end{aligned}$$

so

$$\rho = \frac{\alpha\mu_X + \beta\mu_X^2 + \beta\sigma_X^2 - \mu_X\mu_Y}{\sigma_X\sigma_Y}. \tag{6.22}$$

Since

$$\mu_Y = E[E(Y|X)] = E(\alpha + \beta X) = \alpha + \beta\mu_X, \tag{6.23}$$

it follows that $\alpha = \mu_Y - \beta\mu_X$. Substituting this expression into Eq. (6.22), we obtain $\rho = \beta\sigma_X^2/\sigma_X\sigma_Y$, from which Eq. (6.19) follows. We substitute the expression for β in Eq. (6.19) into Eq. (6.23) and solve for α to obtain Eq. (6.20). Finally, since

$$\begin{aligned}
\sigma_Y^2 &= E[V(Y|X)] + V[E(Y|X)] = E(\sigma^2) + V(\alpha + \beta X) \\
&= \sigma^2 + \beta^2\sigma_X^2 = \sigma^2 + \sigma_Y^2\rho^2
\end{aligned}$$

(from Eq. (6.19), Eq. (6.21) follows.

The joint density of X and Y can now be expressed in terms of the five parameters μ_X, μ_Y, σ_X^2, σ_Y^2, and ρ. By Theorem 6.4, the conditional distribution of Y given x is $N(\alpha + \beta x, \sigma^2)$, which, by Theorem 6.5, is the

$$N\left(\mu_Y - \frac{\sigma_Y}{\sigma_X}\rho\mu_X + \frac{\sigma_Y}{\sigma_X}\rho x, \sigma_Y^2(1 - \rho^2)\right)$$

distribution. The conditional density of Y given $X = x$ is thus

$$f_{Y|X}(y|x) = \frac{1}{\sqrt{2\pi\sigma_Y^2(1 - \rho^2)}}$$

$$\exp\left\{ -\frac{1}{2\sigma_Y^2(1 - \rho^2)} \left[y - \mu_Y - \frac{\sigma_Y}{\sigma_X}\rho(x - \mu_X) \right]^2 \right\}. \tag{6.24}$$

The marginal density of X is the $N(\mu_X, \sigma_X^2)$ density,

$$f_X(x) = \frac{1}{\sqrt{2\pi\sigma_X^2}} \exp\left[-\frac{1}{2\sigma_X^2}(x - \mu_X)^2 \right]. \tag{6.25}$$

The product of these two normal densities is the bivariate normal density of the pair (X, Y). After some algebraic simplification in the exponent, the product may be expressed in the following form (see Exercise 6.76):

Definition 6.3   The *bivariate normal distribution* with parameters μ_X, μ_Y, σ_X^2, σ_Y^2, and ρ is given by the density function

$$f(x, y) = \frac{1}{2\pi\sigma_X\sigma_Y\sqrt{1 - \rho^2}} \exp\left\{ -\frac{1}{2(1 - \rho^2)} \right.$$

$$\times \left[\left(\frac{x - \mu_X}{\sigma_X}\right)^2 - 2\rho\left(\frac{x - \mu_X}{\sigma_X}\right) \right.$$

$$\left. \left. \times \left(\frac{y - \mu_Y}{\sigma_Y}\right) + \left(\frac{y - \mu_Y}{\sigma_Y}\right)^2 \right] \right\}, \tag{6.26}$$

where μ_X and μ_Y are any real numbers, σ_X^2 and σ_Y^2 are positive, and $-1 < \rho < 1$.

A typical "bell-shaped" bivariate normal surface is shown in Fig. 6.5. While the parameters μ_X, μ_Y, σ_X^2, σ_Y^2, and ρ are given in Definition 6.3 simply as constants satisfying certain restrictions, we have seen that they correspond to moments of the bivariate normal random pair. Some insight may be gained by noting that, if the parameter ρ is eliminated by setting it

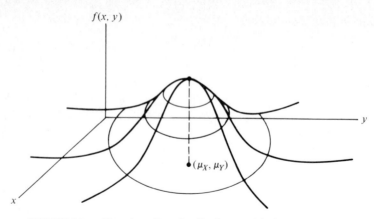

FIGURE 6.5   *Bivariate Density Surface, with Some Contours*

equal to 0, the joint density will be

$$f(x, y) = \frac{1}{2\pi\sigma_X\sigma_Y} \exp\left\{ -\frac{1}{2}\left[\left(\frac{x - \mu_X}{\sigma_X}\right)^2 + \left(\frac{y - \mu_Y}{\sigma_Y}\right)^2 \right] \right\},$$

which could then be written as the product

$$f(x, y) = \frac{1}{\sqrt{2\pi\sigma_X^2}} \exp\left[-\frac{1}{2\sigma_X^2}(x - \mu_X)^2 \right] \\ \cdot \frac{1}{\sqrt{2\pi\sigma_Y^2}} \exp\left[-\frac{1}{2\sigma_Y^2}(y - \mu_Y)^2 \right].$$

(6.27)

This expression is exactly what we would write for the joint density of X and Y if we were told that X is $N(\mu_X, \sigma_X^2)$, Y is $N(\mu_Y, \sigma_Y^2)$, and X and Y are *independent* (in which case, the correlation of X and Y would have to be 0).

Bivariate normal distributions are used in models for a wide variety of phenomena, ranging from the path of a particle under the influence of molecular bombardment to the crop yield and annual rainfall in a certain locality. Very often, the use of a bivariate normal model is prompted by the observation that it would be appropriate to model the outcomes of the two random variables separately with normal distributions. This amounts to saying that, since the marginals are normal, it may be reasonable to assume that the joint is also normal (bivariate normal). While it is true that having normal marginals is a clue tending to support the *assumption* of a joint normal distribution, care must be exercised in making such a conclusion. Recall that we examined an example in Chapter 5 of a nonnormal joint distribution with normal marginals (Example 5.11 in Section 5.3).

The joint moment-generating function for the bivariate normal distribution,

$$M(t_1, t_2) = E(e^{t_1 X + t_2 Y}) = \int_{-\infty}^{\infty} \int_{-\infty}^{\infty} e^{t_1 x + t_2 y} f(x, y) \, dx \, dy,$$

may be found by making the transformation of variables $u = (x - \mu_x)/\sigma_X$ and $v = (y - \mu_Y)/\sigma_Y$, and completing the square in the integrand. After a fair amount of algebraic detail, the joint moment-generating function can be expressed as

$$M(t_1, t_2) = \exp[\mu_X t_1 + \mu_Y t_2 + \tfrac{1}{2}(\sigma_X^2 t_1^2 + \sigma_Y^2 t_2^2 + 2\rho\sigma_X\sigma_Y t_1 t_2)]. \quad (6.28)$$

EXAMPLE 6.21   Use the bivariate normal moment-generating function to find the marginals of X and Y and to verify that the parameters are moments.

SOLUTION   The marginal moment-generating function for X is given by

$$M_X(t) = M_{X,Y}(t, 0) = \exp(\mu_X t + \tfrac{1}{2}\sigma_X^2 t^2).$$

Recall that this expression is the moment-generating function of a random variable that is $N(\mu_X, \sigma_X^2)$. It must be the case, then, that $E(X) = \mu_X$ and $V(X) = \sigma_X^2$. This result could also be verified by taking appropriate partial derivatives of $M(t_1, t_2)$ and evaluating them at $t_1 = t_2 = 0$. In a symmetrical fashion, Y is $N(\mu_Y, \sigma_Y^2)$ from the fact that its marginal moment-generating function is given by $M(0, t) = \exp(\mu_Y t + \tfrac{1}{2}\sigma_Y^2 t^2)$. That identifies four of the parameters. As for the fifth one, note that

$$\frac{\partial^2 M(t_1, t_2)}{\partial t_1 \, \partial t_2} = M(t_1, t_2)(\rho\sigma_X\sigma_Y)$$

$$+ M(t_1, t_2)(\mu_Y + \sigma_Y^2 t_2 + \rho\sigma_X\sigma_Y^2 t_1)(\mu_Y + \sigma_X^2 t_1 + \rho\mu_X\mu_Y t_2)$$

Let $t_1 = t_2 = 0$, and use the fact that $M(0, 0) = 1$ to obtain

$$E(XY) = \frac{\partial^2 M(t_1, t_2)}{\partial t_1 \, \partial t_2}\bigg|_{(0,0)} = \rho\sigma_X\sigma_Y + \mu_Y\mu_X.$$

Then $\text{Cov}(X, Y) = \rho\sigma_X\sigma_Y$ and $\rho(X, Y) = \rho$.

EXAMPLE 6.22   Suppose (X, Y) has moment-generating function

$$M(t_1, t_2) = e^{t_1 + 3t_1^2 + 2t_2^2 - 3t_1 t_2}.$$

Since this expression is of the form of Eq. (6.28), it may be concluded that (X, Y) is bivariate normal with parameters $\mu_X = 1$, $\mu_Y = 0$, $\sigma_X^2 = 6$, $\sigma_Y^2 = 4$, and $\rho = -\tfrac{3}{2}\sqrt{6} \approx -.6124$.

Note that, if $\rho = 0$, Eq. (6.28) may be written as

$$M_{(X,Y)}(t_1, t_2) = e^{t_1\mu_X + t_2\mu_Y + (1/2)(t_1^2\sigma_X^2 + t_2^2\sigma_Y^2)} = e^{t_1\mu_X + (1/2)t_1^2\sigma_X^2}e^{t_2\mu_Y + (1/2)t_2^2\sigma_Y^2}$$

$$= M_X(t_1) \cdot M_Y(t_2);$$

so, in that case, X and Y are independent. Thus the following theorem holds:

Theorem 6.6   If (X, Y) has a bivariate normal distribution, then X and Y are independent if and only if they are uncorrelated.

Remarks: First, recall that, in general, if X and Y are independent, then $\rho(X, Y) = 0$, but the converse is not true in general. But here, for this bivariate normal case, the converse is also true, as we also observed in Eq. (6.27), which shows what happens in the joint density when $\rho = 0$. Second, it is sometimes convenient to allow the limiting cases $\rho = \pm 1$ in the definition of the bivariate normal distribution. In such cases, the density function in Eq. (6.26) is undefined; indeed, the distribution is then said to be *singular,* with the unit of probability distributed along a line in the *xy*-plane. In this case, it may be argued that X and Y are still (marginally) normal, but they are linearly related (with probability 1). For example, if $\rho = 1$,

$$P\left[Y = \frac{\sigma_Y}{\sigma_X}(X - \mu_X) + \mu_Y\right] = 1,$$

as we saw before. Unless specified otherwise, we will henceforth assume that $-1 < \rho < 1$ in our discussion of bivariate normal distributions.

It is interesting to examine the form of "level curves" of the bivariate normal density function. These may be interpreted as contours at fixed elevations on the bivariate density surface (see Fig. 6.5) or as the intersection of this surface with planes parallel to the *xy*-plane. The constraining equation of such a level curve (say at height h') is obtained by letting $f(x, y) = h'$. From Eq. (6.26), this expression is equivalent to

$$\left(\frac{x - \mu_X}{\sigma_X}\right)^2 - 2\left(\frac{x - \mu_X}{\sigma_X}\right)\left(\frac{y - \mu_Y}{\sigma_Y}\right) + \left(\frac{y - \mu_Y}{\sigma_Y}\right)^2 = h, \quad (6.29)$$

where $h = -2(1 - \rho^2) \log(2\pi\sigma_X\sigma_Y \sqrt{1 - \rho^2} \, h')$. For h-values such that Eq. (6.29) is satisfied by points (x, y) in the plane, the graph of all such points is an ellipse with center at (μ_X, μ_Y) and with a semimajor axis that makes an angle α with the *x*-axis, where $\cot 2\alpha = (\sigma_Y^2 - \sigma_X^2)/-2\rho\sigma_X\sigma_Y$. Thus, for example, if $\mu_X = -2$, $\mu_Y = 3$, $\sigma_X^2 = 9$, $\sigma_Y^2 = 4$, and $\rho = .5$, a typical contour is as shown in Fig. 6.6. The semimajor axis will have positive slope if and only if X and Y have a positive correlation.

EXAMPLE 6.23   Suppose $\mu_X = \mu_Y = 0$, $\sigma_X^2 = 1$, $\sigma_Y^2 = 4$, and $\rho = .25$. Then Eq. (6.29) becomes

$$x^2 - \tfrac{1}{4}xy + \tfrac{1}{4}y^2 = h. \tag{6.30}$$

If we now rotate coordinates (to a new $x'y'$-coordinate system) through angle α, where $\cot 2\alpha = 3$, Eq. (6.30) may be written as $a(x')^2 + b(y')^2 = h$, where $a = \tfrac{5}{8} - 1/\sqrt{10}$ and $b = \tfrac{5}{8} + 1/\sqrt{10}$. A graph of the resulting ellipse for $h = 2$ is sketched in Fig. 6.7.

The rotation of coordinates considered in the example above has an interesting and important interpretation in general. A rotation can be made so the contour has no $x'y'$-term in its equation. Since the correlation term enters the bivariate normal density only through the cross product term, its elimination through a rotation may be interpreted as making a linear transformation on the original variables to achieve new random variables with zero correlation, that is, independent normally distributed random variables. Further, by standardizing the resulting independent normal variables, we can take them as standard normal. We summarize this result as follows:

> **Theorem 6.7**   Suppose (X, Y) has the general bivariate normal distribution. There exist constants a, b, c, d, e, and f such that the random variables $U = aX + bY + c$ and $V = dX + eY + f$ are independent and standard-normal-distributed.

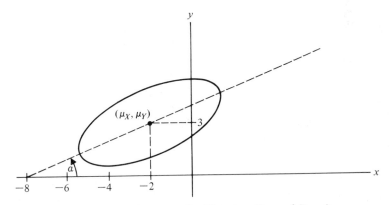

FIGURE 6.6   *Contour of a Bivariate Normal Density*

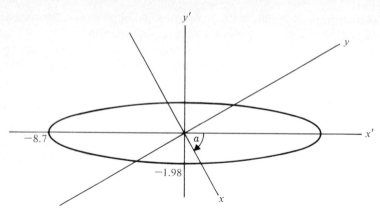

FIGURE 6.7   *Contour Before Rotation (xy-Axes) and After Rotation (x'y'-Axes)*

   In our development of the bivariate normal density, it was seen that the conditional distribution of Y given x is normal with mean $\mu_Y + (\rho\sigma_Y/\sigma_X)$ $(x - \mu_X)$ (linear in x) and variance $\sigma_Y^2(1 - \rho^2)$ (not depending on x). Since the roles of x and y could be interchanged, it follows also that the conditional distribution of X given y is $N(\mu_X + (\rho\sigma_X/\sigma_Y)(y - \mu_Y), \sigma_X^2(1 - \rho^2))$. The conditional means in these distributions are linear functions of the given variable, called *regression functions*. Thus, for example,

$$E(Y|X = x) = \mu_Y + \frac{\rho\sigma_Y}{\sigma_X}(x - \mu_X) = \left(\mu_Y - \frac{\rho\sigma_Y\mu_X}{\sigma_X}\right) + \frac{\rho\sigma_Y}{\sigma_X}x$$

is a linear function of x that is defined to be the regression function of Y on X. It is useful to define the regression function more generally than just in the bivariate normal setting.

Definition 6.4   If X and Y are jointly distributed and $E(Y|x)$ exists for all x in R_X, then $r(x) = E(Y|x)$ is called the *regression function* of Y on X. The graph of r is called the *regression curve* of Y on X.

   Since the regression function of Y on X is a linear function of x when X and Y are jointly normal, we might speak of the *regression line* of Y on X. The regression line $y = \alpha + \beta x$ (where α and β are shown above) is often used to predict the value of Y that will occur in the pair (X, Y), given $X = x$ will occur. Thus, for example, X might represent the score of a randomly selected student on a college entrance examination and Y might be that student's grade point average after the first year of college work. In making decisions about student admissions, the admissions officer may wish to predict Y for a certain applicant from the applicant's score, x. If (X, Y) is bivariate normal, a linear prediction equation could, in principle, be used for this purpose. In general, the regression function of Y on X has the following interesting and useful property:

Theorem 6.8   If X and Y are jointly distributed with finite second moments, then the regression function of Y on X is a function $r(x)$ for which $E\{[Y - r(X)]^2\}$ is minimized.

PROOF   Suppose ϕ is any function [such that $\phi(X)$ is a random variable with finite variance], and let $r(x) = E(Y|x)$. Note, first, that

$$[Y - \phi(X)]^2 = \{[Y - r(X)] + [r(X) - \phi(X)]\}^2$$
$$= [Y - r(X)]^2 + [r(X) - \phi(X)]^2 \qquad (6.31)$$
$$+ 2[Y - r(X)][r(X) - \phi(X)].$$

Now, for any function u such that $u(X)$ has a mean,

$$E\{[Y - r(X)]u(X)\} = E[Yu(X)] - E[r(X)u(X)]$$
$$= E\{E[Yu(X)|X]\} - E[r(X)u(X)]$$
$$= E\{u(X)E[Y|X]\} - E[r(X)u(X)]$$
$$= E[u(X)r(X)] - E[r(X)u(X)] = 0.$$

Thus from Eq. (6.31), with $u(X) = r(X) - \phi(X)$,

$$E\{[Y - \phi(X)]^2\} = E\{[Y - r(X)]^2\} + E\{[r(X) - \phi(X)]^2\}.$$

The right-hand side of this expression is minimized by taking $\phi(X) = r(X)$.

The result of Theorem 6.8 is similar to the fact established earlier that the constant a that minimizes $E[(X - a)^2]$ is $a = E(X)$. Two things should be noted about the regression function in the bivariate normal case:

1. The conditional mean or regression of X on Y is a linear function of y. Its graph, the regression line, has *slope* $\rho(\sigma_X/\sigma_Y)$, which is positive or negative according to whether ρ is positive or negative, and intercept $\mu_X - \rho(\sigma_X/\sigma_Y)\mu_Y$, a constant involving all five parameters.

2. The conditional variance, as a function of y, is a constant; that is, it does not depend on which value of y is given.

The five parameters in the bivariate normal distribution are often grouped so as to resemble the pair of parameters in the univariate normal. Thus the pair (μ_X, μ_Y) is often referred to as the *mean* of (X, Y), and the matrix

$$\begin{pmatrix} \sigma_X^2 & \text{Cov}(X, Y) \\ \text{Cov}(X, Y) & \sigma_Y^2 \end{pmatrix} = \begin{pmatrix} \sigma_X^2 & \rho\sigma_X\sigma_Y \\ \rho\sigma_X\sigma_Y & \sigma_Y^2 \end{pmatrix}$$

is referred to as the *covariance matrix* of (X, Y). This notation and terminology can be extended to multivariate normal distributions of higher dimensions.

EXERCISES

THEORY

6.56 Suppose X and Y are bivariate normal.
   a. Show that any linear function of X and Y, say $Z = aX + bY + c$, is normally distributed, assuming that not both a and b are zero. Find the first two moments of Z in terms of those of X and Y.
   b. Show that $U = aX + Y$ and $V = -X + aY$ have a bivariate normal distribution for any a.

6.57 Prove that, if (X, Y) is bivariate normal, $X \cos \theta + Y \sin \theta$ and $-X \sin \theta + Y \cos \theta$ are independent, where

$$\theta = \tfrac{1}{2} \cot^{-1} \left[\frac{\sigma_X^2 - \sigma_Y^2}{2 \, \text{Cov}(X, Y)} \right],$$

by verifying the following:
   a. The absolute value of the Jacobian of the transformation is 1.
   b. The inverse transformation is given by

$$x = u \cos \theta - v \sin \theta \quad \text{and} \quad y = u \sin \theta + v \cos \theta.$$

   c. $f(x(u, v), y(u, v))|J|$ factors into a function of u only times a function of v only.

6.58 Verify that the semimajor axis in an elliptical contour of the bivariate normal surface has positive slope if $\rho > 0$ and negative slope if $\rho < 0$. *Hint:* Consider the two cases $\sigma_X^2 > \sigma_Y^2$ and $\sigma_X^2 < \sigma_Y^2$.

6.59 The following definition for the general bivariate normal distribution is sometimes given: A random vector (X, Y) is said to have a bivariate normal distribution if X and Y can each be expressed as linear functions of two independent standard normal random variables U and V:

$$X = aU + bV + h \quad \text{and} \quad Y = cU + dV + k.$$

   a. Show that the moments of (X, Y) are as follows:
      i. $\mu_X = h$; $\mu_Y = k$.
      ii. $\sigma_X^2 = a^2 + b^2$; $\sigma_Y^2 = c^2 + d^2$.
      iii. $\sigma_{XY} = \text{Cov}(X, Y) = ac + bd$.

b. Show further that

$$\begin{vmatrix} a & b \\ c & d \end{vmatrix}^2 = (ad - bc)^2 = \sigma_X^2\sigma_Y^2(1 - \rho^2).$$

c. Verify that the above definition is equivalent to Definition 6.3.
d. Discuss the interpretation if $ad = bc$.
e. Use the above definition to find the general bivariate moment-generating function from $M_{(U,V)}$.

6.60 a. Show that the intersection of a bivariate normal density surface with a vertical plane is a normal density, up to a multiplicative constant. *Hint:* First, translate the center to the origin by the transformations $x' = x - \mu_X$ and $y' = y - \mu_Y$.
b. Show that, if the plane in part (a) is parallel to one of the axes, then the section is a marginal density, up to a multiplicative constant.

6.61 Suppose (X, Y) is bivariate normal. Is it possible that X and $Y - \rho X$ are independent if $\rho \neq 0$? Explain.

6.62 Show directly by integration of the joint normal density function that the marginals are normal. *Hint:* To find $f_X(x) = \int_{-\infty}^{\infty} f(x, y)\,dy$, let $v = (y - \mu_Y)/\sigma_Y$. Complete the square on v in the exponent, and then let

$$u = \frac{v - [\rho(x - \mu_X)/\sigma_X]}{\sqrt{1 - \rho^2}}.$$

6.63 Suppose (X, Y) has a singular bivariate normal distribution as discussed in a remark following Theorem 6.6, where $\rho = 1$. Find the probability that $X < 3$ and $Y < 5$, assuming that (a) $\mu_X = \mu_Y = 0$ and $\sigma_X^2 = \sigma_Y^2 = 9$; (b) $\mu_X = 3$, $\mu_Y = 5$, $\sigma_X^2 = 4$, and $\sigma_Y^2 = 25$.

6.64 If the bivariate density pictured in Fig. 6.5 is intersected with a plane parallel to the xy-plane, the equation is given by $f_{X,Y}(x, y) = k$, where k is some constant. Show that the resulting locus of points is an ellipse and that, in particular, if $\rho = 0$ and $\sigma_X = \sigma_Y$, then the ellipse is a circle.

6.65 Suppose X and Y are independent, with X distributed $N(\mu_X, \sigma^2)$ and Y distributed $N(\mu_Y, \sigma^2)$.
a. Find the probability that (X, Y) will fall in a circle of radius r centered at (μ_X, μ_Y). *Hint:* Transform to polar coordinates.
b. The term CEP (circular error probable) is used in artillery applications to denote the radius of a circle centered at (μ_X, μ_Y) such that (X, Y) falls within the circle with probability $\frac{1}{2}$. Express the CEP in terms of σ.

APPLICATION

6.66 Suppose X and Y have a bivariate normal distribution in which $E(X|Y = y) = 3.7 - .15y$, $E(Y|X = x) = .4 - .6x$, and $V(Y|X = x) = 3.64$. Find $E(X)$, $E(Y)$, $V(X)$, $V(Y)$, and $\rho(X, Y)$.

6.67 The center of a target at which a missile is aimed is taken as the origin of a rectangular system of coordinates with reference to which the point of impact has coordinates X and Y. Suppose X and Y have a bivariate normal distribution with $\mu_X = 0$, $\mu_Y = 0$, $\sigma_X = 100$ feet, and $\sigma_Y = 100$ feet, and $\rho = 0$.

a. Find the probability that the point of impact will be inside a square with side 150 feet with center at the origin and sides not necessarily parallel to the coordinate axes.

b. Find the probability that the point of impact will be inside a circle with center at the origin, having the same area.

6.68 Tables of values related to the bivariate normal CDF are available from several sources. For example, the National Bureau of Standards has compiled a volume entitled *Tables of the Bivariate Normal Distribution Function and Related Functions* (volume 50 in the U.S. National Bureau of Standards Applied Mathematics Series, issued in 1959). One of these tables gives values of

$$L(h, k, r) = \int_k^\infty \int_h^\infty g(x, y, r) \, dx \, dy,$$

where $g(x, y, r)$ is the bivariate normal density function with parameters $\mu_X = \mu_Y = 0$, $\sigma_X^2 = \sigma_Y^2 = 1$, and $r = \rho$. Now, suppose (U, V) is bivariate normal with parameters μ_U, μ_V, σ_U^2, σ_V^2, and ρ, and we have available only the tabulated values of L and the standard normal CDF Φ. Find the probability of each of the following events:

a. $P(U > a, V > b)$. *Hint:* Let $X = (U - \mu_U)/\sigma_U$ and $Y = (V - \mu_V)/\sigma_V$.

b. $P(X < h, Y < k)$.

c. $P(X < h, Y > k)$.

d. $P(|X| < h, |Y| < k)$.

e. $P[(X < c_1) \cup (X < c_2, X + Y < c_3)]$. *Hint:* Consider several cases, depending on whether $c_1 > 0$, $c_2 > 0$, $c_3 > 0$; $c_1 < 0$, $c_2 > 0$, $c_3 > 0$; and so on. Probabilities of events such as these are used in connection with certain screening tests in the inspection of drugs.

6.69 Suppose X is distributed $N(\mu, \sigma^2)$ and $Y = X^2$.

a. Show that $\rho(X, Y) \neq \pm 1$.

b. Show that the regression function of Y on X is defined by $r(x) = x^2$.

6.70 If X and Y are bivariate normal, verify that the slope of the regression line of Y on X is σ_Y^2/σ_X^2 times the slope of the regression line of X on Y.

6.71 Suppose the height H and weight W of an individual selected at random from a certain population is such that the mean weight is 140 pounds with standard deviation 12 pounds, and the mean height is 68 inches with standard deviation 4 inches. Assume $\rho(H, W) = \frac{2}{3}$. Given that an individual is 66 inches tall, what is the individual's expected weight?

6.72 Let X and Y be independent and normally distributed.

a. Show that $\rho(X + Y, X - Y) = (\sigma_X^2 - \sigma_Y^2)/(\sigma_X^2 + \sigma_Y^2)$.

b. Conclude that, if $\sigma_X^2 = \sigma_Y^2$, then $X + Y$ and $X - Y$ are independent normal random variables.

6.73 A certain density function is given by

$$f_{X,Y}(x, y) = k e^{-(4x^2 - xy + 2y^2)}.$$

a. Find k.

b. Verify that (X, Y) is bivariate normal, and determine the first and second moments of X and Y.

6.74 Suppose (X, Y) is bivariate normal with parameters $\mu_X = 5$, $\mu_Y = 10$, $\sigma_X^2 = 1$, and $\sigma_Y^2 = 25$.

a. For $\rho > 0$, find ρ if $P(4 < Y < 16|X = 5) = .954$.
b. For $\rho = 0$, find $P[(X - 5)^2 + (Y - 10)^2/25 < 2]$.
c. For $\rho^2 \neq 1$, find $P[X = Y]$.

6.75 Suppose (X, Y) is bivariate normal with mean $(1, -1)$ and covariance matrix

$$\begin{pmatrix} 2 & 1 \\ 1 & 3 \end{pmatrix}.$$

a. Show that $X + Y$ is normal, and give its mean.
b. Show that the variance of $X + Y$ is the sum of the entries in the covariance matrix.

6.76 Verify that the product of the densities in Eqs. (6.24) and (6.25) may be expressed as in Eq. (6.26).

6.6   MULTINOMIAL DISTRIBUTIONS*

We now consider an extremely useful class of discrete joint models. These models have in common the concept of tallying the numbers of times, in a series of independent repeated trials, that outcomes of various defined types occur. It is desired to find the k-variate joint distribution of these tally counts.

EXAMPLE 6.24 Suppose an experiment consists of five independent trials, each of which produces an outcome of a given type, say type 1, type 2, or type 3. We refer to such trials as *multinomial trials*. Suppose N_1, N_2, and N_3 denote the number of outcomes of type 1, 2, and 3, respectively, in the five multinomial trials. Let us denote this result by saying "(N_1, N_2, N_3) is distributed multinomial $(5, p_1, p_2, p_3)$," where $p_1 + p_2 + p_3 = 1$ and the p_i's are positive. The experiment might consist of selecting five registered voters in a certain district and determining whether each is a Republican, Democrat, or Independent. The jointly distributed random variables N_1, N_2, and N_3 then measure the numbers of Republicans, Democrats, and Independents in the sample.

If there are k possible outcome types, each multinomial trial can be imagined to result in an outcome that is a k-tuple of the form $(0, 0, 1, 0, \ldots, 0)$, where all the components are 0s except for a 1 in the position corresponding to the type of outcome that occurs on that trial. The outcome on the multinomial k-tuple $(N_1, N_2, \ldots, N_k)$ is then the sum of the n

---

* This section is optional.

multinomial trials in the experiment. We seek the joint mass function of $(N_1, N_2, \ldots, N_k)$.

There are $k + 1$ parameters in the distribution of $(N_1, N_2, \ldots, N_k)$: n together with the probabilities $p_1, p_2, \ldots, p_k$ that on each single trial the outcome will be of type 1, type 2, ..., type k, respectively. Since the list of outcome types is exhaustive, $\Sigma_{j=1}^{k} p_j = 1$ and $\Sigma_{j=1}^{k} N_j = n$. Thus we could discuss this case as a k-parameter, $(k - 1)$-dimensional distribution, where it could be agreed to take $p_k = 1 - (\Sigma_{j=1}^{k-1} p_j)$ and $N_k = n - \Sigma_{j=1}^{k-1} N_j$, for example. We will continue to view the distribution as a $(k + 1)$-parameter, k-dimensional discrete distribution with all masses concentrated on the $(k - 1)$-dimensional plane defined by $\Sigma n_j = n$; it is convenient to denote the range of $(N_1, N_2, \ldots, N_k)$ by R_N in what follows.

The joint mass function for $N_1, N_2, \ldots, N_k$ is positive only at points $(n_1, n_2, \ldots, n_k)$, where the n_i are nonnegative integers that sum to n. At such a point,

$$p_{N_1, N_2, \ldots, N_k}(n_1, n_2, \ldots, n_k) = P(N_1 = n_1, N_2 = n_2, \ldots, N_k = n_k)$$

$$= P(n_1 \text{ of the trials result in outcome type } 1, \ldots, n_k \text{ of the trials result in type } k)$$

$$= P(\text{the first } n_1 \text{ trials result in outcome type } 1, \ldots, \text{the final } n_k \text{ trials result in outcome type } k), \times (\text{the number of ways of ordering } n \text{ items of which } n_1 \text{ are of type } 1, \ldots, n_k \text{ are of type } k)$$

$$= p_1^{n_1} \cdot p_2^{n_2} \cdots p_k^{n_k} \left(\frac{n!}{n_1! n_2! \cdots n_k!} \right).$$

So $\quad p_{N_1, N_2, \ldots, N_k}(n_1, n_2, \ldots, n_k)$

$$= \binom{n}{n_1, n_2, \ldots, n_k} p_1^{n_1}, p_2^{n_2} \cdots p_k^{n_k} \quad (6.32)$$

for $n_i \in \{0, 1, \ldots, n\}$ and $\Sigma n_i = n$, where

$$\binom{n}{n_1, n_2, \ldots, n_k} = \frac{n!}{n_1! n_2! \cdots n_k!}$$

is the multinomial coefficient.

EXAMPLE 6.25   Here is a practical setting in which the multinomial arises as a model. Suppose you are going to test-fire 20 rocket motors, all of the same configuration, independently of each other. Suppose each one can fail in any one of 5 mutually exclusive ways or else be a success. If we let the respective failure probabilities be denoted by $p_1, p_2, \ldots, p_5$ and let p_6 stand for the probability of success, then it is natural to allow $N_1, N_2, \ldots, N_5$ represent the random number of failures of the respective k types and N_6 the number of successes in the 20 firings. A model for this failure vector would then be given quite naturally by the multinomial distribution, with parameters 20, $p_1, p_2, \ldots, p_6$.

EXAMPLE 6.26 Suppose systems of a certain kind are composed of two components and that these systems are tested, as they come off the assembly line, by testing the components. Each system test can be viewed as a multinomial trial, where the four possible states are "all components up," "component 1 up and component 2 down," "component 1 down and component 2 up," and "both components down." A reasonable model for the numbers of systems found in the various states in testing 50 systems is multinomial (50, p_1, p_2, p_3, p_4).

The marginal mass functions can be found by using the multinomial theorem, which asserts that, for every positive integer n,

$$(a_1 + a_2 + \cdots + a_k)^n$$

$$= \sum_{j_1=0}^{n} \sum_{j_2=0}^{n-j_1} \sum_{j_3=0}^{n-j_1-j_2} \cdots \sum_{j_{k-1}=0}^{n-j_1-\cdots-j_{k-2}} \binom{n}{j_1, j_2, \ldots, n-j_1-\cdots-j_{k-1}}$$

$$\cdot a_1^{j_1} a_2^{j_2} \cdots a_k^{n-j_1-\cdots-j_{k-1}}.$$

Using the multinomial theorem, $p_{N_1}(z)$ is found for $z \in \{0, 1, \ldots, n\}$ as follows:

$$p_{N_1}(z) = \sum_{n_2+\cdots+n_k=n-z} p_{N_1,N_2,\ldots,N_k}(z, n_2, \ldots, n_k)$$

$$= \sum_{n_2+\cdots+n_k=n-z} \frac{n!}{z!n_2! \cdots n_k!} p_1^z p_2^{n_2} \cdots p_k^{n_k}$$

$$= \frac{n!}{z!(n-z)!} p_1^z \sum_{n_2+\cdots+n_k=n-z} \frac{(n-z)!}{n_2! \cdots n_k!} p_2^{n_2} \cdots p_k^{n_k}.$$

Since $\sum_{i=1}^{k} p_i = 1$, it follows that $(1 - p_1)^{n-z} = (p_2 + \cdots + p_k)^{n-z}$. But

$$(p_2 + \cdots + p_k)^{n-z} = \sum_{n_2+\cdots+n_k=n-z} \frac{(n-z)!}{n_2! \cdots n_k!} p_2^{n_2} \cdots p_k^{n_k}$$

by the multinomial theorem. It thus follows that, for any $z \in \{0, 1, \ldots, n\}$,

$$p_{N_1}(z) = \frac{n!}{z!(n-z)!} p_1^z (1 - p_1)^{n-z},$$

which is the binomial (n, p_1) mass function. This marginal is not surprising since N_1 might typically denote the number of outcomes of type 1 in n independent repeated trials, where on each trial the outcome is of type 1 with probability p_1 [and not of type 1 with probability $(1 - p_1)$]. In such cases, a binomial (n, p_1) model is appropriate.

The moment-generating function is easily computed as another consequence of the multinomial theorem. Thus,

$$M_{N_1,N_2,\ldots,N_k}(t_1, t_2, \ldots, t_k)$$

$$= \sum_{R_N} e^{t_1 n_1 + \cdots + t_k n_k} p_{N_1,N_2,\ldots,N_k}(n_1, n_2, \ldots, n_k)$$

$$= \sum_{R_N} \binom{n}{n_1, n_2, \ldots, n_k} (p_1 e^{t_1})^{n_1} (p_2 e^{t_2})^{n_2} \cdots (p_k e^{t_k})^{n_k}$$

$$= (p_1 e^{t_1} + p_2 e^{t_2} + \cdots + p_k e^{t_k})^n.$$

Evaluating $M_{N_1,N_2,\ldots,N_k}(0, 0, \ldots, t_i, 0, 0)$, we find that the marginal moment-generating function of N_i is

$$M_{N_i}(t_i) = \left(p_i e^{t_i} + \sum_{j \neq i} p_j \right)^n,$$

which is recognized as the binomial (n, p_i) moment-generating function (recalling that $\sum_{j \neq i} p_j = 1 - p_i$ from the restriction on the parameters). Thus each marginal is binomial, as was noted earlier.

If $i \neq j$, then

$$M_{(N_i, N_j)}(t_i, t_j) = \left(p_i e^{t_i} + p_j e^{t_j} + \sum_{k \neq i,j} p_k \right)^n$$

(which is not a multinomial moment-generating function). Differentiating yields

$$E(N_i N_j) = \frac{\partial^2 M_{(N_i, N_j)}}{\partial t_i \, \partial t_j} \bigg|_{(0,0)} = n(n - 1) p_i p_j,$$

so $\text{Cov}(N_i, N_j) = -n p_i p_j$ and $\rho(N_i, N_j) = -\sqrt{p_i p_j / (1 - p_i)(1 - p_j)}$. The covariance is negative because types i and j are "competing" for tallies. If we find that n_j is large, then N_i is more likely small because, given large n_j, there are fewer remaining opportunities for outcomes of type i.

In summary, the moments up to order two of the components of N are the following:

$$E(N_i) = n p_i, \qquad V(N_i) = n p_i q_i, \qquad \text{Cov}(N_i, N_j) = -n p_i p_j.$$

The form of M also shows that the multinomial family is reproductive, if all the p_i's are held fixed.

It may be shown that the conditional distribution of some components of $(N_1, N_2, \ldots, N_k)$, say $(N_1, N_2, \ldots, N_j)$, given values of the remaining components, say $n_{j+1}, n_{j+2}, \ldots, n_k$, is multinomial $(n - (n_{j+1} + \cdots + n_k), p_1', p_2', \ldots, p_j')$, where $p_i' = p_i/(p_1 + p_2 + \cdots + p_j)$.

EXERCISES

THEORY

6.77 Suppose $(N_1, N_2, \ldots, N_5)$ is distributed multinomial $(n, p_1, \ldots, p_5)$, where the dimension is $k = 5$.
a. Show that the joint marginal of (N_1, N_2, N_3) is not necessarily multinomial.
b. Define $(W_1, W_2, \ldots, W_4) = (N_1, N_2, N_3, n - (N_1 + N_2 + N_3))$, so that the last component represents a new type of outcome, say "none of the above." Show that $(W_1, W_2, \ldots, W_4)$ is multinomial.

6.78 An urn contains t_1 items of type 1, t_2 of type 2, $\ldots$, t_k of type k. Suppose n items are sampled from the urn with replacement. Let the components of $(N_1, N_2, \ldots, N_k)$ represent the numbers of each type of item drawn.
a. Argue that $(N_1, N_2, \ldots, N_k)$ is multinomial, and state the parameter values.
b. If sampling is without replacement, argue that a multivariate hypergeometric is appropriate, and give a formula for the mass function.

6.79 Suppose $(N_1, N_2, \ldots, N_5)$ is distributed multinomial $(17, .1, .1, .3, .2, .3)$.
a. What is the conditional distribution of (N_2, N_3, N_5), given $N_1 + N_4 = 6$?
b. What is the distribution of $(N_1 + N_4, N_5, N_2 + N_3)$?
c. Use the result of part (b) to find $V(N_1 + N_4)$, and use this answer to find $\text{Cov}(N_1, N_4)$.

6.80 Find the joint moment generating function for a multinomial trial; and use this to obtain the moment generating function for a multinomial distribution.

6.81 Suppose $N = (N_1, N_2, \ldots, N_k)$ is multinomial $(n, p_1, \ldots, p_k)$. We have seen that, for $i \neq j$, $\text{Cov}(N_i, N_j) = -np_ip_j$.
a. What is the largest possible value of $|\rho(N_i, N_j)|$ in this case?
b. What selection of $p_1, p_2, \ldots, p_k$ will make N_i and N_j have the smallest possible correlation (largest possible $|\rho|$)?
c. Discuss the intuitive meaning of your result for part (b).
d. What is the conditional distribution of $(N_{m+1}, N_{m+2}, \ldots, N_k)$, given $N_1 = n_1, N_2 = n_2, \ldots, N_m = n_m$?
e. Find the distribution of $Z = \sum_{i=1}^{k} a_iN_i$, where $a_1, a_2, \ldots, a_k$ are real numbers not all zero. Find the mean and variance of Z.

6.82 Suppose (N_1, N_2) is multinomial (n, p_1, p_2).
a. Find the conditional mass function of N_2 given n_1.
b. Verify directly that $E[E(N_2)|N_1)] = E(N_2)$.
c. What is $\rho(N_1, N_2)$?

6.83 This exercise explores a multinomial approximation of a multivariate hypergeometric. Suppose n_1 items in a box are of type 1, n_2 of type 2, $\ldots$, n_k of type k, and assume that $m < \min\{n_1, n_2, \ldots, n_k\}$ items are drawn at random without replacement from the box. Let Y_i denote the number of items of type i drawn from the box.

  a. Write down an explicit expression for $p_{(Y_1, Y_2, \ldots, Y_k)}$.
  b. Find the probability p_i that the first item drawn is of type i.
  c. Argue that, if m is much smaller than the smallest of the n_i, then the probability of drawing an item of type i is nearly the same for each of the drawings.
  d. Conclude that the distribution of $(Y_1, Y_2, \ldots, Y_k)$ is approximated by a multinomial distribution with parameters $p_1, p_2, \ldots, p_k$ and m.

6.84 Show that the multinomial family is reproductive; that is, if $(X_1, X_2, \ldots, X_k)$ is multinomial $(n, p_1, \ldots, p_k)$ and $(Y_1, Y_2, \ldots, Y_k)$ is multinomial $(m, p_1, \ldots, p_k)$, then $(X_1 + Y_1, X_2 + Y_2, \ldots, X_k + Y_k)$ is multinomial $(m + n, p_1, \ldots, p_k)$ if $(X_1, X_2, \ldots, X_k)$ and $(Y_1, Y_2, \ldots, Y_k)$ are independent.

6.85 A connection between the multinomial and Poisson distributions is investigated here. Suppose $X_1, X_2, \ldots, X_n$ are independent random variables, where X_i is Poisson (λ_i), and let $Y = \Sigma_{i=1}^{n} X_i$. Show that the joint conditional distribution of $(X_1, X_2, \ldots, X_n)$, given $Y = y$, is multinomial with parameters y and $p_i = \lambda_i / \Sigma_{j=1}^{n} \lambda_i$, $i = 1, 2, \ldots, n$.

APPLICATION

6.86 For the two-component, system-testing model described in Example 6.26 suppose the two components fail independently of one another, and component 1 has reliability .9, whereas component 2 has reliability .85.
  a. Find the probability that, in testing 20 systems, the outcome is $(14, 2, 1, 3)$.
  b. Find the probability that not more than three systems have a defect.
  c. Find the expected number of systems with exactly one defect.

6.87 Suppose a loaded die gives outcome j with probability proportional to j, $j = 1, 2, \ldots, 6$. The die is tossed ten times, and $(N_1, N_2, \ldots, N_6)$ represents the numbers of times each outcome occurs.
  a. Find the probability that no 3s occur.
  b. Find the probability that $(0, 1, 0, 2, 3, 4)$ occurs.
  c. Find the probability that $\Sigma_{i=1}^{6} N_i$ is even.
  d. Find the probability that $(0, 1, 0, 2, 3) = (N_1, N_2, N_3, N_4, N_5)$, given $N_6 = 4$.
  e. Find the variance of $N_6 - N_5$.

6.88 Suppose 15% of the students at a certain high school are freshmen, 15% are sophomores, 38% are juniors, and 32% are seniors. Suppose 14 students are selected at random from the school. What is the probability that at most six are freshmen or sophomores?

6.89 Suppose X_1 is Poisson $(.5)$, X_2 is Poisson $(.3)$, X_3 is Poisson (1), and X_4 is Poisson (1.25), and these random variables are independent. Find the joint conditional distribution of (X_1, X_2, X_3, X_4) given that the X's sum to 4. *Hint:* See Exercise 6.85.

6.90 An automobile dealer has three models available, the "economy 4," the "modest 6," and the "outlandish 8." A salesman working for the dealer estimates that, for a given customer entering the showroom, the probability of no sale is .7, of selling an economy 4 is .15, of selling a modest 6 is .10, and of selling an outlandish 8 to .05. Find the probability that, for the next five customers arriving, (a) he makes no sales; (b) he sells at least one economy 4; (c) he sells at least one of each model.

6.91 Three restaurants compete for customers. Suppose that each of 15 customers will choose a restaurant at random, independent of the other customers' choices. At a minimum, how many seats must each restaurant have available so that it can accommodate the customers arriving from this group of 15, with probability of at least .80?

6.92 A class score on a test is assumed to be normal with parameters $\mu = 50$ and $\sigma = 10$. Grades are assigned by the scheme A $= 70+$, B $= 60-70$, C $= 40-60$, D $= 40-$. For a class of 20 students, what is the probability that the letter grades will be evenly distributed?

6.93 Suppose that each of a certain block of 500 storage locations in a computer may be classified as "holding a number," "used in processing," or "not used." Assume that 200 of these locations actually hold numbers and that half of the remaining locations are used in processing. Suppose 30 locations are picked at random in the block of 500.
a. What is the probability that none is used?
b. What is the expected number of those holding numbers?
c. What is the probability that not more than 10 will be used in processing?
   Hint: Recall the normal approximation to the binomial.

6.94 For the example involving test firing of rocket engines (Example 6.25), suppose there are five failure modes with respective probabilities .2, .1, .05, .05, and .03. Suppose there are ten trials.
a. What is the probability of obtaining two type 1 failures and one each of types 2, 3, 4, and 5 (and, hence, also four successes)?
b. What is the probability of obtaining no failures?
c. What is the probability of obtaining no successes?

COMPUTER-ORIENTED

Note: The following exercises exploit a calculator or computer program you are asked to prepare to generate multinomial mass values.

6.95 Write a program to generate multinomial mass values. Allow input of the parameters n and $(p_1, p_2, \ldots, p_k)$; assume $k \le 20$.

6.96 In what follows, suppose (N_1, N_2, N_3, N_4) is multinomial (n, p_1, p_2, p_3, p_4), so that the joint mass function of $(N_1, N_2, \ldots, N_4)$ is

$$p_{N_1, N_2, \ldots, N_4}(n_1, n_2, \ldots, n_4) = \frac{n!}{n_1! n_2! \cdots n_4!} p_1^{n_1} p_2^{n_2} \cdots p_4^{n_4}$$

where $\Sigma_{j=1}^4 n_j = 3$ and $(p_1, p_2, p_3, p_4) = (.1, .1, .3, .5)$. In Exercise 6.95, you developed a program to calculate such multinomial probabilities.
a. Identify the 20 points (in four-dimensional space) at which $p_{N_1, N_2, \ldots, N_4}$ is positive. List them in a column in lexicographical order. Calculate a parallel column of corresponding mass values, using your program.
b. Verify that the 20 mass values sum to 1.
c. Find the marginal mass function of N_4 as follows: For each of the 4 possible values of n_4, sum the joint mass function over n_1, n_2, n_3.

d. Compute the binomial $(3, .5)$ mass function, using your program. Compare the answer with the result in part (c). What do you conclude?

e. Calculate and list the conditional joint mass function of (N_1, N_2, N_3), given $N_4 = 1$.

f. Calculate and list the multinomial mass function with parameters $n = 2$ and $(p_1, p_2, p_3) = (.2, .2, .6)$, and compare the answer with the result in part (e). What do you conclude?

6.97 a. For a univariate discrete distribution with positive masses $p_1, p_2, \ldots, p_{20}$ at the points $n_1, n_2, \ldots, n_{20}$, the mean is given by $\Sigma_{i=1}^{20} n_i p_i$. Use this idea, together with the multinomial mass function you obtained in Exercise 6.96, to calculate $E(N_1, N_2, \ldots, N_4)$ by calculating the sum $\Sigma_{i=1}^{20} n_i p_i$. (Here, each of the 20 n_i's is a 4-tuple in $R_{N_1, N_2, \ldots, N_4}$.)

b. Calculate the 4-tuple of expected values $(E(N_1), E(N_2), \ldots, E(N_4))$, using the fact that each N_j is (marginally) binomially distributed.

c. Compare $E[(N_1, N_2, N_3, N_4)]$ and $(E(N_1), E(N_2), E(N_3), E(N_4))$. What do you conclude?

d. Define the variance-covariance matrix V of the 4-tuple $N = (N_1, N_2, \ldots, N_4)$ as follows:

$$V_{4 \times 4} = E[(\underline{N} - \underline{\mu})(\underline{N} - \underline{\mu})'],$$

where $(\underline{N} - \underline{\mu})' = (N_1 - \mu_1, \ldots, N_4 - \mu_4)$. You could calculate this expected value by forming the twenty 4×4 matrices $(n_i - \underline{\mu})(n_i - \underline{\mu})' = m_i$, and then calculating $\Sigma_{i=1}^{20} m_i p_i$. Think of an easier method, using your results for part (c) above, and use it to calculate V for the present multinomial example.

e. i. Calculate the four conditional mass functions of (N_1, N_2, N_3), given (a) $N_4 = 0$; (b) $N_4 = 1$; (c) $N_4 = 2$; (d) $N_4 = 3$.
   ii. Find the mean of each of the conditional distributions.
   iii. Calculate the sum $\Sigma_{j=1}^{4} E[(N_1, N_2, N_3)|N_4 = j] \cdot p_{N_4}(j)$. *Note:* This sum is $E_{N_4} E[(N_1, N_2, N_3)|N_4]$.
   iv. Calculate $E(N_1, N_2, N_3)$, using your work in part (c).
   v. Compare the result of part (iii) with that of part (iv). What do you conclude?

6.7   THE POISSON PROCESS

In applications, it is frequently desired to devise probability models for phenomena in which occurrences of events are observed over time. For example, it may be desired to model traffic density at a certain intersection in terms of the occurrences of automobile arrivals at the intersection over time. In this section, we will discuss a class of such models that have proved to be useful in a wide variety of applications. These models are approached from the standpoint of making a few fundamental assumptions about the types of phenomena being modeled, assumptions that might seem plausible from a consideration of the phenomena themselves. The implications of these basic assumptions are far-reaching, for they lead to a particular type of model known as the *Poisson process*. This discussion of the Poisson process provides a good example of model building.

Imagine that it is desired to model the arrival, at a Geiger counter, of subatomic particles that result from the radioactive decay of some material. After observing a large number of individual arrivals, the analyst may be willing to assume that, over small intervals of time, the probability of observing one arrival in any preselected interval is roughly proportional to the duration of the interval. It may also seem reasonable to assume that the probability of two or more arrivals in a short time interval is small and that, in fact, this probability goes to zero rapidly as the duration of the interval tends to zero. In addition, the analyst might assume that the number of arrivals in some time interval (t_0, t_1) is independent of the number of arrivals in any nonoverlapping time interval, say (t_2, t_3), where $t_2 \geq t_1$, and that the distribution of the number of arrivals in an interval of time depends only on the *length* of the interval and not on its *location* (time of day or date). We will now show that these assumptions are sufficient to characterize the Poisson process in which, for each $t > 0$, the random variable N_t denotes the number of arrivals of particles at the counter in the time interval $(0, t)$. We are thus considering a set $\{N_t : t \geq 0\}$ of jointly distributed random variables. A typical realization on $\{N_t : t \geq 0\}$ is shown in Fig. 6.8.

Now, let us make more precise the assumptions about the process $\{N_t : t \geq 0\}$ of arrivals of particles at a counter. For any two times t_1 and t_2, with $0 < t_1 < t_2$, the difference $N_{t_2} - N_{t_1}$ is called an *increment* of the process. The increment is a random variable representing the number of particle arrivals in the time interval (t_1, t_2). A process has *independent increments* provided that the random variables $N_{t_2} - N_{t_1}$, $N_{t_4} - N_{t_3}$, $N_{t_6} - N_{t_5}$, ... are independent whenever the time intervals (t_1, t_2), (t_3, t_4), (t_5, t_6), ... are pairwise disjoint. The process has *stationary increments* provided that the distribution of the increments $N_{t_2} - N_{t_1}$ and

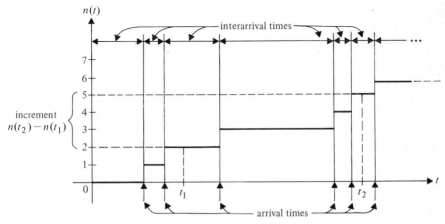

FIGURE 6.8   *A Realization n_t of the Process N_t*

$N_{t_2+h} - N_{t_1+h}$ are identical for any choice of $0 \le t_1 < t_2$ and $h > 0$. Thus in a process with stationary increments, the distribution of an increment depends only on the length of the associated time interval and not on the location of the interval. We are now ready to state formally a set of conditions that leads to the Poisson process $\{N_t : t \ge 0\}$ in which each N_t has a marginal Poisson distribution.

Theorem 6.9   Suppose $\{N_t : t \ge 0\}$ is a process with the following properties:

1. N_0 is degenerate at zero.
2. $\{N_t : t \ge 0\}$ has independent increments.
3. $\{N_t : t \ge 0\}$ has stationary increments.
4. $\lim_{t \to 0} (1/t)P(N_t = 1) = \lambda$.
5. $\lim_{t \to 0} (1/t)P(N_t \ge 2) = 0$.

Then for any $t > 0$, N_t is distributed Poisson (λt).

Let us paraphrase the assumptions in Theorem 6.9 in the language of arrivals of particles at a counter. Thus we can state the following, in loose terms:

1. There are no arrivals at time zero.
2. The numbers of arrivals in nonoverlapping time intervals are independent.
3. The distribution of the number of arrivals in a time interval depends only on the duration of the time interval.
4. For small time intervals, the probability of an arrival is proportional to the duration of the interval (λ is the proportionality constant).
5. There are no simultaneous arrivals.

A process with these properties is a *Poisson process with rate* λ.

Before outlining a proof of Theorem 6.9, we consider an example in which the time parameter is replaced by length along a computer tape.

EXAMPLE 6.27   Suppose splices in a long tape occur "at random" with one splice per 2000 feet, on the average. If a Poisson distribution is assumed for the number of splices in a given length of tape, then the process $\{N_t : t \ge 0\}$, where N_t denotes the number of splices in a randomly selected section of tape t feet long, might reasonably be assumed to be a Poisson process with a rate of 1/2000 per foot. Thus, for example, the probability that a

5000-foot section of tape has no splices is

$$P(N_{5000} = 0) = e^{-(1/2000)t}\Big|_{t=5000} = e^{-2.5} \approx .08,$$

whereas the probability that it has at most two splices is

$$P(N_{5000} \leq 2) = e^{-2.5}\left[1 + 2.5 + \frac{(2.5)^2}{2}\right] \approx .54.$$

We give an outline of a proof of Theorem 6.9 in what follows. You may wish to just skim this material on a first reading. Our argument will center on finding the function $P_n(t)$, which we define to be $P(N_t = n)$, by forming a differential equation that has $P_n(t)$ as its solution. Accordingly, let $P_n(t) = P(N_t = n)$ for any $t > 0$ and $n = 0, 1, 2, \ldots$. It is desired to form the derivative $P'_n(t)$; hence, we seek an expression for $P_n(t + h) - P_n(t)$. Now,

$$P_n(t) = P(N_t - N_0 = n) \quad \text{(by condition 1)}$$

$$= P(N_{t+h} - N_h = n) \quad \text{for any } h > 0 \quad \text{(by condition 3)}.$$

But $P_n(t + h)$ is the probability that we will observe n arrivals in $(0, t)$ and none in $(t, t + h)$, or $n - 1$ arrivals in $(0, t)$ and one in $(t, t + h)$, or $\ldots$, or, finally, no arrivals in $(0, t)$ and n in $(t, t + h)$. More precisely,

$$P_n(t + h) = P(N_t = n, N_{t+h} - N_t = 0)$$

$$+ P(N_t = n - 1, N_{t+h} - N_t = 1) + \cdots$$

$$+ P(N_t = 0, N_{t+h} - N_t = n),$$

since the events $[N_t = n, N_{t+h} - N_t = 0]$, $[N_t = n - 1, N_{t+h} - N_t = 1]$, and so on, are disjoint. But by condition (2), the events $[N_t - N_0 = n - j]$ and $[N_{t+h} - N_t = j]$ are independent, so

$$P(N_t = n - j, N_{t+h} - N_t = j) = P(N_t = n - j) \cdot P(N_{t+h} - N_t = j)$$

$$= P(N_t = n - j) \cdot P(N_h = j)$$

$$= P_{n-j}(t) \cdot P_j(h),$$

by condition (3). Thus

$$P_n(t + h) = \sum_{j=0}^{n} P_{n-j}(t) \cdot P_j(h)$$

$$= P_n(t) \cdot P_0(h) + P_{n-1}(t) \cdot P_1(h) + \sum_{j=2}^{n} P_{n-j}(t) P_j(h),$$

where $P_0(h) = 1 - [P_1(h) + P(N_h \geq 2)]$. It then follows that, for each fixed n and t,

$$P'_n(t) = \lim_{h \to 0} \frac{P_n(t + h) - P_n(t)}{h} = -\lambda P_n(t) + \lambda P_{n-1}(t). \qquad (6.33)$$

It may be shown that the solution to this system of differential difference equations is unique, subject to the natural boundary condition $P_{-1}(t) \equiv 0$ and conditions (4) and (5). (See Exercise 6.98.) But

$$\frac{d}{dt}\left[\frac{e^{-\lambda t}(\lambda t)^n}{n!}\right] = \lambda \left[\frac{e^{-\lambda t}(\lambda t)^n}{n!}\right] + \lambda \left[\frac{e^{-\lambda t}(\lambda t)^{n-1}}{(n-1)!}\right],$$

and thus the Poisson mass function with parameter λt satisfies Eq. (6.33) and the boundary condition. It follows that, for $t > 0$,

$$P_n(t) = \frac{e^{-\lambda t}(\lambda t)^n}{n!}, \qquad n = 0, 1, 2, \ldots . \qquad (6.34)$$

Hence, N_t is Poisson-distributed, as asserted.

The mean value of a variable N_t in a Poisson process with mean rate λ per unit time can be viewed as a function of t, $E(N_t) = \lambda t, t \geq 0$. The covariance of N_s and N_t may be considered to be a function $C(s, t)$, which may be determined as follows: First, note that for any t and s, where $0 < s \leq t$,

$$E(N_s N_t) = E[N_s(N_t - N_s + N_s)] = E[(N_s - N_0)(N_t - N_s) + N_s^2]$$

$$= E(N_s)E(N_t - N_s) + E(N_s^2),$$

by the independent increments assumption. By the stationary increments assumption, $E(N_t - N_s) = E(N_{t-s})$, so that

$$C(s, t) = \text{Cov}(N_s, N_t) = E(N_s)E(N_{t-s}) + E(N_s^2) - E(N_s)E(N_t)$$

$$= (\lambda s)[\lambda(t - s)] + (\lambda s + \lambda^2 s^2) - (\lambda s)(\lambda t)$$

$$= \lambda s = \lambda \min\{s, t\}.$$

Then the correlation between N_t and N_s is given by

$$\rho(N_t, N_s) = \frac{\lambda s}{\sqrt{\lambda t \lambda s}} = \sqrt{\frac{s}{t}}, \qquad 0 < s < t.$$

This quantity is large for any pair (s, t) in which s and t do not differ very much, whereas if t is much larger than s, the correlation between N_t and N_s is nearly zero.

One very important consequence of the fact that a process $\{N_t : t \geq 0\}$ is a Poisson process is that the *interarrival times*, that is, the random waiting times between arrivals, are exponentially distributed (see Fig. 6.8). Indeed, since the exponential distribution is "memoryless" (see Section 5.6), the distribution of time to the next arrival, *measured from any fixed starting time,* is the same as the distribution of time from $t = 0$ to the first arrival in the process.

Theorem 6.10   Suppose $\{N_t : t \geq 0\}$ is a Poisson process with mean rate λ, and let T denote the random waiting time until the first arrival, so that $T = \min\{t : N_t \geq 1\}$. Then T is distributed exponential (λ).

PROOF   Since the waiting time to first arrival is longer than t if and only if there have been no arrivals in the interval $(0, t)$, we have

$$1 - F_T(t) = P(T > t) = P(N_t = 0)$$

$$= e^{-\lambda t}, \qquad t > 0.$$

Thus $F_T(t) = 1 - e^{-\lambda t}$, $t > 0$, which is the exponential CDF.

EXAMPLE 6.28   Reconsider Example 6.27, dealing with splices in a computer tape. If we begin looking for a splice at any point in the tape and search (in one direction) for a splice, the length of tape inspected before the first splice is found is distributed exponential $(1/2000)$. Thus the expected length inspected is $1/\lambda = 2000$ feet, and the variance in this length is $1/\lambda^2 = 4 \times 10^6$ feet.^{2}

Since the only memoryless continuous distribution is the exponential, it follows that, if a process $\{N_t : t \geq 0\}$ has stationary and independent increments, with N_0 degenerate at zero, and if the waiting time to the next arrival does not depend on when one begins to wait, then the process is a Poisson process. Of course, this remark presupposes that the Poisson process is the only such process with exponentially distributed interarrival times (see Exercise 6.99). It also follows that interarrival times between adjoining arrivals are independent, as you are asked to show in Exercise 6.105.

EXERCISES

THEORY

6.98   This exercise involves solution of a differential difference equation. Verify that the Poisson mass function given in Eq. (6.34) is the unique solution to the differential difference equation in (6.33), subject to the constraint $P_{-1}(t) \equiv 0$ and to conditions (4) and (5) of Theorem 6.9. *Hint:* For $n = 0$, Eq. (6.33) is $P_0'(t) = -\lambda P_0(t)$, so that $P_0'(t)/P_0(t) = -\lambda$, and thus $\log P_0(t) = -\lambda t + \log c$. Then

$$P_0(t) = ce^{-\lambda t} = c\left(1 - \lambda t + \frac{\lambda^2 t^2}{2} - \cdots\right).$$

But by conditions (4) and (5), $P_0(t) \approx 1 - \lambda t$ for small t. The only choice of the real number c is thus $c = 1$. Next, solve the equation for $n = 1$:

$$P_1'(t) = -\lambda P_1(t) + \lambda P_0(t) = -\lambda P_1(t) + \lambda e^{-\lambda t}.$$

After P_1 has been found, P_2 is obtained by solving the corresponding differential equation; then P_3, then P_4, and so on, may be found. [You may use induction, for example, to show that Eq. (6.34) follows in general.]

6.99   a. Show that, if $\{N_t : t \geq 0\}$ is a process with stationary and independent increments, with N_0 degenerate at zero, and if the interarrival times T_1, $T_2, \ldots$ are independent exponential (λ) random variables, then $\{N_t : t \geq 0\}$ is a Poisson process. *Hint:* It follows that the waiting time to the nth arrival, $W_n = \Sigma_{i=1}^n T_i$, is gamma-distributed. Now, observe that $N_t \leq n$ if and only if $W_n \geq t$ and derive the CDF of N_t from that of W_n. By Exercise 6.98, it suffices to show that N_t is distributed exponential (λt) for any $t > 0$.

   b. Does the fact that $T_1, T_2, \ldots$ are independent imply that $\{N_t : t \geq 0\}$ has independent increments?

6.100 Suppose $\{X_t : t \geq 0\}$ is a stochastic process with stationary and independent increments, where X_0 is degenerate at zero.

   a. Show that the mean value function $m(t) = E(N_t)$ is linear. *Hint:*

$$m(t + s) = E(X_{t+s}) = E(X_{t+s} - X_s + X_s) = E(X_t) + E(X_s)$$

$$= m(t) + m(s).$$

   b. Show that the covariance function C is of the form $C(t, s) = \sigma^2 \min\{t, s\}$, where $\sigma^2 = V(X_1)$. *Hint:* Use an argument similar to the one used for the Poisson process.

6.101 Let $\{N_t : t \geq 0\}$ and $\{M_t : t \geq 0\}$ be independent Poisson processes with mean rates λ and v, respectively. For each $t \geq 0$, let $Q_t = N_t + M_t$. Show that $\{Q_t : t \geq 0\}$ is a Poisson process with mean rate $\lambda + v$.

6.102 This exercise involves the distribution of distance to the nearest neighbor. Suppose the trees in a certain region are distributed in a manner such that the number of trees in a randomly selected plot of area A is a Poisson-distributed random variable that is independent of the number of trees in any other nonoverlapping plot. In addition, assume that the distribution of the number of trees in a randomly selected plot depends only on the area of the plot. Now suppose a fixed location is selected in the region. What is the distribution of the distance to the nearest tree?

6.103 Derive Theorem 6.10 by using the law of the mean for integrals, as follows: Assume T has a density function f. Then for small $dt > 0$, and for $t > 0$,

$$f(t) \, dt \approx P(t \leq T \leq t + dt) = P(N_t = 0, N_{t+dt} - N_t \geq 1)$$

$$= P(N_t = 0)P(N_{dt} \geq 1) = e^{-\lambda t}(\lambda \, dt).$$

Now find $f(t) = \lim_{dt \to 0} f(t) \, dt/dt$.

6.104 Let W_r denote the waiting time until the rth arrival in a Poisson process.

   a. Define W_r explicitly in terms of $\{N_t : t \geq 0\}$.

   b. Show that W_r is gamma-distributed.

6.105   Suppose T_i denotes the ith interarrival time in a Poisson process; that is, T_i is the length of time for which $N_t = i$.
   a. Show that the T_i are independent and identically distributed.
   b. Verify that the random variable W_r of Exercise 6.104 is given by $W_r = \Sigma_{i=1}^{r} T_i$.
   c. Use the results of part (a) and of Exercise 6.104 to conclude that T_i is exponentially distributed.

6.106   This exercise deals with a connection between the binomial distribution and the Poisson process. Suppose $\{N_t : t \geq 0\}$ is a Poisson process. Show that

$$P(N_s = j | N_t = k) = \binom{k}{j} p^j q^{k-j},$$

where $0 < s < t$ and $p = s/t$.

6.107   A compound Poisson process is explored here. A process $\{X_t : t \geq 0\}$ is said to be a *compound Poisson process* if, for any $t \geq 0$, $X_t = \Sigma_{j=1}^{N_t} U_j$, where $\{N_t : t \geq 0\}$ is a Poisson process and $\{U_j : j = 1, 2, \ldots\}$ is a sequence of independent, identically distributed random variables, independent of $\{N_t : t \geq 0\}$.
   a. Verify, at least intuitively, that $\{X_t : t \geq 0\}$ has stationary and independent increments.
   b. Verify that X_0 is degenerate at zero. *Note:* the empty sum $\Sigma_{j=1}^{0} U_j$ is defined to be zero.
   c. Show that the moment-generating function of X_t can be given in terms of those for N_t and U as follows:

$$M_{X_t}(\alpha) = e^{\lambda t M_U(\alpha) - 1} = M_{N_t}(\log M_U(\alpha)),$$

   where λ is the rate of $\{N_t : t \geq 0\}$.
   d. Assuming the moments exist, show that (i) $\{X_t : t \geq 0\}$ has mean value function $m(t) = \lambda t E(Y)$ and (ii) $\{X_t : t \geq 0\}$ has covariance function $C(s, t) = \lambda E(Y^2)\min\{s, t\}$. *Hint:* Consider, first, the conditional moment-generating function of X_i given N_t, and then take its expectation with respect to the distribution of N_t.

6.108   Suppose the number of orders received in a day for a certain type of material stored at a warehouse is a Poisson process with rate v per day. Assume that the quantity of material of this type requested on each order is gamma-distributed with parameters r and λ, and that the quantity requested on any order is independent of the quantities requested on other orders and is independent of the number of orders received. Show that the total quantity of items requested in d days is negative binomial distributed, and find the mean and variance of the quantity requested. *Hint:* See Exercise 6.107.

APPLICATION

6.109   a. Explain why, for a Poisson process, it is true that N_α and $N_{t+\alpha} - N_t$ are identically distributed.
   b. Under what conditions (on t) are the random variables N_α and $(N_{t+\alpha} - N_t)$ independent?

c. Explain why specifying the marginal distributions of the N_t for $t > 0$ is sufficient to determine the joint distribution of any finite collection of the random variables, say $N_{t_1}, N_{t_2}, \ldots, N_{t_k}$, where $0 < t_1 < t_2 < \cdots < t_k$.

6.110   In Example 6.28, explain why it is unnecessary for it to be stipulated that the search is conducted in only one direction along the tape, as long as the total length of tape inspected is properly specified.

6.111   In a certain book of 520 pages, 390 typographical errors occur. What is the probability that 4 pages selected at random in the book contain no errors? *Hint:* Assume errors occur in the book in such a way that they can be modeled as occurrences of a Poisson process.

6.112   Ships of a certain type destined for port A arrive at an average rate of four per day. It takes one day for a ship to be serviced at port A. There are five docks at port A that can accommodate these ships. If the docks are all in service when a ship of this type arrives, the ship immediately departs for a second (presumably less desirable) port (say port B). Assume "day" means "during daylight hours," and assume the docks are all available at the beginning of each day.
a. What is the probability that a given ship will have to go to port B?
b. How many docks must port A have so that a given ship can be serviced there with probability of at least .95?
c. Suppose port A has the number of docks found in part (b). What is the expected percentage of its docks that are used in a given day?

6.113   a. Each of the delivery trucks of a certain company breaks down on the average of once a month, in accordance with a Poisson process. Assuming that the company has 10 delivery trucks and that the breakdowns of any one of the trucks are independent of the breakdowns of any other, describe the process that counts the arrivals of trucks at the company garage for repair.
b. Suppose it takes the company mechanic half a workday to repair each breakdown. What is the probability that a truck will arrive at the garage for repair while the mechanic is occupied with a repair?
c. What is the average amount of time per month the mechanic spends making repairs? *Hint:* Settle for a reasonable close upper bound here.

6.114   This exercise deals with selective shoppers. Suppose shoppers pass a store in accordance with a Poisson process, with a rate of five per minute. Assume that each shopper passing the store enters the store with probability .1, independent of what other shoppers decide to do. (Is this a reasonable assumption, in practice?) Show that shoppers enter the store in accordance with a Poisson process, with rate of .5 per minute. *Hint:* Given that k shoppers pass the store in a given time interval, the number entering is distributed binomial $(k, .1)$. Now, find the unconditional distribution of the number entering.

6.115   Let T denote the number of shoppers entering the store of Exercise 6.114 in a 10-minute period. Find $E(T)$, $V(T)$, and $P[T > E(T)]$.

6.116   Errors occur in the transmission of data over a certain communication link on the average of one per 1000 characters. What is the probability that a message 850 characters long has no transmission errors? At least two transmission errors? What is the expected number of errors?

6.117   A certain recruiting station recruits, on the average, four men per day. In recent weeks, it has also recruited about two women per week.
   a. What is the probability that at most two people will be recruited at the station on a randomly selected day next week?
   b. What is the probability that one man and one woman will be recruited?

6.118   A dam was constructed on the Allegheny River in the late 1960s. When the project was proposed, the Army Corps of Engineers told the residents of a town upriver that, on the basis of hydrological data, the river would flood only once in 50 years. Two years after the project was completed, the Northeast was hit with heavy rains and the river flooded.
   a. What is the probability of this event happening?
   b. What is the probability of another flood in the next 25 years?

6.8   MARKOV CHAINS

There are many applications of probability models for phenomena in which there is "evolution" with time, or perhaps change with respect to some other index such as volume or distance. For example, it may be desired to model the depth $d(t)$ of a river at time t, as measured from some fixed origin. If the analyst is interested in future depths (e.g., in an attempt to predict the frequency and duration of floods), it is natural to consider the depths at (future) times t_1, t_2, t_3, and so on as observed values of random variables X_{t_1}, X_{t_2}, X_{t_3}, and so on. Since the set of possible future times is large, perhaps the interval $(0, \infty)$, the analyst may be concerned with infinitely many jointly distributed random variables, say those in the class $\{X_t : t > 0\}$. A set $\{X_t : t \in T\}$ of random variables is called a *stochastic process*. The set T is called the *index set* of the stochastic process. If the index set is finite, say $T = \{t_1, t_2, \ldots, t_n\}$, then the stochastic process $\{X_t : t \in T\}$ is simply a set of n jointly distributed random variables. In the following discussion, we will be primarily interested in stochastic processes with infinite index sets.

   In this section, a class of relatively simple stochastic processes are considered, those having a countably infinite index set. They are called *random walks*. In order to get an idea of the kinds of problems we wish to model, consider the following example (we'll return with a solution later on).

EXAMPLE 6.29   Consider a system (rocket, computer, bicycle, or whatever) that is imagined to have a single failure mode, a "weak link" whose presence may make the system fail. Suppose, on any test of the system, the failure mode fails with probability .1. When there is a failure, an attempt is made to remove the failure mode by redesigning the system. Suppose the failure mode is successfully removed (immediately after it has failed) with probability .4; otherwise, the failure mode remains unrepaired for the next system trial. It is desired to compute the reliability of the system after a sequence of trials and to determine the nature of the "reliability growth" as testing proceeds.

Stochastic processes are used as models for a wide variety of phenomena. They are often used in connection with phenomena whose characteristics evolve in time, so that a separate random variable X_t is needed at each time t. In practice, we wish to take advantage of this dependence on time of the random variables in the stochastic process, so as to be able to gain inference about the future from observation of the past. In our model, this means nothing more than trying to take advantage of observed values of X_t for $t \leq t_0$ to gain some idea of the accompanying outcomes that will be observed for X_t when $t > t_0$. To do so, we might use conditional distributions of random variables X_t with $t > t_0$, given outcomes on random variables X_t for $t \leq t_0$, where t_0 is viewed as the "present" in our analogy.

EXAMPLE 6.30 In many inventory situations, we can view demands for various items in each time period (say each day) as observed values of random variables. It is of great interest to be able to take advantage of the demand history for various types of items held in stock in some storage facility, such as a warehouse. It is desired to hold no more items than necessary in stock, since the storage of items and the investment of capital in the inventory represent expenses. On the other hand, if, in a given day, more items of a certain type are demanded than there are in stock, then expense is again incurred due to the shortage. Thus understanding the demand process may be helpful in determining the most efficient stock levels and reorder policies.

EXAMPLE 6.31 In the transmission of signals, such as telephone conversations or telegraph messages, the signal received may be viewed as a stochastic process in which X_t may represent the strength of the signal at time t. Or if we imagine that the signal is composed of a series of 0s and 1s, then X_t might denote the tth symbol received.

In some of these examples, the "future" depends on the "past" only through the "most recent" observation (let us call this observation the "present" observation). That is, the conditional distribution of X_t for any $t > t_0$, given $X_{t_1}, X_{t_2}, \ldots, X_{t_n}$, where $t_1 < t_2 < \cdots < t_n \leq t_0$, may depend only on the value of the most recent observation X_{t_n}, or

$$P(X_t \leq x | X_{t_1}, X_{t_2}, \ldots, X_{t_n}) = P(X_t \leq x | X_{t_n}).$$

If this is the case for all choices of $t_1 < t_2 < \cdots < t_n < t$, the process $\{X_t : t \in T\}$ is a *Markov process*.

EXAMPLE 6.32   Consider a *random walk* on the real line, as follows: A particle starts at position 1 at time period zero. At each succeeding time period (1, 2, 3, . . .), the walk jumps one unit to the right with probability .3, and it jumps one unit to the left with probability .1; it remains stationary with probability .6. Assume the jump at each period is independent of the jumps at other periods. Since the position of the particle at period $t > t_n$ evolves from its position at period t_n (it may jump one unit right or left of that position), and it does not depend on *how* the process managed to get where it was at period t_n, this random walk is a Markov process.

Now, let us specialize to Markov processes with denumerable parameter spaces $T = \{0, 1, 2, 3, . . .\}$, and assume that the random variables in $\{X_t : t \in T\}$ have as their range a common finite set A of real numbers. Let us call A the *state space* of the process. Since the state space of the Markov process is finite, the process is said to be a *finite* Markov process, and since, in addition, T is discrete, the process is called a *finite Markov chain*.

EXAMPLE 6.33   The random walk of Example 6.32 is a *denumerable* Markov chain because the state space is the denumerable set $\{. . . , -1, 0, 1, . . .\}$. It can be modified by the introduction of reflecting barriers, so that the modified chain is finite. Suppose the point is *reflected* from 0 and 6 to the adjoining points in the opposite direction. That is, whenever the original walk would have entered the state 0, the modified walk enters 2 instead; whenever it would have entered 6, it is reflected to 4 instead. This situation may be modeled with a finite Markov chain with state space $A = \{1, 2, 3, 4, 5\}$. Stochastic processes like this one are called *random walks with reflecting barriers*.

Many of the properties of finite Markov chains are characterized by "one-step transition probabilities."

Definition 6.5   Suppose $\{X_t : t = 0, 1, 2, . . .\}$ is a finite Markov chain with state space $A = \{1, 2, . . . , n\}$. The probability

$$p_{ij}(t) = P(X_t = j | X_{t-1} = i), \qquad i, j \in A,$$

is the *transition probability* from i to j. If the $p_{ij}(t)$'s are constant with respect to t, say $p_{ij}(t) \equiv p_{ij}$, the Markov chain is *homogeneous,* and the p_{ij}'s are called *one-step transition probabilities*. The $n \times n$ matrix $\mathbf{P} = (p_{ij})$ is then called the *one-step transition probability matrix* of the Markov chain.

Since X_t is a random variable with range A, it follows that $\Sigma_{j=1}^{n} p_{ij}$ = 1 for each i. This result is interpreted as asserting that the process will, with probability 1, enter *some* state at time t, given that it is in state i at time $t - 1$. Thus the rows of **P** consist of nonnegative real numbers that sum to 1. Such a matrix is said to be a *stochastic matrix*. It is assumed in the following discussion that the Markov chains under consideration are homogeneous.

EXAMPLE 6.34   For the random walk with reflecting barriers discussed in Examples 6.32 and 6.33, it is easy to verify that the Markov chain is homogeneous and that

$$\mathbf{P} = \begin{pmatrix} .6 & .4 & 0 & 0 & 0 \\ .1 & .6 & .3 & 0 & 0 \\ 0 & .1 & .6 & .3 & 0 \\ 0 & 0 & .1 & .6 & .3 \\ 0 & 0 & 0 & .4 & .6 \end{pmatrix}.$$

For example, $p_{5,5} = .6$ since the walk stays in state 5 with the "remains stationary" probability. Otherwise, the walk enters 4 from 5, so $p_{5,4} = .4$.

Our discussion of Markov chains has concentrated on the one-step transition probabilities. It is also of interest to determine the *m-step transition probabilities of homogeneous Markov chains,* defined by

$$p_{ij}^{(m)} = P(X_{t+m} = j|X_t = i). \tag{6.35}$$

Let us denote the matrix of *m*-step transition probabilities by $\mathbf{P}^m$. The following theorem shows that this notation is consistent (you may skip this theorem and proof if you aren't familiar with matrix multiplication).

> **Theorem 6.11**   If **P** and $\mathbf{P}^m$ are, respectively, the one-step and *m*-step transition probabilities of a Markov chain with finite state space, then $\mathbf{P}^m = (\mathbf{P})^m$.

PROOF   We use mathematical induction on *m*. Clearly, the theorem is true for $m = 1$. As another example, and to provide insight for the general case, consider $m = 2$. Then for any fixed element $p_{ij}^{(2)}$ of $\mathbf{P}^2$, assuming the state space has n elements,

$$p_{ij}^{(2)} = \sum_{\alpha=1}^{n} p_{i\alpha} \cdot p_{\alpha j}. \tag{6.36}$$

This equation can be interpreted as follows: The probability that the process goes from state i to state j in two steps is the sum (over α) of all the probabilities of going from i to α in one step and then from α to j in one step. But by Eq. (6.36), $\mathbf{P}^2 = (p_{ij}^{(2)}) = \mathbf{P} \cdot \mathbf{P} = (\mathbf{P})^2$. Now, assuming $\mathbf{P}^k = (\mathbf{P})^k$, we verify that $\mathbf{P}^{k+1} = (\mathbf{P})^{k+1}$. Arguing in a manner similar to the case for $m = 2$ above, we have, for any fixed element $p_{ij}^{(k+1)}$ of $\mathbf{P}^{k+1}$,

$$p_{ij}^{(k+1)} = \sum_{\alpha=1}^{n} p_{i\alpha}^{(k)} p_{\alpha j},$$

so

$$\mathbf{P}^{k+1} = \mathbf{P}^k \cdot \mathbf{P} = (\mathbf{P})^k \mathbf{P} = (\mathbf{P})^{k+1}.$$

The m-step transition probabilities are conditional probabilities. In many applications, it is of interest to determine the marginal distribution of X_t for some fixed t. The marginal distribution of X_0 is called the *initial distribution* of the process. If the mass function p_{X_0} is known, along with $\mathbf{P}$ (and, hence, in theory, $\mathbf{P}^m$ for any m), then the marginal distribution of X_t, called the *distribution of states occupied at time t,* is, in principle, known. For convenience, let p_t denote the mass function at time t, p_{X_t}. Given the initial distribution p_0 and the one-step transition probability matrix $\mathbf{P}$, the marginal distribution of X_t is given by

$$p_t(j) = \sum_{i=1}^{n} p_{ij}^{(t)} p_{X_0}(i), \qquad j = 1, 2, \ldots, n, \tag{6.37}$$

since

$$p_t(j) = P(X_t = j) = \sum_{i=1}^{n} P(X_t = j | X_0 = i) \cdot P(X_0 = i).$$

EXAMPLE 6.35   Suppose that, in selecting a brand of soap in a supermarket for her jth purchase of soap, a typical housewife makes selections among brands 1, 2, and 3 with the probabilities shown in the matrix

$$\mathbf{P} = \begin{pmatrix} \frac{1}{4} & \frac{1}{2} & \frac{1}{4} \\ 0 & \frac{3}{4} & \frac{1}{4} \\ 0 & \frac{1}{2} & \frac{1}{2} \end{pmatrix}.$$

Thus, for example, given that the housewife selected brand 2 last time, she will again select brand 2 with probability $\frac{3}{4}$; otherwise, she will select brand 3. Assume that, when the supermarket first opened ($t = 0$), the housewife selected the brands with initial probabilities $p_0(j) = \frac{1}{3}, j = 1, 2, 3$. Then the distribution of brand preference at the time of her third purchase is given by

$$p_3(1) = \frac{1}{64} \cdot \frac{1}{3} + 0 \cdot \frac{1}{3} + 0 \cdot \frac{1}{3} = \frac{1}{192},$$

$$p_3(2) = \frac{127}{192}, \qquad p_3(3) = \frac{64}{192}.$$

EXAMPLE 6.36   Consider the reliability growth situation of Example 6.29. Suppose a system starts with a simple failure mode, which fails with probability q unless the system has been (permanently) repaired. On each trial in which the failure mode fails, an attempt at repair is made. With probability π, the mode is repaired immediately after it has failed; otherwise, the mode remains unrepaired for the next trial. Identify state 1 with an unrepaired system and state 2 with a repaired system, so the situation described can be viewed as a Markov chain. The chain starts in state 1 [i.e., $p_0(0) = 1$] and has the one-step transition matrix

$$\mathbf{P} = \begin{pmatrix} 1 - q\pi & q\pi \\ 0 & 1 \end{pmatrix}.$$

For example,

p_{12} = P(system fails and is repaired|system has not been previously repaired)

     = P(unrepaired system fails) $\cdot$ P(system is repaired immediately after a failure is observed)

     = $q\pi$,

where we have obviously made certain independence assumptions. It follows that the reliability R_m of the system after m trials, with repair attempts on those trials in which a failure occurred, is given by $R_m = p_{11}^{(m)}(1 - q) + p_{12}^{(m)}$. Now (see Exercise 6.128),

$$\mathbf{P}^m = \begin{pmatrix} (1 - q\pi)^m & 1 - (1 - q\pi)^m \\ 0 & 1 \end{pmatrix},$$

so $R_m = 1 - q(1 - q\pi)^m$.

In general, we are interested in several aspects of Markov chains, such as, "Given $X_t = j$, what are the mean and variance of the number of steps required to first reach state k? How many times will state i be visited, on the average, when the walk goes from state j to state k? What is the limit of $\mathbf{P}^m$ as $m \to \infty$, and what is the marginal distribution of X_t as t gets very large? Under what conditions is the limiting distribution of X_t as $t \to \infty$, if it exists, not dependent on the initial distribution?" We will not consider these questions here, except for a brief comment concerning the last one.

     A homogeneous Markov chain with state space $\{1, 2, \ldots, n\}$ is *ergodic* provided that there is some matrix $\mathbf{P}^*$ such that

$$\lim_{n \to \infty} \mathbf{P}^m = \mathbf{P}^*, \tag{6.38}$$

where the rows of $\mathbf{P}^*$ are all the same, say of the form $(p^*(1),$

$p^*(2), \ldots, p^*(n))$. It then follows that

$$\lim_{t \to \infty} p_t(j) = p^*(j), \tag{6.39}$$

and p^* is said to be the *limiting*, or *steady-state distribution* of the process. The interpretation is that the distribution of states occupied by the walk eventually settles down to steady state; this steady-state condition does not depend on where the walk started.

Several questions remain concerning the stationary probabilities of an ergodic Markov chain. One is, "How can we tell whether or not a chain is ergodic?" Although we will not prove it here, it may be shown that a Markov chain is ergodic provided that, for some m, $\mathbf{P}^m$ has positive components. There are, in addition, other criteria that can be used to determine whether or not a chain is ergodic. (For more details, consult the classic work of Kemeny and Snell.*) Another interesting problem is that of determining p^*, once the chain is known to be ergodic. Since

$$p_t(j) = \sum_{k=1}^{n} p_{t-1}(k)p_{kj},$$

and since the limit in Eq. (6.39) does not depend on the initial distribution p_0, then

$$p^*(j) = \lim_{t \to \infty} p_t(j) = \lim_{t \to \infty} \sum_{k=1}^{n} p_{t-1}(k)p_{kj} = \sum_{k=1}^{n} p^*(k)p_{kj}. \tag{6.40}$$

Thus p^* is a solution of the system of n linear equations, formed in Eq. (6.40) by taking $j = 1, 2, \ldots, n$, subject to the conditions that $p^*(j) \geq 0$ for all j and $\sum_{j=1}^{n} p^*(j) = 1$. It can be shown that, if the chain is ergodic, such a solution exists, is unique, and satisfies the limit condition of Eq. (6.39).

EXAMPLE 6.37   Suppose a random walk behaves in accordance with the transition matrix

$$\mathbf{P} = \begin{pmatrix} \frac{1}{2} & \frac{1}{4} & \frac{1}{4} \\ \frac{1}{3} & \frac{1}{3} & \frac{1}{3} \\ \frac{1}{8} & \frac{1}{4} & \frac{5}{8} \end{pmatrix}.$$

The steady-state probabilities are given by the solutions to the system

$$p^*(1) + p^*(2) + p^*(3) = 1$$

$$\tfrac{1}{2}p^*(1) + \tfrac{1}{3}p^*(2) + \tfrac{1}{8}p^*(3) = p^*(1)$$

$$\tfrac{1}{4}p^*(1) + \tfrac{1}{3}p^*(2) + \tfrac{1}{4}p^*(3) = p^*(2)$$

$$\tfrac{1}{4}p^*(1) + \tfrac{1}{3}p^*(2) + \tfrac{5}{8}p^*(3) = p^*(3)$$

---

* J. G. Kemeny and J. L. Snell, *Finite Markov Chains* (Princeton, N.J.: D. Van Nostrand, 1960).

which is $p^*(1) = \dfrac{16}{55}$, $p^*(2) = \dfrac{15}{55}$, $p^*(3) = \dfrac{24}{55}$.

Thus, roughly speaking, after a long time t (a large number of transitions), the marginal distribution of the states occupied at time t is approximated well by the steady-state solution p^*. It then follows that the proportions of time spent by the walk in each of the three states are, in the long run, $p^*(1)$, $p^*(2)$, and $p^*(3)$, respectively.

As a final example, we consider a random walk with barriers that are *absorbing* rather than reflecting. This type of model is used for random walks that have states from which the walk never exits; for all intents, the walk stops when it enters these absorbing states. The example is a "gambler's ruin" situation.

EXAMPLE 6.38   —*Gambler's Ruin*   Two opponents, say A and B, engage in a gambling game in which each wagers $1 at each "play." The outcome of each play is random, with A winning the pot (and hence winning $1) with probability p, independent of previous plays. Suppose A begins the series of plays with a stake of i and B starts with a stake of $(n - i)$. The game ends when A or B is broke ("ruined"). Then A's position after t plays forms a random walk over the state space $\{0, 1, \ldots, n\}$, where states 0 and n are absorbing states. The walk is a Markov process; at each play, it steps one unit to the right with probability p or one unit to the left with probability $q = 1 - p$. It is desired to answer questions about the probability that A is eventually ruined, the expected duration of the game, and the roles of A's relative skill (p) and stake (i) in these answers. It is not difficult to show that the walk will eventually terminate with probability 1; this result is tacitly assumed as we proceed.

SOLUTION   Let $r(i)$ denote the probability that A is ruined, as a function of his initial stake i, where $0 < i < n$. After the first play of the game, A has either $(i - 1)$ (with probability q) or $(i + 1)$ (with probability p), from which he commences the next play. Thus $r(i)$ satisfies the difference equation

$$r(i) = r(i + 1) \cdot p + r(i - 1) \cdot q, \qquad 0 < i < n, \qquad (6.41)$$

with boundary conditions $r(0) = 1$ and $r(n) = 0$. In the exercises, you are asked to verify that the solution to this system is

$$r(i) = \begin{cases} \dfrac{(q/p)^n - (q/p)^i}{(q/p)^n - 1} & \text{if } p \neq q, \\[3mm] 1 - \dfrac{i}{n} & \text{if } p = q. \end{cases} \qquad (6.42)$$

Since the walk eventually terminates with probability 1, it follows that B is ruined with probability $1 - r(i)$.

To find the expected duration $E(i)$ of the game, again consider the situation after one play of the game, at which time A's pool (the state occupied by the walk) is either $i - 1$ or $i + 1$. Then $E(i)$ satisfies the equation

$$E(i) = E(i + 1) \cdot p + E(i - 1) \cdot q, \, 0 < i < n, \qquad (6.43)$$

with boundary conditions $E(0) = E(n) = 0$. The solution to this system is

$$E(i) = \begin{cases} \dfrac{i}{q - p} - \dfrac{n}{q - p} \left[\dfrac{1 - (q/p)^i}{1 - (q/p)^n} \right] & \text{if } p \neq q, \\[2ex] i(n - i) & \text{if } p = q. \end{cases} \qquad (6.44)$$

From Eqs. (6.42) and (6.44), it can be seen that, if A and B are equally skillful ($p = q$), the probability that B is ruined increases linearly with A's relative stake, and the expected duration of the game is greatest if the players start with equal stakes.

EXERCISES

THEORY

6.119 Discuss whether a random variable is a stochastic process.

6.120 Verify that, if $\lim_{t \to \infty} p_{ij}^{(t)} = p_j^*$ for all i and j, then $\lim_{t \to \infty} p_t(j) = p_j^*$ for any initial distribution p_0.

6.121 Verify that a product of $n \times n$ stochastic matrices is stochastic.

6.122 Prove that, if a Markov chain is homogeneous, then the m-step transition probabilities do not depend on the particular value of $t \in T$ being considered.

6.123 Verify that Eq. (6.42) is a solution to Eq. (6.41).

6.124 Verify that Eq. (6.44) is a solution to Eq. (6.43).

6.125 Suppose
$$\mathbf{P} = \begin{pmatrix} 1 & 0 & 0 \\ .2 & .8 & 0 \\ .3 & .4 & .3 \end{pmatrix}$$

    a. Show that $\mathbf{P}$ is positive definite. *Hint:* Show that the characteristic values of $\mathbf{P}$ are the elements along the main diagonal.
    b. Find a nonsingular matrix $\mathbf{Q}$ such that $\mathbf{QPQ}^{-1} = \mathbf{D}$, where $\mathbf{D} = (\lambda_i \delta_{ij})$.
    c. Find $\mathbf{P}^{16} = \mathbf{Q}^{-1} \mathbf{D}^{16} \mathbf{Q}$.
    d. Suppose the initial distribution of X_0 is given by $p_0(1) = .3$, $p_0(2) = .6$, and $p_0(3) = .1$. Find the distribution of X_{16}.

6.126 a. Suppose $\mathbf{P}^m$ is given for some integer $m > 1$. Is $\mathbf{P}$ determined?
    b. Suppose the marginal distribution of X_k is known, where k is some fixed integer (not necessarily zero). In addition, assume that $\mathbf{P}$ is known. De-

termine whether these may be used to find the marginal distribution of X_t for any t, and, if possible, describe a procedure for finding such distributions.

6.127   Suppose $\{X_t : t = 0, 1, 2, \ldots\}$ is a homogeneous finite Markov chain with one-step transition matrix $\mathbf{P}$ and initial distribution p_{20}. Let M_t denote the moment-generating function of X_t. Find a recursion relation between M_t and M_{t+1} (i.e., given $\mathbf{P}$ and M_{X_t}, find $M_{X_{t+1}}$).

6.128   Suppose

$$\mathbf{P} = \begin{pmatrix} \alpha & 1 - \alpha \\ 0 & 1 \end{pmatrix},$$

where $\alpha \in (0, 1)$. Show that

$$\mathbf{P}^m = \begin{pmatrix} \alpha^m & 1 - \alpha^m \\ 0 & 1 \end{pmatrix}.$$

Hint: A product of upper triangular matrices is upper triangular, and a product of stochastic matrices is stochastic.

6.129   a. A stochastic matrix $\mathbf{P}$ is said to be *doubly stochastic* if $\mathbf{P}'$ is also stochastic. Show that, if such a matrix is the one-step transition probability matrix of an ergodic Markov chain with state space $\{1, 2, \ldots, n\}$, then the stationary probabilities are equal (and hence are all $1/n$).
        b. Show that the converse holds.
        c. Is a product of doubly stochastic matrices necessarily doubly stochastic?

6.130   Suppose $Y_1, Y_2, Y_3, \ldots$ is a sequence of independent, identically distributed random variables, and define the random walk process $\{X_t : t = 1, 2, 3, \ldots\}$ by $X_t = \Sigma_{t=1}^t Y_i$. Show that the process is a Markov process.

6.131   The following questions concern the gambler's ruin walk (Example 6.38).
        a. Suppose A is more skillful than B, so $p > \frac{1}{2}$. Is it possible for B to compensate by having a large stake?
        b. Suppose $p = q$ and A can stop the game whenever he wishes. Also, suppose A wants to win $\$m$ before stopping. Show that the probability that he can do so is $i/(i + m)$. What happens if A is greedy (m is large)?
        c. Suppose the bet size is increased from $\$1$ to $\$d$ per play. Show that this scheme favors the more skillful player. *Hint:* This procedure is like playing for $\$1$ with A's initial stake $\$i/d$ and ruin for B when the walk reaches n/d.

APPLICATION

6.132   Discuss the meaning of the statement "The present depends only on the most recent past" in connection with the Markov property.

6.133   Verify that the 5×5 matrix of Example 6.34 is stochastic.

6.134   Suppose a stochastic process has the one-step transition matrix

$$\mathbf{P} = \begin{pmatrix} \frac{1}{2} & 0 & \frac{1}{2} \\ 0 & \frac{1}{2} & \frac{1}{2} \\ 0 & 0 & 1 \end{pmatrix}.$$

    a. Argue that state 3 is absorbing in the sense that, regardless of the initial distribution, the process eventually enters state 3 and remains there with probability 1. *Hint:* Argue that $\lim_{t \to \infty} p_t(3) = 1$.

    b. Suppose $p_0(0) = 1$. What is the probability that state 3 is first entered from state 1?

    c. With the same p_0 as in part (b), what is the expected number of transitions until state 3 is first entered?

    d. With the same p_0 as in part (b), what is p_3?

6.135   Two players, A and B, match pennies. Player A starts with 3 cents and B starts with 4. The game ends whenever one of the players has all seven pennies. What is the probability that A wins?

6.136   Urn I originally contains three red balls and urn II contains three white balls. A ball is selected from urn I and put into urn II. A ball is drawn from urn II (at random): if it is white, it is put in urn I; otherwise, the process terminates. If a white ball is transferred to urn I, a ball is then selected at random from urn I; if it is red, it is transferred to urn II; otherwise, the process terminates. This procedure is carried out until either a white ball is drawn from urn I or a red ball is drawn from urn II.

    a. What is the expected duration of the process?

    b. What is the distribution of balls in urn II when the process terminates?

6.137   Suppose a random walk has the one-step transition matrix

$$\mathbf{P} = \begin{pmatrix} 0 & p & 0 & q \\ q & 0 & p & 0 \\ 0 & q & 0 & p \\ p & 0 & q & 0 \end{pmatrix}.$$

    a. Describe the behavior of the walk in terms of tossing a biased coin at each transition.

    b. Show that the walk may eventually visit any state from any given state.

    c. The chain is ergodic. Find the steady-state distribution p^*.

    d. Suppose $p_0(1) = \frac{1}{2}$ and $p_0(2) = p_0(3) = \frac{1}{4}$. Find the expected number of transitions until the walk first enters state 4.

6.138   a. In Example 6.35, where

$$\mathbf{P} = \begin{pmatrix} \frac{1}{4} & \frac{1}{2} & \frac{1}{4} \\ 0 & \frac{3}{4} & \frac{1}{4} \\ 0 & \frac{1}{2} & \frac{1}{2} \end{pmatrix},$$

    compute $\mathbf{P}^2$ and show that $\mathbf{P}^3$ is the same when computed either as $\mathbf{P}^2 \cdot \mathbf{P}$ or $\mathbf{P} \cdot \mathbf{P}^2$.

    b. Conjecture what form $\mathbf{P}^m$ will have as $m \to \infty$.

6.139   In Example 6.36, let N denote the number of trials required until the failure mode is repaired. Find the mean and variance of N as functions of q and π, noting that both moments of N decrease with increasing q and π. Discuss why this result should be true.

6.9 SUMMARY

To find the distribution of a function of two random variables, say $Y = g(X_1, X_2)$, several approaches are possible: First, if (X_1, X_2) is discrete, so is Y; and p_Y is found at each $y_0 \in R_Y$ by summing p_{X_1, X_2} over points (x_1, x_2) such that $g(x_1, x_2) = y_0$. Second, the CDF method involves writing $F_Y(y)$ as $P[g(X_1, X_2) \leq y]$ and then "solving" the inequality so that the probability can be obtained directly from the known distribution of (X_1, X_2). Third, if (X_1, X_2) is continuous and $g(X_1, X_2)$ is continuous, the change-of-variable technique may be used, where the system $y_1 = g(x_1, x_2)$ and $y_2 = x_1$ (or some other simple equation) is solved for x_1 and x_2 in terms of y_1 and y_2. Then the joint density of (Y_1, Y_2) is found utilizing the Jacobian of the transformation. Finally, the moment-generating function of $g(X_1, X_2)$ may be calculated and recognized as belonging to a known family.

In particular, if $Y = X_1 + X_2$, and (X_1, X_2) is continuous, the density of Y is the convolution

$$f_Y(y) = \int_{-\infty}^{\infty} f(x_1, y - x_1)\, dx_1,$$

with a similar formula for the discrete case. If X_1 and X_2 are independent and have marginal distributions belonging to a common family, the distribution of $X_1 + X_2$ may also belong to that family. In this case, the family has the reproductive property. Reproductive families include the normal, the Poisson, the gamma (with common λ), the binomial (with common p), and the chi-square.

A joint moment of (X_1, X_2) of order $\alpha_1 + \alpha_2$ is

$$\mu_{\alpha_1, \alpha_2} = E(X_1^{\alpha_1} \cdot X_2^{\alpha_2}).$$

A joint central moment of order $\alpha_1 + \alpha_2$ is

$$\mu_{\alpha_1, \alpha_2} = E[(X_1 - \mu_1)^{\alpha_1} \cdot (X_2 - \mu_2)^{\alpha_2}].$$

The covariance of X_1 and X_2 is thus $\text{Cov}(X_1, X_2) = \mu_{1,1}$. The correlation coefficient of X_1 and X_2 is $\rho(X_1, X_2) = \mu_{1,1}/\sigma_{X_1}\sigma_{X_2}$; this quantity is always between -1 and $+1$. If X_1 and X_2 are linearly related ($X_2 = \alpha X_1 + \beta$), then $|\rho| = 1$; if $|\rho(X_1, X_2)| = 1$, then there exist constants α and β such that $P(X_2 = \alpha X_1 + \beta) = 1$. The variance of a sum of n random variables is given by

$$V\left(\sum_{i=1}^{n} X_i\right) = \sum_{i=1}^{n} V(X_i) + \sum_{\substack{i=1 \\ i \neq j}}^{n} \sum_{j=1}^{n} \text{Cov}(X_i, X_j);$$

if the X's are independent, this expression reduces to $\sum_{i=1}^{n} \sigma_{X_i}^2$.

The moment-generating function of a sum of independent random variables is the product of their marginal moment-generating functions. This result is useful in determining whether a family has the reproductive property.

The joint moment-generating function of a random pair (X_1, X_2) is

$$E(e^{t_1 X_1 + t_2 X_2}) = M_{X_1, X_2}(t_1, t_2).$$

Joint moments can be found by taking partial derivatives of M_{X_1, X_2} and evaluating at $t_1 = 0$, $t_2 = 0$.

The bivariate normal family is a five-parameter family of distributions; for any μ_1 and μ_2, for $\sigma_1^2 > 0$ and $\sigma_2^2 > 0$, and for ρ between -1 and $+1$, the joint density is

$$f_{X,Y}(x, y) = \frac{1}{2\pi\sigma_1\sigma_2 \sqrt{1 - \rho^2}} \exp\left\{ - \frac{1}{2(1 - \rho^2)} \right.$$

$$\left. \times \left[\left(\frac{x - \mu_1}{\sigma_1} \right)^2 - 2\rho \left(\frac{x - \mu_1}{\sigma_1} \right) \left(\frac{y - \mu_2}{\sigma_2} \right) + \left(\frac{y - \mu_2}{\sigma_2} \right)^2 \right] \right\}.$$

The following relationships of moments to parameters exist:

$$E(X) = \mu_1, \qquad E(Y) = \mu_2, \qquad V(X) = \sigma_1^2, \qquad V(Y) = \sigma_2^2,$$

$$\rho(X, Y) = \rho, \qquad \text{Cov}(X, Y) = \sigma_1\sigma_2\rho,$$

$$E(Y|X = x) = \mu_2 + \frac{\sigma_2}{\sigma_1} \rho(x - \mu_1), \qquad V(Y|X = x) = \sigma_2^2(1 - \rho^2),$$

with symmetric conditional moments for X given y. The conditional distribution of Y given x is normal, and the marginal distributions of X and Y are normal. $E(Y|X = x)$ is the regression line of Y on X; similarly, $E(X|Y = y)$ is the regression of X on Y.

The multinomial distribution is a discrete distribution with $k + 1$ parameters: positive integer n and probabilities $p_1, p_2, \ldots, p_k$, which sum to 1. The multinomial mass function is

$$p_{N_1, N_2, \ldots, N_k}(n_1, n_2, \ldots, n_k) = \binom{n}{n_1, n_2, \ldots, n_k} p_1^{n_1} p_2^{n_2} \cdots p_k^{n_k},$$

which is positive over points $(n_1, n_2, \ldots, n_k)$ with nonnegative integer components such that $\sum_{i=1}^{k} n_i = n$. This model is one for the counts N_i of occurrences of type i in n independent multinomial trials, where in each trial a type i outcome occurs with probability p_i. Some moments of the N_i's are

$$E(N_i) = np_i, \qquad V(N_i) = np_iq_i, \qquad \text{Cov}(N_i, N_j) = -np_ip_j.$$

The Poisson process is a stochastic process $\{N_t : t \geq 0\}$ in which N_t, the number of arrivals in t units of time, is distributed Poisson (λt). The parameter λ is the arrival rate per unit of time. The interarrival times in a Poisson process are independent, exponential (λ) random variables.

A finite Markov chain is a stochastic process $\{X_t : t = 0, 1, \ldots\}$ with finite state space $A = R_{X_t}$. The random behavior of a homogeneous Markov chain is determined by the initial distribution of states occupied, p_0, together with the one-step transition matrix $\mathbf{P}$. If the chain is ergodic, there is a steady-state distribution $p^*(i)$ of states occupied, defined by $p^*(i) = \lim_{t \to \infty} p_t(i)$, where $p_t(i)$ is the probability that the process occupies state i at time t. The steady-state probabilities may be interpreted as the long-term proportions of times the process occupies the various states.

Chapter 7

SAMPLING AND ASYMPTOTIC DISTRIBUTIONS

7.1 SAMPLES AND SAMPLE MOMENTS

Much of this chapter involves the determination of distributions of certain functions of jointly distributed random variables. While this determination is nothing new, the particular functions involved (to be called "statistics") are of special interest for applications in the field of statistics. Generally speaking, these functions are selected so as to allow the statistician to make inferences about a population based on a sample of outcomes on random variables whose distribution is not completely known. For example, an inspector may sample 100 items from a shipment of 10,000 items in order to estimate the proportion defective in the population (shipment). The concepts of statistics will not be pursued here, although some of the basic notions are mentioned briefly in order to motivate the probabilistic results.

You have already encountered the notion of sampling in connection with reports on public opinion polls, or the Nielsen ratings of TV programs. A basic notion in sampling is that each observation should be from a common source or *population*. The objective in taking samples is to use the data to make inferences about the entire population sampled, using only the observations of the sample values. To cast this in a probabilistic framework, imagine that each sample value is the result of a component experiment and that taking the entire sample is the performance of a compound experiment. In Chapter 2, we considered experiments involving sampling from an urn. Sampling with replacement was associated with the notion of independent repeated trials and with the binomial and multinomial models. Sampling without replacement led to dependent trials and hypergeometric models.

Let us now describe compound experiments in terms of random variables. The notion of a "parent population" (urn) that is sampled can be modeled as the CDF F of a random variable X. A *sample of size n* from the population is modeled as a collection $X_1, X_2, \ldots, X_n$ of n jointly distributed random variables all having the same marginal CDF F. That is, the X_i are identically distributed as X. As we have seen, this result is not in itself enough to determine the joint distribution of $X_1, X_2, \ldots, X_n$. When used in this way, X is often referred to as the *population* or *parent* random variable, and, similarly, F is called the *parent distribution* (as is also the density f or mass function p).

The parent random variable X is defined on some basic sample space S of possible outcomes of a random experiment. When the experiment is performed, resulting in an outcome o, the sampler records the real number $X(o)$. Such a performance and recording of the value of X is referred to as a *trial* of the experiment. A sample $X_1, X_2, \ldots, X_n$ of size n on X may be viewed as constituting n repeated trials of the same basic experiment. The result of performing these trials is to record an n-tuple $(x_1, x_2, \ldots, x_n)$ of respective values of the trials $(X_1, X_2, \ldots, X_n)$. The components of this n-tuple of real numbers are then referred to as the *sample values* or *sample data*. It is this information that furnishes the basis for statistical analysis.

Suppose special care is exercised to perform the experiment in such a way that the outcome of one trial does not influence the outcome of any other. Then in a structuring of a model for these successive trials, it seems reasonable to assume that the random variables $X_1, X_2, \ldots, X_n$ of the sample are mutually independent. A sample having this additional property of independence is called a *random sample*.* The clear-cut advantage of having a random sample is that the joint distribution of the sample is then determined from the specified common marginal distributions by multiplication. Independence is a strong assumption; if it should happen that the outcome of a given trial does, in fact, provide some information concerning another outcome, then inferences based on the assumption of independence may very well be in error. Later, we discuss how data from random samples of specified parent distributions can be constructed, and we discuss some applications of such values to simulation.

EXAMPLE 7.1   Consider an urn containing v white and $u - v$ red balls. The (component) experiment of drawing a ball and noting its color can be viewed as observing the outcome on a Bernoulli random variable X with parameter $p = v/u$ (so $X = 1$ if the ball drawn is white; otherwise, $X = 0$). If a sample of size n is taken *with replacement*, the experiment may be modeled with random variables $X_1, X_2, \ldots, X_n$, which are independent and identically Bernoulli (v/u) distributed. In this case, $(X_1, X_2, \ldots, X_n)$ has the joint mass function

$$p(x_1, x_2, \ldots, x_n) = \left(\frac{v}{u}\right)^{\Sigma x_i} \left(1 - \frac{v}{u}\right)^{n - \Sigma x_i}, \qquad x \in \{0, 1\}, i = 1, 2, \ldots, n,$$

---

* The term "random sample" is also used when every possible sample of n items from a finite population has the same probability of being selected.

and $X_1, X_2, \ldots, X_n$ is a random sample consisting of n independent Bernoulli trials. We have noted before that the number $\Sigma\, X_i$ of white balls in the n trials is binomial $(n, v/u)$. On the other hand, if sampling is conducted without replacement, and if $n < v < u$, then the trials are dependent and the total number of white balls drawn has the hypergeometric distribution. It is still the case, however, that each of the random variables in the sample has a Bernoulli (v/u) distribution; so this sample is one of size n of X, where X is Bernoulli (v/u), but it is not a random sample of size n.

In both types of schemes of sampling from an urn, the central point of investigation is the total number of white balls (or defectives, etc.) in the sample; the order in which these outcomes occur is irrelevant. The data furnished by the sample are thus summarized quite adequately by the sum of the observed values. Keeping track of this one piece of information rather than the n numbers $x_1, x_2, \ldots, x_n$ that are recorded for the sample is more convenient and serves the purpose just as well. Such functions of the sample data, summarizing the data for a particular purpose, are the basis for statistical inference.

> **Definition 7.1**   Let $X_1, X_2, \ldots, X_n$ be a sample, and suppose ϕ is a function of n variables that does not depend on unknown parameters. Then $\phi(X_1, X_2, \ldots, X_n)$ is called a *statistic*.

A statistic is a random variable and hence has a probability distribution, moments, and other characteristics. The reason for the added requirement that no unknown parameters occur in the description of a statistic is that the statistician often uses its value to summarize the sample data, so that ultimately decisions can be based on this value. If this value were not definitive because it depended on unknown quantities, it would hardly be useful for this purpose.

EXAMPLE 7.2   The random variable $\Sigma_{i=1}^{n} X_i$ is a statistic but $\Sigma_{i=1}^{n} (X_i - \theta)$ is not, assuming that θ is unknown. Similarly, if

$$\phi(x_1, x_2) = \begin{cases} x_1 \text{ if } x_1 - x_2 > \theta, \\ x_2 \text{ if } x_1 - x_2 \le \theta, \end{cases}$$

then $\phi(X_1, X_2)$ is not a statistic, since ϕ depends on θ through its range.

Note that we do not mean to say that the probability *distribution* of a statistic will be free of parameters. For example, in both sampling with and without replacement from an urn, the sum of the Bernoulli random variables involved is a statistic, but in both cases its probability mass function is dependent on all the parameters involved. The point is that the outcome on a statistic T can actually be calculated when the experiment is performed; that is, T is observable. The size n of the sample making up a statistic usually occurs itself as a parameter in the probability distribution of the statistic. Normally, however, the sample size is considered to be fixed and known, though generally it is left arbitrary in appearance for the sake of generality. Usually, in applications, the sample underlying a statistic is a random sample, although Definition 7.1 makes no such formal requirement. One of the most common statistics is the sum of the random variables in the sample. Another very important class of statistics are sample analogues of the moments of a probability distribution.

Definition 7.2   Let $X_1, X_2, \ldots, X_n$ be a sample, and, for each positive integer k, let $M_k = (1/n) \Sigma_{i=1}^n X_i^k$. The statistic M_k is the kth *sample moment* of the sample.

If the respective values of the sample random variables $X_1, X_2, \ldots, X_n$ are $x_1, x_2, \ldots, x_n$, then the value of M_k is simply the arithmetic average of the real numbers $x_1^k, x_2^k \ldots, x_n^k$. The first sample moment is usually denoted $\overline{X}$ and is called the *sample mean*. When it is important to observe the dependence of $\overline{X}$ on n, it is denoted $\overline{X}_n$. The prefix "sample" in these definitions is very important. The moments of the parent random variable X, when they exist, are not random variables, as are these sample moments. The moments of X are generally functions of the parameters involved in the parent distribution and are thus numerical characteristics associated with that distribution. To distinguish moments of X from sample moments, we call moments of the parent distribution *population moments*.

An important distinction exists between these two types of moments, and it must be carefully maintained. Notice the difference, for example, between the sample mean $\overline{X}$, the population mean $\mu = E(X)$, and the ordinary (arithmetic) mean, or average,

$$\overline{x} = \frac{1}{n} \sum_{i=1}^n x_i,$$

where x_i is the outcome on X_i. The first is a random variable (statistic); the second is a characteristic of the parent probability distribution; the third is the *value* of the random variable $\overline{X}$, a real number obtained *after* the experiment is performed. The *function* $\overline{X}$ is never the same as μ, and even $\overline{x}$, the value of $\overline{X}$, is rarely the same as μ. Indeed, when $\overline{X}$ is continuous, the event $(\overline{X} = \mu)$ has probability zero. Neither, of course, is $\overline{X}$ the same as $\overline{x}$ (a random variable and one of its values cannot be the same object); $\overline{x}$ has no probability distribution and represents the outcome of a single performance of the experiment. A second performance of the experiment would usually result in a different value in the range of $\overline{X}$.

One of the most commonly treated sample moments is the first sample moment, $\overline{X}$. Being a *linear* combination $(1/n)X_1 + (1/n)X_2 + \cdots + (1/n)X_n$ of the sample random variables, the probability distribution of $\overline{X}$ is one of the easier ones to derive by using the methods discussed in Chapter 6. In particular, if the sample random variables are mutually independent—that is, if the sample is a random sample—the distribution of $\overline{X}$ can be found by convoluting the densities, or multiplying the generating functions, of the $(1/n) \cdot (X_i)$'s. Thus if M is the common moment-generating function of each of the independent, identically distributed random variables $X_1, X_2, \ldots,$ X_n, then the distribution of $\overline{X}$ is given by

$$M_X(t) = \left[M\left(\frac{t}{n}\right) \right]^n.$$

EXAMPLE 7.3   Suppose the parent random variable X is $N(\mu, \sigma^2)$. Then for a random sample of size n,

$$M_{\overline{X}}(t) = \left[\exp\left(\frac{\mu t}{n} + \frac{1}{2}\sigma^2\frac{t^2}{n^2} \right) \right]^n = \exp\left(\mu t + \frac{1}{2}\frac{\sigma^2}{n}t^2 \right).$$

Thus $\overline{X}$ is again normal, in fact, $N(\mu, \sigma^2/n)$, so that $E(\overline{X}) = \mu$ and $V(\overline{X}) = \sigma^2/n$. That is, the expected value of the sample mean is the population mean, while its variance is *smaller* than the population variance. This fact has profound implications, which we will pursue later. The relationship between the population distribution and that of the statistic $\overline{X}$ is shown in Fig. 7.1.

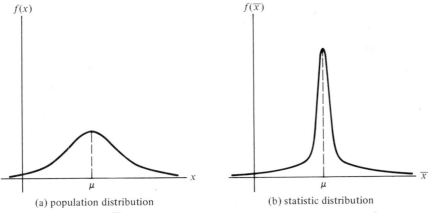

(a) population distribution   (b) statistic distribution

FIGURE 7.1 $\overline{X}$ *Has Same Mean as X but Smaller Variance, by a Factor of* $1/n$

The last two facts in the example above concerning the moments of $\overline{X}$ can be established quite independently of the normality of the parent random variable.

Theorem 7.1   Let $\overline{X}$ be the sample mean of a random sample $X_1, X_2, \ldots, X_n$ of a population random variable X having population mean μ and population variance σ^2. Then

$$E(\overline{X}) = \mu \qquad \text{and} \qquad V(\overline{X}) = \frac{\sigma^2}{n}.$$

PROOF   Using the linearity of expectation yields

$$E(\overline{X}) = E\left(\frac{1}{n} \sum_{i=1}^{n} X_i\right) = \frac{1}{n} \sum_{i=1}^{n} E(X_i) = \frac{1}{n}\left(\sum_{i=1}^{n} \mu\right) = \frac{1}{n} \cdot n\mu = \mu.$$

Next, using the independence of the X_i's yields

$$V(\overline{X}) = V\left(\frac{1}{n} \sum_{i=1}^{n} X_i\right) = \frac{1}{n^2} \sum_{i=1}^{n} V(X_i) = \frac{n\sigma^2}{n^2} = \frac{\sigma^2}{n}.$$

Definition 7.3   Let $X_1, X_2, \ldots, X_n$ be a sample, and, for each positive integer k, let

$$S^k = \frac{1}{n} \sum_{i=1}^{n} (X_i - \overline{X})^k.$$

The statistic S^k is the kth *central sample moment* of the sample.

Note that S^1 is degenerate, since

$$\frac{1}{n} \sum_{i=1}^{n} (X_i - \overline{X}) = \frac{1}{n} \sum_{i=1}^{n} X_i - \frac{n\overline{X}}{n} = \overline{X} - \overline{X} = 0.$$

Our main interest is in S^2, which is called the *sample variance* and is sometimes denoted as S_X^2. The positive square root of S_X^2 is denoted by S_X; it is the *sample standard deviation*. Naturally, the same distinction among S_X^2, its value s_X^2, and the population variance σ^2 applies here as it did for $\overline{X}$, $\overline{x}$, and μ.

Generally speaking, the probability distribution of S_X^2 is more difficult to find than that of $\overline{X}$ because of the presence of quadratic terms. Also, even when the sample is random, $(X_1 - \overline{X})^2, \ldots, (X_n - \overline{X})^2$ are not necessarily independent, so S_X^2 is not, in general, a linear combination of independent random variables. Thus derivation of the distribution of S_X^2 is generally more difficult than that of $\overline{X}$. One special circumstance in which the distribution of S_X^2 is tractable is when the parent distribution is normal. In this case, a most interesting and valuable result arises: S_X^2, although a function of $\overline{X}$, is statistically independent of $\overline{X}$. We now state this result as a theorem and give proof for the case of a sample of size 2.

Theorem 7.2   Suppose X is $N(\mu, \sigma^2)$ and X_1, X_2, $\ldots$, X_n is a random sample of X (with $n > 1$). Then the statistics $\overline{X}$ and S_X^2 are independent random variables, and, moreover, nS_X^2/σ^2 has a $\chi_{(n-1)}^2$ distribution and $\overline{X}$ is $N(\mu, \sigma^2/n)$.

PROOF   For $n = 2$: First, examine the expression for nS_X^2. For $n = 2$,

$$2S_X^2 = (X_1 - \overline{X})^2 + (X_2 - \overline{X})^2$$

$$= \left(X_1 - \frac{X_1 + X_2}{2} \right)^2 + \left(X_2 - \frac{X_1 + X_2}{2} \right)^2$$

$$= \left(\frac{X_1 - X_2}{2} \right)^2 + \left(\frac{X_2 - X_1}{2} \right)^2 = \frac{1}{2}(X_1 - X_2)^2.$$

The independence of $\overline{X}$ and S_X^2, in this case, is implied by the independence of $X_1 + X_2$ and $X_1 - X_2$, since $\overline{X}$ is a function of the former and S_X^2 is a function of the latter. That $X_1 + X_2$ and $X_1 - X_2$ are independent is easily seen by considering their joint moment-generating function:

$$m_{X_1+X_2, X_1-X_2}(t_1, t_2) = E[e^{t_1(X_1+X_2)+t_2(X_1-X_2)}] = E[e^{(t_1+t_2)X_1 + (t_1-t_2)X_2}]$$

$$= m_{X_1, X_2}(t_1 + t_2, t_1 - t_2) = m_{X_1}(t_1 + t_2) \cdot m_{X_2}(t_1 - t_2),$$

since X_1 and X_2 are assumed to be independent. But the latter product is

$$m_{X_1}(t_1 + t_2) \cdot m_{X_2}(t_1 - t_2) = \exp[\mu(t_1 + t_2) + \tfrac{1}{2}\sigma^2(t_1 + t_2)^2]$$

$$\times \exp[\mu(t_1 - t_2) + \tfrac{1}{2}\sigma^2(t_1 - t_2)^2]$$

$$= \exp[2\mu t_1 + \tfrac{1}{2}(2\sigma^2)t_1^2] \times \exp[\tfrac{1}{2}(2\sigma^2)t_2^2].$$

Since $m_{X_1+X_2,X_1-X_2}(t_1, t_2)$ factors into a product of a function of t_1 times a function of t_2, it follows that $X_1 + X_2$ and $X_1 - X_2$ are independent. Indeed, you can recognize these functions as the $N(2\mu, 2\sigma^2)$ and $N(0, 2\sigma^2)$ moment-generating functions, respectively. Since $X_1 - X_2$ is distributed $N(0, 2\sigma^2)$, $(X_1 - X_2)/\sqrt{2}\,\sigma$ is $N(0, 1)$, so $2S_X^2/\sigma^2 = (X_1 - X_2)^2/2\sigma^2$ is the square of a standard normal random variable, which we have seen is $\chi^2_{(1)}$. It was seen in Example 7.3 that $\overline{X}_n$ is $N(\mu, \sigma^2/n)$.

The fact that, under the conditions of Theorem 7.2, $\overline{X}$ and S_X^2 are independent, even though $\overline{X}$ appears in the expression for S_X^2, is somewhat surprising. This independence does not happen for other parent distributions; it is a characterization of the normal family. We state this formally, without proof.

Theorem 7.3   Suppose $X_1, X_2, \ldots, X_n$ is a random sample of a continuous parent random variable X, and suppose $\overline{X}$ and S_X^2 are independent. Then X is normally distributed.

In statistics, problems such as the following are considered: It is known that the parent random variable X has a normal distribution (say), but the parameters in this distribution are not known. So that some idea of these parameter values is gained, a random sample of X is observed, and these values are used to *estimate* μ and σ^2. It seems reasonable to estimate μ by the value $\overline{X}$ of the sample mean. The fact that $\overline{X}$ is $N(\mu, \sigma^2/n)$ gives some idea of how well we might expect $\overline{X}$ to estimate μ. The *estimator* $\overline{X}$ (the random variable whose value is the estimate), in this case, has the property that its *expected value* is the parameter being estimated, $E(\overline{X}) = \mu$. Such estimators are said to be *unbiased*.

Similarly, it seems reasonable to estimate σ^2 by S_X^2. As an estimator for σ^2, however, S_X^2 is biased, since, when nS_X^2/σ^2 is $\chi^2_{(n-1)}$, it has been shown that $E(nS_X^2/\sigma^2) = n - 1$, so

$$E(S_X^2) = \frac{(n - 1)\sigma^2}{n}. \tag{7.1}$$

It is left for you to show that Eq. (7.1) holds for *any* parent distribution having a variance σ^2 (Exercises 7.1 and 7.2).

Let us now consider a second example of an estimation problem.

EXAMPLE 7.4   Suppose certain electronic components are assumed to have exponentially distributed lifetimes; that is, a randomly selected component of this type operates in an acceptable manner for T hours, where T is exponential (λ). A component is judged to be "successful" in its mission if it operates for more than 5 hours (say). Thus the reliability $R(\lambda)$ of the component is the probability that a randomly selected component has a lifetime in excess of 5 hours, or

$$R(\lambda) = P(T > 5) = 1 - F_T(5) = e^{-5\lambda}.$$

Suppose λ is unknown and it is desired to estimate λ, using data obtained by putting n components on test and observing the times $T_1, T_2, \ldots, T_n$ at which they fail.

SOLUTION   Since the mean time to failure is $1/\lambda = E(T)$, it seems reasonable to estimate $1/\lambda$ by $\overline{T}$ and, hence, λ by $1/\overline{T}$. Thus we may estimate the unknown reliability $R(\lambda)$ of the components by using the estimator $R(1/\overline{T}) = e^{-(5/\overline{T})}$. For example, if four items were put on test, and these failed at the times 30, 60, 40, and 50, then $\overline{T} = 180/4 = 45$; so the estimated reliability is $e^{-(5/45)} \approx .90$.

EXERCISES

THEORY

7.1   a. Show that, for any real numbers $z_1, z_2, \ldots, z_n$,

$$\sum_{i=1}^{n} (z_i - \overline{z})^2 = \sum_{i=1}^{n} z_i^2 - n\overline{z}^2$$

and $$\sum_{i=1}^{n} (z_i - \overline{z})^2 = \sum_{i=1}^{n} (z_i - \theta)^2 - n(\overline{z} - \theta)^2.$$

    b. Use part (a) to show that $E(S_X^2) = [(n - 1)/n]\sigma^2$, where $\sigma^2 = V(X)$, the population variance. *Note:* It is not assumed that X is normal.

7.2   Let $X_1, X_2, \ldots, X_n$ be a random sample of an exponential (λ) population. Show that $n\overline{X}$ is gamma (n, λ). Is $\overline{X}$ an unbiased estimator for $1/\lambda$?

7.3   Show that the variance of S_X^2 is given by $2(n - 1)\sigma^4/n^2$ when the parent distribution of the random sample is $N(\mu, \sigma^2)$.

7.4   Show, subject to existence, that $E(M_k) = \mu_k'$ for every positive integer k.

7.5   Show that the kth factorial moment of Y is given by $(M)_k(n)_k/(N)_k$ when Y is hypergeometric with parameters M, N, and n.

7.6   Let $A_1, A_2, \ldots, A_n$ be independent events, and suppose $P(A_i) = p_i$, $i = 1$, $2, \ldots, n$. What is the probability that, in a single performance of the experiment, exactly k of these events will occur? *Hint:* Let $Y = \sum_{i=1}^{n} I_{A_i}$.

7.7   Show that the variance for the binomial ($n, M/N$) random variable is greater than the variance of the hypergeometric (n, M, N) random variable. Explain why this result should be true, in terms of sampling from an urn.

7.8   This exercise employs an approach to the hypergeometric by using conditional distributions. Suppose we sample balls, without replacement, from an urn containing M white and $N - M$ red balls. Let $X_1, X_2, \ldots, X_n$ denote the outcomes of a sample of size n, where $(X_i = 1)$ is the event "A white ball is drawn on the ith trial."

a. Show that the conditional distribution of X_2, given $x_1 = 1$, is binomial $(1, (M - 1)/(N - 1))$, whereas the conditional distribution of X_2, given $x_1 = 0$, is binomial $(1, M/(N - 1))$.

b. Suppose $x_1, x_2, \ldots, x_k$ are the outcomes of the first k trials. Show that the conditional distribution of X_{k+1}, given $y_k = \Sigma_{i=1}^{k} x_i$, is binomial $(1, p_k)$, where $p_k = (M - y_k)/(N - k)$.

c. Conclude that the mass function for $(X_1, X_2, \ldots, X_n)$ is given by

$$p_{X_1, X_2, \ldots, X_n}(x_1, x_2, \ldots, x_n) = \prod_{j=1}^{n} p_{j-1}^{x_j}(1 - p_{j-1})^{1-x_j}$$

$$= \prod_{j=1}^{n} \frac{(M - y_{j-1})^{x_j}(N - M + y_{j-1} - j + 1)^{1-x_j}}{(N - j + 1)}.$$

d. Prove that the joint distribution of $X_1, X_2, \ldots, X_k$ $(k \le n)$ may be written as

$$p_{(X_1, X_2, \ldots, X_k)}(x_1, x_2, \ldots, x_k) = \frac{(M)_{y_k}(N - M)_{k - y_k}}{(N)_{y_k}}.$$

Hint: Use induction.

e. Show that, for any k in $\{1, 2, \ldots, n - 1\}$, the marginal distribution of X_{k+1} is binomial $(1, M/N)$. Hint: Verify that

$$p_{X_{k+1}}(1) = \sum_{x_1=0}^{1} \sum_{x_2=0}^{1} \cdots \sum_{x_k=0}^{1} p(x_1, x_2, \ldots, x_k, 1) = \frac{M}{N}.$$

f. Let $Y = \Sigma_{i=1}^{n} X_i$. Show that

$$p_Y(y) = \frac{\binom{M}{y}\binom{N-M}{n-y}}{\binom{N}{n}}, \qquad y = 0, 1, \ldots, n.$$

Hint: $p_Y(y) = \Sigma_{x_1} \Sigma_{x_2} \cdots \Sigma_{x_n} p(x_1, x_2, \ldots, x_n)$, $(x_1, x_2, \ldots, x_n)$ in B_Y, where $B_y = \{(x_1, x_2, \ldots, x_n): \Sigma_{i=1}^{n} x_i = y\}$. If $(x_1, x_2, \ldots, x_n)$ is in B_y,

$$p_{X_1, X_2, \ldots, X_n}(x_1, x_2, \ldots, x_n) = \frac{(M)_y(N - M)_{n-y}}{(N)_y}.$$

7.9   Use the results in Exercise 7.8 to show that the mean and variance of a hypergeometric random variable Y are nM/N and $nM(N - n)(N - M)/$

$N^2(N - 1)$, respectively. *Hint:*

$$E(Y) = \sum_{i=1}^{n} E(X_i),$$

$$V(Y) = \sum_{i=1}^{n} V(X_i) + 2 \sum_{\substack{i,j \\ i<j}} \text{Cov}(X_i, X_j),$$

$$\text{Cov}(X_i, X_j) = \frac{M(N - M)}{N^2(N - 1)}.$$

7.10 Let $X_1, X_2, \ldots, X_n$ be a sample of a population for which μ_k exists. Show that $E[(1/n) \sum_{i=1}^{n} (X_i - \mu)^k] = \mu_k$, $k = 1, 2, \ldots$.

7.11 Let $X_1, X_2, \ldots, X_n$ be a random sample of X. Let

$$\overline{X}_1 = \frac{1}{k} \sum_{i=1}^{k} X_i$$

and

$$\overline{X}_2 = \frac{1}{n - k} \sum_{i=k+1}^{n} X_i,$$

where $1 < k < n$. Show that $\overline{X}_1$ and $\overline{X}_2$ are independent. Find formulas for $E(\overline{X}_1 \pm \overline{X}_2)$ and $V(\overline{X}_1 \pm \overline{X}_2)$ in terms of $\mu = E(X)$ and $\sigma^2 = V(X)$. What is the probability distribution of $\overline{X}_1 - \overline{X}_2$ when X is normal?

7.12 a. Verify that, if $z_1, z_2, \ldots, z_n$ are real numbers, then

$$\frac{1}{n} \sum_{j=1}^{n} (z_i - \bar{z})^2 = \frac{(n - 1)}{n^2} \sum_{j=1}^{n} z_j^2 - \frac{2}{n^2} \sum_{i=1}^{n-1} \sum_{j=1}^{n-1} z_i z_{j+1}.$$

b. Use the expansion in part (a) to show that Eq. (7.1) holds.

APPLICATION

7.13 Determine the variance of $\overline{X}$ when $X_1, X_2, \ldots, X_n$ are independent Bernoulli (M/N) trials.

7.14 Suppose X is binomial $(1, p)$, and let $X_1, X_2, \ldots, X_n$ be a random sample of X. Write out the mass function for $\overline{X}$. Find $P(\overline{X} \leq z)$ by using the fact that $\overline{X} = Y/n$, where Y is binomial (n, p). Is $\overline{X}$ an unbiased estimator for p?

7.15 Find a sample size n in Exercise 7.14 such that, with probability of at least .8, $\overline{X}$ does not differ from p by more than .2.

7.16 Suppose $\overline{X}_n$ is based on a sample of an $N(\mu, \sigma^2)$ population. Determine a formula for the 100αth percentile of the distribution of $\overline{X}_n$; that is, determine ξ_α so that $P(\overline{X}_n < \xi_\alpha) = \alpha$.

7.17 When the population is $N(0, 16)$, evaluate $P(S_X^2 \leq 1)$ for $n = 2$ and $n = 5$.

7.18 Suppose X is $N(50, 90)$, and $X_1, X_2, \ldots, X_{10}$ is a random sample. Evaluate $P(-55 < \overline{X} < 55, S_X^2 > 80)$.

7.19 Let $\overline{X}$ be based on a sample from some population having mean μ and variance σ^2, and let $Y = \sqrt{n}(\overline{X} - \mu)/\sigma$. Show that $E(Y) = 0$ and $V(Y) = 1$. (If the population is normal, we know that Y is standard normal.)

7.20 A population is sampled and results in the five values 2.1, 1.4, 3.2, 6.8, 4.4. Determine the outcomes on $\overline{X}$ and S_X^2. *Note:* The equations in Exercise 7.1 are often helpful for computing the value of S_X^2.

7.21 A supplier of electronic tubes normally supplies tubes in lots of 1000, which are only .5% defective. A prospective buyer is allowed to sample 10 items of the lot before purchasing. What is the probability that the sampler will find at least one defective item in the lot if the lot is up to par?

7.22 A bag of 50 items is known to have 10 defective ones. A sampler selects 8 items without replacement and finds them all good. What is the conditional probability that the ninth item selected will be defective?

7.23 A pair of unbiased dice are to be tossed ten times. What is the probability of throwing a seven exactly four times?

7.24 An opinion poll is to be conducted by sampling n persons selected at random. Each person will be asked whether he or she is for ($X = 1$) or against ($X = 0$) a certain candidate.
   a. State the problem in terms of estimation of p in a Bernoulli (p) population.
   b. State why $\overline{X}_n$ is a reasonable estimator for p.
   c. For what value of p will this estimator have largest variance?
   d. How large would n have to be, with the "worst case" p found in part (c), in order that the estimate be within .1 of the true value (p) with probability .8?

7.2   LIMITING DISTRIBUTIONS

In Chapter 1, it was stated that attempting to define probability as relative frequency is not satisfactory, and yet interpreting probability in applications as relative frequency is quite useful. To see why this statement is so, we now examine the law of large numbers. Suppose $X_1, X_2, \ldots, X_n$ is a random sample of X having mean μ and variance σ^2. Regardless of the underlying probability distribution, the sample mean $\overline{X}_n$ has mean μ and variance σ^2/n. Then by the Chebyshev inequality, for any $\sigma > 0$,

$$0 \leq P(|\overline{X}_n - \mu| \geq \delta) \leq \frac{\sigma^2}{\delta^2 n}.$$

With σ^2 and δ fixed,

$$\lim_{n \to \infty} P(|\overline{X}_n - \mu| \geq \delta) = 0.$$

The significance of this result is that, if we take the sample size n sufficiently large, the likelihood that the sample mean $\overline{X}_n$ will differ from the population mean μ by an amount δ or more can itself be made very small. This result does not mean that $\overline{X}_n$ is ever necessarily equal to μ; it is a statement that, for $n = 1, 2, \ldots$, the sequence $\overline{X}_1, \overline{X}_2, \ldots$ converges to μ in a certain sense. This type of convergence is different from the usual pointwise convergence encountered in calculus. This type of convergence is called *convergence in probability* to μ, written $\overline{X}_n \overset{P}{\to} \mu$. The result that $\overline{X}_n \overset{P}{\to} \mu$ is referred to as the *law of large numbers*.

Theorem 7.4   —*The Law of Large Numbers*   Let $\overline{X}_n$ be the sample mean of a random sample of size n on a parent random variable X, where $E(X) = \mu$ and $V(X)$ exists. Then $\overline{X}_n$ converges in probability to μ, that is, $\lim_{n \to \infty} P(|\overline{X}_n - \mu| \geq \delta) = 0$ for any $\delta > 0$.

EXAMPLE 7.5   Imagine a quality control situation in which items coming off the assembly line are sampled at random and tested. Assume a Bernoulli model for these individual tests, where the success probability p is the reliability of the items produced. The reliability is estimated by the relative frequency of successes with the items sampled. The law of large numbers ensures that, for large sample sizes, the observed relative frequency of success (the estimate of reliability) is very likely to be close to the true (unknown) reliability.

Let us review the significance of the law of large numbers for the relative frequency concept of probability. Suppose A is a given event, and let $X = I_A$, the indicator function of A. Then X is a Bernoulli random variable with parameter $p = P(A)$ and $E(X) = p$. Performing the experiment n times and observing each time whether A occurs is equivalent to an outcome on n Bernoulli trials $X_1, X_2, \ldots, X_n$. Assuming these trials are independent, $Y = \Sigma_{i=1}^n X_i$ has a binomial (n, p) distribution, and the value of Y is just the number of times A occurs. Then the sample mean Y/n is the relative frequency n_A/n of the occurrence of the event A. Since $E(\overline{X}_n) = p = P(A)$, the law of large numbers asserts that $n_A/n \overset{P}{\to} P(A)$. This sense is the one in which relative frequency may be used to approximate the value of $P(A)$. The reason it is not satisfactory for *defining* $P(A)$ is that the convergence mode already involves P, and a circular definition would result.

There is a second concept of convergence of sequences of random variables, called *convergence in distribution*. This concept enables us to establish approximations of various intractable distributions. The concept of convergence in distribution is introduced in the following example.

EXAMPLE 7.6   Suppose X is $N(0, 1)$ and $\overline{X}_n$ is the sample mean of a random sample. We know $\overline{X}_n$ is $N(0, 1/n)$ and $\overline{X}_n \xrightarrow{P} 0$. Since $V(\overline{X}_n) \to 0$, it seems plausible that, for large n, $\overline{X}_n$ will essentially be degenerate at zero. Let F_n be the CDF of $\overline{X}_n$, so $F_n(x) = \Phi(\sqrt{n}\, x)$ for every x, and let F be the CDF of a random variable T degenerate at zero. It is easy to establish that, for $x \neq 0$, $\lim_{n \to \infty} F_n(x) = F(x)$. Thus for $x < 0$,

$$\lim_{n \to \infty} F_n(x) = \lim_{u \to \infty} \Phi(u) = 0,$$

while for $x > 0$,

$$\lim_{n \to \infty} F_n(x) = \lim_{u \to -\infty} \Phi(u) = 1.$$

We say, in this case, that $\overline{X}_n$ converges in distribution to the degenerate random variable T.

Definition 7.4   Suppose X_1, X_2, . . . is a sequence of random variables with corresponding CDFs F_1, F_2, ... Let T be a random variable with distribution function F. The sequence X_n *converges in distribution* to T provided that $\lim_{n \to \infty} F_n(x) = F(x)$ at each point x where F is continuous.

The significance of convergence in distribution is that, if n is sufficiently large, $F(x)$ is approximately $F_n(x)$ at any continuity point x. This approximation is useful when F_n is unknown or difficult to compute while the limiting distribution F is known and is computationally tractable. The use of a limiting distribution F to approximate F_n is discussed in Section 7.5.

In Definition 7.4, suppose each X_n has a corresponding moment-generating function M_n and T has moment-generating function M. Since moment-generating functions characterize the distributions, it seems plausible that, if $M_n(t)$ converges to $M(t)$ for all t in some neighborhood of the origin, then X_n converges in distribution to T. This result is true, and we will use this fact in what follows.

EXAMPLE 7.7   Consider again the sample mean based on a random sample of an $N(0, 1)$ population. Since $\overline{X}_n$ is $N(0, 1/n)$, its moment-generating function is

$$M_n(t) = e^{(1/2n)t^2}.$$

For any real t, $\lim_{n \to \infty} M_n(t) = e^{0t} = 1$, which is the moment-generating function of a random variable degenerate at zero. Thus, again, $\overline{X}_n$ converges in distribution to a random variable degenerate at 0 (and, hence, $\overline{X}_n \xrightarrow{P} 0$; see Exercise 7.25).

One of the most celebrated theorems in probability expresses the fact that the $N(0, 1)$ CDF occurs as a limiting distribution in a wide variety of cases. The proof we give assumes the existence of moment-generating functions, and this assumption limits the class of random variables for which the conclusion is valid. Even so, this class is large enough to make the result surprising and extremely useful. A more general theorem can be proved, but it is beyond the scope of our treatment.

> **Theorem 7.5**   —*The Central Limit Theorem*   Let X_1, $X_2, \ldots, X_n$ be a random sample of a parent random variable X whose moment-generating function M exists. Let $Z_n = \sqrt{n}(\overline{X}_n - \mu)/\sigma$, where $\mu = E(X)$ and $\sigma^2 = V(X)$. Then Z_n converges in distribution to Z, where Z is $N(0, 1)$.*

PROOF   Let $Y = (X - \mu)/\sigma$, so $M_Y(t) = e^{-(\mu/\sigma)t}M(t/\sigma)$. Expand M_Y in a Taylor expansion with remainder to obtain

$$M_Y(t) = 1 + \frac{t^2}{2} + \frac{M_Y'''(\theta)t^3}{6},$$

where $0 < \theta < t$. If M_n is the moment-generating function of Z_n, then

$$M_n(t) = e^{-(\sqrt{n}\mu/\sigma)t}M_{\overline{X}}\left(\frac{t\sqrt{n}}{\sigma}\right)$$

$$= e^{-(\sqrt{n}\mu/\sigma)t}\left[M\left(\frac{t}{\sqrt{n}\,\sigma}\right)\right]^n$$

$$= e^{-(\sqrt{n}\mu/\sigma)t}\left[e^{(\mu/\sqrt{n}\sigma)t}M_Y\left(\frac{t}{\sqrt{n}}\right)\right]^n$$

$$= \left[M_Y\left(\frac{t}{\sqrt{n}}\right)\right]^n$$

$$= \left[1 + \frac{t^2}{2n} + \frac{M_Y''(\theta)t^3}{6n^{3/2}}\right]^n.$$

---

* This result is often expressed by saying Z_n is *asymptotically normal*.

Hence,

$$\log M_n(t) = n \log\left[1 + \frac{t^2}{2n} + \frac{M_Y'''(\theta)t^3}{6n^{3/2}}\right]$$

$$= na_n(t)\,\frac{\log[1 + a_n(t)]}{a_n(t)},$$

where

$$a_n(t) = \frac{t^2}{2n} + \frac{M_Y'''(\theta)t^3}{6n^{3/2}}.$$

As $n \to \theta$, with θ and t fixed,

$$a_n(t) \to 0 \qquad \text{and} \qquad na_n(t) = \frac{t^2}{2} + \frac{M_Y'''(\theta)t^3}{n^{1/2}} \to \frac{t^2}{2}.$$

Since $\lim_{x \to 0} \log[(1 + x)/x] = 1$,

$$\lim_{n \to \infty} \log M_n(t) = \frac{t^2}{2} = \log \lim_{n \to \infty} M_n(t),$$

so $\lim_{n \to \infty} M_n(t) = e^{t^2/2}$. But this expression is the moment-generating function of an $N(0, 1)$ random variable.

The central limit theorem is a very important result, particularly for statistical inference. It is often desired to evaluate $P(a < \overline{X} < b)$, where $\overline{X}$ is the sample average of a random sample of size n of some distribution having mean μ and variance σ^2. According to the central limit theorem, for large n,

$$P(a < \overline{X} \le b) = P\left[\frac{\sqrt{n}(a - \mu)}{\sigma} < \frac{\sqrt{n}(\overline{X} - \mu)}{\sigma} \le \frac{\sqrt{n}(b - \mu)}{\sigma}\right]$$

$$\approx \Phi\left(\frac{\sqrt{n}(b - \mu)}{\sigma}\right) - \Phi\left(\frac{\sqrt{n}(a - \mu)}{\sigma}\right).$$

Thus if n is sufficiently large, we may use standard normal tables to evaluate $P(a < \overline{X} \le b)$, *regardless of the underlying probability distribution* (subject to the existence of σ^2). The central limit theorem can also be used to evaluate probabilities of events involving the sum $\Sigma_{i=1}^{n} X_i$, using an $N(n\mu, n\sigma^2)$ approximation.

The central limit theorem is quite useful in constructing probability models. For example, to model the motion of a particle that is subject to collisions with a great many smaller particles, we may imagine that the overall displacement in a given time is the sum of a great many smaller displacements; consequently, we may be led to adopt a *normal* distribution for the displacement of the larger particle. This result leads to a useful class of stochastic models for Brownian motion.

The central limit theorem may be interpreted roughly as stating that the sample mean based on a random sample of size *n* has a distribution approaching a *normal distribution* as *n* increases, regardless of the parent distribution. In order to show how rapid this convergence may be, even for parent distributions quite unlike the normal, we have sketched the densities of $\overline{X}_n$ in Fig. 7.2 for several values of *n* and several parent density shapes.

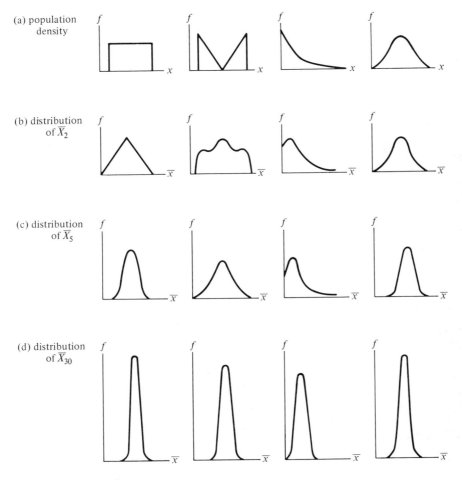

FIGURE 7.2   *Densities of $\overline{X}_n$, n = 2, 5, 30, for Various Population Densities*

Even for n as small as 5, the density of $\overline{X}$ may look nearly bell-shaped, even though the parent density does not.

EXERCISES

THEORY

7.25 The concept of the law of large numbers can be extended to any sequence $\{X_n\}_{n=1}^{\infty}$ of random variables. Such a sequence is said to *converge in probability* to the constant a, written as $X_n \overset{P}{\rightarrow} a$, if $\lim_{n \to \infty} P(|X_n - a| \geq \delta) = 0$ for every $\delta > 0$.
  a. Show that a sequence $X_1, X_2, \ldots$ of random variables converges in probability to a constant a if and only if the corresponding sequence of CDFs converges to the CDF for a random variable degenerate at a, at all points other than a itself, that is, if and only if X_n converges in distribution to a random variable X degenerate at a.
  b. Show that the condition in part (a) is equivalent to $\lim_{n \to \infty} P(|X_n - a| < \delta) = 1$ for any $\delta > 0$.

7.26 Derive the result of the central limit theorem directly when each X_i in the theorem is $\chi^2_{(1)}$.

7.27 Suppose X is uniform over $(0, b)$, and let $X_1, X_2, \ldots, X_n$ be a random sample of X.
  a. If $Y_n = \max\{X_1, X_2, \ldots, X_n\}$ is the largest value in the sample, show that $Y_n \overset{P}{\rightarrow} b$.
  b. Let $Z_n = n(b - Y_n)$. Show that Z_n converges in distribution to an exponential random variable whose parameter is $\lambda = 1/b$.

7.28 Let X_n be degenerate at n for $n = 1, 2, 3, \ldots$. Show that the sequence X_1, $X_2, \ldots$ has no limiting distribution.

7.29 Suppose X has mean μ and variance σ^2, and let $Y = |(X - \mu)/\sigma|$. The value of Y is a measure of the *relative deviation* of a random variable from its mean.
  a. Use Chebyshev's inequality to show that, for any $\delta > 0$, $P(Y < \delta) \leq \sigma^2/\delta^2\mu^2$.
  b. If X is normally distributed, show that $P(Y > \delta) \leq \alpha$ for $\delta = K_{1-\alpha}\sigma/|\mu|$, where $K_{1-\alpha} = \Phi^{-1}(1 - \alpha)$; that is, $P(Z > K_{1-\alpha}) = \alpha$, where Z is $N(0, 1)$.
  c. Evaluate δ for $\alpha = .01, .02, .05, .10$. *Note:* The quantity σ^2/μ^2 is often called the *coefficient of variation* in applied statistical work. The smaller this coefficient of variation is, the more restricted (for given α) are the values of Y, that is, the closer the values of X are to the mean μ. The square root of the reciprocal of the coefficient of variation, $|\mu|/\sigma$, arises in engineering applications and is called the *measurement signal-to-noise ratio*.

7.30 This exercise involves convergence in probability to a random variable. Suppose $X_1, X_2, \ldots$ is a sequence of random variables, and X is a random variable. The sequence X_n converges to X in probability, written $X_n \overset{P}{\rightarrow} X$, provided that $X_n - X$ converges in probability to zero, that is, provided that $\lim_{n \to \infty} P(|X_n - X| \geq \delta) = 0$ for every $\delta \geq 0$.

a. Suppose Y and Z are independent $N(0, 1)$ random variables, and define

$$X_n = \left(\frac{1}{n}\right) Y + \left(\frac{n - 1}{n}\right) Z.$$

Show that $X_n \overset{P}{\to} Z$ and that X_n converges in distribution to z.

b. Suppose X is binomial $(1, .5)$ and $X_n = X$, $n = 1, 2, \ldots$. Let $Y = 1 - X$. Show that X_n converges in distribution to Y and that X_n does *not* converge in probability to Y.

APPLICATION

7.31 If X_n is binomial (n, p) and $Y_n = n - X_n$, show that $Y_n/n \overset{P}{\to} q$, where $q = 1 - p$.

7.32 Let $\overline{X}_n$ be the sample mean of a population random variable whose distribution is unknown. Approximately how large should n be to guarantee that, with probability .95, $\overline{X}_n$ is within $.2\sigma$ units of μ, where μ and σ^2 are the population mean and variance? *Hint:* Try using the Chebyshev inequality and the central limit theorem.

7.33 A scale is used to weigh samples of items. The differences between observed and true weights are assumed to be the observed values of a random variable X, whose mean is 0 and whose standard deviation is .5. What is the probability that a reading is within one unit of the true weight? What is the probability the mean of 30 readings is within one unit of the true weight?

7.34 Assume that telephone conversations have a mean duration of 4.5 minutes with a variance of 4 minutes. It is desired to establish a probability model for the combined duration of 30 conversations. Discuss why a normal model might be appropriate, and suggest a choice for the parameters in the normal distribution involved. What is the probability that the duration will not exceed 150 minutes?

7.35 Add 100 real numbers, each of which is rounded off to the nearest integer. Assume that each round-off error is a random variable uniformly distributed between $-.5$ and $.5$ and that the 100 round-off errors are independent. Find (approximately) the probability that the error in the sum will be between -3 and 3. Find a quantity A such that the probability is approximately .99 that the error in the sum will be less than A in absolute terms.

7.36 A standard drug is known to be effective in about 80% of the cases in which it is used to treat infections. A new drug has been found effective in 85 of the first 100 cases tried.
a. Is the superiority of the new drug well established?
b. Would 63 successes out of 70 cases be stronger evidence (than 85 out of 100) in support of the new drug?

7.37 Numbers are selected at random from the interval $(0, 1)$.
a. If 10 numbers are selected, what is the probability that exactly 5 are less than $\frac{1}{2}$?
b. If 10 numbers are selected, how many are less than $\frac{1}{2}$, on the average?
c. If 20 numbers are selected, what is the probability that their average is less than $\frac{1}{2}$?

   d. If 50 numbers are selected, what is the probability that their average is
      between .48 and .52?

7.38 A machine fills cereal boxes with cornflakes. Each box is advertised to contain
      at least 16 ounces of cereal. The machine is calibrated to put 16.5 ounces in
      each box, but since this calibration is based on a volumetric measurement,
      the weight of cereal actually put in the box has a variance of $\frac{1}{4}$ ounce^2.
      a. What is the distribution of the weight of a carton of 24 boxes of cereal?
      b. Could a reasonable test for shortages of boxes in the carton be made by
         weighing the carton?
      c. What is the probability that a carton weighs less than 390 ounces?

7.39 An aircraft has 80 passenger seats and can carry a passenger load of 10,000
      pounds.
      a. If passegers weigh 130 pounds, on the average, with a standard deviation
         of 15 pounds, what is the probability that all seats can be loaded?
      b. If the airline runs a half-fare special for children, the mean weight of pas-
         sengers drops to 111 pounds, with a standard deviation of 19 pounds. Now
         what is the probability that all seats can be loaded?

7.40 An insurance company has long-term records indicating that the probability
      that a married male 30 years of age will survive to at least age 60 is .66. Of
      a given set of 100 such males, what is the probability that at least 75 survive
      to age 60?

7.41 It is sometimes argued that a student's grade point average cannot be a measure
      more precise than the precision with which individual grades are determined.
      "A chain is no stronger than its weakest link," it is argued. Comment on this
      argument, especially as the number of courses increases.

7.3 THE t- AND F-DISTRIBUTIONS*

In practice, it is frequently assumed that a random sample is taken from a
normal population. Consequently, there is a great deal of interest in the
distributions of various functions of independent normal random variables.
We have already discussed several particular cases; for example, a linear
function of jointly normal random variables is normally distributed (hence,
$\bar{X}$ is normal), and a sum of squares of independent standard normal random
variables has a chi-square distribution [so $nS_{\bar{X}}^2/\sigma^2$ is $\chi^2_{(n-1)}$]. We will now
discuss two other important families of continuous distributions that arise
in connection with sampling from a normal population, the t- and F-distri-
butions.

   The t-distribution was discovered by the English statistician W. Gossett, who published his results under the pseudonym "Student." For this
reason, it has come to be known as Student's t-distribution. Interest in this
one-parameter family of distributions arises in statistics in connection with

---

* This section is optional.

an attempt to standardize an $N(0, \sigma^2)$ random variable X when σ^2 is unknown. This standardization involves replacing σ in X/σ by its estimate, s_X. Thus we become interested in deriving the distribution of a quotient of the random variables X and S_X. The family of F-distributions also arises in statistics as the distribution of a quotient of certain random variables. Members of this two-parameter family of distributions are called variously the F-distributions (in honor of the great English statistician R. A. Fisher) or Snedecor's F-distributions (in honor of the American statistician G. W. Snedecor). Let us begin by considering how the t-distribution arises as a sampling distribution.

Suppose X is $N(0, 1)$ and Y is $\chi^2_{(n)}$, and X and Y are independent random variables. The probability density of $W = X/\sqrt{Y/n}$ is a t-density with parameter n (called "degrees of freedom"). Since independence is assumed, the joint density of X and Y is

$$f(x, y) = \frac{1}{\sqrt{2\pi}} e^{-(1/2)x^2} \cdot \frac{1}{\Gamma(n/2)2^{n/2}} y^{(n/2)-1} e^{-(y/2)},$$

for $-\infty < x < \infty$, $0 < y < \infty$. Now, we employ the change-of-variable technique, using the system of equations $w = x/\sqrt{y/n}$ and $z = y$. The inverse of this transformation is $x = w\sqrt{z/n}$ and $y = z$. Hence, the Jacobian of the transformation is

$$J = \begin{vmatrix} \sqrt{\dfrac{z}{n}} & \left(\dfrac{w}{2n}\right)\left(\dfrac{z}{n}\right)^{-(1/2)} \\ 0 & 1 \end{vmatrix} = \sqrt{\dfrac{z}{n}}.$$

Accordingly, the joint density g of W and Z is

$$g(w, z) = \sqrt{\frac{z}{n}} f\left(w\sqrt{\frac{z}{n}}, z\right)$$

$$= \frac{1}{\sqrt{2\pi n}\,\Gamma(n/2)2^{n/2}} z^{(n-1)/2} e^{-(z/2)[1+(w^2/n)]}, \quad -\infty < w < \infty, 0 < z < \infty.$$

To find the required marginal density of W, we integrate the joint with respect to z. Letting $u = (z/2)[1 + (w^2/n)]$, we see that, for fixed w,

$$\int_0^\infty z^{(n-1)/2} e^{-(z/2)[1+(w^2/n)]}\, dz = \frac{2^{(n+1)/2}}{[1 + (w^2/n)]^{(n+1)/2}} \int_0^\infty u^{[(n+1)/2]-1} e^{-u}\, du.$$

But the latter integral is $\Gamma((n + 1)/2)$, so the density h_n of W is

$$h_n(w) = \frac{\Gamma((n + 1)/2)}{\sqrt{n\pi}\,\Gamma(n/2)} \cdot \left(1 + \frac{w^2}{n}\right)^{-(n+1)/2}, \quad -\infty < w < \infty. \quad (7.2)$$

This expression is the *t–density function* with *n degrees of freedom*, and we may state that W is $t_{(n)}$. The degrees of freedom correspond to the parameter in the chi-square variable in the denominator of W. If $n = 1$, W is Cauchy. For $n > 2$, the mean of W is zero. In Exercise 7.55, you are asked to show that $V(W) = n/(n - 2)$ for $n > 2$. Higher central moments of W may be found by using the additional results of Exercise 7.55. The moment-generating function for the *t*-family does not exist, and the cumulative distribution function for W is not simply expressed. Yet the *t*-distribution is so important in statistical analysis that the *t*–CDF has been extensively tabulated; a short table is given in Table 4 of the Appendix.

From functional considerations of the *t*-density in Eq. (7.2), it can be verified that the graph of h_n, like the special case when $n = 1$ (the Cauchy), has the general appearance of a bell-shaped curve resembling the $N(0, 1)$ density function. Indeed, the similarity of appearance of the *t*-densities and the standard normal density misled many early investigators into the impression that ratios of the form X/S_X were standard normal. Gossett's discovery of the correct distribution of this statistic was, therefore, an important achievement.

We may now employ the independence results of Theorem 7.2 to yield a *t*-distributed random variable that is of great significance in the field of statistics.

Theorem 7.6   Suppose X is $N(\mu, \sigma^2)$, and let $X_1, X_2, \ldots, X_n$ be a random sample of X with $n > 1$. Then the random variable $\sqrt{n - 1}\,(\bar{X} - \mu)/S_X$ has a *t*-distribution with $n - 1$ degrees of freedom.

PROOF   The standardized variable $\sqrt{n}(\bar{X} - \mu)/\sigma$ is $N(0, 1)$, and (from Theorem 7.2) nS_X^2/σ^2 is $\chi^2_{(n-1)}$. Moreover, by Theorem 7.2, S_X^2 and $\bar{X}$ are independent, so $\sqrt{n}(\bar{X} - \mu)/\sigma$ and nS_X^2/σ^2 are also independent. Thus

$$\frac{\sqrt{n}(\bar{X} - \mu)/\sigma}{\sqrt{nS_X^2/\sigma^2(n - 1)}} = \frac{\sqrt{n - 1}(\bar{X} - \mu)}{\sqrt{S_X^2}}$$

has a *t*-distribution with $n - 1$ degrees of freedom.

Let us now turn to the *F*-family of distributions mentioned earlier. Members of this family, like members of the *t*-family, arise in a natural way as to the distributions of quotients of two random variables. If X is $\chi^2_{(n)}$ and Y is $\chi^2_{(m)}$, where X and Y are independent, then $W = (m/n)(X/Y)$ is *F*-distributed with parameters n and m. The density of W is derived as follows:

The joint distribution of X and Y is the product of the respective χ^2 marginals,

$$f_{X,Y}(x, y) = \frac{1}{2^{n/2}\Gamma(n/2)} x^{(n/2)-1} e^{-x/2} \frac{1}{2^{m/2}\Gamma(m/2)} y^{(m/2)-1} e^{-y/2}, \quad x > 0, y > 0.$$

Now make the transformation $W = (X/n)/(Y/m)$ and $Z = Y$. The Jacobian of this transformation is $(n/m)z$, so

$$f_{W,Z}(w, z) = \frac{1}{\Gamma(n/2)\Gamma(m/2)2^{(n+m)/2}} \left(\frac{n}{m} \cdot wz\right)^{(n-2)/2} z^{(m-2)/2} e^{-(1/2)[(n/m)wz+z]},$$

for $w > 0$ and $z > 0$; and the marginal of W is

$$f_W(w) = \left[\frac{(n/m)^{n/2}\Gamma((n+m)/2)}{\Gamma(n/2)\Gamma(m/2)}\right] \left\{\frac{w^{(n/2)-1}}{[1 + (n/m)w]^{(n+m)/2}}\right\}, \quad w > 0. \quad (7.3)$$

This two-parameter family of density functions is called the *F-family*. The parameters n and m are called *degrees of freedom*, and when a random variable W has density function f_W, we say that W is an $F_{(n,m)}$ *random variable*. In some contexts, W is written $W_{(n,m)}$ in order to emphasize dependence on the parameter values. The order in which the parameters are read is very important, for Eq. (7.3) is not symmetric in n and m. For obvious reasons, and so that the distinction between the parameters is clear, n is called the *numerator* degrees of freedom and m the *denominator* degrees of freedom in Eq. (7.3); W is sometimes called a *variance ratio*.

Like the *t*-family, neither the moment-generating function nor the cumulative distribution function for the *F*-distributions is tractable. The CDF is related to the incomplete beta function (see Exercise 7.44). This fact is sometimes used in tabulation of the CDF of W; for instance, it was used to obtain the short table of F CDFs given in Table 5 of the Appendix. As for moments, they may be computed directly from the density function by a convenient change of variable that relates the integral to the beta function. Thus

$$\int_0^\infty \frac{x^{n/2}\, dx}{[1 + (n/m)x]^{(n+m)/2}}$$

$$= \left(\frac{m}{n}\right)^{(n/2)+1} \int_0^1 u^{[(n/2)+1]-1}(1 - u)^{[(m/2)-1]-1}\, du, \quad (7.4)$$

under the change of variable

$$u = \frac{(n/m)x}{1 + (n/m)x}.$$

But if $m > 2$, this integral is precisely

$$\beta\left(\frac{n}{2} + 1, \frac{m}{2} - 1\right) = \frac{\Gamma((n/2) + 1)\Gamma((m/2) - 1)}{\Gamma((m + n)/2)} = \frac{(n/2)\Gamma(n/2)\Gamma(m/2)}{\Gamma((m + n)/2)[(m/2) - 1]}$$

$$= \frac{n\Gamma(n/2)\Gamma(m/2)}{(m - 2)\Gamma((m + n)/2)}.$$

Multiplying Eq. (7.4) by the constant in Eq. (7.3) yields

$$E[W_{(n,m)}] = \frac{m}{n} \cdot \frac{n}{m - 2} = \frac{m}{m - 2}, \qquad \text{if } m > 2.$$

In Exercise 7.46, you are asked to show that

$$V[W_{(n,m)}] = \frac{m^2(2m + 2n - 4)}{n(m - 2)^2(m - 4)}, \qquad \text{if } m > 4. \tag{7.5}$$

One important way in which the F-distribution arises in statistics is embodied in the next theorem.

> **Theorem 7.7**   Let X be $N(\mu_X, \sigma_X^2)$ and Y be $N(\mu_Y, \sigma_Y^2)$. Suppose $X_1, X_2, \ldots, X_n$ is a random sample of X and $Y_1, Y_2, \ldots, Y_m$ is a random sample of Y; moreover, assume that all $m + n$ random variables are independent. Then
>
> $$\frac{n(m - 1)}{m(n - 1)}\left(\frac{\sigma_Y^2}{\sigma_X^2}\right)\left(\frac{S_X^2}{S_Y^2}\right)$$
>
> has an $F_{(n-1, m-1)}$ distribution.

PROOF   By Theorem 7.2, nS_X^2/σ_X^2 and mS_Y^2/σ_Y^2 are chi-square random variables with $n - 1$ and $m - 1$ degrees of freedom, respectively. Also, S_X^2 and S_Y^2 are independent because S_X^2 is a function of $X_1, X_2, \ldots, X_n$, while S_Y^2 is a function of $Y_1, Y_2, \ldots, Y_m$. It follows that

$$\frac{m - 1}{n - 1}\left(\frac{nS_X^2}{\sigma_X^2}\right)\left(\frac{\sigma_Y^2}{mS_Y^2}\right)$$

has an $F_{(n-1, m-1)}$ distribution.

As a special case of Theorem 7.7 with $\sigma_X^2 = \sigma_Y^2$, $n(m - 1)/m(n - 1)(s_X^2/S_Y^2)$ is a statistic having an $F_{(n-1,m-1)}$ distribution. This statistic is a (sample) variance ratio, up to a multiplicative constant.

EXAMPLE 7.8   A machine produces steel balls for use in ball bearings. The balls are supposed to have a diameter of .2497 inch each, but due to rolling tolerances in the machine, the actual diameters vary. In a check of whether the machine is within acceptable accuracy and precision specifications, a sample of 15 balls is drawn at random from a day's production and the diameters carefully measured. The observed diameters are .2490, .2485, .2481, .2590, .2491, .2506, .2511, .2489, .2512, .2500, .2510, .2489, .2491, .2480, and .2501.

   a. What are the sample mean and variance?

   b. Assuming the population mean is .2497, what is the value of the t-statistic $(\overline{X} - .2497)\sqrt{14}/S_X$?

   c. What is the probability that a $t_{(14)}$ random variable would be this large or larger?

   d. Suppose, in an independent random sample on Y of size 11 from the same population, it is found that $s_Y^2 = 2.88 \times 10^{-6}$. What is the value of the F-statistic defined in Theorem 7.7?

   e. What is the probability that an $F_{(14,10)}$ random variable would be larger than the outcome in part (d)?

SOLUTION

   a. $\overline{x}_{15} = .2502$ and $s_X^2 = 6.631 \times 10^{-6}$, using the definitions of $\overline{X}$ and S_X^2.

   b. $t = .6878$.

   c. $P(T \geq .6878)$ is approximately .25, using a t-table.

   d. 2.898, by computation with $\sigma_Y^2 = \sigma_X^2$.

   e. $P(W \geq 2.898)$ is approximately .05, using an F-table.

EXERCISES

THEORY

7.42   Suppose X is $N(0, 1)$, Y is exponential (1), and X and Y are independent. Show that $X/\sqrt{Y}$ has a $t_{(2)}$ distribution.

7.43   Suppose T is $t_{(2)}$. Verify that $V(T)$ does not exist.

7.44   Suppose X is an $F_{(n,m)}$ random variable, and let $Y = 1/[1 + (m/nX)]$.
   a. Show that Y has a beta $(n/2, m/2)$ distribution.
   b. Show that, consequently, $F_X(x) = F_Y[nx/(m + nx)]$ for each $x > 0$.

7.45   Suppose X has the $F_{(n,m)}$ distribution, and let $Y = 1/X$. Show directly that Y is $F_{(m,n)}$ (i.e., do not assume that X is the ratio of chi-square random variables). Use this fact to verify the identity $F_{\alpha;m,n}F_{1-\alpha;n,m} = 1$, where, if W is $F_{(n,m)}$, then $P(W > F_{\alpha;n,m}) = \alpha$.

7.46 Show that, if X is $F_{(n,m)}$, then

$$E(X^2) = \frac{m^2(n + 2)}{n(m - 2)(m - 4)} \qquad \text{if } m > 4.$$

Hence, verify Eq. (7.5).

7.47 a. Suppose X has a $t_{(n)}$ distribution, and let $Y = X^2$. Show that Y has an $F_{(1,n)}$ distribution.
  b. Explain why it does not follow that $\sqrt{Y}$ is $t_{(n)}$.

7.48 Let $X_1, X_2, \ldots, X_{2n}$ be a random sample from an $N(\mu, \sigma^2)$ distribution. Let

$$W_1 = \frac{1}{n} \sum_{i=1}^{n} (X_i - \overline{X}_1)^2 \qquad \text{and} \qquad W_2 = \frac{1}{n} \sum_{i=n+1}^{2n} (X_i - \overline{X}_2)^2,$$

where $\qquad \overline{X}_1 = \frac{1}{n} \sum_{i=1}^{n} X_i \qquad$ and $\qquad \overline{X}_2 = \frac{1}{n} \sum_{i=n+1}^{2n} X_i.$

Show that W_1/W_2 has an F-distribution.

7.49 Let X be exponential (λ_1) and Y be exponential (λ_2) (X and Y might be "times to failure" for two different electronic components). Suppose $X_1, X_2, \ldots, X_n$ is a random sample of X and that $Y_1, Y_2, \ldots, Y_m$ is a random sample of Y, with the X's and Y's independent. Show that $\lambda_1\overline{X}/\lambda_2\overline{Y}$ has an $F_{(2n,2m)}$ distribution.

7.50 Suppose $X_1, X_2, \ldots, X_n$ are independent, where X_i is $\chi^2_{(n_i)}$. Show that $n\overline{X}$ is $\chi^2_{(\Sigma n_i)}$.

7.51 Suppose (X_1, X_2) is bivariate normal with mean $(0, 0)$ and covariance matrix I.
  a. Show that $Y_1 = X_1^2 + X_2^2$ and $Y_2 = X_1/X_2$ are independent.
  b. Find the marginal distributions of Y_1 and Y_2.

7.52 Suppose X and Y have a bivariate normal distribution, where X and Y are identically $N(0, 9)$ distributed and $\rho(X, Y) = .2$. Find the equation of an ellipse bounding a region A such that $P[(X, Y) \in A] = .5$.

7.53 Suppose $X_1, X_2, \ldots, X_n$ are independent, where X_i is $N(\mu, \sigma_i^2)$.
  a. Show that

$$Y = \frac{\sum_{i=1}^{n} X_i/\sigma_i^2}{\sum_{i=1}^{n} 1/\sigma_i^2} \qquad \text{and} \qquad V = \sum_{i=1}^{n} \frac{(X_i - \mu)^2}{\sigma_i^2}$$

are independent.
  b. Show that Y is normally distributed.
  c. Show that V is $\chi^2_{(n)}$.

7.54 Suppose X is $N(\mu_X, \sigma_X^2)$, Y is $N(\mu_Y, \sigma_Y^2)$, and $X_1, X_2, \ldots, X_n$ is a random sample of X while $Y_1, Y_2, \ldots, Y_m$ is a random sample of Y. Find the joint density of

$$U = \frac{\sqrt{m}(\bar{Y} - \mu_Y)}{S_Y} \quad \text{and} \quad V = \frac{S_X^2}{S_Y^2}.$$

(Assume the X's and Y's are independent.)

7.55 This exercise concerns moments of t-distributions. The family of t-distributions, where each distribution is associated with a parameter $n > 0$, is characterized by the density function

$$f(x) = c \cdot \left(1 + \frac{x^2}{n}\right)^{-(1/2)(n+1)},$$

where n is called the degrees of freedom.

a. Show that the "normalizing constant" is given by

$$c = \frac{1}{\sqrt{n\pi}} \left[\frac{\Gamma((n+1)/2)}{\Gamma(n/2)}\right].$$

b. Verify that the Cauchy distribution is a t-distribution with one degree of freedom.

c. Show that the kth moment exists only for $k < n$.

d. Show that, if $k < n$ and k is odd, then the kth moment is zero. *Hint:* Consider the integral of an odd function.

e. Show that, if $k < n$ and k is even, then the kth moment is

$$\frac{n^{k/2}\Gamma((k+1)/2)\Gamma((n-k)/2)}{\Gamma(1/2)\Gamma(n/2)}.$$

Hint: If X is $t_{(n)}$, then

$$E(X^k) = c\int_{-\infty}^{\infty} \frac{x^k}{[1 + (x^2/n)]^{(n+1)/2}}\,dx = 2c\int_0^{\infty} \frac{x^n}{[1 + (x^2/n)]^{(n+1)/2}}\,dx,$$

since n is even. Let $y = x^2/n$, so that $dx = (\sqrt{n}/2\sqrt{y})\,dy$. Then

$$E(X^k) = 2c\int_0^{\infty} \frac{(ny)^{k/2}}{(1+y)^{(n+1)/2}} \cdot \frac{\sqrt{n}}{2\sqrt{y}}\,dy$$

$$= cn^{(k+1)/2}\int_0^{\infty} \frac{y^{[(k+1)/2]-1}}{(1+y)^{[(k+1)/2]+[(n+1)/2]-[(k+1)/2]}}\,dy$$

$$= cn^{(k+1)/2}B\left(\frac{k+1}{2}, \frac{n-k}{2}\right),$$

where B is the beta function whose value at $((k + 1)/2, (n - k)/2)$ is

$$\frac{\Gamma((k + 1)/2)\Gamma((n - k)/2)}{\Gamma((n + 1)/2)}.$$

APPLICATION

7.56 Sketch the graph of the t-density having four degrees of freedom.

7.57 Suppose $X_1, X_2, \ldots, X_{10}$ is a random sample from an $N(0, 4)$ distribution. Find $P(\overline{X} > .3S_X)$ in terms of a t-distribution function. Use a t-table to approximate this value.

7.58 Suppose X_1 and X_2 are independent and identically distributed, each uniform over $(0, 1)$. [That is, X_1, X_2 is a random sample of size 2 from a uniform over $(0, 1)$ population.] Find the distribution of $\overline{X}$, and show directly that $E(\overline{X})$ is the mean of the population.

7.59 Suppose $X_1, X_2, \ldots, X_n$ is a random sample from an $N(0, 1)$ population. What is the distribution of each of the following?

a. $\displaystyle\sum_{i=1}^{n} X_i$ b. $\displaystyle\sum_{i=1}^{n} X_i^2$ c. $\overline{X}_n$ d. $\Sigma (X_i - \overline{X})^2$

e. $\dfrac{X_1^2}{X_2^2}$ f. $\dfrac{X_1}{\sqrt{X_2^2}}$ g. $\dfrac{X_1 - X_2}{\sqrt{2}}$

h. $\dfrac{(X_1 - X_2)^2}{(X_1 + X_2)^2}$ i. $\dfrac{X_1 - X_2}{\sqrt{(X_1 + X_2)^2}}$ j. $\dfrac{\overline{X}}{\sqrt{nS_X^2/(n - 1)}}$

7.60 a. Find $P(X > 15.101)$, where X is $F_{(4,3)}$.
b. Find $P(Y < 1.65)$, where Y is $t_{(6)}$.

7.4   ORDER STATISTICS*

In a sampling of a population, the respective values of the random variables $X_1, X_2, \ldots, X_n$ obtained after the experiment has been performed are the components of an n-tuple $(x_1, x_2, \ldots, x_n)$ of real numbers, the sample data. In some cases, such as sampling without replacement and the development of the hypergeometric distribution, the *order* in which the values are observed may be quite critical. On the other hand, when the sample is *random*, the order in which these numbers are obtained is immaterial. To see this idea, suppose X_1, X_2, X_3 is a random sample of X. If (say) $Y_1 = X_2$, $Y_2 = X_3$, and $Y_3 = X_1$, then Y_1, Y_2, Y_3 is also a random sample of X. Indeed, the joint distributions of (X_1, X_2, X_3) and (Y_1, Y_2, Y_3) are identical. However,

---

* This section is optional.

if the sample data are rearranged according to some rule involving the sample values themselves, such as listing the values in increasing order of magnitude, then the resulting n-tuple of data is no longer the result of a random sample. Rather, the result may be thought of as respective values of the components of a 3-tuple of statistics, called *order statistics*. Such statistics are very useful in the development of stochastic models and in statistical inference.

The objective in what follows is to make the notions of order statistics more precise and to develop their distributions. There are two possible approaches. We could write the order statistic $(Y_1, Y_2, \ldots, Y_n)$ as a function of the random sample $(X_1, X_2, \ldots, X_n)$ and use the change-of-variable technique to find the distribution of $(Y_1, Y_2, \ldots, Y_n)$. The approach we will use, however, is to consider a multinomial tally of outcomes in small intervals along the line and use the mean value theorem for integrals.

Consider a random sample $(X_1, X_2, \ldots, X_n)$ of size n on a continuous parent random variable X having density $f(x)$ and CDF $F(x)$. The order statistic $(Y_1, Y_2, \ldots, Y_n)$ is a rearrangement of the components of $(X_1, X_2, \ldots, X_n)$ into ascending order, so Y_1 is the smallest component in $(X_1, X_2, \ldots, X_n)$, while $Y_n = \max\{X_1, X_2, \ldots, X_n\}$. Since X is continuous, there is zero probability that two or more of the X_i's will have precisely the same outcome (i.e., that there are "ties"). The joint density of $(Y_1, Y_2, \ldots, Y_n)$ will be positive only within the region $y_1 < y_2 < \cdots < y_n$ in n-dimensional space. Now, for small positive $\Delta y_1, \Delta y_2, \ldots, \Delta y_n$, and for $y_1 < y_2 < \cdots < y_n$,

$$f_{Y_1, Y_2, \ldots, Y_n}(y_1, y_2, \ldots, y_n) \, \Delta y_1 \, \Delta y_2 \cdots \Delta y_n \approx P(y_1 < Y_1 \le y_1$$
$$+ \Delta y_1, y_2 < Y_2 \le y_2 + \Delta y_2, \ldots, y_n < Y_n \le y_n + \Delta y_n). \quad (7.6)$$

The latter may be described as the probability that one of the X's is in $(y_1, y_1 + \Delta y_1]$, another is in $(y_2, y_2 + \Delta y_2]$, and so on. We obtain this probability by multiplying the probability that a specific X, say X_1, is in $(y_1, y_1 + \Delta y_1]$ and (say) X_2 is in $(y_2, y_2 + \Delta y_2]$, and so on, by the number $n!$ of possible arrangements of the X_i's. Thus Eq. (7.6) may be written as

$$f_{Y_1, Y_2, \ldots, Y_n}(y_1, y_2, \ldots, y_n) \, \Delta y_1 \, \Delta y_2 \cdots \Delta y_n$$
$$\approx n! f(y_1) \, \Delta y_1 \cdot f(y_2) \, \Delta y_2 \cdots f(y_n) \, \Delta y_n.$$

Dividing both sides by $\Delta y_1 \, \Delta y_2 \cdots \Delta y_n$, then taking the limit as the Δy_i's tend to zero, yields the equality

$$f_{Y_1, Y_2, \ldots, Y_n}(y_1, y_2, \ldots, y_n) = n! f(y_1) f(y_2) \cdots f(y_n),$$
$$y_1 < y_2 < \cdots < y_n. \quad (7.7)$$

EXAMPLE 7.9   Consider the order statistic (Y_1, Y_2, Y_3, Y_4) of a random sample X_1, X_2, X_3, X_4 from a uniform $(0, 1)$ population. The joint density of the order statistic is

$$f_{Y_1, Y_2, Y_3, Y_4}(y_1, y_2, y_3, y_4) = 24, \qquad 0 < y_1 < y_2 < y_3 < y_4 < 1.$$

The Y_i's are not independent since the range of (Y_1, Y_2, Y_3, Y_4) is not a product set.

EXAMPLE 7.10 It is easy to use the CDF method to find the marginal distributions of the extreme order statistics $Y_1 = \min\{X_1, X_2, \ldots, X_n\}$ and $Y_n = \max\{X_1, X_2, \ldots, X_n\}$. For example,

$$F_{Y_n}(y) = P(Y_n \le y) = P(X_1 \le y, X_2 \le y, \ldots, X_n \le y)$$

$$= P(X_1 \le y)P(X_2 \le y) \cdots P(X_n \le y)$$

$$= [F_X(y)]^n.$$

Thus the density of the largest order statistic is

$$f_{Y_n}(y) = n[F_X(y)]^{n-1} f_X(y).$$

If the parent distribution is uniform $(0, 1)$, for example, then $f_{Y_n}(y) = ny^{n-1}$, $0 < y < 1$. This expression is recognized to be the beta $(n, 1)$ distribution. A similar argument shows that Y_1 is beta $(1, n)$, in this case (see Exercise 7.74).

Let us examine the distributions of the individual order statistics and some simple functions of the order statistics. The order statistics of a random sample $X_1, X_2, \ldots, X_n$ are usually denoted $X_{(1)}, X_{(2)}, \ldots, X_{(n)}$ in the probability literature. Thus $X_{(i)}$ is the ith smallest value in a random sample of size n on X.

The marginal distributions of $X_{(1)} = \min\{X_1, X_2, \ldots, X_n\}$ and $X_{(n)} = \max\{X_1, X_2, \ldots, X_n\}$ have already been examined. The random variable $R = X_{(n)} - X_{(1)}$ is called the *range* of the random sample. Let us obtain the distribution of the sample range from the joint density of $X_{(1)}$ and $X_{(n)}$. For $y_1 < y_n$, the joint density, g_{1n}, of $X_{(1)}$ and $X_{(n)}$ is the appropriate bivariate marginal in the joint distribution of $X_{(1)}, X_{(2)}, \ldots, X_{(n)}$ given by Eq. (7.7). Letting h denote the joint density of the order statistics, we have

$$g_{1n}(y_1, y_n) = n! \int_{y_1}^{y_n} \int_{y_1}^{y_{n-1}} \cdots \int_{y_1}^{y_4} \int_{y_1}^{y_3} h(y_1, y_2, \ldots, y_n)$$

$$\times \, dy_2 \, dy_3 \cdots dy_{n-2} \, dy_{n-1}$$

$$= n! f(y_1) f(y_n) \int_{y_1}^{y_n} \cdots \int_{y_1}^{y_3} f(y_2) \cdots f(y_{n-1})$$

$$\times \, dy_2 \cdots dy_{n-1}. \tag{7.8}$$

These limits of integration are determined by the restrictions on

$y_1, y_2, \ldots, y_n$. Note that $\int_{y_1}^{y_3} f(y_2)\,dy_2 = F(y_3) - f(y_1)$. Remembering that y_1 is fixed, we have

$$\int_{y_1}^{y_4} \int_{y_1}^{y_3} f(y_2)f(y_3)\,dy_2\,dy_3 = \int_{y_1}^{y_4} f(y_3)[F(y_3) - F(y_1)]\,dy_3$$

$$= \frac{[F(y_3) - F(y_1)]^2}{2} \Bigg|_{y_1}^{y_4} = \frac{[F(y_4) - F(y_1)]^2}{2}.$$

Then

$$\int_{y_1}^{y_5} \int_{y_1}^{y_4} \int_{y_1}^{y_3} f(y_2)f(y_3)f(y_4)\,dy_2\,dy_3\,dy_4$$

$$= \int_{y_1}^{y_5} f(y_4)\frac{[F(y_4) - F(Y_1)]^2}{2}\,dy_4$$

$$= \frac{[F(y_4) - F(y_1)]^3}{2 \cdot 3} \Bigg|_{y_1}^{y_5} = \frac{[F(y_5) - F(y_1)]^3}{3!}.$$

Proceeding in this fashion, we obtain

$$g_{1n}(y_1, y_n) = n(n - 1)f(y_1)f(y_n)[F(y_n) - F(y_1)]^{n-2}, \qquad y_1 < y_n. \quad (7.9)$$

EXAMPLE 7.11 The marginal density g_1 of the smallest order statistic $X_{(1)}$ can be computed by integrating the joint density of $X_{(1)}$ and $X_{(n)}$ given in Eq. (7.9). Thus

$$g_1(y_1) = n(n - 1)f(y_1) \int_{y_1}^{\infty} f(y_n)[F(y_n) - F(y_1)]^{n-2}\,dy_n$$

$$= n(n - 1)f(y_1)\frac{[F(y_n) - F(y_1)]^{n-1}}{n - 1} \Bigg|_{y_1}^{\infty} = nf(y_1)[1 - F(y_1)]^{n-1}.$$

To find the density function of the sample range R, we use the change-of-variable technique. Let $z = y_1$ and $r = y_n - y_1$, so $y_1 = z$, $y_n = z + r$, and the Jacobian is

$$J = \begin{vmatrix} 1 & 0 \\ 1 & 1 \end{vmatrix} = 1.$$

Then the joint density g of $R = X_{(n)} - X_{(1)}$ and $X_{(1)}$ is

$$g(z, r) = g_{1n}(z, z + r) = n(n - 1)f(z)f(z + r)[F(z + r) - F(z)]^{n-2},$$

provided $z < z + r$. The density g_R of the sample range is found by integrating g with respect to z. How explicitly this integration can be done depends very much on the form of f and F. We use numerical methods in those cases in which F cannot be given in explicit form, such as the normal CDF.

EXAMPLE 7.12   Suppose X is uniform over (a, b). Then $f(x) = 1/(b - a)$ and $F(x) = (x - a)/(b - a)$ for $a < x < b$, so that $X_{(1)}$, $X_{(n)}$, and $R = X_{(n)} - X_{(1)}$, from a random sample of size n, have respective densities

$$g_1(y_1) = \frac{n(b - y_1)^{n-1}}{(b - a)^n}, \qquad a < y_1 < b;$$

$$g_n(y_n) = \frac{n(y_n - a)^{n-1}}{(b - a)^n}, \qquad a < y_n < b;$$

and

$$g_R(r) = \frac{n(n - 1)}{(b - a)^n} r^{n-2}(b - a - r), \qquad 0 < r < b - a. \qquad (7.10)$$

In a manner similar to the derivation of g_{1n}, and, hence, g_1, you are invited to establish the following result (see Exercise 7.66).

Theorem 7.8   Suppose X has continuous CDF F, and suppose $X_{(i)}$ and $X_{(j)}$ $(i < j)$ are the ith and jth order statistics from a random sample of X of size n. Then the joint density g_{ij} of $X_{(i)}$ and $X_{(j)}$ is given by

$$g_{ij}(y_i, y_j) = \frac{n!f(y_i)f(y_j)}{(i - 1)!(j - i - 1)!(n - j)!}$$

$$\times F^{i-1}(y_i)[F(y_j) - F(y_i)]^{j-i-1}$$

$$\times [1 - F(y_j)]^{n-j}, \qquad y_i < y_j, \quad (7.11)$$

and the marginal density g_i of $X_{(i)}$ is given by

$$g_i(y_i) = n \binom{n - 1}{i - 1} f(y_i)F^{i-1}(y_i)[1 - F(y_i)]^{n-1}. \qquad (7.12)$$

We will now consider several examples that suggest how order statistics may be useful in various applications.

EXAMPLE 7.13   —*Order Statistics from a Uniform* $(0, 1)$ *Population Are Beta-Distributed*   If $X_{(i)}$ is the ith order statistic from a random sample of size n from a uniform $(0, 1)$ population, then the density of $X_{(i)}$ is

$$f_{X_{(i)}}(x) = n \binom{n-1}{i-1} x^{i-1}(1-x)^{n-i}, \qquad x \in (0, 1).$$

This expression is a beta density with parameters i and $n - i + 1$. It follows that

$$E[X_{(i)}] = \frac{i}{i + (n - i + 1)} = \frac{i}{n + 1}.$$

In general, the kth moment of $X_{(i)}$ is

$$E[X_{(i)}^k] = \frac{\beta(i + k, n - i + 1)}{\beta(i, n - i + 1)}.$$

(See Exercise 7.68.)

EXAMPLE 7.14 —*Tolerance Intervals in the Uniform $(0, 1)$ Case*   Suppose it is desired to observe a sample sufficiently large so that $P[X_{(n)} - X_{(1)} \geq .99] = .95$. By Eq. (7.10) (with $a = 0$ and $b = 1$), the density of $X_{(n)} - X_{(1)} = R$ is

$$g_R(r) = n(n - 1)(1 - r)r^{n-2}, \qquad 0 < r < 1.$$

Thus

$$P(R \geq .99) = n(n - 1) \int_{.99}^1 (1 - r)r^{n-2}\, dr = n(n - 1)\left[\frac{r^{n-1}}{n-1} - \frac{r^n}{n}\right]_{.99}^1$$

$$= 1 - (.99)^{n-1}(.01n + .99).$$

Setting this value equal to .95 and solving for n yields a solution of about 475. Thus in a large number of random samples of size 475 from a uniform $(0, 1)$ population, about 95% of the sample ranges will exceed .99. The pair $(X_{(1)}, X_{(n)})$ is sometimes called a *tolerance interval* for the population in question. In general, statistics L_1 and L_2, based on a random sample from a population with density f and having the property that the area between L_1 and L_2 under f is at least β with probability at least p, are called *tolerance limits* for f.

EXAMPLE 7.15 —*Estimation of One Parameter in a Uniform Distribution Within a Prescribed Percentage of Error*   Suppose the parent distribution is uniform $(0, b)$, where b is unknown, and suppose $X_{(n)}$, the largest observation in a random sample of size n, is taken as an estimator for b. (In statistics, it is shown that this estimator is, indeed, a "good" estimator for b.) It is desired to find a sample size n such that, with some given probability, say p, the *relative error* $[b - X_{(n)}]/b$ is less than some fixed number $\alpha \in$ (0, 1), that is,

$$p = P\left[\frac{b - X_{(n)}}{b} < \alpha\right] = P\left[\frac{X_{(n)}}{b} > 1 - \alpha\right].$$

Since X/b is uniform (0, 1) it follows that $X_{(n)}/b$ is the largest observation in a random sample of size n from a uniform (0, 1) population, and, hence, it has a CDF described by $F(x) = x^n$, $0 < x < 1$. Thus

$$P\left[\frac{X_{(n)}}{b} > 1 - \alpha\right] = 1 - P\left[\frac{X_{(n)}}{b} \leq 1 - \alpha\right]$$

$$= 1 - F(1 - \alpha) = 1 - (1 - \alpha)^n.$$

But $1 - (1 - \alpha)^n \geq p$ if and only if $n \geq \log(1 - p)/\log(1 - \alpha)$; thus for fixed p and α, the sample size may be taken as the smallest integer n that satisfies the inequality.

EXAMPLE 7.16   —*Testing n Items Until r Fail*   Suppose items of a certain type have exponentially distributed lifetimes with unknown parameter $1/\theta$. [Thus if L denotes the time to failure of a randomly selected item, the mean time to failure is $E(L) = \theta$.] Suppose n items are randomly selected and placed on test simultaneously. The test continues until r of the n items have failed. Let $L_{(1)}, L_{(2)}, \ldots, L_{(r)}$ denote the r ordered failure times, noting that these are the first r order statistics in a random sample of size n from L. It is desired to estimate θ.

SOLUTION   The joint density of $L_{(1)}, L_{(2)}, \ldots, L_{(r)}$ is

$$f_{L_{(1)}L_{(2)},\ldots,L_{(r)}}(y_1, y_2, \ldots, y_r) = \frac{n!}{(n - r)!}\left(\frac{1}{\theta^r}\right)$$

$$\exp\left(-\frac{1}{\theta}\sum_{i=1}^{r} y_i\right) \exp\left[\left(-\frac{1}{\theta}\right) y_r^{(n-r)}\right],$$

for $0 < y_1 < y_2 < \cdots < y_r$. A reasonable estimator T for θ is the average test time to failure,

$$T = \frac{1}{r}\left[\sum_{i=1}^{r} L_{(i)} + (n - r)L_{(r)}\right].$$

In Exercise 7.63, you are asked to find the distribution (apart from a constant) of T and the first two moments of this estimator.

We end this section with a discussion of sample analogues of the (population) median. The "middle" order statistic of a random sample of odd size is the *sample median*; for even sample size, the sample median is defined to be the average of the middle pair of order statistics. That is, the sample median $\tilde{X}$, from a random sample of size n on X, is defined by

$$\tilde{X} = \begin{cases} X_{(n+1)/2)} \text{ if } n \text{ is odd}, \\ \\ \dfrac{X_{(n/2)} + X_{((n/2)+1)}}{2} \text{ if } n \text{ is even}. \end{cases}$$

The sample median is sometimes used in inferential statistics in a role similar to that played by the sample mean. The sample median is an attractive alternative to the sample mean in situations where some of the sample values may be "wild" (very small or very large), because the extreme sample values might unreasonably affect the sample mean. The sample median is *robust* (insensitive) with respect to the presence of a few extreme values in the data. Such extreme values are sometimes called "outliers" in the statistical literature.

EXERCISES

THEORY

7.61   a. Show that the distribution of the range R of a random sample of size n from an $N(\mu, \sigma^2)$ population does not depend on μ.
   b. Find an expression for the density of R/σ, noting that it does not depend on the parameters of the parent distribution. *Note:* The random variable R/σ is called the *standardized range*. Tables of the CDF of R/σ are available for various sample sizes n.

7.62   This exercise deals with estimation of the mean of an exponential distribution within a prescribed percentage of error. Suppose $X_1, X_2, \ldots, X_n$ are independent and exponential (λ), where λ is unknown, and suppose $\overline{X}_n$ is used as an estimator for the mean $1/\lambda$ of the parent distribution.
   a. Show that a table of the $\chi^2_{(2n)}$ CDF can be used to find a sample of size n such that, for a given α, $p \in (0, 1)$,

$$P\left(\lambda \left| \overline{X} - \frac{1}{\lambda} \right| < \alpha\right) = p.$$

   Hint: This expression is equivalent to $P(|n\lambda\overline{X} - n| < n\alpha) = p$. Show that $2n\lambda\overline{X}$ is $\chi_{(2n)}$.
   b. Show that the Chebyshev inequality can be used to find a value of n sufficiently large so that the condition in part (a) is met.
   c. Of the two approaches, parts (a) and (b), which yields the smaller value of n?

7.63 Consider the situation described in Example 7.16, in which n exponential-lived items are put on test until r fail. Let T denote total test time divided by r.
  a. Show that $2rT/\theta$ is $\chi^2_{(2r)}$. *Hint:* Let

$$X_1 = nL_{(1)},$$

$$X_2 = (n - 1)[L_{(2)} - L_{(1)}],$$

$$X_3 = (n - 2)[L_{(3)} - L_{(2)}],$$

$$\vdots$$

$$X_r = (n - r + 1)[L_{(r)} - L_{(r-1)}].$$

  b. Show that $E(T) = \theta$ (i.e., T is an unbiased estimator for the mean time to failure θ).
  c. Show that $E[L_{(r)}] = \theta \sum_{j=1}^{r} 1/(n - j + 1)$. *Hint:*

$$E(X_j) = (n - j + 1)E[L_{(j)} - L_{(j-1)}] = \theta.$$

  d. Find $V(T)$ and $V[L_{(r)}]$.

7.64 Suppose the time T to failure of a certain type of item has the density function $f_T(t) = \lambda \alpha t^{\alpha-1} e^{-\lambda t^\alpha}$, $t > 0$, where λ and α are positive parameters. This expression is the *Weibull distribution* with parameters λ and α.
  a. Show that $F_T(t) = 1 - e^{-\lambda t^\alpha}$, $t \ge 0$.
  b. Show that the mean time to failure is $\Gamma[(1/\alpha) + 1]\lambda^{-(1/\alpha)}$.
  c. Show that T^α is exponential (λ).
  d. Verify that, if $\alpha = 1$, the Weibull distribution is exponential.

7.65 Suppose X is exponential (λ), and let $X_1, X_2, \ldots, X_n$ be a random sample. Show that the first order statistic $X_{(1)}$ is exponential ($n\lambda$). Conversely, show that, if this condition is true, X must be exponential (λ).

7.66 Use the following approach to prove Theorem 7.8: First, establish Eq. (7.11) by an argument analogous to that used for Eq. (7.9). Then integrate Eq. (7.11) with respect to y_j to establish Eq. (7.12).

7.67 For X distributed uniform over (a, b), find the first two moments of the order statistics $X_{(1)}$ and $X_{(n)}$. Also, determine $\rho(X_{(1)}, X_{(n)})$.

7.68 Find the expected value of the range of a random sample of size n from a uniform (a, b) population. Show that, if $a = 0$ and $b = 1$, then

$$E[X_{(i)}^k] = \frac{\beta(i + k, n - i + 1)}{\beta(i, n - i + 1)}.$$

7.69 Suppose F is strictly increasing, and let $X_1, X_2, \ldots, X_n$ be a random sample of F with order statistics $X_{(1)}, X_{(2)}, \ldots, X_{(n)}$. For each i, let $V_i = F(X_{(i)})$, $i = 1, 2, \ldots, n$.
  a. Show that $(V_1, V_2, \ldots, V_n)$ has the same distribution as the n-tuple of order statistics from a uniform $(0, 1)$ distribution.
  b. Show that V_i has a beta distribution, and find the corresponding parameters.
  c. Determine $E(V_j)$ for $j = 1, 2, \ldots, n$.
  d. Show that $E(V_j - V_{j-1}) = 1/(n + 1)$ for $j = 2, 3, \ldots, n$.
  e. Find $V(V_j)$, $j = 1, 2, \ldots, n$.

7.70   This exercise involves tolerance intervals. Suppose X has a CDF that is strictly increasing on its range $R_X = (a^*, b^*)$, and suppose $X_1, X_2, \ldots, X_n$ is a random sample of X. Show that, for given p and α (both between 0 and 1), there is a sample size n such that, with probability of at least α, the proportion of the population between $X_{(1)}$ and $X_{(n)}$ is at least p. Hint: This amounts to showing that, for some n, $P[F(X_{(n)}) - F(X_{(1)}) \geq p] \geq \alpha$. Compare this result with that in Exercise 7.69 and in Example 7.14.

7.71   Suppose X has a continuous distribution and $X_{(n)}$ is the largest order statistic from a random sample of size n. Show that, if sampling is resumed, a value greater than $x_{(n)}$ will eventually be observed with probability 1.

7.72   When will the next record rainfall occur? Suppose the annual rainfall in a given location in the ith year of record keeping is an observation of the continuous random variable X_i, $i = 1, 2, \ldots, j$ (so that X_j is the rainfall in the most recent year, last year, say). Suppose the record rainfall to date occurred last year [so that $X_{(j)} = X_j$] and that $X_1, X_2, \ldots, X_j, X_{j+1}, \ldots, X_{j+N}$ are independent and identically distributed, where $j + N$ is the (random) year at which X_j is first exceeded. (That is, N additional years of observations are required to first observe a rainfall greater than last year's.)

a. Show that the distribution of N (for given j) is given by the mass function

$$p_n(n) = \frac{j}{(j + n)(j + n - 1)}, \qquad n = 1, 2, \ldots.$$

Hint:

$$p_N(n) = P(N = n | X_j = \max_{k \leq j}\{X_k\})$$

$$= P(X_{j+1}, \ldots, X_{j+n-1} < X_j < X_{j+n} | X_j > X_1, X_2, \ldots, X_{j-1})$$

$$= \frac{P(X_1, X_2, \ldots, X_{j-1}, X_{j+1}, \ldots, X_{j+n-1} < X_j < X_{j+n})}{P(X_1, X_2, \ldots, X_{j-1} < X_j)}. \qquad (7.13)$$

Under the assumption that $X_1, X_2, \ldots, X_{j+N}$ are independent and identically distributed, it follows that each possible ordering of a collection of these random variables is equally likely. (With probability 1, there will be no ties among the X_k's since their distribution is assumed to be continuous.) Since there are $j!$ possible orderings of $\{X_1, X_2, \ldots, X_j\}$, the denominator in Eq. (7.13) is $(j - 1)!/j!$. A similar argument may be used with the numerator of Eq. (7.13).

b. Show that the CDF of N is $F_N(n) = n/(j + n)$. Hint:

$$F_j(n) = P(N \leq n | X_j = \max_{k \leq j}\{X_k\}) = \sum_{i=1}^{n} p_j(i)$$

$$= \sum_{i=1}^{n} \left(\frac{i}{j + 1} - \frac{i - 1}{j + i - 1} \right) = \frac{n}{j + n}.$$

c. Show that N has no mean.

d. Show that the median time until the present record is broken is $F_N^{-1}(\tfrac{1}{2}) + \tfrac{1}{2} = j + \tfrac{1}{2}$.

e. Show that the most probable length of time until the present record is broken is one year.

7.73 Consider the order statistics of a random sample of size n from a uniform $(0, 1)$ population.

a. Compute the correlation between $X_{(n-1)}$ and $X_{(n)}$. What happens as n gets large?

b. Compute the correlation between $X_{(1)}$ and $X_{(n)}$. What happens as n gets large?

c. Explain, in intuitive terms, the limiting results in parts (a) and (b).

APPLICATION

7.74 Suppose Y_1 is the smallest order statistic of a random sample of size 5 from a uniform $(0, 1)$ population.

a. Use an argument like that in Example 7.10 to show that Y_1 is beta $(1, 5)$.

b. Find the distribution of the sample median, $\check{Y}$.

7.75 Given an intuitive explanation of why the ith order statistic of a random sample of size n on a uniform $(0, 1)$ population has expected value $i/(n + 1)$. *Hint:* Think about n points drawn at random in the interval $(0, 1)$.

7.76 Give an intuitive explanation of why $X_{(n)}$ is a reasonable estimator for b in the uniform $(0, b)$ population, as described in Example 7.15. Explain why this estimator should get better as n increases. Is $X_{(n)}$ an unbiased estimator for b? Show that $X_{(n)} \xrightarrow{P} b$.

7.77 Suppose X is uniform (a, b), and let $X_1, X_2, \ldots, X_n$ be a random sample of X. Find the joint density function for $X_{(1)}$ and $X_{(n)}$.

7.78 Suppose X is uniform $(0, 1)$ and $X_1, X_2, \ldots, X_5$ is a random sample on X. What is the probability that all the sample values will exceed .5 when the experiment is performed? *Hint:* $x_i > .5$, $i = 1, 2, 3, 4, 5$, if and only if min $\{x_1, x_2, \ldots, x_5\} > .5$.

7.79 Suppose a random sample of X yields the values 1.2, -4.0, 2.1, .7, $-.1$, and 56.

a. What is the outcome on the order statistic?

b. What is the sample median?

c. What would the sample median have been if the value recorded as 56 was supposed to be 5.6?

7.5   APPROXIMATIONS

The central limit theorem furnishes us with one method of approximating the values of distribution functions. Although the theorem is stated for sample averages, it is also useful for approximating sums. If $X_1, X_2, \ldots, X_n$

is a random sample from a population with mean μ and variance σ^2, then

$$P\left(\sum_{i=1}^{n} X_i \le y\right) = P\left(\frac{\sum_{i=1}^{n} X_i - n\mu}{\sqrt{n}\,\sigma} \le \frac{y - n\mu}{\sqrt{n}\,\sigma}\right)$$

$$= P\left(\frac{\bar{X}_n - \mu}{\sigma/\sqrt{n}} \le \frac{y - n\mu}{\sqrt{n}\,\sigma}\right) \approx \Phi\left(\frac{y - n\mu}{\sqrt{n}\,\sigma}\right). \quad (7.14)$$

EXAMPLE 7.17   One special instance of the result above may be seen in the case of the chi-square distributions. The chi-square CDFs tabulated in Table 6 in the Appendix are limited to $n = 40$ degrees of freedom. In practice, chi-square tables for larger degrees of freedom are not needed, for the following reason. If $X_1, X_2, \ldots, X_n$ are independent and $\chi^2_{(1)}$, then $Y_n = \sum_{i=1}^{n} X_i$ is distributed $\chi^2_{(n)}$. On the other hand, Y_n is a sum of independent, identically distributed random variables with $\mu = E(X_i) = 1$ and $\sigma^2 = V(X_i) = 2$ for $i = 1, 2, \ldots, n$. Then by Eq. (7.14),

$$P(Y_n \le y) \approx \Phi\left(\frac{y - n}{\sqrt{2n}}\right). \quad (7.15)$$

Hence, for $n > 40$, we might use Eq. (7.15) in place of entries in a χ^2-table. For example, if Y is $\chi^2_{(40)}$,

$$P(T \le 60) \approx \Phi\left(\frac{60 - 40}{\sqrt{80}}\right) = \Phi(2.22) \approx .987.$$

Actually, in this case, the rate of convergence of Y_n to normality is rather slow and should not be used unless n is quite large. It has been shown that, if Y is $\chi^2_{(n)}$, then $\sqrt{2Y} - \sqrt{2n - 1}$ is also asymptotic $N(0, 1)$, and this expression gives a better approximation than does Eq. (7.15). Thus for $n > 30$, $P(Y \le y) \approx \Phi(\sqrt{2y} - \sqrt{2n - 1})$. For our example, with Y distributed $\chi^2_{(40)}$,

$$P(Y \le 60) \approx \Phi(\sqrt{120} - \sqrt{80}) = \Phi(1.95) = .974,$$

which makes a slight difference and a better approximation to the correct value, .978.

EXAMPLE 7.18   Suppose each X_i is Bernoulli (p) and $Y_n = \sum_{i=1}^{n} X_i$. Then

$$\frac{Y_n - n\mu}{\sqrt{n}\,\sigma} = \frac{Y_n - np}{\sqrt{npq}}$$

has a limiting $N(0, 1)$ distribution. On the other hand, we know Y_n is binomial (n, p). So for large n (when the binomial tables become inadequate), binomial probabilities may be approximated by the normal distribution, a fact noted earlier without proof. Thus from Eq. (7.14),

$$\sum_{x \le y} \binom{n}{x} p^x (1 - p)^{n-x} \approx \Phi\left(\frac{y - np}{\sqrt{npq}}\right). \tag{7.16}$$

The approximation may be fairly good even for "moderate" values of n. For example, if $n = 10$ and $p = .40$, then

$$\sum_{x \le 4} \binom{10}{x} (.4)^x (.6)^{10-x} = .63.$$

while

$$\Phi\left(\frac{4 - 4}{2.4}\right) = \Phi(0) \approx .50.$$

For a slightly better approximation, a "continuity correction" is often given. When $y = k$, adjust Eq. (7.16) slightly to take

$$\sum_{x=0}^{k} \binom{n}{x} p^x (1 - p)^{n-x} \approx \Phi\left(\frac{k + \frac{1}{2} - np}{\sqrt{npq}}\right).$$

For the example values above, this equation gives the better approximation $\Phi((4.5 - 4)/2.4) \approx .58$.

Similarly, for $k_1 \le k_2$,

$$\sum_{x=k_1}^{k_2} \binom{n}{x} p^x (1 - p)^{n-x} \approx \Phi\left(\frac{k_2 + \frac{1}{2} - np}{\sqrt{npq}}\right) - \Phi\left(\frac{k_1 - \frac{1}{2} - np}{\sqrt{npq}}\right).$$

The continuity correction can be motivated by considering graphs of the binomial mass function and the approximating normal density, such as those shown in Fig. 7.3. The binomial mass at k is approximated by the area under the normal density and above the interval $(k - \frac{1}{2}, k + \frac{1}{2})$. Summing such values leads to the approximations shown above.

On any occasion when limiting results are used to approximate quantities, the question of what constitutes "sufficiently large" values of n is an important one, but one that is difficult to answer in general. Studies have been carried out that show that approximations obtained via the central limit theorem are, in general, remarkably good for n as small as 5, when the population distributions are "well behaved." (See Fig. 7.2.) As a general rough guide only, we usually think of $n \ge 30$ as a reasonable range of values of n for applying approximations in most asymptotic results.

The normal approximation to the binomial just discussed is reasonably good for values of p near $\frac{1}{2}$, even for moderate values of n. For p nearer the extremes of 0 and 1, the corresponding value of n required to achieve a given degree of accuracy increases rapidly. As we saw in Section 3.5, an

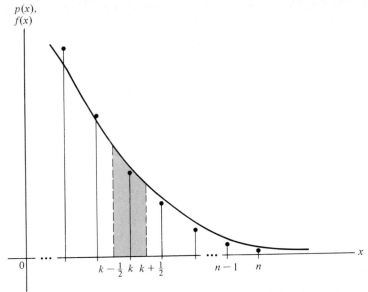

FIGURE 7.3   *Binomial Mass $p(k)$ Approximated by Area Under Normal Curve $f(x)$ Between $k - \frac{1}{2}$ and $k + \frac{1}{2}$*

alternative approximation is available, using the Poisson distribution for cases in which p is near one of these extremes. Rules of thumb for sample size requirements for these approximations were also discussed in Section 3.5.

There are other ways, quite apart from the central limit theorem, in which the normal distribution arises as an approximation. For example, consider the t-density with n degrees of freedom,

$$g_n(x) = \frac{\Gamma((n+1)/2)}{\sqrt{n\pi}\,\Gamma(n/2)} \left(1 + \frac{x^2}{n}\right)^{-(n+1)/2}.$$

Now,

$$\left(1 + \frac{x^2}{n}\right)^{-(n+1)/2} = \left[\left(1 + \frac{x^2}{n}\right)^n\right]^{-(1/2)} \left(1 + \frac{x^2}{n}\right)^{-(1/2)}.$$

But

$$\lim_{n\to\infty} \left(1 + \frac{x^2}{n}\right)^n = e^{x^2},$$

and

$$\lim_{n\to\infty} \left(1 + \frac{x^2}{n}\right)^{-(1/2)} = 1.$$

Hence,

$$\lim_{n\to\infty} \left(1 + \frac{x^2}{n}\right)^{-(n+1)/2} = e^{-x^2/2}.$$

With some algebraic detail, using Stirling's formula (see Exercise 7.87), it may be shown that $\Gamma((n+1)/2)/\sqrt{n\pi}\,\Gamma(n/2)$ is, for large values of n, approximately $1/\sqrt{2\pi}$. Hence, for n sufficiently large (say greater than 30),

$$g_n(x) \approx \frac{1}{\sqrt{2\pi}} e^{-(1/2)x^2}.$$

You can verify that the t-distribution is asymptotically normal by observing the tabled t–CDF as degrees of freedom increase. Scan down a column of the t-table, Table 4 in the Appendix, and observe the values approach those of the standard normal CDF.

EXAMPLE 7.19 The binomial mass function may be used to approximate the hypergeometric mass function when the parameter N (corresponding to the lot size in sampling applications) is large. This notion was discussed intuitively in Section 2.3; you are asked to establish the result in Exercise 7.82. Generally speaking, the approximation is good for practical purposes when $n/N < .1$ and the fraction m/N of items of one type is between .1 and .9.

So far, the discussion has concentrated on limiting distributions of sums or averages. In the applications, however, there is also interest in sequences of statistics other than sums or sample means. An example was seen previously, in which the composition $\sqrt{2Y} - \sqrt{2n-1}$ gives a better asymptotic normal approximation of the chi-square distribution than does the central limit approach with standardized variable $(Y - n)/\sqrt{2n}$. Often, a transformation of the sample data is suggested (e.g., by an examination of experimental data) that will help satisfy certain desirable assumptions, perhaps other than explicit distribution assumptions, in a model. For example, suppose it is desired to compare several (say k) populations through a comparison of only their means. Assume that, from past experience with these populations (either through sampling or through theoretical considerations of the phenomenon being observed), the mean μ_i and standard deviation σ_i of the ith population are known to be functionally related:

$$\sigma_i = h(\mu_i), \qquad i = 1, 2, \ldots, k. \tag{7.17}$$

It may be desired to find a transformation g such that the standard deviation of $g(X_i)$ is approximately the same for each i, where X_i denotes the outcome of a sample of size 1 from the ith population. Let us now consider how to find candidate transformations to accomplish such *variance stabilization*.

Suppose $g(x_i)$ has a Taylor series expansion about μ_i given approximately by $g(x_i) \approx g(\mu_i) + (x_i - \mu_i)g'(\mu_i)$, where the quadratic and higher-order terms have been omitted. Then it follows (see Exercise 7.83) that

$$E[g(X_i)] \approx g(\mu_i),$$

$$E[g^2(X_i)] \approx g^2(\mu_i) + [g'(\mu_i)]^2 E(X_i - \mu_i)^2,$$

so
$$V[g(X_i)] \approx [g'(\mu_i)]^2 \sigma_i^2. \tag{7.18}$$

Thus from Eqs. (7.17) and (7.18), it is desired to find a transformation g such that $g'(\mu_i)h(\mu_i) = c$, where c is a constant. This result suggests the transformation

$$g(x) = c \int \frac{1}{h(x)} \, dx. \tag{7.19}$$

EXAMPLE 7.20   —*The arc sin Transformation*   If $X_1, X_2, \ldots, X_n$ are independent and Bernoulli (p), then $\overline{X}_n$ has mean p and standard deviation $\sqrt{[p(1-p)]/n}$. Let us attempt to find a function g such that $g(\overline{X}_n)$ has variance independent of p.

SOLUTION Applying the foregoing remarks, we have $h(p) = \sqrt{[p(1-p)]/n}$; so by Eq. (7.19), with $c = 1/\sqrt{n}$,

$$g(\overline{x}) = c \int \frac{\sqrt{n}}{\sqrt{\overline{x}(1-\overline{x})}} \, d\overline{x} = 2 \sin^{-1} \sqrt{\overline{x}}.$$

Then the mean and variance of $g(\overline{X}_n)$ are, by Eq. (7.18), approximately

$$2 \sin^{-1} \sqrt{p} \qquad \text{and} \qquad \frac{[g'(p)]^2 p(1-p)}{n} = \frac{1}{n},$$

respectively. Indeed, it may be shown that, since $\overline{X}_n$ is asymptotically normal with mean p and variance $[p(1-p)]/n$, $Y_n = 2 \sin^{-1} \sqrt{\overline{X}_n}$ is also asymptotically normal with mean $2 \sin^{-1} \sqrt{p}$ and variance $1/n$. This transformation was suggested by R. A. Fisher and has been extremely useful in various applications.

As another type of approximation, we consider very briefly the basic notions of *simulation*. Very often, the mathematical structure involved in a model is so complicated that it is not feasible to analytically determine answers to the questions of interest. For example, the analyst might be interested in modeling objects lining up (forming a queue) for service of some sort, say cars queueing up at a toll booth. Suppose the analyst wishes to determine the mean length of the queue (i.e., the average number of cars at a given time whose drivers are waiting to pay their toll) and the mean duration of the waiting period for a given car. Even if the analyst assumes that the cars arrive in accordance with some known distribution and that the amount of time the attendant is occupied with a given car is a random variable with a second known distribution, the questions of interest are still very difficult. Indeed, only special cases (such as Poisson arrival and independent exponential service times, independent of the arrivals) have been solved analytically.

So that useful answers to questions about the queueing process are obtained, it is possible to *simulate* the process with a computer. Thus the analyst may generate arrivals of "cars" and "service times" in accordance with any distributions desired. By keeping track of the length of the queue and the waiting time for each car over a period of time as the simulation progresses, the analyst can estimate the mean queue length and waiting time. But with a computer, the analyst may "observe," in a small amount of running time, thousands of "car arrivals" and associated "service times." The law of large numbers guarantees that, with very high probability, these averages are extremely close to the true means being sought. This process amounts to building a digital model, within the computer, of a phenomenon (the queue) about which there are questions. The simulation of the queueing process described involves the generation of outcomes of random variables with the given distributions. We consider next the generation of such sample data.

Suppose the analyst has available only a "random number generator" for the simulation. It is desired to find ways of using outcomes of the uniform, over $(0, 1)$, random variable X to further generate outcomes of a general random variable Y having a specified distribution. If the analyst can analytically specify F as a one-to-one CDF for Y, then it follows that $F^{-1}(X)$ has the given distribution (see Exercise 7.85). Thus if $X_1, X_2, \ldots, X_n$ is a random sample of X, then $Y_1 = F^{-1}(X_1)$, $Y_2 = F^{-1}(X_2)$, $\ldots$, $Y_n = F^{-1}(X_n)$ is a random sample of Y. By generating uniform $(0, 1)$ outcomes and mapping them through F^{-1}, the analyst obtains outcomes from a population with distribution F.

But suppose F_Y cannot be given in closed form. For example, suppose it is desired to generate $N(\frac{1}{2}, 1)$ outcomes, so $F_Y(y) = \Phi(y - \frac{1}{2})$. In this case, the analyst might use the law of large numbers, taking the average of (say) 12 uniform $(0, 1)$ outcomes as an outcome from the normal population.* Actually, this result is only approximate, but the simulation model itself is not exact, so the results may be sufficient for the intended purposes.

---

* Note that, if $X_1, X_2, \ldots, X_{12}$ are independent and uniform $(0, 1)$, then

$$Y_1 = \sum_{i=1}^{12} X_i - \tfrac{11}{2}$$

has mean $6 - \tfrac{11}{2} = \tfrac{1}{2}$ and variance $\tfrac{12}{12} = 1$. By taking k random samples of size 12 from the uniform $(0, 1)$ population, the analyst may thus generate k outcomes from an $N(\frac{1}{2}, 1)$ population. But this approach requires 12 calls to the random number generator for each normal outcome. More efficient normal generators are actually used in practice, such as one outlined in Exercise 7.90, which returns a pair of normal outcomes for each two calls to the random number generator.

To generate $N(\mu, \sigma^2)$ outcomes for any specified μ and σ^2, take

$$Y_j = \sigma \left(\sum_{i=1}^{12} X_i - 6 \right) + \mu,$$

where $\sum_{i=1}^{12} X_i$ is determined by random samples of size 12 from the uniform $(0, 1)$ distribution. In the exercises that follow, you are asked to consider the generation of samples from other distributions by using a computer routine for generating outcomes from the uniform $(0, 1)$ distribution. Such a routine is called a "random number generator."

EXERCISES

THEORY

7.80   This exercise involves an alternate proof of the Poisson approximation to the binomial. Suppose $0 < \mu < 1$, and let $p_n = \mu/n$, $n = 1, 2, \ldots$. Let X_1, $X_2, \ldots, X_n$ be independent random variables such that X_k is binomial (k, p_k), $k = 1, 2, \ldots, n$.
   a. Prove that $\lim_{n \to \infty} M_n(t) = e^{\mu(e^t - 1)}$, where M_n is the moment-generating function of X_n. Hint:

$$M_n(t) = (p_n e^t + 1 - p_n)^n \quad \text{and} \quad \lim_{n \to \infty} \left(1 + \frac{\lambda}{n} \right)^n = e^\lambda.$$

   b. Conclude that X_n has a limiting Poisson (μ) distribution, so

$$P(X_n \leq y) \approx \sum_{x \leq y} \frac{e^{-\mu} \mu^x}{x!} .$$

   c. Extend these results to the case in which $\mu \geq 1$.
   d. Show that, if the approximation is "good" for $p \leq .1$, then it is also "good" for $p \geq .9$. Hint: Consider $(n - X_n)$, which is binomial (n, q_n).

7.81   Suppose X is binomial $(10,000, .04)$ and it is desired to evaluate $P(X \leq 450)$. Since $np = 400$, the Poisson approximation of Exercise 7.80, while applicable in theory, is not practical since the Poisson tables do not extend to $\lambda = 400$. Prove that a normal approximation to the (approximating) Poisson may be used, by the following argument:
   a. If X is Poisson (λ), show that the standardized random variable $Y = (X - \lambda)/\sqrt{\lambda}$ has moment-generating function

$$M_Y(t) = \exp[\lambda e^{(t/\sqrt{\lambda})} - \lambda - \sqrt{\lambda}t].$$

   b. Show that

$$M_Y(t) = \exp \left(\frac{t^2}{2} + \frac{e^\theta t^3}{6\lambda^{3/2}} \right),$$

where $0 < \theta < t/\sqrt{\lambda}$.

c. Conclude that $\lim_{\lambda \to \infty} M_Y(t) = e^{t^2/2}$, so Y is asymptotically $N(0, 1)$.

d. Use the appropriate approximating normal distribution, with a continuity correction, to find $P(X \le 450)$.

7.82   If $\{a_n\}_{n=1}^{\infty}$ and $\{b_n\}_{n=1}^{\infty}$ are two real sequences, a_n and b_n are *asymptotically equivalent*, written $a_n \sim b_n$, provided $\lim_{n \to \infty} a_n/b_n = 1$. In that case, $a_n \approx b_n$ for all n sufficiently large. Show that the binomial and hypergeometric mass functions are asymptotically equivalent. Assume the proportion M/N of items in a lot is defective, and let p_1 denote the binomial (M/N) mass function

$$p_1(x) = \binom{n}{x} \left(\frac{M}{N}\right)^x \left(1 - \frac{M}{N}\right)^{n-x}, \qquad x = 1, 2, \ldots, n.$$

Let p_2 be the hypergeometric mass function.

a. Show that

$$p_2(x) = \binom{n}{x} \left(\frac{M}{N}\right)^x \left(1 - \frac{M}{N}\right)^{n-x}$$

$$\times \frac{(1 - 1/M) \cdots [1 - (x - 1)/M][1 - 1/(N - M) \cdots [1 - (n - X - 1)/(N - M)]}{(1 - 1/N) \cdots [1 - (n - 1)/N]}.$$

b. Show that, if M and N both increase without bound in such a way that $M/N = p$ (the proportion of defectives in the lot remains constant), then $N - M = Nq$ also increases without bound, and $p_2(x)/p_1(x) \to 1$. Thus for fixed x and for large values of N and M, $p_2(x) \approx p_1(x)$.

7.83   Suppose X has mean μ and variance σ^2, and suppose g is a function such that $g(x) = g(\mu) + (x - \mu)g'(\mu)$. Show that $E[g(X)] = g(\mu)$ and $V[g(X)] = [g'(\mu)]^2\sigma^2$.

7.84   Suppose $S = \{o_1, o_2, \ldots, o_k\}$ is an unknown finite sample space such that $P(\{o\}) > 0$ for all $o \in S$.

a. Show that, for all simple events $\{o\}$, the probability of $\{o\}$ being observed at least once converges to 1 as the size n of the random sample from this population gets large.

b. Conclude that the sample space S will be determined through continued sampling, with probability 1.

c. Show that, for any k, $\{o\}$ will, with probability 1, be observed at least k times for sufficiently large sample size.

d. Show that, as sampling continues, P may be determined arbitrarily closely with high probability. *Hint:* It is necessary only to determine $P(\{o\})$ for each $o \in S$. Use the law of large numbers for the proportion of times $\{o\}$ occurs.

e. Let X_i denote the number of times $\{o_i\}$ occurs in n independent repeated trials of the experiment. Discuss the distribution of $(X_1, X_2, \ldots, X_k)$.

7.85   If F is a continuous and strictly increasing CDF, and X is uniform $(0, 1)$, show that $F^{-1}(X)$ has CDF F. *Hint:* Let $Z = F^{-1}(X)$, so for any z,

$$F_Z(z) = P(Z \le z) = P[F^{-1}(X) \le z] = P[X \le F(z)] = F(z).$$

7.86   Extend the results of Exercise 7.80 to the case in which μ_n is a sequence of real numbers such that

$$\lim_{n \to \infty} \mu_n = \mu \quad \text{and} \quad p_n = \frac{\mu_n}{n}, \quad n = 1, 2, 3, \ldots .$$

Hint: First, show that

$$\lim_{n \to \infty} \left(1 - \frac{\mu_n}{n} \right)^n = e^{-\mu}.$$

7.87   *Stirling's formula* states that, for large n,

$$\Gamma(n + 1) \approx \sqrt{2\pi} \, e^{-n} n^{n+(1/2)}.$$

Using this formula, show that

$$\frac{\Gamma((n + 1)/2)}{\Gamma(n/2)} \approx \sqrt{\frac{(n - 2)}{2}} \, e^{-(1/2)} \left[\left(1 + \frac{1}{n - 2} \right)^n \right]^{(1/2)},$$

and, hence, for large values of n,

$$\frac{\Gamma((n + 1)/2)}{\Gamma(n/2)} \approx \sqrt{\frac{n}{2}}.$$

7.88   Suppose X is $\chi_{(100)}$ (chi distribution with parameter 100). Find an approximate expression for $E(X)$.

7.89   This exercise involves the limiting volume of a 1-ball. Let $V_n(r^2)$ denote the "volume" of an n-dimensional sphere of radius r, that is,

$$V_n(r^2) = \int_0^{r} \cdots \int_{\{(x_1,x_2,\ldots,x_n):x_1^2+x_2^2+\ldots+x_n^2\leq1\}} dx_1 \, dx_2 \cdots dx_n.$$

Thus, for example, $V_1(1) = 2$ [length of the interval $(-1, 1)$], $V_2(1) = \pi$ (area of a circle of radius 1), $V_3(1) = \frac{4}{3}\pi$ (volume of a sphere of radius 1).
a. Find $V_n(1)$. *Hint:*

$$V_n(r^2) = \frac{\pi^{(n/2)} r^n}{\Gamma((n/2) + 1)}.$$

This result may be seen as follows: Suppose $X_1, X_2, \ldots, X_n$ are independent and $N(0, 1)$. Then

$$P(X_1^2 + X_2^2 + \cdots + X_n^2 \leq r^2)$$

$$= \int_0^{r^2} \int \cdots \int_{\{(x_1,x_2,\ldots,x_n):\Sigma x_i^2=y\}} \frac{1}{(2\pi)^{n/2}} e^{-(1/2)\Sigma x_i^2} \, dx_1 \cdots dx_{n-1} \, dy$$

$$= \int_0^{r^2} \frac{1}{2^{n/2}\Gamma(n/2)} \, y^{(n/2)-1} e^{-(y/2)} \, dy,$$

since ΣX_i^2 is $\chi_{(n)}^2$. Now,

$$P(\Sigma X_i^2 \leq r^2)$$

$$= \int_0^{r^2} \frac{1}{(2\pi)^{n/2}} e^{-(1/2)y} \int \cdots \int_{\{(x_1, x_2, \ldots, x_n): \Sigma x_i^2 = y\}} dx_1 \cdots dx_{n-1} \, dy$$

$$= \int_0^{r^2} \frac{1}{2^{n/2}\Gamma(n/2)} y^{(n/2)-1} e^{-(y/2)} \, dy,$$

and taking derivatives with respect to r^2 yields

$$\frac{1}{(2\pi)^{n/2}} e^{-(1/2)r^2} V_n'(r^2) = \frac{1}{2^{n/2}\Gamma(n/2)} (r^2)^{(n/2)-1} e^{-(r^2/2)}.$$

The last step depends on the fact that, if $V_{(n)}(r^2)$ is the volume of an n-dimensional sphere of radius r, then $V_{(n)}'(r^2)$ is the surface area of that sphere. Finally,

$$V_n'(r^2) = \frac{(r^2)^{(n/2)-1} e^{(r^2/2)}(2\pi)^{(n/2)}}{2^{(n/2)}\Gamma(n/2) e^{(r^2/2)}}$$

$$= \frac{(r^2)^{(n/2)-1} \pi^{(n/2)}}{\Gamma(n/2)},$$

so

$$V_n(r^2) = \int V_n'(r^2) \, dr^2 = \frac{(r^2)^{(n/2)} \pi^{(n/2)}}{\Gamma(n/2) + 1)}.$$

b. Find $\lim_{n\to\infty} V_n(1)$. *Hint:* Use Exercise 7.87.
c. Find the surface area $S_n(1)$ of the n-dimensional sphere, and find $\lim_{n\to\infty} S_n(1)$.
d. Show that the maximal volume of a ball of radius 1 occurs for $n = 5$ dimensions.
e. Imagine the ball is inscribed in a box with side 2. Find the volume of the circumscribed box in n-dimensional space.
f. Note that the volume of the box becomes unbounded as n increases. In view of the result for part (b), it follows that nearly all the volume of a box in high-dimensional space is in its corners. Discuss.

7.90 It is desired to generate two independent $N(0, 1)$ outcomes, say X and Y. Imagine (X, Y) is plotted on the plane, and express the point in polar form as (R, Θ). Since $R^2 = X^2 + Y^2$ is $\chi_{(2)}^2$, it is easy to generate R as the square root of an exponential $(\frac{1}{2})$ random variable, since the exponential CDF and its inverse can be given in closed form. Since Θ is uniform over $(0, 2\pi)$, it is also easy to generate. Θ and R are independent.

a. Explain how this procedure can be used to generate outcomes on X and Y, using only two uniform over $(0, 1)$ outcomes. This method of generating normal outcomes is especially convenient for implementation on calcu-

lators that have a hard-wired, polar-to-rectangular coordinate transformation.
b. Describe how $N(\mu, \sigma^2)$ outcomes can be generated.

7.91   It is desired to generate points at random (uniform distribution) on the surface
of a sphere, using a uniform over $(0, 1)$ generator.
a. Suppose spherical coordinates are adopted and the angles θ and ϕ are
generated as uniform over $(0, 2\pi)$ and uniform over $(-\pi, \pi)$, respectively.
Show that the resulting point on the sphere is *not* uniform on the sphere.
b. Discuss how uniform points on the surface of a sphere can be generated.

APPLICATION

7.92   Suppose X is binomial $(100, \frac{1}{3})$. Find an approximate probability that $Y = [X/10]$ (the greatest integer in $X/10$) will take on one of the values 4, 5, 6, 7, 8.

7.93   Suppose $X_1, X_2, \ldots, X_{50}$ is a random sample from an $N(\mu, \sigma^2)$ population. Determine $P(S_X^2 \leq \sigma^2)$.

7.94   Suppose a fair $(p = \frac{1}{2})$ coin is tossed independently 499 times. What is the likelihood that the tosses will result in 260 heads? In at most 260 heads? In at least 260 heads? In an even number of heads?

7.95   In Exercise 7.94, how many tosses are needed to ensure that the probability of seeing at most 52% heads is at least .99?

7.96   Suppose it is found that the variance σ^2 of a random variable X is linearly related to its mean μ, say $\sigma^2 = a\mu$ for some $a > 0$.
a. Suggest a transformation g such that $g(X)$ has a variance not dependent on μ.
b. Give an example of a parent distribution that would give rise to this relationship between μ and σ^2.
c. Find an approximate expression for the variance of $g(X)$.

7.97   Same as Exercise 7.96, except that the relationship between μ and σ^2 is (a) $\sigma^2 = b\mu^2$, $b > 0$; (b) $\sigma^2 = b\mu(\mu - 1)$, $b > 0$.

7.98   Show how one could use a random number generator [uniform $(0, 1)$ random variables] to generate outcomes from the following populations:
a. Exponential (λ).
b. *Truncated normal* with parameters μ, σ^2, α; that is, a random variable X such that

$$F_X(x) = \begin{cases} c \cdot \Phi\left(\dfrac{x - \mu}{\sigma}\right) & \text{if } x \leq \alpha, \\[2mm] 1 & \text{if } x \geq \alpha. \end{cases}$$

Hint: First, determine c in terms of α; then consider generating $N(\mu, \sigma^2)$ outcomes, omitting those falling above α.
c. Binomial (n, p). *Hint:* First, consider generating a Bernoulli (p); then sum to get binomial (n, p) outcomes. To get a Bernoulli random variable,

consider

$$Y = \begin{cases} 0 \text{ if } X \le 1 - p, \\ 1 \text{ if } X > 1 - p, \end{cases}$$

where X is uniform $(0, 1)$.
d. Poisson (λ).
e. Gamma with integer parameters.
f. Bivariate normal with mean (μ_1, μ_2) and covariance matrix V.
g. $t_{(n)}$.
h. Uniform (a, b).
i. Equal likelihood over the set $\{a_1, a_2, \ldots, a_n\}$.

7.99 This exercise deals with estimation of σ^2 in an $N(\mu, \sigma^2)$ distribution to within a prescribed percentage of error. Suppose S_X^2, based on a random sample of size n from an $N(\mu, \sigma^2)$ population, where μ and σ^2 are unknown, is used as an estimator for σ^2. It is desired to find a sample size such that $P(|S_X^2 - \sigma^2| \le \alpha\sigma^2) \ge p$ for given α and $p \in (0, 1)$.
  a. Show that an upper bound on n, obtained by using Chebyshev's inequality, is given by the smallest integer larger than $2/[\alpha^2(1 - p)]$.
  b. Show that an "exact" value of n can be found for each α, p by using the chi-square tables and a normal approximation to the chi-square distribution.

7.100 A sample of size 10 is taken with replacement from a lot having .5% defective items.
  a. Find the exact probability of getting no more than one defective item, and compare it with the probability obtained by using a Poisson approximation.
  b. Determine the answer if the sampling is done without replacement and the lot contains 100 items. Compare the result with that of part (a).

7.101 Accidents in a certain industrial plant occur according to the Poisson probability law with a mean rate of 5 per year. What is the probability of fewer than 40 accidents in a ten-year period?

7.102 A random sample of size 100 is taken from an $N(0, 4)$ population. What is the probability that the sample mean will be less than the sample standard deviation?

7.103 Suppose X is binomial $(100, .4)$. Find $P(2 < X \le 51)$.

7.104 Suppose X is binomial $(100, .04)$. Find $P(2 < X \le 51)$.

7.105 Suppose Y is Poisson (50). Find $P(Y > 60)$.

7.6   SUMMARY

A sample of size n is a set of n identically distributed random variables, X_1, $X_2, \ldots, X_n$. The common distribution is called the parent distribution or the population distribution. If the sample variables (X_i's) are independent, the sample is a random sample. A function $\phi(X_1, X_2, \ldots, X_n)$ of the sample variables whose value $\phi(x_1, x_2, \ldots, x_n)$ does not depend on unknown parameter values is a statistic. Some common statistics include the sample moments and central sample moments,

$$M_k = \frac{1}{n} \sum_{i=1}^{n} X_i^k \qquad \text{and} \qquad S^k = \frac{1}{n} \sum_{i=1}^{n} (X_i - \overline{X})^k.$$

M_1 is the sample mean, $\overline{X}$. On the basis of a random sample from a population with mean μ and variance σ^2, $\overline{X}$ has mean μ and variance σ^2/n. S^2 is the sample variance and $S = \sqrt{S^2}$ is the sample standard deviation. If the population is $N(\mu, \sigma^2)$, then $\overline{X}$ is distributed $N(\mu, \sigma^2/n)$, nS^2/σ^2 is distributed $\chi^2_{(n-1)}$, and the statistics $\overline{X}$ and S^2 are independent.

If $\overline{X}_n$ is based on a random sample of size n from a population having mean μ and variance σ^2, the law of large numbers asserts that, for any positive distance δ, the probability that $\overline{X}_n$ falls more than δ units from μ is close to zero for n sufficiently large. That is,

$$\lim_{n \to \infty} P(|\overline{X}_n - \mu| > \delta) = 0,$$

or $\overline{X}_n \xrightarrow{P} \mu$ (the sample mean converges in probability to the population mean).

If $X_1, X_2, \ldots$ have corresponding CDFs $F_1, F_2, \ldots$, then the sequence of random variables converges in distribution to a random variable T having CDF F, provided that $\lim_{n \to \infty} F_n(x) = F(x)$ at all points where F is continuous. In such a case, F can be used as an approximation of the F_n's for large n. The central limit theorem states that, if $X_1, X_2, \ldots$ is a random sample from a population with mean μ and variance σ^2, and if the population has some higher-order moments, then $\sqrt{n}(\overline{X}_n - \mu)/\sigma$ converges in distribution to an $N(0, 1)$ random variable. This result is often used to provide normal approximations for sums and averages of random samples.

A t-distribution with n degrees of freedom has density

$$f(x) = \frac{\Gamma((n+1)/2)}{\sqrt{n\pi}\,\Gamma(n/2)} \left(1 + \frac{x^2}{n}\right)^{-(n+1)/2}.$$

If $\overline{X}$ and S^2 are based on a random sample of size n from an $N(\mu, \sigma^2)$ population, the ratio $\sqrt{n-1}(\overline{X} - \mu)/S$ is distributed $t_{(n-1)}$. As degrees of freedom get large, $t_{(n)}$ converges in distribution to $N(0, 1)$. An F-distribution with n and m degrees of freedom has density

$$f(x) = \frac{(n/m)^{n/2}\Gamma((n+m)/2)}{\Gamma(n/2)\Gamma(m/2)} \cdot \frac{x^{(n/2)-1}}{[1 + (n/m)x]^{(n+m)/2}}, \qquad x > 0.$$

If S_X^2 is based on a random sample of size n from an $N(\mu_X, \sigma_X^2)$ population and S_Y^2 is based on an independent random sample of size m from an $N(\mu_Y, \sigma_Y^2)$ population, then the ratio $n(m-1)\sigma_Y^2 S_X^2 m(n-1)\sigma_X^2 S_Y^2$ is distributed $F_{n-1, m-1}$.

If X_1, X_2, . . . , X_n is a random sample from a population having continuous CDF F, and the sample values are to be reordered in order of increasing magnitude, the ordered values are order statistics, $X_{(1)}$, $X_{(2)}$, . . . , $X_{(n)}$. The marginal density of $X_{(i)}$ is

$$g_i(x) = n \binom{n-1}{i-1} f(x) F^{i-1}(x)[1 - F(x)]^{n-i}.$$

If the X_i's are uniform over $(0, 1)$, the $X_{(i)}$'s are beta-distributed, and $E[X_{(i)}] = i/(n + 1)$.

APPENDIX: TABLES OF COMMON DISTRIBUTIONS

TABLE 1   BINOMIAL MASS FUNCTIONS b(n, p)

This table gives mass function values $F_Y(y) = \binom{n}{y} p^y q^{n-y}$ for some Y distributions in the binomial family. For values of p larger than .50, use the fact that

$$\binom{n}{y} p^y q^{n-y} = \binom{n}{n-y} q^{n-y} p^y$$

(where $q < .50$). For example, if Y is $b(5, .7)$, then

$$P(Y = 3) = F_Y(3) = \binom{5}{3}(.7)^3(.3)^2$$

$$= \binom{5}{2}(.3)^2(.7)^3 = .3087,$$

the latter value being tabulated under $n = 5$, $y = 2$, $p = .30$.

					P			
n	**y**	**.01**	**.05**	**.10**	**.20**	**.30**	**.40**	**.50**
2	0	9801	9025	8100	6400	4900	3600	2500
	1	0198	0950	1800	3200	4200	4800	5000
	2	0001	0025	0100	0400	0900	1600	2500
3	0	9703	8574	7290	5120	3430	2160	1250
	1	0294	1354	2430	3840	4410	4320	3750
	2	0003	0071	0270	0960	1890	2880	3750
	3	0000	0001	0010	0080	0270	0640	1250
4	0	9606	8145	6561	4096	2401	1296	0625
	1	0388	1715	2916	4096	4116	3456	2500
	2	0006	0135	0486	1536	2646	3456	3750
	3	0000	0005	0036	0256	0756	1536	2500
	4	0000	0000	0001	0016	0081	0256	0625

TABLE 1   (*CONTINUED*)

n	y	.01	.05	.10	.20	.30	.40	.50
					P			
5	0	9510	7738	5905	3277	1681	0778	0312
	1	0480	2036	3280	4096	3602	2592	1562
	2	0010	0214	0729	2048	3087	3456	3125
	3	0000	0011	0081	0512	1323	2304	3125
	4	0000	0000	0004	0064	0283	0768	1562
	5	0000	0000	0000	0003	0024	0102	0312
6	0	9415	7351	5314	2621	1176	0467	0156
	1	0570	2321	3543	3932	3025	1866	0937
	2	0014	0305	0984	2458	3241	3110	2344
	3	0000	0021	0146	0819	1852	2765	3125
	4	0000	0001	0012	0154	0595	1382	2344
	5	0000	0000	0001	0015	0102	0369	0937
	6	0000	0000	0000	0001	0007	0041	0156
7	0	9321	6983	4783	2097	0824	0280	0078
	1	0660	2572	3720	3670	2474	1306	0547
	2	0020	0406	1240	2753	3177	2613	1641
	3	0000	0036	0230	1147	2269	2903	2734
	4	0000	0002	0026	0287	0972	1935	2734
	5	0000	0000	0002	0043	0250	0774	1641
	6	0000	0000	0000	0004	0036	0172	0547
	7	0000	0000	0000	0000	0002	0016	0078
8	0	9227	6634	4305	1678	0576	0168	0039
	1	0747	2793	3826	3355	1976	0896	0312
	2	0026	0515	1488	2936	2965	2090	1094
	3	0001	0054	0331	1468	2541	2787	2187
	4	0000	0004	0046	0459	1361	2322	2734
	5	0000	0000	0004	0092	0467	1239	2187
	6	0000	0000	0000	0011	0400	0413	1094
	7	0000	0000	0000	0001	0012	0079	0312
	8	0000	0000	0000	0000	0001	0007	0039
9	0	9135	6302	3874	1342	0404	0101	0020
	1	0830	2985	3874	3020	1556	0605	0176
	2	0034	0629	1722	3020	2668	1612	0703
	3	0001	0077	0446	1762	2668	2508	1641
	4	0000	0006	0074	0661	1715	2508	2461
	5	0000	0000	0008	0165	0735	1672	2461
	6	0000	0000	0001	0028	0210	0743	1641
	7	0000	0000	0000	0003	0039	0212	0703
	8	0000	0000	0000	0000	0004	0035	0176
	9	0000	0000	0000	0000	0000	0003	0020

TABLE 1   (CONTINUED)

					P			
n	*y*	.01	.05	.10	.20	.30	.40	.50
10	0	9044	5987	3487	1074	0282	0060	0010
	1	0914	3151	3874	2684	1211	0403	0098
	2	0042	0746	1937	3020	2335	1209	0439
	3	0001	0105	0574	2013	2668	2150	1172
	4	0000	0010	0112	0881	2001	2508	2051
	5	0000	0001	0015	0264	1029	2007	2461
	6	0000	0000	0001	0055	0368	1115	2051
	7	0000	0000	0000	0008	0090	0425	1172
	8	0000	0000	0000	0001	0014	0106	0439
	9	0000	0000	0000	0000	0001	0016	0098
	10	0000	0000	0000	0000	0000	0001	0010
11	0	8953	5688	3138	0859	0198	0036	0005
	1	0995	3293	3835	2362	0932	0266	0054
	2	0050	0867	2131	2953	1998	0887	0269
	3	0002	0137	0710	2215	2568	1774	0806
	4	0000	0014	0158	1107	2201	2365	1611
	5	0000	0001	0025	0388	1321	2207	2256
	6	0000	0000	0003	0097	0566	1471	2256
	7	0000	0000	0000	0017	0173	0701	1611
	8	0000	0000	0000	0002	0037	0234	0806
	9	0000	0000	0000	0000	0005	0052	0269
	10	0000	0000	0000	0000	0000	0007	0054
	11	0000	0000	0000	0000	0000	0000	0005
12	0	8864	5404	2824	0687	0138	0022	0002
	1	1074	3413	3766	2062	0712	0174	0029
	2	0060	0988	2301	2835	1678	0639	0161
	3	0002	0173	0852	2362	2397	1419	0537
	4	0000	0021	0213	1329	2311	2128	1208
	5	0000	0002	0038	0532	1585	2270	1934
	6	0000	0000	0005	0155	0792	1766	2256
	7	0000	0000	0000	0033	0291	1009	1934
	8	0000	0000	0000	0005	0078	0420	1208
	9	0000	0000	0000	0001	0015	0125	0537
	10	0000	0000	0000	0000	0002	0025	0161
	11	0000	0000	0000	0000	0000	0003	0029
	12	0000	0000	0000	0000	0000	0000	0002
13	0	8775	5133	2542	0550	0097	0013	0001
	1	1152	3512	3672	1787	0540	0113	0016
	2	0070	1109	2448	2680	1388	0453	0095
	3	0003	0214	0997	2457	2181	1107	0349
	4	0000	0028	0277	1535	2337	1845	0873

TABLE 1   (*CONTINUED*)

n	y	.01	.05	.10	.20	.30	.40	.50
	5	0000	0003	0055	0691	1803	2214	1571
	6	0000	0000	0008	0230	1030	1968	2095
	7	0000	0000	0001	0058	0442	1312	2095
	8	0000	0000	0000	0011	0142	0656	1571
	9	0000	0000	0000	0001	0034	0243	0873
	10	0000	0000	0000	0000	0006	0065	0350
	11	0000	0000	0000	0000	0001	0012	0095
	12	0000	0000	0000	0000	0000	0001	0016
	13	0000	0000	0000	0000	0000	0000	0001
14	0	8687	4877	2288	0440	0068	0008	0001
	1	1229	3593	3559	1539	0407	0073	0009
	2	0081	1230	2570	2501	1134	0317	0056
	3	0003	0259	1142	2501	1943	0845	0222
	4	0000	0037	0349	1720	2290	1549	0611
	5	0000	0004	0078	0860	1963	2066	1222
	6	0000	0000	0013	0322	1262	2066	1833
	7	0000	0000	0002	0092	0618	1574	2095
	8	0000	0000	0000	0020	0232	0918	1832
	9	0000	0000	0000	0003	0066	0408	1222
	10	0000	0000	0000	0000	0014	0136	0611
	11	0000	0000	0000	0000	0002	0033	0222
	12	0000	0000	0000	0000	0000	0005	0056
	13	0000	0000	0000	0000	0000	0001	0009
	14	0000	0000	0000	0000	0000	0000	0001
15	0	8601	4633	2059	0352	0047	0005	0000
	1	1303	3658	3432	1319	0305	0047	0005
	2	0092	1348	2669	2309	0916	0219	0032
	3	0004	0307	1285	2501	1700	0634	0139
	4	0000	0049	0428	1876	2186	1268	0417
	5	0000	0006	0105	1032	2061	1859	0916
	6	0000	0000	0019	0430	1472	2066	1527
	7	0000	0000	0003	0138	0811	1771	1964
	8	0000	0000	0000	0035	0348	1181	1964
	9	0000	0000	0000	0007	0116	0612	1527
	10	0000	0000	0000	0001	0030	0245	0916
	11	0000	0000	0000	0000	0006	0074	0417
	12	0000	0000	0000	0000	0001	0016	0139
	13	0000	0000	0000	0000	0000	0003	0032
	14	0000	0000	0000	0000	0000	0000	0005
	15	0000	0000	0000	0000	0000	0000	0000

TABLE 1   (*CONTINUED*)

n	y	.01	.05	.10	.20	.30	.40	.50
16	0	8515	4401	1853	0281	0033	0003	0000
	1	1376	3706	3294	1126	0228	0030	0002
	2	0104	1463	2745	2111	0732	0150	0018
	3	0005	0359	1423	2463	1465	0468	0085
	4	0000	0061	0514	2001	2040	1014	0278
	5	0000	0008	0137	1201	2099	1623	0667
	6	0000	0001	0028	0550	1649	1983	1222
	7	0000	0000	0004	0197	1010	1889	1746
	8	0000	0000	0001	0055	0487	1417	1964
	9	0000	0000	0000	0012	0185	0839	1746
	10	0000	0000	0000	0002	0056	0392	1222
	11	0000	0000	0000	0000	0013	0142	0667
	12	0000	0000	0000	0000	0002	0040	0278
	13	0000	0000	0000	0000	0000	0008	0085
	14	0000	0000	0000	0000	0000	0001	0018
	15	0000	0000	0000	0000	0000	0000	0002
	16	0000	0000	0000	0000	0000	0000	0000
17	0	8429	4181	1668	0225	0023	0002	0000
	1	1447	3741	3150	0957	0169	0019	0001
	2	0117	1575	2800	1914	0581	0102	0010
	3	0006	0415	1556	2393	1245	0341	0052
	4	0002	0076	0605	2093	1868	0796	0182
	5	0000	0010	0175	1361	2081	1379	0472
	6	0000	0001	0039	0680	1784	1839	0944
	7	0000	0000	0007	0267	1201	1927	1484
	8	0000	0000	0001	0084	0644	1606	1855
	9	0000	0000	0000	0021	0276	1070	1855
	10	0000	0000	0000	0004	0095	0571	1484
	11	0000	0000	0000	0001	0026	0242	0944
	12	0000	0000	0000	0000	0006	0081	0472
	13	0000	0000	0000	0000	0001	0021	0182
	14	0000	0000	0000	0000	0000	0004	0052
	15	0000	0000	0000	0000	0000	0001	0010
	16	0000	0000	0000	0000	0000	0000	0001
	17	0000	0000	0000	0000	0000	0000	0000
18	0	8345	3972	1501	0180	0016	0001	0000
	1	1517	3763	3002	0811	0126	0012	0001
	2	0130	1683	2835	1723	0458	0069	0006
	3	0007	0473	1680	2297	1046	0246	0031
	4	0000	0093	0700	2153	1681	0614	0117

TABLE 1   (CONTINUED)

n	*y*	.01	.05	.10	.20	.30	.40	.50
					P			
	5	0000	0014	0218	1507	2017	1146	0327
	6	0000	0002	0052	0816	1873	1655	0708
	7	0000	0000	0010	0350	1376	1892	1214
	8	0000	0000	0002	0120	0811	1734	1669
	9	0000	0000	0000	0033	0386	1284	1855
	10	0000	0000	0000	0008	0149	0771	1669
	11	0000	0000	0000	0001	0046	0374	1214
	12	0000	0000	0000	0000	0012	0145	0708
	13	0000	0000	0000	0000	0002	0045	0327
	14	0000	0000	0000	0000	0000	0011	0117
	15	0000	0000	0000	0000	0000	0002	0031
	16	0000	0000	0000	0000	0000	0000	0006
	17	0000	0000	0000	0000	0000	0000	0001
	18	0000	0000	0000	0000	0000	0000	0000
19	0	8262	3774	1351	0144	0011	0001	0000
	1	1586	3774	2852	0685	0093	0008	0000
	2	0144	1787	2852	1540	0358	0046	0003
	3	0008	0533	1796	2182	0869	0175	0018
	4	0000	0112	0798	2182	1491	0467	0074
	5	0000	0018	0266	1636	1916	0933	0222
	6	0000	0002	0069	0955	1916	1451	0518
	7	0000	0000	0014	0443	1525	1797	0961
	8	0000	0000	0002	0166	0980	1797	1442
	9	0000	0000	0000	0051	0514	1464	1762
	10	0000	0000	0000	0013	0220	0976	1762
	11	0000	0000	0000	0003	0077	0532	1442
	12	0000	0000	0000	0000	0022	0237	0961
	13	0000	0000	0000	0000	0005	0085	0518
	14	0000	0000	0000	0000	0001	0024	0222
	15	0000	0000	0000	0000	0000	0005	0074
	16	0000	0000	0000	0000	0000	0001	0018
	17	0000	0000	0000	0000	0000	0000	0003
	18	0000	0000	0000	0000	0000	0000	0000
	19	0000	0000	0000	0000	0000	0000	0000
20	0	8179	3583	1216	0115	0008	0000	0000
	1	1652	3774	2702	0576	0068	0005	0000
	2	0159	1887	2852	1369	0278	0031	0002
	3	0010	0596	1901	2054	0716	0123	0011
	4	0000	0133	0898	2182	1304	0350	0046
	5	0000	0022	0319	1746	1789	0746	0148
	6	0000	0003	0089	1091	1916	1244	0370

TABLE 1   (*CONCLUDED*)

n	y	.01	.05	.10	.20	.30	.40	.50
					P			
	7	0000	0000	0020	0545	1643	1659	0739
	8	0000	0000	0004	0222	1144	1797	1201
	9	0000	0000	0001	0074	0654	1597	1602
	10	0000	0000	0000	0020	0308	1171	1762
	11	0000	0000	0000	0005	0120	0710	1602
	12	0000	0000	0000	0001	0039	0355	1201
	13	0000	0000	0000	0000	0010	0146	0739
	14	0000	0000	0000	0000	0002	0049	0370
	15	0000	0000	0000	0000	0000	0013	0148
	16	0000	0000	0000	0000	0000	0003	0046
	17	0000	0000	0000	0000	0000	0000	0011
	18	0000	0000	0000	0000	0000	0000	0002
	19	0000	0000	0000	0000	0000	0000	0000
	20	0000	0000	0000	0000	0000	0000	0000

TABLE 2   VALUES OF THE POISSON MASS FUNCTION

Entries are values of the Poisson mass function

$$p(x) = \frac{e^{-\lambda}\lambda^x}{x!}$$

for values of λ and x shown. The entries are to be read with a decimal point preceding the digits.

					λ					
x	.05	.10	.15	.20	.25	.30	.35	.40	.45	.50
0	9512	9048	8607	8187	7788	7408	7047	6703	6376	6065
1	0476	0905	1291	1637	1947	2222	2466	2681	2869	3033
2	0012	0045	0097	0164	0243	0333	0432	0536	0646	0758
3	0000	0002	0005	0011	0020	0033	0050	0072	0097	0126
4	0000	0000	0000	0001	0001	0003	0004	0007	0011	0016
5	0000	0000	0000	0000	0000	0000	0000	0001	0001	0002

					λ					
x	.55	.60	.65	.70	.75	.80	.85	.90	.95	1.00
0	5769	5488	5220	4966	4724	4493	4274	4066	3867	3679
1	3173	3293	3393	3476	3543	3595	3633	3659	3674	3679
2	0873	0988	1103	1217	1329	1438	1544	1647	1745	1839
3	0160	0198	0239	0284	0332	0383	0437	0494	0553	0613
4	0022	0030	0039	0050	0062	0077	0093	0111	0131	0153
5	0002	0004	0005	0007	0009	0012	0016	0020	0025	0031
6	0000	0000	0001	0001	0001	0002	0002	0003	0004	0005

					λ					
x	1.05	1.10	1.15	1.20	1.25	1.30	1.35	1.40	1.45	1.50
0	3499	3329	3166	3012	2865	2725	2592	2466	2346	2231
1	3674	3662	3641	3614	3581	3543	3500	3452	3401	3347
2	1929	2014	2094	2169	2238	2303	2362	2417	2466	2510
3	0675	0738	0803	0867	0933	0998	1063	1128	1192	1255
4	0177	0203	0231	0260	0291	0324	0359	0395	0432	0471
5	0037	0045	0053	0062	0073	0084	0097	0111	0125	0141
6	0007	0008	0010	0012	0015	0018	0022	0026	0030	0035
7	0001	0001	0002	0002	0003	0003	0004	0005	0006	0008
8	0000	0000	0000	0000	0000	0001	0001	0001	0001	0001

TABLE 2   (CONTINUED)

x	1.55	1.60	1.65	1.70	1.75	1.80	1.85	1.90	1.95	2.00
					λ					
0	2122	2019	1920	1827	1738	1653	1572	1496	1423	1353
1	3290	3230	3169	3106	3041	2975	2909	2842	2774	2707
2	2550	2584	2614	2640	2661	2678	2691	2700	2705	2707
3	1317	1378	1438	1496	1552	1607	1659	1710	1758	1804
4	0510	0551	0593	0636	0679	0723	0767	0812	0857	0902
5	0158	0176	0196	0216	0238	0260	0284	0309	0334	0361
6	0041	0047	0054	0061	0069	0078	0088	0098	0109	0120
7	0009	0011	0013	0015	0017	0020	0023	0027	0030	0034
8	0002	0002	0003	0003	0004	0005	0005	0006	0007	0009
9	0000	0000	0000	0001	0001	0001	0001	0001	0002	0002

x	2.05	2.10	2.15	2.20	2.25	2.30	2.35	2.40	2.45	2.50
					λ					
0	1287	1225	1165	1108	1054	1003	0954	0907	0863	0821
1	2639	2572	2504	2438	2371	2306	2241	2177	2114	2052
2	2705	2700	2692	2681	2668	2652	2633	2613	2590	2565
3	1848	1890	1929	1966	2001	2033	2063	2090	2115	2138
4	0947	0992	1037	1082	1126	1169	1212	1254	1295	1336
5	0388	0417	0446	0476	0506	0538	0570	0602	0635	0668
6	0133	0146	0160	0174	0190	0206	0223	0241	0259	0278
7	0039	0044	0049	0055	0061	0068	0075	0083	0091	0099
8	0010	0011	0013	0015	0017	0019	0022	0025	0028	0031
9	0002	0003	0003	0004	0004	0005	0006	0007	0008	0009
10	0000	0001	0001	0001	0001	0001	0001	0002	0002	0002

x	2.55	2.60	2.65	2.70	2.75	2.80	2.85	2.90	2.95	3.00
					λ					
0	0781	0743	0707	0672	0639	0608	0578	0550	0523	0498
1	1991	1931	1872	1815	1758	1703	1649	1596	1544	1494
2	2539	2510	2481	2450	2417	2384	2349	2314	2277	2240
3	2158	2176	2191	2205	2216	2225	2232	2237	2239	2240
4	1376	1414	1452	1488	1523	1557	1590	1622	1652	1680
5	0702	0735	0769	0804	0838	0872	0906	0940	0974	1008
6	0298	0319	0340	0362	0384	0407	0431	0455	0479	0504
7	0109	0118	0129	0139	0151	0163	0175	0188	0202	0216
8	0035	0038	0043	0047	0052	0057	0062	0068	0074	0081
9	0010	0011	0013	0014	0016	0018	0020	0022	0024	0027
10	0003	0003	0003	0004	0004	0005	0006	0006	0007	0008
11	0001	0001	0001	0001	0001	0001	0001	0002	0002	0002

TABLE 2   (*CONTINUED*)

x	λ 3.05	3.10	3.15	3.20	3.25	3.30	3.35	3.40	3.45	3.50
0	0474	0450	4029	0408	0388	0369	0351	0334	0316	0302
1	1444	1397	1350	1304	1260	1217	1175	1135	1095	1057
2	2203	2165	2126	2087	2048	2008	1969	1929	1889	1850
3	2239	2237	2232	2226	2218	2209	2198	2186	2173	2158
4	1708	1733	1758	1781	1802	1823	1841	1858	1874	1888
5	1042	1075	1108	1140	1172	1203	1234	1264	1293	1322
6	0530	0555	0581	0608	0635	0662	0689	0716	0743	0771
7	0231	0246	0262	0278	0295	0312	0330	0348	0366	0385
8	0088	0095	0103	0111	0120	0129	0138	0148	0158	0169
9	0030	0033	0036	0040	0043	0047	0051	0056	0061	0066
10	0009	0010	0011	0013	0014	0016	0017	0019	0021	0023
11	0003	0003	0003	0004	0004	0005	0005	0006	0007	0007
12	0001	0001	0001	0001	0001	0001	0001	0002	0002	0002

x	λ 3.55	3.60	3.65	3.70	3.75	3.80	3.85	3.90	3.95	4.00
0	0287	0273	0260	0247	0235	0224	0213	0202	0193	0183
1	1020	0984	0949	0915	0882	0850	0819	0789	0781	0733
2	1810	1771	1731	1692	1654	1615	1577	1539	1502	1465
3	2142	2125	2106	2087	2067	2046	2024	2001	1987	1954
4	1901	1912	1922	1931	1938	1944	1948	1951	1953	1954
5	1350	1377	1403	1429	1453	1477	1500	1522	1543	1563
6	0799	0826	0854	0881	0908	0936	0962	0989	1016	1042
7	0405	0425	0445	0466	0487	0508	0529	0551	0573	0595
8	0180	0191	0203	0215	0228	0241	0255	0269	2083	2098
9	0071	0076	0082	0089	0095	0102	0109	0116	0124	0132
10	0025	0028	0030	0033	0036	0039	0042	0045	0049	0053
11	0008	0009	0010	0011	0012	0013	0015	0016	0018	0019
12	0002	0003	0003	0003	0004	0004	0005	0005	0006	0006
13	0001	0001	0001	0001	0001	0001	0001	0002	0002	0002

TABLE 2   (*CONTINUED*)

x					λ					
	4.05	**4.10**	**4.15**	**4.20**	**4.25**	**4.30**	**4.35**	**4.40**	**4.45**	**4.50**
0	0174	0166	0158	0150	0143	0136	0129	0123	0117	0111
1	0706	0679	0654	0630	0606	0583	0561	0540	0520	0500
2	1429	1393	1358	1323	1288	1254	1221	1188	1156	1125
3	1929	1904	1878	1852	1825	1798	1771	1743	1715	1687
4	1953	1951	1948	1944	1939	1933	1926	1917	1908	1898
5	1582	1600	1617	1633	1648	1662	1675	1687	1698	1708
6	1068	1093	1118	1143	1167	1191	1215	1237	1260	1281
7	0618	0640	0663	0686	0709	0732	0755	0778	0801	0824
8	0313	0328	0344	0360	0377	0393	0410	0428	0445	0463
9	0141	0150	0159	0168	0178	0188	0198	0209	0220	0232
10	0057	0061	0066	0071	0076	0081	0086	0092	0098	0104
11	0021	0023	0025	0027	0029	0032	0034	0037	0040	0043
12	0007	0008	0009	0009	0010	0011	0012	0013	0015	0016
13	0002	0002	0003	0003	0003	0004	0004	0005	0005	0006
14	0001	0001	0001	0001	0001	0001	0001	0001	0002	0002

x					λ					
	4.55	**4.60**	**4.65**	**4.70**	**4.75**	**4.80**	**4.85**	**4.90**	**4.95**	**5.00**
0	0106	0101	0096	0091	0087	0082	0078	0074	0071	0067
1	0481	0462	0445	0427	0411	0395	0380	0365	0351	0337
2	1094	1063	1034	1005	0976	0948	0921	0894	0868	0842
3	1659	1631	1602	1574	1545	1517	1488	1460	1432	1404
4	1887	1875	1863	1849	1835	1820	1805	1789	1772	1755
5	1717	1725	1732	1738	1743	1747	1751	1753	1754	1755
6	1302	1323	1343	1362	1380	1398	1415	1432	1447	1462
7	0846	0869	0892	0914	0937	0959	0980	1002	1023	1044
8	0481	0500	0518	0537	0556	0575	0594	0614	0633	0653
9	0243	0255	0268	0281	0293	0307	0320	0334	0348	0363
10	0111	0118	0125	0132	0139	0147	0155	0164	0172	0181
11	0046	0049	0053	0056	0060	0064	0068	0073	0078	0082
12	0017	0019	0020	0022	0024	0026	0028	0030	0032	0034
13	0006	0007	0007	0008	0009	0009	0010	0011	0012	0013
14	0002	0002	0002	0003	0003	0003	0004	0004	0004	0005
15	0001	0001	0001	0001	0001	0001	0001	0001	0001	0002

TABLE 2   (CONTINUED)

					λ					
x	5.10	5.20	5.30	5.40	5.50	5.60	5.70	5.80	5.90	6.00
0	0061	0055	0050	0045	0041	0037	0033	0030	0027	0025
1	0311	0287	0265	0244	0225	0207	0191	0176	0162	0149
2	0793	0746	0701	0659	0618	0580	0544	0509	0477	0446
3	1348	1293	1239	1185	1133	1082	1033	0985	0938	0892
4	1719	1681	1641	1600	1558	1515	1472	1428	1383	1339
5	1753	1748	1740	1728	1714	1697	1678	1656	1632	1606
6	1490	1515	1537	1555	1571	1584	1594	1601	1605	1606
7	1086	1125	1163	1200	1234	1267	1298	1326	1353	1377
8	0692	0731	0771	0810	0849	0887	0925	0962	0998	1033
9	0392	0423	0454	0486	0519	0552	0586	0620	0654	0688
10	0200	0220	0241	0262	0285	0309	0334	0359	0386	0413
11	0093	0104	0116	0129	0143	0157	0173	0190	0207	0225
12	0039	0045	0051	0058	0065	0073	0082	0092	0102	0113
13	0015	0018	0021	0024	0028	0032	0036	0041	0046	0052
14	0006	0007	0008	0009	0011	0013	0015	0017	0019	0022
15	0002	0002	0003	0003	0004	0005	0006	0007	0008	0009
16	0001	0001	0001	0001	0001	0002	0002	0002	0003	0003
17	0000	0000	0000	0000	0000	0001	0001	0001	0001	0001

					λ					
x	6.10	6.20	6.30	6.40	6.50	6.60	6.70	6.80	6.90	7.00
0	0022	0020	0018	0017	0015	0014	0012	0011	0010	0009
1	0137	0126	0116	0106	0098	0090	0082	0076	0070	0064
2	0417	0390	0364	0340	0318	0296	0276	0258	0240	0223
3	0848	0806	0765	0726	0688	0652	0617	0584	0552	0521
4	1294	1249	1205	1162	1118	1076	1034	0992	0952	0912
5	1579	1549	1519	1487	1454	1420	1385	1349	1314	1277
6	1605	1601	1595	1586	1575	1562	1546	1529	1511	1490
7	1399	1418	1435	1450	1462	1472	1480	1486	1489	1490
8	1066	1099	1130	1160	1188	1215	1240	1263	1284	1304
9	0723	0757	0791	0825	0858	0891	0923	0954	0985	1014
10	0441	0469	0498	0528	0558	0588	0618	0649	0679	0710
11	0244	0265	0285	0307	0330	0353	0377	0401	0426	0452
12	0124	0137	0150	0164	0179	0194	0210	0227	0245	0263
13	0058	0065	0073	0081	0089	0099	0108	0119	0130	0142
14	0025	0029	0033	0037	0041	0046	0052	0058	0064	0071
15	0010	0012	0014	0016	0018	0020	0023	0026	0029	0033
16	0004	0005	0005	0006	0007	0008	0010	0011	0013	0014
17	0001	0002	0002	0002	0003	0003	0004	0004	0005	0006
18	0000	0001	0001	0001	0001	0001	0001	0002	0002	0002
19	0000	0000	0000	0000	0000	0000	0001	0001	0001	0001

TABLE 2   (*CONTINUED*)

x	λ 7.10	7.20	7.30	7.40	7.50	7.60	7.70	7.80	7.90	8.00
0	0008	0007	0007	0006	0006	0005	0005	0004	0004	0003
1	0059	0054	0049	0045	0041	0038	0035	0032	0029	0027
2	0208	0194	0180	0167	0156	0145	0134	0125	0116	0107
3	0492	0464	0438	0413	0389	0366	0345	0324	0305	0286
4	0874	0836	0799	0764	0729	0696	0663	0632	0602	0573
5	1241	1204	1167	1130	1094	1057	1021	0986	0951	0916
6	1468	1445	1420	1394	1367	1339	1311	1282	1252	1221
7	1489	1486	1481	1474	1465	1454	1442	1428	1413	1396
8	1321	1337	1351	1363	1373	1381	1388	1392	1395	1396
9	1042	1070	1096	1121	1144	1167	1187	1207	1224	1241
10	0740	0770	0800	0829	0858	0887	0914	0941	0967	0993
11	0478	0504	0531	0558	0585	0613	0640	0667	0695	0722
12	0283	0303	0323	0344	0366	0388	0411	0434	0457	0481
13	0154	0168	0181	0196	0211	0227	0243	0260	0278	0296
14	0078	0086	0095	0104	0113	0123	0134	0145	0157	0169
15	0037	0041	0046	0051	0057	0062	0069	0075	0083	0090
16	0016	0019	0021	0024	0026	0030	0033	0037	0041	0045
17	0007	0008	0009	0010	0012	0013	0015	0017	0019	0021
18	0003	0003	0004	0004	0005	0006	0006	0007	0008	0009
19	0001	0001	0001	0002	0002	0002	0003	0003	0003	0004
20	0000	0000	0001	0001	0001	0001	0001	0001	0001	0002

TABLE 2   (*CONTINUED*)

x	λ									
	8.10	8.20	8.30	8.40	8.50	8.60	8.70	8.80	8.90	9.00
0	0003	0003	0002	0002	0002	0002	0002	0002	0001	0001
1	0025	0023	0021	0019	0017	0016	0014	0013	0012	0011
2	0100	0092	0086	0079	0074	0068	0063	0058	0054	0050
3	0269	0252	0237	0222	0208	0195	0183	0171	0160	0150
4	0544	0517	0491	0466	0443	0420	0398	0377	0357	0337
5	0882	0849	0816	0784	0752	0722	0692	0663	0635	0607
6	1191	1160	1128	1097	1066	1034	1003	0972	0941	0911
7	1378	1358	1338	1317	1294	1271	1247	1222	1197	1171
8	1395	1392	1388	1382	1375	1366	1356	1344	1332	1318
9	1256	1269	1280	1290	1299	1306	1311	1315	1317	1318
10	1017	1040	1063	1084	1104	1123	1140	1157	1172	1186
11	0749	0776	0802	0828	0853	0878	0902	0925	0948	0970
12	0505	0530	0555	0579	0604	0629	0654	0679	0703	0728
13	0315	0334	0354	0374	0395	0416	0438	0459	0481	0504
14	0182	0196	0210	0225	0240	0256	0272	0289	0306	0324
15	0098	0107	0116	0126	0136	0147	0158	0169	0182	0194
16	0050	0055	0060	0066	0072	0079	0086	0093	0101	0109
17	0024	0026	0029	0033	0036	0040	0044	0048	0053	0058
18	0011	0012	0014	0015	0017	0019	0021	0024	0026	0029
19	0005	0005	0006	0007	0008	0009	0010	0011	0012	0014
20	0002	0002	0002	0003	0003	0004	0004	0005	0005	0006
21	0001	0001	0001	0001	0001	0002	0002	0002	0002	0003
22	0000	0000	0000	0000	0000	0001	0001	0001	0001	0001

TABLE 2   (*CONTINUED*)

x	9.10	9.20	9.30	9.40	9.50	9.60	9.70	9.80	9.90	10.00
					λ					
0	0001	0001	0001	0001	0001	0001	0001	0001	0001	0000
1	0010	0009	0009	0008	0007	0007	0006	0005	0005	0005
2	0046	0043	0040	0037	0034	0031	0029	0027	0025	0023
3	0140	0131	0123	0115	0107	0100	0093	0087	0081	0076
4	0319	0302	0285	0269	0254	0240	0226	0213	0201	0189
5	0581	0555	0530	0506	0483	0460	0439	0418	0398	0378
6	0881	0851	0822	0793	0764	0736	0709	0682	0656	0631
7	1145	1118	1091	1064	1037	1010	0982	0955	0928	0901
8	1302	1286	1269	1251	1232	1212	1191	1170	1148	1126
9	1317	1315	1311	1306	1300	1293	1284	1274	1263	1251
10	1198	1210	1219	1228	1235	1241	1245	1249	1250	1251
11	0991	1012	1031	1049	1067	1083	1098	1112	1125	1137
12	0752	0776	0799	0822	0844	0866	0888	0908	0928	0948
13	0526	0549	0572	0594	0617	0640	0662	0685	0707	0729
14	0342	0361	0380	0399	0419	0439	0459	0479	0500	0521
15	0208	0221	0235	0250	0265	0281	0297	0313	0330	0347
16	0118	0127	0137	0147	0157	0168	0180	0192	0204	0217
17	0063	0069	0075	0081	0088	0095	0103	0111	0119	0128
18	0032	0035	0039	0042	0046	0051	0055	0060	0065	0071
19	0015	0017	0019	0021	0023	0026	0028	0031	0034	0037
20	0007	0008	0009	0010	0011	0012	0014	0015	0017	0019
21	0003	0003	0004	0004	0005	0006	0006	0007	0008	0009
22	0001	0001	0002	0002	0002	0002	0003	0003	0004	0004
23	0000	0001	0001	0001	0001	0001	0001	0001	0002	0002
24	0000	0000	0000	0000	0000	0000	0000	0001	0001	0001

TABLE 2   (CONCLUDED)

x	10.5	11.0	11.5	12.0	12.5	13.0	13.5	14.0	14.5	15.0
0	0000	0002	0000	0000	0000	0000	0000	0000	0000	0000
1	0003	0010	0001	0001	0000	0000	0000	0000	0000	0000
2	0015	0037	0007	0004	0003	0002	0001	0001	0001	0000
3	0053	0102	0026	0018	0012	0008	0006	0004	0003	0002
4	0139	0224	0074	0053	0038	0027	0019	0013	0009	0006
5	0293	0411	0170	0127	0095	0070	0051	0037	0027	0019
6	0513	0646	0325	0255	0197	0152	0115	0087	0065	0048
7	0769	0888	0535	0437	0353	0281	0222	0174	0135	0104
8	1009	1085	0769	0655	0551	0457	0375	0304	0244	0194
9	1177	1194	0982	0874	0756	0661	0563	0473	0394	0324
10	1236	1194	1129	1048	0956	0859	0760	0663	0571	0486
11	1180	1094	1181	1144	1087	1015	0932	0844	0753	0663
12	1032	0926	1131	1144	1132	1099	1049	0984	0910	0829
13	0834	0728	1001	1056	1089	1099	1089	1060	1014	0956
14	0625	0534	0822	0905	0972	1021	1050	1060	1051	1024
15	0438	0367	0630	0724	0810	0885	0945	0989	1016	1024
16	0287	0237	0453	0543	0633	0719	0798	0866	0920	0960
17	0177	0145	0306	0383	0465	0550	0633	0713	0785	0847
18	0104	0084	0196	0255	0323	0397	0475	0554	0632	0706
19	0057	0046	0119	0161	0213	0272	0337	0409	0483	0557
20	0030	0024	0068	0097	0133	0177	0228	0286	0350	0418
21	0015	0012	0037	0055	0079	0109	0146	0191	0242	0299
22	0007	0006	0020	0030	0045	0065	0090	0121	0159	0204
23	0003	0003	0010	0016	0024	0037	0053	0074	0100	0133
24	0001	0001	0005	0008	0013	0020	0030	0043	0061	0083
25	0000	0000	0002	0004	0006	0010	0016	0024	0035	0050
26	0000	0000	0001	0002	0003	0005	0008	0013	0020	0029
27	0000	0000	0000	0001	0001	0002	0004	0007	0011	0016
28	0000	0000	0000	0000	0001	0001	0002	0003	0005	0009
29	0000	0000	0000	0000	0000	0000	0001	0002	0003	0004
30	0000	0000	0000	0000	0000	0000	0000	0001	0001	0002
31	0000	0000	0000	0000	0000	0000	0000	0000	0001	0001

λ

TABLE 3   NORMAL CDF

Entries are values of

$$\Phi(x) = \int_{-\infty}^{x} \frac{1}{\sqrt{2\pi}} e^{-(1/2)z^2} dz$$

for the values of $x \geq 0$ shown. For $x < 0$, use the identity $\Phi(x) = 1 - \Phi(-x)$. The entries are to be read with a decimal point preceding the digits.

x	0	1	2	3	4	5	6	7	8	9
.00	5000	5004	5008	5012	5016	5020	5024	5028	5032	5036
.01	5040	5044	5048	5052	5056	5060	5064	5068	5072	6076
.02	5080	5084	5088	5092	5096	5100	5104	5108	5112	5116
.03	5120	5124	5128	5132	5136	5140	5144	5148	5152	5156
.04	5160	5164	5168	5171	5175	5179	5183	5187	5191	5195
.05	5199	5203	5207	5211	5215	5219	5223	5227	5231	5235
.06	5239	5243	5247	5251	5255	5259	5263	5267	5271	5275
.07	5279	5283	5287	5291	5295	5299	5305	5307	5311	5315
.08	5319	5323	5327	5331	5335	5339	5343	5347	5351	5355
.09	5359	5363	5367	5370	5374	5378	5382	5386	5390	5394
.10	5398	5402	5406	5410	5414	5418	5422	5426	5430	5434
.11	5438	5442	5446	5450	5454	5458	5462	5466	5470	5474
.12	5478	5482	5486	5489	5493	5497	5501	5505	5509	5513
.13	5517	5521	5525	5529	5533	5537	5541	5545	5549	5553
.14	5557	5561	5565	5569	5572	5576	5580	5584	5588	5592
.15	5596	5600	5604	5608	5612	5616	5620	5624	5628	5632
.16	5636	5640	5643	5647	5651	5655	5659	5663	5667	5671
.17	5675	5679	5683	5687	5691	5695	5699	5702	5706	5710
.18	5714	5718	5722	5726	5730	5734	5738	5742	5746	5750
.19	5753	5757	5761	5765	5769	5773	5777	5781	5785	5789
.20	5793	5797	5800	5804	5808	5812	5816	5820	5824	5828
.21	5832	5836	5839	5843	5847	5851	5855	5859	5863	5867
.22	5871	5875	5878	5882	5886	5890	5894	5898	5902	5906
.23	5910	5913	5917	5921	5925	5929	5933	5937	5941	5944
.24	5948	5952	5956	5960	5964	5968	5972	5975	5979	5983
.25	5987	5991	5995	5999	6003	6006	6010	6014	6018	6022
.26	6026	6030	6033	6037	6041	6045	6049	6053	6057	6060
.27	6064	6068	6072	6076	6080	6083	6087	6091	6095	6099
.28	6103	6106	6110	6114	6118	6122	6126	6129	6133	6137
.29	6141	6145	6149	6152	6156	6160	6164	6168	6171	6175

TABLE 3   (CONTINUED)

x	0	1	2	3	4	5	6	7	8	9
.30	6179	6183	6187	6191	6194	6198	6202	6206	6210	6213
.31	6217	6221	6225	6229	6232	6236	6240	6244	6248	6251
.32	6255	6259	6263	6267	6270	6274	6278	6282	6285	6289
.33	6293	6297	6301	6304	6308	6312	6316	6319	6323	6327
.34	6331	6334	6338	6342	6346	6350	6353	6357	6361	6365
.35	6368	6372	6376	6380	6383	6387	6391	6395	6398	6402
.36	6406	6410	6413	6417	6421	6424	6428	6432	6436	6439
.37	6443	6447	6451	6454	6458	6462	6465	6469	6473	6477
.38	6480	6484	6488	6491	6495	6499	6503	6506	6510	6514
.39	6517	6521	6525	6528	6532	6536	6539	6543	6547	6551
.40	6554	6558	6562	6565	6569	6573	6576	6580	6584	6587
.41	6591	6595	6598	6602	6606	6609	6613	6617	6620	6624
.42	6628	6631	6635	6639	6642	6646	6649	6653	6657	6660
.43	6664	6668	6671	6675	6679	6682	6686	6689	6693	6697
.44	6700	6704	6708	6711	6715	6718	6722	6726	6729	6733
.45	6736	6740	6744	6747	6751	6754	6758	6762	6765	6769
.46	6772	6776	6780	6783	6787	6790	6794	6798	6801	6805
.47	6808	6812	6815	6819	6823	6826	6830	6833	6837	6840
.48	6844	6847	6851	6855	6858	6862	6865	6869	6872	6876
.49	6879	6883	6886	6890	6893	6897	6901	6904	6908	6911
.50	6915	6918	6922	6925	6929	6932	6936	6939	6943	6946
.51	6950	6953	6957	6960	6964	6967	6971	6974	6978	6981
.52	6985	6988	6992	6995	6999	7002	7006	7009	7013	7016
.53	7019	7023	7026	7030	7033	7037	7040	7044	7047	7051
.54	7054	7057	7061	7064	7068	7071	7075	7078	7082	7085
.55	7088	7092	7095	7099	7102	7106	7109	7112	7116	7119
.56	7123	7126	7129	7133	7136	7140	7143	7146	7150	7153
.57	7157	7160	7163	7167	7170	7174	7177	7180	7184	7187
.58	7190	7194	7197	7201	7204	7207	7211	7214	7217	7221
.59	7224	7227	7231	7234	7237	7241	7244	7247	7251	7254
.60	7257	7261	7264	7267	7271	7274	7277	7281	7284	7287
.61	7291	7294	7297	7301	7304	7307	7311	7314	7317	7320
.62	7324	7327	7330	7334	7337	7340	7343	7347	7350	7353
.63	7357	7360	7363	7366	7370	7373	7376	7379	7883	7386
.64	7389	7392	7396	7399	7402	7405	7409	7412	7415	7418
.65	7422	7425	7428	7431	7434	7438	7441	7444	7447	7451
.66	7454	7457	7460	7463	7467	7470	7473	7476	7479	7483
.67	7486	7489	7492	7495	7498	7502	7505	7508	7511	7514
.68	7517	7521	7524	7527	7530	7533	7536	7540	7543	7546
.69	7549	7552	7555	7558	7562	7565	7568	7571	7574	7577

TABLE 3   (CONTINUED)

x	0	1	2	3	4	5	6	7	8	9
.70	7580	7583	7587	7590	7593	7596	7599	7602	7605	7608
.71	7611	7615	7618	7621	7624	7627	7630	7633	7636	7639
.72	7642	7645	7649	7652	7655	7658	7661	7664	7667	7670
.73	7673	7676	7679	7682	7685	7688	7691	7694	7697	7700
.74	7703	7707	7710	7713	7716	7719	7722	7725	7728	7731
.75	7734	7737	7740	7743	7746	7749	7752	7755	7758	7761
.76	7764	7767	7770	7773	7776	7779	7782	7785	7788	7791
.77	7793	7796	7799	7802	7805	7808	7811	7814	7817	7820
.78	7823	7826	7829	7832	7835	7838	7841	7844	7847	7849
.79	7852	7855	7858	7861	7864	7867	7870	7873	7876	7879
.80	7881	7884	7887	7890	7893	7896	7899	7902	7905	7907
.81	7910	7913	7916	7919	7922	7925	7927	7930	7933	7936
.82	7939	7942	7945	7947	7950	7953	7956	7959	7962	7964
.83	7967	7970	7973	7976	7979	7981	7984	7987	7990	7993
.84	7995	7998	8001	8004	8007	8009	8012	8015	8018	8021
.85	8023	8026	8029	8032	8034	8037	8040	8043	8046	8048
.86	8051	8054	8057	8059	8062	8065	8068	8070	8073	8076
.87	8078	8081	8084	8087	8089	8092	8095	8098	8100	8103
.88	8106	8108	8111	8114	8117	8119	8122	8125	8127	8130
.89	8133	8135	8138	8141	8143	8146	8149	8151	8154	8157
.90	8159	8162	8165	8167	8170	8173	8175	8178	8181	8183
.91	8186	8189	8191	8194	8196	8199	8202	8204	8207	8210
.92	8212	8215	8217	8220	8223	8225	8228	8230	8233	8236
.93	8238	8241	8243	8246	8248	8251	8254	8256	8259	8261
.94	8264	8266	8269	8272	8274	8277	8279	8282	8284	8287
.95	8289	8292	8295	8297	8300	8302	8305	8307	8310	8312
.96	8315	8317	8320	8322	8325	8327	8330	8332	8335	8337
.97	8340	8342	8345	8347	8350	8352	8355	8357	8360	8362
.98	8365	8367	8370	8372	8374	8377	8379	8382	8384	8387
.99	8389	8392	8394	8396	8399	8401	8404	8406	8409	8311
1.00	8413	8416	8418	8421	8423	8426	8428	8430	8433	8435
1.01	8438	8440	8442	8445	8447	8449	8452	8454	8457	8459
1.02	8461	8464	8466	8468	8471	8473	8476	8478	8480	8483
1.03	8485	8487	8490	8492	8494	8497	8499	8501	8504	8506
1.04	8508	8511	8513	8515	8518	8520	8522	8525	8527	8529
1.05	8531	8534	8536	8538	8541	8543	8545	8547	8550	8552
1.06	8554	8557	8559	8561	8563	8566	8568	8570	8572	8575
1.07	8577	8579	8581	8584	8586	8588	8590	8593	8595	8597
1.08	8599	8602	8604	8606	8608	8610	8613	8615	8617	8619
1.09	8621	8624	8626	8628	8630	8632	8635	8637	8639	8641

TABLE 3   (CONTINUED)

x	0	1	2	3	4	5	6	7	8	9
1.10	8643	8646	8648	8650	8652	8654	8656	8659	8661	8663
1.11	8665	8667	8669	8671	8674	8676	8678	8680	8682	8684
1.12	8686	8689	8691	8693	8695	8697	8699	8701	8703	8706
1.13	8708	8710	8712	8714	8716	8718	8720	8722	8724	8726
1.14	8729	8731	8733	8735	8737	8739	8741	8743	8745	8747
1.15	8749	8751	8753	8755	8757	8760	8762	8764	8766	8768
1.16	8770	8772	8774	8776	8778	8780	8782	8784	8786	8788
1.17	8790	8792	8794	8796	8798	8800	8802	8804	8806	8808
1.18	8810	8812	8814	8816	8818	8820	8822	8824	8826	8828
1.19	8830	8832	8834	8836	8838	8840	8842	8843	8845	8847
1.20	8849	8851	8853	8855	8857	8859	8861	8863	8865	8867
1.21	8869	8871	8872	8874	8876	8878	8880	8882	8884	8886
1.22	8888	8890	8891	8893	8895	8897	8899	8901	8903	8905
1.23	8907	8908	8910	8912	8914	8916	8918	8920	8921	8923
1.24	8925	8927	8929	8931	8933	8934	8936	8938	8940	8942
1.25	8944	8945	8647	8949	8951	8953	8954	8956	8958	8960
1.26	8962	8963	8965	8967	8969	8971	8972	8974	8976	8978
1.27	8980	8981	8983	8985	8987	8988	8990	8992	8994	8996
1.28	8997	8999	9001	9003	9004	9006	9008	9010	9011	9013
1.29	9015	9016	9018	9020	9022	9023	9025	9027	9029	9030
1.30	9032	9034	9035	9037	9039	9041	9042	9044	9046	9047
1.31	9049	9051	9052	9054	9056	9057	9059	9061	9062	9064
1.32	9066	9067	9069	9071	9072	9074	9076	9077	9079	9081
1.33	9082	9084	9086	9087	9089	9091	9092	9094	9096	9097
1.34	9099	9100	9102	9104	9105	9107	9108	9110	9112	9113
1.35	9115	9117	9118	9120	9121	9123	9125	9126	9128	9129
1.36	9131	9132	9134	9136	9137	9139	9140	9142	9143	9145
1.37	9147	9148	9150	9151	9153	9154	9156	9157	9159	9161
1.38	9162	9164	9165	9167	9168	9170	9171	9173	9174	9176
1.39	9177	9179	9180	9182	9183	9185	9186	9188	9189	9191
1.40	9192	9194	9195	9197	9198	9200	9201	9203	9204	9206
1.41	9207	9209	9210	9212	9213	9215	9216	9218	9219	9221
1.42	9222	9223	9225	9226	9228	9229	9231	9232	9234	9235
1.43	9236	9238	9239	9241	9242	9244	9245	9246	9248	9249
1.44	9251	9252	9253	9255	9256	9258	9259	9261	9262	9263
1.45	9265	9266	9267	9269	9270	9272	9273	9274	9276	9277
1.46	9279	9280	9281	9283	9284	9285	9287	9288	9289	9291
1.47	9292	9294	9295	9296	9298	9299	9300	9302	9303	9304
1.48	9306	9307	9308	9310	9311	9312	9314	9315	9316	9318
1.49	9319	9320	9322	9323	9324	9325	9327	9328	9329	9331

TABLE 3   (CONTINUED)

x	0	1	2	3	4	5	6	7	8	9
1.50	9332	9333	9335	9336	9337	9338	9340	9341	9342	9344
1.51	9345	9346	9347	9349	9350	9351	9352	9354	9355	9356
1.52	9357	9359	9360	9361	9362	9364	9365	9366	9367	9369
1.53	9370	9371	9372	9374	9375	9376	9377	9379	9380	9381
1.54	9382	9383	9385	9386	9387	9388	9389	9391	9392	9393
1.55	9394	9395	9397	9398	9399	9400	9401	9403	9404	9405
1.56	9406	9407	9409	9410	9411	9412	9413	9414	9416	9417
1.57	9418	9419	9420	9421	9423	9424	9425	9426	9427	9428
1.58	9429	9431	9432	9433	9434	9435	9436	9437	9439	9440
1.59	9441	9442	9443	9444	9445	9446	9448	9449	9450	9451
1.60	9452	9453	9454	9455	9456	9458	9459	9460	9461	9462
1.61	9463	9464	9465	9466	9467	9468	9470	9471	9472	9473
1.62	9474	9475	9476	9477	9478	9479	9480	9481	9482	9483
1.63	9484	9486	9487	9488	9489	9490	9491	9492	9493	9494
1.64	9495	9496	9497	9498	9499	9500	9501	9502	9503	9504
1.6	9452	9463	9474	9484	9495	9505	9515	9525	9535	9545
1.7	9554	9564	9573	9582	9591	9599	9608	9616	9625	9633
1.8	9641	9649	9656	9664	9671	9678	9686	9693	9699	9706
1.9	9713	9719	9726	9732	9738	9744	9750	9756	9761	9767
2.0	9772	9778	9783	9788	9793	9798	9803	9808	9812	9817
2.1	9821	9826	9830	9834	9838	9842	9846	9850	9854	9857
2.2	9861	9864	9868	9871	9875	9878	9881	9884	9887	9890
2.3	9893	9896	9898	9901	9904	9906	9909	9911	9913	9916
2.4	9918	9920	9922	9925	9927	9929	9931	9932	9934	9936
2.5	9938	9940	9941	9943	9945	9946	9948	9949	9951	9952
2.6	9953	9955	9956	9957	9959	9960	9961	9962	9963	9964
2.7	9965	9966	9967	9968	9969	9970	9971	9972	9973	9974
2.8	9974	9975	9976	9977	9977	9978	9979	9979	9980	9981
2.9	9981	9982	9982	9983	9984	9984	9985	9985	9986	9986
3.0	9987									
3.1	9990									
3.2	9993									
3.3	9995									
3.4	9997									
3.5	9998									
3.6	9998									
3.7	9999									
3.8	9999									

TABLE 4   VALUES OF THE t–CDF

Entries are values of x such that

$$\alpha = \int_{-\infty}^{x} \frac{\Gamma((n+1)/2)}{\sqrt{n\pi}\ \Gamma(n/2)} \left(1 + \frac{w^2}{n}\right)^{-(n+1)/2} dw$$

for various values of α and degrees of freedom n. For values of $\alpha < .5$, use the identity $t_\alpha = -t_{1-\alpha}$. These values were obtained from Table 5, using the identity $t_\gamma = \sqrt{F_{2\gamma-1;1,n}}$.

					α				
n	.80	.85	.90	.925	.950	.975	.980	.990	.995
1	1.376	1.962	3.078	4.165	6.314	12.73	15.87	31.75	64.00
2	1.061	1.386	1.886	2.282	2.920	4.303	4.849	6.964	9.925
3	0.978	1.250	1.638	1.924	2.353	3.182	3.482	4.541	5.841
4	0.941	1.190	1.533	1.778	2.132	2.776	2.998	3.747	4.604
5	0.920	1.156	1.476	1.699	2.015	2.571	2.756	3.365	4.032
6	0.906	1.134	1.440	1.650	1.943	2.447	2.612	3.143	3.707
7	0.896	1.119	1.415	1.617	1.895	2.365	2.517	2.998	3.499
8	0.889	1.108	1.397	1.592	1.860	2.306	2.449	2.896	3.355
9	0.883	1.100	1.383	1.574	1.833	2.262	2.398	2.821	3.250
10	0.879	1.093	1.372	1.559	1.812	2.228	2.359	2.764	3.169
11	0.876	1.088	1.363	1.548	1.796	2.201	2.328	2.718	3.106
12	0.873	1.083	1.356	1.538	1.782	2.179	2.303	2.681	3.054
13	0.870	1.080	1.350	1.530	1.771	2.160	2.282	2.650	3.012
14	0.868	1.076	1.345	1.523	1.761	2.145	2.264	2.624	2.977
15	0.866	1.074	1.341	1.517	1.753	2.132	2.248	2.602	2.947
16	0.865	1.071	1.337	1.512	1.746	2.120	2.235	2.584	2.921
17	0.863	1.069	1.333	1.508	1.740	2.110	2.224	2.567	2.898
18	0.862	1.067	1.330	1.504	1.734	2.101	2.214	2.552	2.878
19	0.861	1.066	1.328	1.500	1.729	2.093	2.205	2.540	2.861
20	0.860	1.064	1.325	1.497	1.725	2.086	2.197	2.528	2.845
21	0.859	1.063	1.323	1.494	1.721	2.080	2.189	2.518	2.831
22	0.858	1.061	1.321	1.492	1.717	2.074	2.183	2.508	2.819
23	0.858	1.060	1.320	1.489	1.714	2.069	2.177	2.500	2.807
24	0.857	1.059	1.318	1.487	1.711	2.064	2.172	2.492	2.797
25	0.856	1.058	1.316	1.485	1.708	2.060	2.167	2.485	2.788
26	0.856	1.058	1.315	1.483	1.706	2.056	2.162	2.479	2.779
27	0.855	1.057	1.314	1.482	1.703	2.052	2.158	2.473	2.771
28	0.855	1.056	1.312	1.480	1.701	2.048	2.154	2.467	2.763
29	0.854	1.055	1.311	1.479	1.699	2.045	2.150	2.462	2.756
30	0.854	1.055	1.310	1.477	1.697	2.042	2.147	2.457	2.750

TABLE 5 VALUES OF THE F-CDF

Entries are values of x such that

$$\alpha = \int_0^x \frac{\Gamma((n+m)/2)\,(n/m)^{n/2}\, t^{(n/2)-1}}{\Gamma(n/2)\,\Gamma(m/2)\,[1+(n/m)t]^{(n+m)/2}}\, dt$$

for various values of α and degrees of freedom n and m. With x denoting $F_{\alpha;n,m}$, the table may be used for values corresponding to $1-\alpha$ by means of the identity $F_{\alpha;n,m} \cdot F_{1-\alpha;m,n} = 1$.

									n								
m	α	1	2	3	4	5	6	7	8	9	10	15	20	30	50	100	200
1	.800	9.5	12.0	13.1	13.6	14.0	14.3	14.4	14.6	14.7	14.8	15.0	15.2	15.3	15.4	15.5	15.5
	.825	12.6	15.8	17.2	17.9	18.4	18.7	19.0	19.2	19.3	19.4	19.7	19.9	20.1	20.2	20.3	20.4
	.850	17.3	21.7	23.6	24.6	25.2	25.7	26.0	26.2	26.4	26.5	27.0	27.2	27.5	27.7	27.8	27.9
	.875	25.3	31.5	34.1	35.6	36.5	37.1	37.6	37.9	38.2	38.4	39.1	39.4	39.7	40.0	40.2	40.3
	.900	39.9	49.5	53.6	55.8	57.2	58.2	58.9	59.4	59.9	60.2	61.2	61.7	62.3	62.7	63.0	63.2
	.925	71.4	88.4	95.6	99.6	102.1	103.8	105.0	106.0	106.7	107.3	109.1	110.0	111.0	111.7	112.3	112.6
	.950	161.4	199.5	215.7	224.6	230.2	234.0	236.8	238.9	240.5	241.9	245.9	248.0	250.1	251.8	253.0	253.7
	.975	647.8	799.5	864.2	899.6	921.9	937.1	948.2	956.7	963.3	968.6	984.9	993.1	1001.4	1008.1	1013.2	1015.7
	.990	4052.2	4999.5	5403.3	5624.5	5763.6	5859.0	5928.3	5981.2	6022.5	6055.9	6157.3	6208.6	6260.7	6302.5	6334.0	6349.9
	.995	16211.	20000.	21615.	22499.	23056.	23438.	23716.	23925.	24091.	24225.	24629.	24837.	25043.	25211.	25338.	25401.
2	.800	3.5556	4.0000	4.1563	4.2361	4.2844	4.3168	4.3401	4.3576	4.3712	4.3822	4.4151	4.4316	4.4482	4.4614	4.4714	4.4764
	.825	4.2622	4.7143	4.8721	4.9523	5.0008	5.0334	5.0567	5.0743	5.0880	5.0989	5.1319	5.1484	5.1650	5.1783	5.1883	5.1933
	.850	5.2072	5.6667	5.8258	5.9065	5.9553	5.9880	6.0114	6.0290	6.0427	6.0537	6.0867	6.1033	6.1199	6.1332	6.1431	6.1481
	.875	6.5333	7.0000	7.1605	7.2416	7.2907	7.3234	7.3469	7.3646	7.3783	7.3893	7.4224	7.4390	7.4556	7.4689	7.4789	7.4839
	.900	8.5263	9.0000	9.1618	9.2434	9.2926	9.3255	9.3491	9.3668	9.3805	9.3916	9.4247	9.4413	9.4579	9.4712	9.4812	9.4862
	.925	11.853	12.333	12.496	12.578	12.628	12.661	12.684	12.702	12.716	12.727	12.760	12.777	12.793	12.807	12.817	12.822
	.950	18.513	19.000	19.164	19.274	19.297	19.329	19.354	19.371	19.385	19.396	19.429	19.445	19.463	19.476	19.486	19.491
	.975	38.506	39.000	39.167	39.249	39.299	39.331	39.355	39.373	39.388	39.396	39.432	39.447	39.465	39.476	39.488	39.491
	.990	98.502	98.999	99.165	99.250	99.299	99.332	99.356	99.375	99.389	99.399	99.432	99.450	99.466	99.478	99.491	99.495
	.995	198.50	199.00	199.17	199.25	199.30	199.33	199.36	199.38	199.39	199.40	199.43	199.45	199.46	199.48	199.49	199.49

TABLE 5   (CONTINUED)

m	α	1	2	3	4	5	6	7	8	9	10	15	20	30	50	100	200
3	.800	2.6822	2.8860	2.9359	2.9555	2.9652	2.9707	2.9741	2.9763	2.9779	2.9791	2.9819	2.9830	2.9838	2.9842	2.9844	2.9845
	.825	3.1302	3.2944	3.3260	3.3351	3.3381	3.3389	3.3389	3.3385	3.3380	3.3375	3.3351	3.3335	3.3316	3.3298	3.3284	3.3276
	.850	3.7030	3.8133	3.8209	3.8166	3.8109	3.8058	3.8013	3.7976	3.7945	3.7918	3.7828	3.7778	3.7724	3.7678	3.7642	3.7624
	.875	4.4651	4.5000	4.4750	4.4526	4.4354	4.4223	4.4120	4.4038	4.3971	4.3915	4.3738	4.3643	4.3543	4.3460	4.3396	4.3363
	.900	5.5383	5.4624	5.3908	5.3426	5.3092	5.2847	5.2662	5.2517	5.2400	5.2304	5.2003	5.1845	5.1681	5.1546	5.1443	5.1390
	.925	7.1865	6.9343	6.7901	6.7021	6.6435	6.6017	6.5705	6.5463	6.5269	6.5112	6.4622	6.4367	6.4105	6.3891	6.3727	6.3644
	.950	10.128	9.5521	9.2766	9.1172	9.0135	8.9406	8.8867	8.8452	8.8123	8.7855	8.7029	8.6602	8.6166	8.5810	8.5539	8.5402
	.975	17.443	16.044	15.439	15.101	14.885	14.735	14.624	14.540	14.473	14.419	14.253	14.167	14.080	14.010	13.956	13.929
	.990	34.116	30.816	29.457	28.710	28.237	27.911	27.672	27.489	27.345	27.229	26.872	26.690	26.505	26.354	26.240	26.183
	.995	55.552	49.799	47.467	46.195	45.392	44.839	44.434	44.126	43.883	43.686	43.085	42.778	42.466	42.213	42.022	41.925
4	.800	2.3507	2.4721	2.4847	2.4826	2.4780	2.4733	2.4691	2.4654	2.4623	2.4596	2.4503	2.4450	2.4392	2.4342	2.4302	2.4281
	.825	2.7111	2.7809	2.7702	2.7549	2.7418	2.7311	2.7225	2.7155	2.7097	2.7048	2.6888	2.6801	2.6707	2.6629	2.6567	2.6535
	.850	3.1620	3.1640	3.1236	3.0916	3.0678	3.0497	3.0357	3.0245	3.0153	3.0078	2.9835	2.9704	2.9567	2.9453	2.9364	2.9313
	.875	3.7468	3.6568	3.5773	3.5236	3.4859	3.4582	3.4371	3.4204	3.4070	3.3959	3.3609	3.3424	3.3231	3.3070	3.2947	3.288
	.900	4.5448	4.3246	4.1909	4.1072	4.0506	4.0098	3.9790	3.9549	3.9357	3.9199	3.8704	3.8443	3.8174	3.7952	3.7782	3.7695
	.925	5.7219	5.3030	5.0883	4.9604	4.8756	4.8154	4.7704	4.7355	4.7077	4.6850	4.6142	4.5772	4.5392	4.5079	4.4840	4.4719
	.950	7.7086	6.9443	6.5914	6.3882	6.2561	6.1631	6.0942	6.0410	5.9988	5.9644	5.8578	5.8026	5.7459	5.6995	5.6641	5.6461
	.975	12.218	10.649	9.9792	9.6045	9.3645	9.1973	9.0741	8.9796	8.9047	8.8439	8.6565	8.5599	8.4613	8.3808	8.3195	8.2885
	.990	21.198	18.000	16.694	15.977	15.522	15.207	14.976	14.799	14.659	14.546	14.198	14.020	13.838	13.690	13.577	13.520
	.995	31.333	26.284	24.259	23.154	22.456	21.974	21.622	21.352	21.139	20.967	20.438	20.167	19.891	19.667	19.497	19.411
5	.800	2.1782	2.2591	2.2530	2.2397	2.2275	2.2174	2.2090	2.2021	2.1963	2.1914	2.1751	2.1660	2.1562	2.1479	2.1413	2.1379
	.825	2.4956	2.5202	2.4889	2.4612	2.4397	2.4231	2.4099	2.3993	2.3906	2.3833	2.3596	2.3467	2.3330	2.3214	2.3124	2.3077
	.850	2.8878	2.8395	2.7764	2.7309	2.6980	2.6733	2.6543	2.6391	2.6268	2.6165	2.5838	2.5662	2.5477	2.5322	2.5202	2.5140
	.875	3.3890	3.2435	3.1392	3.0708	3.0232	2.9884	2.9618	2.9408	2.9239	2.9100	2.8658	2.8423	2.8177	2.7973	2.7815	2.7734
	.900	4.0604	3.7797	3.6195	3.5202	3.4530	3.4045	3.3679	3.3393	3.3163	3.2974	3.2380	3.2066	3.1741	3.1471	3.1263	3.1157
	.925	5.0278	4.5456	4.3037	4.1598	4.0642	3.9961	3.9451	3.9055	3.8738	3.8478	3.7667	3.7242	3.6802	3.6439	3.6160	3.6019
	.950	6.6079	5.7861	5.4094	5.1922	5.0503	4.9503	4.8759	4.8183	4.7725	4.7351	4.6188	4.5581	4.4957	4.4444	4.4051	4.3851
	.975	10.007	8.4336	7.7636	7.3879	7.1464	6.9777	6.8531	6.7572	6.6810	6.6192	6.4277	6.3286	6.2269	6.1436	6.0800	6.0478
	.990	16.258	13.274	12.060	11.392	10.967	10.672	10.456	10.289	10.158	10.051	9.7222	9.5527	9.3794	9.2378	9.1299	9.0754
	.995	22.785	18.314	16.530	15.556	14.940	14.513	14.200	13.961	13.772	13.618	13.146	12.904	12.656	12.454	12.300	12.222

n

TABLE 5   (CONTINUED)

m	α	1	2	3	4	5	6	7	8	9	10	15	20	30	50	100	200
6	.800	2.0729	2.1299	2.1126	2.0924	2.0755	2.0619	2.0508	2.0417	2.0342	2.0278	2.0068	1.9951	1.9825	1.9717	1.9632	1.9588
	.825	2.3650	2.3634	2.3198	2.2847	2.2581	2.2377	2.2216	2.2086	2.1980	2.1891	2.1602	2.1444	2.1277	2.1135	2.1023	2.0966
	.850	2.7231	2.6462	2.5699	2.5164	2.4780	2.4493	2.4270	2.4093	2.3949	2.3830	2.3446	2.3239	2.3021	2.2837	2.2694	2.2620
	.875	3.1761	3.0000	2.8817	2.8048	2.7514	2.7122	2.6823	2.6587	2.6396	2.6238	2.5736	2.5467	2.5186	2.4950	2.4768	2.4674
	.900	3.7760	3.4633	3.2888	3.1808	3.1075	3.0546	3.0145	2.9830	2.9577	2.9369	2.8712	2.8363	2.8000	2.7697	2.7463	2.7343
	.925	4.6269	4.1138	3.8584	3.7061	3.6047	3.5323	3.4779	3.4354	3.4015	3.3736	3.2861	3.2400	3.1921	3.1524	3.1218	3.1062
	.950	5.9874	5.1433	4.7571	4.5337	4.3874	4.2839	4.2067	4.1468	4.0990	4.0600	3.9381	3.8742	3.8082	3.7537	3.7117	3.6904
	.975	8.8131	7.2598	6.5988	6.2272	5.9876	5.8198	5.6955	5.5996	5.5234	5.4613	5.2687	5.1684	5.0652	4.9804	4.9154	4.8824
	.990	13.745	10.925	9.7796	9.1483	8.7459	8.4661	8.2600	8.1017	7.9761	7.8741	7.5590	7.3958	7.2285	7.0915	6.9867	6.9336
	.995	18.635	14.544	12.917	12.027	11.464	11.073	10.786	10.566	10.392	10.250	9.8140	9.5888	9.3583	9.1697	9.0257	8.9529
7	.800	2.0020	2.0434	2.0186	1.9937	1.9736	1.9575	1.9445	1.9339	1.9251	1.9176	1.8930	1.8793	1.8646	1.8519	1.8419	1.8367
	.825	2.2776	2.2589	2.2073	2.1672	2.1371	2.1140	2.0959	2.0812	2.0692	2.0592	2.0264	2.0085	1.9893	1.9730	1.9602	1.9536
	.850	2.6134	2.5183	2.4334	2.3746	2.3324	2.3009	2.2764	2.2570	2.2411	2.2279	2.1854	2.1623	2.1379	2.1173	2.1012	2.0929
	.875	3.0354	2.8401	2.7129	2.6305	2.5732	2.5310	2.4988	2.4733	2.4526	2.4355	2.3809	2.3516	2.3206	2.2947	2.2745	2.2641
	.900	3.5894	3.2574	3.0741	2.9605	2.8833	2.8274	2.7849	2.7516	2.7247	2.7025	2.6322	2.5947	2.5555	2.5226	2.4971	2.4841
	.925	4.3670	3.8363	3.5731	3.4157	3.3107	3.2354	3.1788	3.1345	3.0989	3.0697	2.9777	2.9290	2.8782	2.8359	2.8031	2.7863
	.950	5.5915	4.7374	4.3468	4.1203	3.9715	3.8660	3.7870	3.7257	3.6767	3.6365	3.5107	3.4445	3.3758	3.3189	3.2749	3.2525
	.975	8.0727	6.5415	5.8898	5.5226	5.2852	5.1186	4.9949	4.8993	4.8232	4.7611	4.5678	4.4667	4.3624	4.2763	4.2101	4.1764
	.990	12.246	9.5466	8.4513	7.8466	7.4604	7.1914	6.9928	6.8401	6.7188	6.6200	6.3143	6.1554	5.9920	5.8577	5.7547	5.7024
	.995	16.236	12.404	10.882	10.050	9.5220	9.1553	8.8853	8.6781	8.5138	8.3803	7.9678	7.7540	7.5345	7.3544	7.2165	7.1466
8	.800	1.9511	1.9814	1.9512	1.9230	1.9005	1.8826	1.8682	1.8564	1.8466	1.8383	1.8109	1.7956	1.7791	1.7648	1.7535	1.7476
	.825	2.2150	2.1844	2.1271	2.0834	2.0507	2.0257	2.0060	1.9901	1.9770	1.9660	1.9303	1.9106	1.8895	1.8715	1.8573	1.8499
	.850	2.5352	2.4274	2.3366	2.2740	2.2291	2.1955	2.1694	2.1486	2.1316	2.1174	2.0717	2.0467	2.0202	1.9978	1.9801	1.9710
	.875	2.9356	2.7272	2.5939	2.5076	2.4474	2.4031	2.3692	2.3423	2.3204	2.3023	2.2443	2.2129	2.1798	2.1518	2.1300	2.1187
	.900	3.4579	3.1131	2.9238	2.8064	2.7264	2.6683	2.6241	2.5893	2.5612	2.5380	2.4642	2.4246	2.3830	2.3481	2.3208	2.3068
	.925	4.1851	3.6435	3.3752	3.2145	3.1070	3.0297	2.9714	2.9258	2.8891	2.8589	2.7634	2.7125	2.6593	2.6149	2.5803	2.5626
	.950	5.3177	4.4590	4.0662	3.8379	3.6875	3.5806	3.5005	3.4381	3.3881	3.3472	3.2184	3.1503	3.0794	3.0204	2.9747	2.9513
	.975	7.5709	6.0595	5.4160	5.0526	4.8173	4.6517	4.5286	4.4333	4.3572	4.2951	4.1012	3.9995	3.8940	3.8067	3.7393	3.7050
	.990	11.259	8.6491	7.5910	7.0061	6.6318	6.3707	6.1776	6.0289	5.9106	5.8143	5.5143	5.3591	5.1981	5.0654	4.9633	4.9114
	.995	14.688	11.042	9.5965	8.8051	8.3018	7.9520	7.6942	7.4959	7.3386	7.2107	6.8142	6.6082	6.3961	6.2215	6.0875	6.0194

TABLE 5   (CONTINUED)

n

m	α	1	2	3	4	5	6	7	8	9	10	15	20	30	50	100	200
9	.800	1.9128	1.9349	1.9007	1.8699	1.8455	1.8262	1.8107	1.7979	1.7874	1.7784	1.7488	1.7321	1.7141	1.6985	1.6860	1.6795
	.825	2.1680	2.1287	2.0671	2.0206	1.9860	1.9594	1.9385	1.9216	1.9077	1.8960	1.8578	1.8367	1.8140	1.7945	1.7790	1.7710
	.850	2.4766	2.3597	2.2644	2.1989	2.1520	2.1168	2.0894	2.0675	2.0496	2.0347	1.9863	1.9598	1.9316	1.9075	1.8885	1.8786
	.875	2.8611	2.6433	2.5056	2.4163	2.3540	2.3081	2.2728	2.2448	2.2220	2.2031	2.1423	2.1093	2.0742	2.0446	2.0213	2.0092
	.900	3.3603	3.0064	2.8129	2.6927	2.6106	2.5509	2.5053	2.4694	2.4403	2.4163	2.3396	2.2983	2.2547	2.2180	2.1892	2.1744
	.925	4.0510	3.5020	3.2302	3.0671	2.9578	2.8790	2.8194	2.7727	2.7351	2.7041	2.6058	2.5532	2.4980	2.4516	2.4155	2.3969
	.950	5.1174	4.2565	3.8625	3.6331	3.4817	3.3738	3.2927	3.2296	3.1789	3.1373	3.0061	2.9365	2.8637	2.8028	2.7556	2.7313
	.975	7.2093	5.7147	5.0781	4.7181	4.4844	4.3197	4.1970	4.1020	4.0260	3.9639	3.7694	3.6669	3.5604	3.4719	3.4034	3.3684
	.990	10.561	8.0215	6.9919	6.4221	6.0569	5.8018	5.6129	5.4671	5.3511	5.2565	4.9621	4.8080	4.6486	4.5167	4.4150	4.3631
	.995	13.614	10.107	8.7171	7.9559	7.4712	7.1338	6.8849	6.6933	6.5411	6.4172	6.0325	5.8318	5.6248	5.4539	5.3223	5.2553
10	.800	1.8829	1.8986	1.8614	1.8286	1.8027	1.7823	1.7658	1.7523	1.7411	1.7316	1.7000	1.6823	1.6629	1.6461	1.6327	1.6256
	.825	2.1314	2.0854	2.0205	1.9719	1.9357	1.9079	1.8860	1.8683	1.8537	1.8414	1.8012	1.7788	1.7547	1.7339	1.7173	1.7087
	.850	2.4312	2.3072	2.2085	2.1408	2.0922	2.0557	2.0273	2.0046	1.9860	1.9704	1.9198	1.8920	1.8622	1.8368	1.8166	1.8061
	.875	2.8035	2.5786	2.4374	2.3459	2.2820	2.2347	2.1983	2.1694	2.1459	2.1263	2.0632	2.0288	1.9921	1.9609	1.9363	1.9236
	.900	3.2850	2.9245	2.7277	2.6053	2.5216	2.4606	2.4140	2.3772	2.3473	2.3226	2.2435	2.2007	2.1554	2.1171	2.0869	2.0713
	.925	3.9480	3.3938	3.1195	2.9546	2.8438	2.7639	2.7033	2.6558	2.6174	2.5858	2.4851	2.4310	2.3740	2.3260	2.2884	2.2691
	.950	4.9646	4.1028	3.7083	3.4780	3.3258	3.2172	3.1355	3.0717	3.0204	2.9782	2.8450	2.7740	2.6996	2.6371	2.5884	2.5634
	.975	6.9367	5.4564	4.8256	4.4683	4.2361	4.0721	3.9498	3.8549	3.7790	3.7168	3.5217	3.4185	3.3110	3.2214	3.1517	3.1161
	.990	10.044	7.5594	6.5523	5.9943	5.6363	5.3858	5.2001	5.0567	4.9424	4.8492	4.5581	4.4054	4.2469	4.1155	4.0137	3.9617
	.995	12.826	9.4270	8.0808	7.3428	6.8724	6.5446	6.3025	6.1159	5.9676	5.8467	5.4707	5.2740	5.0705	4.9022	4.7722	4.7058
15	.800	1.7972	1.7952	1.7490	1.7103	1.6801	1.6561	1.6368	1.6209	1.6076	1.5964	1.5584	1.5367	1.5127	1.4914	1.4741	1.4649
	.825	2.0269	1.9621	1.8879	1.8331	1.7923	1.7608	1.7359	1.7156	1.6988	1.6846	1.6375	1.6108	1.5816	1.5560	1.5352	1.5242
	.850	2.3020	2.1586	2.0504	1.9763	1.9228	1.8825	1.8509	1.8254	1.8044	1.7868	1.7289	1.6965	1.6613	1.6305	1.6057	1.5926
	.875	2.6404	2.3963	2.2457	2.1478	2.0790	2.0278	1.9881	1.9563	1.9303	1.9086	1.8377	1.7984	1.7559	1.7191	1.6895	1.6739
	.900	3.0732	2.6952	2.4898	2.3614	2.2730	2.2081	2.1582	2.1185	2.0862	2.0593	1.9722	1.9243	1.8728	1.8284	1.7929	1.7743
	.925	3.6605	3.0938	2.8132	2.6436	2.5287	2.4453	2.3818	2.3316	2.2908	2.2570	2.1484	2.0891	2.0257	1.9714	1.9281	1.9055
	.950	4.5431	3.6823	3.2874	3.0556	2.9013	2.7905	2.7066	2.6408	2.5876	2.5437	2.4034	2.3275	2.2468	2.1780	2.1234	2.0950
	.975	6.1995	4.7650	4.1528	3.8043	3.5764	3.4147	3.2934	3.1987	3.1227	3.0602	2.8621	2.7559	2.6437	2.5488	2.4739	2.4352
	.990	8.6831	6.3589	5.4170	4.8932	4.5556	4.3183	4.1415	4.0045	3.8948	3.8049	3.5222	3.3719	3.2141	3.0814	2.9772	2.9235
	.995	10.798	7.7007	6.4760	5.8029	5.3721	5.0708	4.8472	4.6743	4.5364	4.4235	4.0698	3.8826	3.6867	3.5225	3.3941	3.3279

TABLE 5   (CONTINUED)

m	α								*n* = 1	2	3	4	5	6	7	8	9	10	15	20	30	50	100	200
20	.800								1.7565	1.7462	1.6958	1.6543	1.6218	1.5960	1.5752	1.5581	1.5436	1.5313	1.4897	1.4656	1.4385	1.4143	1.3941	1.3833
	.825								1.9775	1.9041	1.8255	1.7677	1.7245	1.6912	1.6646	1.6429	1.6249	1.6096	1.5585	1.5292	1.4967	1.4678	1.4439	1.4311
	.850								2.2411	2.0890	1.9764	1.8991	1.8433	1.8010	1.7677	1.7407	1.7185	1.6997	1.6376	1.6023	1.5635	1.5291	1.5010	1.4859
	.875								2.5640	2.3114	2.1566	2.0556	1.9844	1.9311	1.8897	1.8565	1.8291	1.8062	1.7308	1.6885	1.6421	1.6014	1.5681	1.5504
	.900								2.9747	2.5893	2.3801	2.2489	2.1582	2.0913	2.0397	1.9985	1.9649	1.9367	1.8449	1.7938	1.7382	1.6896	1.6501	1.6292
	.925								3.5280	2.9567	2.6735	2.5017	2.3850	2.2999	2.2347	2.1831	2.1410	2.1061	1.9927	1.9301	1.8624	1.8036	1.7560	1.7309
	.950								4.3512	3.4928	3.0984	2.8661	2.7109	2.5990	2.5140	2.4471	2.3928	2.3479	2.2033	2.1242	2.0391	1.9656	1.9066	1.8755
	.975								5.8715	4.4613	3.8587	3.5147	3.2891	3.1283	3.0074	2.9128	2.8365	2.7737	2.5731	2.4645	2.3486	2.2493	2.1699	2.1284
	.990								8.0960	5.8489	4.9382	4.4307	4.1027	3.8714	3.6987	3.5644	3.4567	3.3682	3.0880	2.9377	2.7785	2.6430	2.5353	2.4792
	.995								9.9439	6.9865	5.8177	5.1743	4.7616	4.4721	4.2569	4.0900	3.9564	3.8470	3.5020	3.3178	3.1234	2.9586	2.8282	2.7603
30	.800								1.7172	1.6990	1.6445	1.6001	1.5654	1.5378	1.5154	1.4968	1.4812	1.4678	1.4220	1.3949	1.3641	1.3358	1.3116	1.2983
	.825								1.9299	1.8483	1.7655	1.7047	1.6592	1.6239	1.5956	1.5725	1.5531	1.5367	1.4810	1.4486	1.4119	1.3785	1.3502	1.3347
	.850								2.1826	2.0223	1.9054	1.8252	1.7669	1.7226	1.6875	1.6590	1.6354	1.6154	1.5483	1.5097	1.4663	1.4271	1.3941	1.3761
	.875								2.4908	2.2305	2.0716	1.9677	1.8940	1.8387	1.7955	1.7606	1.7318	1.7076	1.6270	1.5810	1.5298	1.4838	1.4452	1.4243
	.900								2.8807	2.4887	2.2761	2.1422	2.0492	1.9803	1.9269	1.8841	1.8490	1.8195	1.7223	1.6673	1.6065	1.5522	1.5069	1.4824
	.925								3.4023	2.8274	2.5421	2.3682	2.2496	2.1627	2.0959	2.0427	1.9992	1.9629	1.8441	1.7775	1.7042	1.6393	1.5855	1.5564
	.950								4.1709	3.3158	2.9223	2.6896	2.5336	2.4205	2.3343	2.2662	2.2107	2.1646	2.0148	1.9317	1.8409	1.7609	1.6950	1.6597
	.975								5.5675	4.1821	3.5894	3.2499	3.0265	2.8667	2.7460	2.6513	2.5746	2.5112	2.3072	2.1952	2.0739	1.9681	1.8816	1.8354
	.990								7.5625	5.3903	4.5097	4.0179	3.6990	3.4735	3.3045	3.1726	3.0665	2.9791	2.7002	2.5487	2.3860	2.2450	2.1307	2.0700
	.995								9.1797	6.3547	5.2388	4.6234	4.2276	3.9492	3.7416	3.5801	3.4505	3.3440	3.0057	2.8230	2.6278	2.4594	2.3234	2.2514
50	.800								1.6867	1.6624	1.6048	1.5581	1.5216	1.4924	1.4687	1.4490	1.4323	1.4179	1.3682	1.3383	1.3035	1.2706	1.2415	1.2249
	.825								1.8930	1.8052	1.7191	1.6560	1.6086	1.5716	1.5420	1.5176	1.4971	1.4796	1.4197	1.3842	1.3432	1.3048	1.2710	1.2518
	.850								2.1374	1.9710	1.8508	1.7682	1.7080	1.6620	1.6254	1.5956	1.5707	1.5496	1.4780	1.4361	1.3880	1.3433	1.3043	1.2822
	.875								2.4345	2.1684	2.0064	1.9002	1.8246	1.7676	1.7228	1.6866	1.6565	1.6311	1.5457	1.4961	1.4398	1.3878	1.3427	1.3172
	.900								2.8087	2.4120	2.1967	2.0608	1.9660	1.8954	1.8405	1.7963	1.7598	1.7291	1.6269	1.5681	1.5018	1.4409	1.3885	1.3590
	.925								3.3065	2.7292	2.4424	2.2671	2.1469	2.0586	1.9904	1.9358	1.8911	1.8356	1.7296	1.6589	1.5798	1.5078	1.4460	1.4115
	.950								4.0343	3.1826	2.7900	2.5572	2.4004	2.2864	2.1992	2.1299	2.0733	2.0261	1.8714	1.7841	1.6872	1.5995	1.5249	1.4835
	.975								5.3403	3.9749	3.3902	3.0544	2.8326	2.6736	2.5530	2.4579	2.3808	2.3168	2.1090	1.9933	1.8659	1.7520	1.6558	1.6029
	.990								7.1706	5.0566	4.1993	3.7195	3.4077	3.1864	3.0202	2.8900	2.7850	2.6981	2.4190	2.2652	2.0976	1.9490	1.8248	1.7567
	.995								8.6258	5.9016	4.8259	4.2316	3.8486	3.5785	3.3764	3.2189	3.0920	2.9875	2.6531	2.4702	2.2717	2.0967	1.9512	1.8719

TABLE 5   (CONTINUED)

α	m	n=1	2	3	4	5	6	7	8	9	10	15	20	30	50	100	200
.800	100	1.6643	1.6356	1.5757	1.5273	1.4894	1.4591	1.4343	1.4136	1.3961	1.3809	1.3278	1.2954	1.2569	1.2191	1.1839	1.1626
.825		1.8660	1.7737	1.6853	1.6204	1.5715	1.5334	1.5026	1.4772	1.4557	1.4374	1.3739	1.3356	1.2905	1.2467	1.2063	1.1819
.850		2.1045	1.9336	1.8111	1.7267	1.6650	1.6177	1.5300	1.5491	1.5232	1.5012	1.4257	1.3807	1.3282	1.2776	1.2312	1.2034
.875		2.3935	2.1233	1.9591	1.8512	1.7741	1.7158	1.6698	1.6325	1.6014	1.5750	1.4855	1.4326	1.3714	1.3130	1.2597	1.2280
.900		2.7564	2.3564	2.1394	2.0019	1.9057	1.8339	1.7778	1.7324	1.6949	1.6632	1.5566	1.4943	1.4227	1.3548	1.2934	1.2571
.925		3.2372	2.6585	2.3707	2.1943	2.0703	1.9835	1.9142	1.8586	1.8128	1.7743	1.6458	1.5715	1.4866	1.4068	1.3353	1.2931
.950		3.9361	3.0873	2.6955	2.4626	2.3053	2.1906	2.1025	2.0323	1.9748	1.9267	1.7675	1.6764	1.5733	1.4772	1.3917	1.3416
.975		5.1786	3.8284	3.2496	2.9166	2.6961	2.5374	2.4168	2.3215	2.2439	2.1793	1.9679	1.8486	1.7148	1.5917	1.4833	1.4203
.990		6.8953	4.8239	3.9837	3.5127	3.2059	2.9877	2.8233	2.6943	2.5898	2.5033	2.2230	2.0666	1.8933	1.7353	1.5977	1.5184
.995		8.2407	5.5892	4.5424	3.9634	3.5895	3.3252	3.1271	2.9722	2.8472	2.7440	2.4113	2.2270	2.0239	1.8400	1.6809	1.5897
.800	200	1.6533	1.6225	1.5614	1.5122	1.4736	1.4427	1.4173	1.3961	1.3781	1.3625	1.3076	1.2738	1.2329	1.1919	1.1521	1.1266
.825		1.8527	1.7582	1.6686	1.6029	1.5533	1.5145	1.4831	1.4572	1.4353	1.4164	1.3510	1.3111	1.2635	1.2161	1.1706	1.1415
.850		2.0882	1.9152	1.7916	1.7063	1.6439	1.5959	1.5576	1.5261	1.4997	1.4772	1.3996	1.3529	1.2976	1.2431	1.1911	1.1581
.875		2.3734	2.1012	1.9360	1.8272	1.7494	1.6904	1.6438	1.6059	1.5742	1.5473	1.4555	1.4008	1.3366	1.2738	1.2144	1.1770
.900		2.7308	2.3293	2.1114	1.9732	1.8763	1.8038	1.7470	1.7011	1.6630	1.6308	1.5218	1.4575	1.3826	1.3100	1.2418	1.1991
.925		3.2034	2.6241	2.3358	2.1589	2.0370	1.9470	1.8771	1.8209	1.7745	1.7354	1.6045	1.5280	1.4396	1.3547	1.2755	1.2263
.950		3.8884	3.0411	2.6497	2.4168	2.2592	2.1441	2.0556	1.9849	1.9269	1.8783	1.7166	1.6233	1.5164	1.4146	1.3206	1.2626
.975		5.1004	3.7578	3.1820	2.8503	2.6304	2.4720	2.3513	2.2558	2.1780	2.1130	1.8996	1.7780	1.6403	1.5108	1.3927	1.3205
.990		6.7633	4.7128	3.8810	3.4143	3.1100	2.8933	2.7298	2.6012	2.4971	2.4106	2.1294	1.9713	1.7941	1.6295	1.4811	1.3912
.995		8.0572	5.4412	4.4085	3.8368	3.4674	3.2059	3.0097	2.8560	2.7319	2.6292	2.2970	2.1116	1.9051	1.7147	1.5442	1.4416

TABLE 6   CHI-SQUARE CDF

Entries are values of x such that

$$\alpha = \int_0^x \frac{1}{2^{n/2}\,\Gamma(n/2)}\, t^{(n/2)-1} e^{-(t/2)}\, dt$$

for various α and degrees of freedom $n = 1, 2, \ldots, 40$.

n	α								
	.005	.010	.025	.050	.100	.150	.200	.250	.300
1	0.000	0.000	0.001	0.004	0.016	0.036	0.064	0.102	0.148
2	0.010	0.020	0.051	0.103	0.211	0.325	0.446	0.575	0.713
3	0.072	0.115	0.216	0.352	0.584	0.798	1.005	1.212	1.424
4	0.207	0.297	0.484	0.711	1.064	1.366	1.649	1.923	2.195
5	0.412	0.554	0.831	1.145	1.610	1.994	2.342	2.674	3.000
6	0.676	0.872	1.237	1.635	2.204	2.661	3.070	3.454	3.828
7	0.989	1.239	1.690	2.167	2.833	3.358	3.822	4.255	4.671
8	1.344	1.646	2.180	2.733	3.490	4.078	4.594	5.071	5.527
9	1.735	2.088	2.700	3.325	4.168	4.816	5.380	5.899	6.393
10	2.156	2.558	3.247	3.940	4.865	5.570	6.179	6.737	7.267
11	2.603	3.053	3.816	4.575	5.578	6.336	6.989	7.584	8.148
12	3.074	3.571	4.404	5.226	6.304	7.114	7.807	8.438	9.034
13	3.565	4.107	5.009	5.892	7.041	7.901	8.634	9.299	9.926
14	4.075	4.660	5.629	6.571	7.790	8.696	9.467	10.165	10.822
15	4.600	5.229	6.262	7.261	8.547	9.499	10.307	11.036	11.721
16	5.142	5.812	6.908	7.962	9.312	10.309	11.152	11.912	12.624
17	5.697	6.407	7.564	8.682	10.085	11.125	12.002	12.792	13.531
18	6.265	7.015	8.231	9.390	10.865	11.946	12.857	13.675	14.440
19	6.843	7.623	8.906	10.117	11.651	12.773	13.716	14.562	15.352
20	7.434	8.260	9.591	10.851	12.443	13.604	14.578	15.452	16.266
21	8.033	8.897	10.283	11.591	13.240	14.439	15.444	16.344	17.182
22	8.643	9.542	10.982	12.338	14.042	15.279	16.314	17.240	18.101
23	9.260	10.195	11.688	13.090	14.848	16.122	17.186	18.137	19.021
24	9.886	10.856	12.401	13.848	15.659	16.969	18.062	19.037	19.943
25	10.519	11.523	13.120	14.611	16.473	17.818	18.940	19.939	20.867
26	11.160	12.198	13.844	15.379	17.292	18.671	19.820	20.843	21.792
27	11.807	12.878	14.573	16.151	18.114	19.527	20.703	21.749	22.719
28	12.461	13.565	15.308	16.928	18.939	20.386	21.588	22.657	23.648
29	13.120	14.256	16.047	17.708	19.768	21.247	22.475	23.566	24.577
30	13.787	14.954	16.791	18.493	20.599	22.110	23.364	24.478	25.508
31	14.457	15.655	17.538	19.280	21.433	22.976	24.255	25.390	26.440
32	15.134	16.362	18.291	20.072	22.271	23.844	25.148	26.304	27.373
33	15.814	17.073	19.046	20.866	23.110	24.714	26.042	27.219	28.307
34	16.501	17.789	19.806	21.664	23.952	25.586	26.938	28.136	29.242
35	17.191	18.508	20.569	22.465	24.796	26.460	27.836	29.054	30.178
36	17.887	19.233	21.336	23.269	25.643	27.336	28.735	29.973	31.115
37	18.584	19.960	22.105	24.075	26.492	28.214	29.635	30.893	32.053
38	19.289	20.691	22.878	24.884	27.343	29.093	30.538	31.815	32.992
39	19.994	21.425	23.654	25.695	28.196	29.974	31.440	32.737	33.932
40	20.706	22.164	24.433	26.509	29.050	30.856	62.345	33.660	34.872

TABLE 6 (CONTINUED)

					α				
n	.350	.400	.450	.500	.550	.600	.650	.700	.750
1	0.206	0.275	0.357	0.455	0.571	0.708	0.874	1.074	1.324
2	0.862	1.022	1.196	1.386	1.597	1.833	2.100	2.408	2.773
3	1.642	1.869	2.109	2.366	2.643	2.946	3.283	3.665	4.108
4	2.470	2.753	3.047	3.357	3.687	4.045	4.438	4.878	5.385
5	3.325	3.656	3.996	4.351	4.728	5.132	5.573	6.064	6.626
6	4.197	4.570	4.952	5.348	5.765	6.211	6.695	7.231	7.841
7	5.082	5.493	5.912	6.346	6.800	7.283	7.806	8.383	9.037
8	5.975	6.423	6.877	7.344	7.832	8.350	8.909	9.524	10.219
9	6.876	7.357	7.843	8.343	8.863	9.414	10.006	10.656	11.389
10	7.783	8.296	8.812	9.342	9.892	10.473	11.097	11.781	12.549
11	8.695	9.237	9.783	10.341	10.920	11.530	12.184	12.899	13.701
12	9.612	10.182	10.755	11.340	11.946	12.584	13.266	14.011	14.845
13	10.532	11.129	11.729	12.340	12.972	13.636	14.345	15.119	15.984
14	11.455	12.079	12.703	13.339	13.996	14.685	15.421	16.222	17.117
15	12.381	13.030	13.679	14.339	15.020	15.733	16.494	17.322	18.245
16	13.310	13.983	14.656	15.338	16.042	16.780	17.565	18.418	19.369
17	14.241	14.937	15.633	16.338	17.064	17.824	18.633	19.511	20.489
18	15.174	15.893	16.611	17.338	18.086	18.868	19.699	20.601	21.605
19	16.109	16.850	17.589	18.338	19.107	19.910	20.764	21.689	22.718
20	17.046	17.809	18.569	19.337	20.127	20.951	21.826	22.774	23.828
21	17.984	18.768	19.548	20.337	21.147	21.991	22.888	23.858	24.935
22	18.924	19.729	20.529	21.337	22.166	23.031	23.947	24.939	26.039
23	19.866	20.690	21.510	22.337	23.185	24.069	25.006	26.018	27.141
24	20.808	21.652	22.491	23.337	24.204	25.106	26.062	27.096	28.241
25	21.752	22.616	23.472	24.336	25.222	26.143	27.118	28.172	29.339
26	22.698	23.579	24.454	25.336	26.240	27.179	28.173	29.246	30.435
27	22.644	24.544	25.437	26.336	27.257	28.214	29.226	30.319	31.528
28	24.591	25.509	26.420	27.336	28.274	29.249	30.279	31.391	32.620
29	25.539	26.475	27.402	28.336	29.291	30.282	31.331	32.461	33.711
30	26.488	27.442	28.386	29.336	30.307	31.316	32.382	33.530	34.800
31	27.438	28.409	29.369	30.336	31.323	32.349	33.431	34.598	35.887
32	29.389	29.376	30.353	31.336	32.339	33.381	34.480	35.665	36.973
33	29.340	30.344	31.337	32.336	33.355	34.412	35.529	36.731	38.057
34	30.293	31.313	32.322	33.336	34.371	35.444	36.576	37.795	39.141
35	31.246	32.282	33.306	34.336	35.386	36.474	37.623	38.859	40.223
36	32.200	33.252	34.291	35.336	36.401	37.505	38.669	39.922	41.303
37	33.154	34.222	35.276	36.335	37.416	38.535	39.715	40.984	42.383
38	34.109	35.192	36.262	37.335	38.430	39.564	40.760	42.045	43.462
39	35.064	36.163	37.247	38.335	39.445	40.593	41.804	43.105	44.539
40	36.021	37.134	38.233	39.335	40.459	41.622	42.848	44.165	45.616

TABLE 6 **(CONCLUDED)**

n	.800	.850	.900	.950	.975	.990	.995
1	1.643	2.073	2.706	3.843	5.025	6.637	7.882
2	3.219	3.794	4.605	5.992	7.378	9.210	10.597
3	4.642	5.317	6.251	7.815	9.348	11.344	12.837
4	5.989	6.745	7.779	9.488	11.143	13.277	14.860
5	7.289	8.115	9.236	11.070	12.832	15.085	16.748
6	8.558	9.446	10.645	12.592	14.449	16.812	18.548
7	9.803	10.748	12.017	14.067	16.012	18.474	20.276
8	11.030	12.027	13.362	15.507	17.534	20.090	21.954
9	12.242	13.288	14.684	16.919	19.022	21.665	23.587
10	13.442	14.534	15.987	18.307	20.483	23.209	25.188
11	14.631	15.767	17.275	19.675	21.920	24.724	26.755
12	15.812	16.989	18.549	21.026	23.337	24.217	28.300
13	16.985	18.202	19.812	22.362	24.735	27.687	29.817
14	18.151	19.406	21.064	23.685	26.119	29.141	31.319
15	19.311	20.603	22.307	24.996	27.488	30.577	32.799
16	20.465	21.793	23.542	26.296	28.845	32.000	34.267
17	21.614	22.977	24.769	27.587	30.190	33.408	35.716
18	22.760	24.156	25.989	28.869	31.526	34.805	37.156
19	23.900	25.329	27.203	30.143	32.852	36.190	38.580
20	25.038	26.498	28.412	31.410	34.170	37.566	39.997
21	26.171	27.662	29.615	32.670	35.478	38.930	41.399
22	27.302	28.822	30.813	33.924	36.781	40.289	42.796
23	28.429	29.979	32.007	35.172	38.075	41.637	44.179
24	29.553	31.132	33.196	36.415	39.364	42.980	45.558
25	30.675	32.282	34.381	37.652	40.646	44.313	46.925
26	31.795	33.430	35.563	38.885	41.923	45.642	48.290
27	32.912	34.573	36.741	40.113	43.194	46.962	49.642
28	34.027	35.715	37.916	41.337	44.461	48.278	50.993
29	35.139	36.854	39.087	42.557	45.722	49.586	52.333
30	36.250	37.990	40.256	43.773	46.979	50.892	53.672
31	37.359	39.124	41.422	44.985	48.231	52.190	55.000
32	38.466	40.256	42.585	46.194	49.480	53.486	56.328
33	39.572	41.386	43.745	47.400	50.724	54.774	57.646
34	40.676	42.514	44.903	48.602	51.966	56.061	58.964
35	41.778	43.640	46.059	49.802	53.203	57.340	60.272
36	42.879	44.764	47.212	50.998	54.437	58.619	61.581
37	43.978	45.886	48.363	52.192	55.667	59.891	62.880
38	45.076	47.007	49.513	53.384	56.896	61.162	64.181
39	46.173	48.126	50.660	54.572	58.119	62.426	65.473
40	47.268	49.244	51.805	55.758	59.342	63.691	66.766

SELECTED APPLICATION ANSWERS TO EXERCISES

Note: Most numerical answers have been rounded to two or three significant digits.

Section 1.2
1.2 $\{s, d\}$ **1.4 a** ordered triples whose components are H or T **b** same as 4(a)
c $S = \{(T, T, T), (T, T, H), (T, H, T), (H, T, T), (T, H, H), (H, T, H), (H, H, T),$
$(H, H, H)\}$ **1.6** $S = \{\alpha\beta : \alpha \text{ and } \beta \text{ are letters of the alphabet}\}; 26 \cdot 26$ **1.8** $\{0, 1,$
$2, \ldots\}$ **1.10** $\{a, e, i, o, u\}$ **1.12** $S = \{(\alpha, \beta) : \alpha \text{ and } \beta \text{ are names of people attending}$
the party and $\alpha \neq \beta\}$ **1.14** $S = \{0, 1, 2, \ldots\}$

Section 1.3
1.23 $\emptyset, \{1\}, \{2\}, \{3\}, \{4\}, \{1, 2\}, \{1, 3\} \{1, 4\}, \{2, 3\} \{2, 4\}, \{3, 4\}, \{1, 2, 3\}, \{1, 2, 4\},$
$\{1, 3, 4\}, \{2, 3, 4\}, \{1, 2, 3, 4\};$ when $n(S) = 5$, there are 2^5 subsets **1.25 a** F **b** T
c F **1.27 a** 4 **b** 9 **c** 10 **d** 11 **e** 6 **f** 5 **g** 12 **h** 7 **i** 1 **j** 8 **1.29 a** $S =$
$\{\{\alpha, \beta\} : \alpha \text{ and } \beta \text{ are denominations of cards from deck and } \alpha \neq \beta\}$ **b** $\{\{x, y\} \in S$
$: x$ is a heart or diamond and y is a heart or diamond$\}$ **1.31 a** $\{0, 1, 2, 3\}$ **b** A
$= \{3\}; B = \{0\}$

Section 1.4
1.48 a $\frac{9}{16}$ **b** $\frac{13}{16}$ **c** $\frac{7}{16}$ **d** $\frac{1}{16}$ **e** $\frac{7}{16}$ **f** $\frac{3}{4}$ **g** $\frac{13}{16}$ **h** $\frac{3}{16}$ **1.50 a** $\frac{2}{3}$ **b** $\frac{23}{30}$ **c** $\frac{7}{30}$ **d** $\frac{9}{10}$
1.52 a 0 **b** $\frac{2}{5}$ **c** $\frac{3}{5}$ **1.54** no **1.56 a** .8 **b** .1 **c** .1 **1.58 a** 30% **b** 35%
c 5% **1.60 a** .25 **b** .15 **c** .9 **d** .1 **e** .45 **f** 1.0 **g** 0

Section 1.5
1.73 .67 **1.75** .05 **1.77** .54 **1.79** Let R denote the event "A is to be released"
and S denote the event "the guard *says* B is to be released." In the case where B
and C are to be released, suppose the guard *says* "B is to be released" with prob-
ability .5. Then $P(S) = P(A \text{ and } B \text{ released}) + .5 \times P(B \text{ and } C \text{ released}) = .5$, so
$P(R|S) = P(R \cap S)/P(S) = (\frac{1}{3})/(\frac{1}{2}) = \frac{2}{3}$, as before. **1.81** .14 **1.83** $(\frac{4}{50})(\frac{3}{49})(\frac{2}{48})$
1.85 .9 **1.87** .92, assuming the prior probabilities are given by the proportions in
the sample

Section 1.6
1.97 assuming independence, .59 **1.99 a** .53 **b** .076 **1.101** .51 **1.103** .5 **b** .3
c no **1.105** yes

Section 1.7
1.108 $\frac{1}{3}$ **1.110** .594 **b** .208 **c** .945 **d** no **1.112** 24.2 **1.114** $\frac{1}{9}$ **b** 1.074

Section 2.1
2.6 take $S = \{y, n, u\}$ and $P(y) = \frac{2}{9}$, $P(n) = \frac{1}{9}$, and $P(u) = \frac{6}{9}$ **2.8** .345; .655
2.10 .432 **2.12** $\frac{1}{2}, \frac{1}{2}, \frac{1}{2}$ **2.14** Let $p(j|k)$ denote the conditional probability in ques-
tion. Then $p(0|1) = p(1|1) = \frac{1}{2}$; $p(0|2) = p(1|2) = p(2|2) = \frac{1}{3}$; $p(0|3) = p(1|3) = p(2|3) = p(3|3) = \frac{1}{4}$; $p(2|1) = p(3|1) = p(3|2) = 0$. **b** Let $q(k|j)$ denote the
conditional probability. Then $q(1|0) = q(1|1) = .46$; $q(2|0) = q(2|1) = q(2|2) = .31$; $q(3|0) = q(3|1) = q(3|2) = q(3|3) = .23$; $q(1|2) = q(1|3) = q(2|3) = 0$. **c** .19
d .27 **e** .92 **2.16 a** .8 **b** .4

Section 2.2
2.30 a .027 **b** .18 **2.32 a** .28 **b** .42 **c** .028 **d** $1 - .028 = .972$ **e** .66
2.34 .33, .19, .10, .04 **2.36** $4/\binom{52}{13}$; $1/\binom{52}{13}$ **2.38** .19, .54 **2.40** 49 to 1
2.42 a .5 **b** One way would be to put all the nickels in one bag, all the dimes in
another bag, and all the quarters in a third bag; then pick a coin from a randomly
selected bag. **2.44 a** .42 **b** .19 **c** .65 **d** .21 **e** .086 **f** .055 **2.46** 116,396,280
unordered subsets, 24 times this many ordered subsets **2.48** .6 **2.50 a** 7! **b**
$6 \cdot 6!$ **2.52 a** $\frac{1}{6}$ **b** $\frac{4}{6}$ **c** $\frac{1}{2}$ **2.54 a** .72 **b** .01 **c** .27

Section 2.3
2.61 .00, .79 **2.63** .50 **2.65** .23 **2.67** .68 **2.69** .55 **2.71** .42 **2.73** $\frac{9}{99}$ **2.75 a** .31
b .50 **c** 2 and 3 have equal likelihoods of .31 **d** .25 **2.77** $P(A|B) = (.9)^3/(.9)^2 = P(A)$; $P(C|A) = .18$, whereas $P(C) = .24$ **2.79** $P(A|B) = P(A) = \frac{1}{2}$;
$P(C|A) = .5 \neq .375 = P(C)$

Section 2.4
2.85 a .015 **b** .022 **2.87** .015 **2.89 a** .16 **b** .042 **2.91** .47

Section 3.2
3.15 Let W denote the winnings, so $p(1) = \frac{8}{36}$, $p(-1) = \frac{2}{36}$, and $p(0) = \frac{26}{36}$. **3.19** $p(3) = \frac{1}{35}$, $p(4) = \frac{3}{35}$, $p(5) = \frac{6}{35}$, $p(6) = \frac{25}{35}$ **3.21** $\{0, 1, 2, 3, 4, 5\}$ **b** $p(0) = p(6) = .031$;
$p(1) = p(4) = .156$; $p(2) = p(3) = .312$ **c** .312 **d** .812 **e** 0 **3.23 b** no, since
$\Sigma f(x) \neq 1$ **3.25 a** $p(0) = .812$, $p(1) = .158$, $p(2) = .0028$, $p(3) = .00024$,
$p(4) \approx 0$ **b** .189 **3.27 a** .7792 **b** .58 **3.29 a** $p(x) = \frac{1}{6}$ for $x = 1, 2, \ldots, 6$;
$F(x) = [x]/6$ for $1 \leq x \leq 6$; $F(x) = 0$ for $x < 1$; $F(x) = 1$ for $x > 6$ **b** $\frac{5}{6}$
3.31 a $\{1, 2, 3, 4\}$ **b** $F(z) = 0$ for $z < 1$; $F(z) = \frac{1}{5}$ for $1 \leq z < 2$; $F(z) = \frac{2}{5}$ for
$2 \leq z < 3$; $F(z) = \frac{4}{5}$ for $3 \leq z < 4$; $F(z) = 1$ for $z \geq 4$ **3.33 a** R_z is finite
b $F(z) = 0$ for $z < 6$; $F(z) = \frac{1}{6}$ for $6 \leq z < 7$; $F(z) = \frac{2}{6}$ for $7 \leq z < 8$; $F(z) = \frac{3}{6}$ for $8 \leq z < 9$; $F(z) = \frac{5}{6}$ for $9 \leq z < 10$; $F(z) = 1$ for $z \geq 10$ **d** $\frac{4}{6}, \frac{3}{6}, \frac{2}{6}, \frac{1}{6}$

Section 3.3

3.44 a .92 **b** .65 **3.46 a** binomial with $n = 1000$ and $p = .005$ **b** $P(X \leq 3)$ $\approx .26$ **3.48 a** .2 **b** .67 **c** .62 **3.50 a** hypergeometric with $N = 100$, $M = 3$, and $n = 20$ **b** .09 **3.52 a** .77 **b** .02 **3.54 a** .13 **b** .36 **3.56 a** .81 **b** .37 **3.58 a** binomial with $n = 200$ and $p = .01$ **b** .41 **3.60 a** geometric with $p = .6$ **b** .94 **c** negative binomial with $r = 3$ and $p = .6$, translated 3 units so that the range is $\{3, 4, 5, \ldots\}$ **3.62** .37, .26 **3.64** \$12.50 **3.66 a** binomial with $n = 3000$ and p 1/6000, which can be approximated as Poisson with $\lambda = .5$ **b** .61 **c** .0003

Section 3.4

3.75 a 3.707 **b** $F(x) = 0$ if $x < -1$; $F(x) = (x + 1)^2/2$ if $-1 \leq x < 0$; $F(x) = .5$ if $0 \leq x < 3$; $F(x) = (x - 3)^2 + .5$ if $3 \leq x < 3.707$; $F(x) = 1$ if $x \geq 3.707$ **c** .5 **3.77** .47; $a^2/2$ if $0 \leq a \leq 2$; $1 - (2 - a)^2/2$ if $1 < a \leq 2$ **3.79** .44, .25 **3.81** .26 **3.83** .32, .42 **3.85 a** 0 **b** $\frac{1}{4}$ **3.87 a** $\frac{1}{3}$ **b** $\frac{1}{3}$ **c** 0 **3.89 a** $F(z) = 0$ for $z < 1$; $F(z) = 2(1 - 1/z)$ for $1 \leq z < 2$; $F(z) = 1$ for $z \geq 2$ **c** .54, .31 **3.91 b** $f(t) = 1/t$ for $l \leq t \leq e$ **3.93 a** .5 **b** $F(w) = 0$ for $w < -1$; $F(w) = (w + 1)^2/4$ for $-1 \leq w < 1$; $F(w) = 1$ for $w \geq 1$ **c** .56, .5, 1 **3.95 a** $f(w) = 1$ for $-.5 < w < .5$ **b** .4 **3.97 a** .0001 **b** .05

Section 3.5

3.109 $$P(a < Z < b) = P\left(\frac{a - \mu}{\sigma} < \frac{Z - \mu}{\sigma} < \frac{b - \mu}{\sigma}\right)$$

$$= \Phi\left(\frac{b - \mu}{\sigma}\right) - \left[1 - \Phi\left(-\frac{a - \mu}{\sigma}\right)\right]$$

3.111

	a	b	c	d	e	f	g
Binomial	.0013	.016	.107	.984	.172	.812	1.00
Normal Approximation	.0009	.011	.072	.980	.103	.647	.96

3.113 a .009 **b** .07 **c** .93 **3.115** 10 **3.117 a** .31 **b** .44

Section 4.1

4.12 .5 **4.14** 7 **4.16** 2 **4.18** $60r$ **4.20** .147, 1.77

Section 4.2

4.28

z	-50	-25	0	25	50	75	$E(Z) = 52¢$
$p(z)$	.01	.01	.02	.18	.41	.37	

4.30 Bernoulli with $p = .3$ **4.32 a** $f_Y(y) = (\lambda/y^2) \exp(-\lambda/y)$ for $y > 0$ **b** integral for $E(Y)$ diverges to ∞, whereas $1/E(X) = \lambda$ **4.34** $F_Y(y) = (y - a)/(b - a)$ is linear function of y for $a < y < b$ **4.36 a** $2/(\pi\sqrt{1 - z^2})$ for $0 < z < 1$ **b** $2/\pi$ **c** $(2/\pi) \sin^{-1} z$ for $0 < z < 1$ **4.38** $p(100) = .135$, $p(200) = .233$, $p(500) = .632$ **4.40** $2/\pi\sqrt{1 - y^2}$ for $0 < y < 1$ **4.42** $f(y) = y^{-2}$ for $y < 1$; $F(y) = 1 - y^{-1}$ for

$y > 1$; $E(Y)$ doesn't exist, whereas $\exp[E(X)] = e$ **4.44** $f(y) = [(y/2)^{1/2} + (y/2)^{-1/2}]/48$ for $0 < y < 18$; $F(y) = \frac{1}{12}[\frac{1}{3}(y/2)^{3/2} + (y/2)^{1/2}]$ for $0 < y < 18$

Section 4.3

4.66 a 0 **b** 0 **c** 2 **d** −1 **4.68 a** $W = (T - .75)/.1936$; $F(w) = (.1936w + .75)^3$ for $-.75/.1936 < w < .25/.1936$ **b** 10.5 **4.70 a** 2.625 **b** −1.875 **c** .515 **d** 12.5 (not unique) **e** 10 (not unique) **4.72 a** $W = (T - \frac{1}{3})/\sqrt{\frac{2}{9}}$ **b** 1 **4.74 a** 8.17 **b** 1.81 **c** 1.83 **d** 9 **4.76 a** 111, 16.3 **b** .5 **c** .85 **4.78 a** $f(u) = \frac{2}{9}$ for $-.25 < u < 4.25$ **b** 1.30 **4.80 a** 500.2, 99.5 **b** 373 **4.82 a** .93 **b** .002 **c** .080 **d** 12 **4.84 a** .44 **b** 5, 5 **c** .26 **4.88** $P(|X - 120| < 10) \geq .16$ **4.90** at least $\frac{3}{4}$ **4.92 a** $\sigma \leq .31$ **b** .60

Section 4.4

4.110 $e^t/(2 - e^t)$ **4.112** $E(X) = m = V(X)$; gamma with parameters m and 1 **4.114** $p(0) = .4$, $p(1) = .3$, $p(2) = .2$, $p(3) = .1$ **4.116** $M_Y(t) = e^{5t}M_X(-t) = .01 + .02e^t + .13e^{2t} + .31e^{3t} + .36e^{4t} + .17e^{5t}$, so $p(0) = .01$, $p(1) = .02$, $p(2) = .13$, $p(3) = .31$, $p(4) = .36$, $p(5) = .17$ **4.118 a** $p(0) = \frac{1}{8}$, $p(1) = \frac{1}{2}$, $p(2) = \frac{1}{4}$, $p(3) = \frac{1}{8}$ **b** 1.375, .734 **c** .625 **4.120** $M(t) = \frac{1}{2}[1/(t + 1) - 1/(t - 1)]$ for $-1 < t < 1$, so $M'(0) = 0$ **4.122 a** $.8e^t/(1 - .2e^t)$ **b** 1.25 **4.124 a** $M_X(t) = .05 + .1e^t + .2e^{2t} + .4e^{3t} + .25e^{4t}$; $N_X(t) = .05 + .1t + .2t^2 + .4t^3 + .25t^4$ **b** $E(X) = 2.7$; $E(X^2) = 8.5$; $E(X^3) = 28.5$ **4.126** $1/(1 - 2t)$; 2, 4 **4.128 a** binomial with $n = 3$ and $p = .1$ **b** .3, 2.7 **4.130 a** $.25 + .5e^t + .25e^{2t}$ **b** $.25 + .5t + .25t^2$ **c** 1, 1.5 **4.132** $F(y) = \sqrt{y}$ for $0 < y < 1$ **4.134** There is no $\lambda > 0$ for which $(\ln 2)/\lambda = 1/\lambda$. **4.136** exponential with $\lambda = .5$

Section 5.2

5.7 b $F(1, 1) = .25$ **c** .25 **d** .5 **e** .25 **f** $\frac{49}{64}$ **5.9** number of B's is binomial with $n = 5$ and $p = .75$, so probability of all B's is .24, $p(2) = .226$, $p(3) = .774$, $p(4) = .00032$ **5.11 a** missing value is .1 **b** .3 **c** .3 **d** .5 **e** .4 **5.13 a** $p(1, 2) = \frac{2}{11}$, $p(1, 3) = \frac{3}{11}$, $p(2, 3) = \frac{6}{11}$ **b** 1 **c** $\frac{2}{11}$ **d** $\frac{6}{11}$ **5.15** $P(X_1 \leq .5, X_2 \leq 100) \approx P(X_1 \leq .5)$

Section 5.3

5.25 a $p(x, y) = p^{x-1}r^{y-1}(1 - p)(1 - r)$ for $x = 1, 2, \ldots$ and $y = 1, 2, \ldots$ **b** $p(x, y) = \binom{n}{x}p^x(1 - p)^{n-x}e^{-\lambda}\lambda^y/y!$ for $x = 0, 1, \ldots, n$ and $y = 0, 1, 2, \ldots$ **5.27** $f_{X_1}(x_1) = \int_0^\infty e^{-x_1-x_2} dx_2 = e^{-x_1}$ for $x_1 > 0$, so $f_{X_1}(x) = f_{X_2}(x)$ **5.29 a** $f(y) = \int_0^1 1 \, dx = 1$ for $0 < y < 1$ **b** $F(x, y) = xy$ for $0 < x < 1$ and $0 < y < 1$ **5.31 a** X and Y identically distributed Bernoulli with $p = \frac{1}{4}$ **b** $E(X) = E(Y) = \frac{1}{4}$

5.33 a.

x_2	x_1 = 0	x_1 = 1		x_3	x_2 = 0	x_2 = 1
1	$\frac{3}{16}$	$\frac{1}{16}$		1	$\frac{3}{16}$	$\frac{1}{16}$
0	$\frac{9}{16}$	$\frac{3}{16}$		0	$\frac{9}{16}$	$\frac{3}{16}$

b $p_{X_1}(0) = \frac{3}{4}$, $p_{X_1}(1) = \frac{1}{4}$; X_3 and X_1 identically distributed **5.35 a** $\frac{1}{2}$, $\frac{1}{10}$ **b** .82 **5.37 c** = 8, $f_X(x) = 8x$ for $0 < x < \frac{1}{2}$, $f_Y(y) = 4 - 8y$ for $0 < y < \frac{1}{2}$

Section 5.4
5.45 a no **b** $\frac{1}{6}$ **c** .25

d

y \ x	0	1
6	0	$\frac{1}{12}$
5	0	$\frac{1}{12}$
4	0	$\frac{1}{12}$
3	0	$\frac{1}{12}$
2	0	$\frac{1}{12}$
1	$\frac{1}{4}$	$\frac{1}{12}$
0	$\frac{1}{4}$	0

5.49 assuming independence, $p(x, y) = e^{-2\lambda}\lambda^{x+y}/x!y!$ for $x = 0, 1, 2, \ldots$ and $y = 0, 1, 2, \ldots$ **5.51 b** .39 **c** .68 **5.53 a** no **b** no **c** yes? **d** yes

Section 5.5
5.64 a $\frac{1}{18}$ **b** $\frac{1}{6}$ **c** $e - 1$ **d** .185 **e** $\binom{3}{k}\alpha^k(1 - \alpha)^{3-k}$ **5.66 a** $ab\rho\sigma_X\sigma_Y$ **b** $\sigma_X^2 - \sigma_Y^2$ **c** $\sigma_X^2 + \mu_X^2 + 2(\rho\sigma_X\sigma_Y + \mu_X\mu_Y) + \sigma_Y^2 + \mu_Y^2$ **5.68 a** .028 **b** .504 **c** .167 **5.70** 0

Section 5.6
5.85 a $\frac{1}{5}$ **b** $\frac{3}{5}$ **c** $\frac{2}{5}$ **5.87 a** $x_1/2$; $(x_2 - 1)/\log x_2$ **b** .415 **c** $(-x_2^2/2 + 2x_2 - \frac{3}{2})/\log x_2$ **5.89 a** $1/y$ for $0 < x < y < 1$ **b** $y/2$ **5.91 a** $\mu_X + \rho\sigma_X(y - \mu_Y)/\sigma_Y$ **b** $(1 - \rho^2)\sigma_X^2$ **5.93 a** $p_X(-2)$·$p_Y(1) \neq 0 = p(-2, 1)$ **b** $p(-2|2) = \frac{1}{3}, p(2|2) = \frac{2}{3}$ **c** $\frac{2}{3}$ **d** 0 for $x < -2$, $\frac{1}{2}$ for $-2 \le x < 2$, 1 for $x \ge 2$ **e** 0 **f** .89 **5.95 a** valid for any x-values, although conditional mass function is positive only for nonnegative integer x such that $z - m \le x \le \min\{n, z\}$

5.99 a

p(y) \\ y	x	1	2	3
$\frac{1}{3}$	3	$\frac{1}{9}$	$\frac{1}{9}$	$\frac{1}{9}$
$\frac{1}{3}$	2	$\frac{1}{9}$	$\frac{1}{9}$	$\frac{1}{9}$
$\frac{1}{3}$	1	$\frac{1}{9}$	$\frac{1}{9}$	$\frac{1}{9}$
p(x)		$\frac{1}{3}$	$\frac{1}{3}$	$\frac{1}{3}$
p(x|1)		$\frac{1}{3}$	$\frac{1}{3}$	$\frac{1}{3}$

b

p(y) \\ y	x	1	2	3
$\frac{1}{3}$	3	$\frac{1}{6}$	$\frac{1}{6}$	0
$\frac{1}{3}$	2	$\frac{1}{6}$	0	$\frac{1}{6}$
$\frac{1}{3}$	1	0	$\frac{1}{6}$	$\frac{1}{6}$
p(x)		$\frac{1}{3}$	$\frac{1}{3}$	$\frac{1}{3}$
p(x|1)		0	$\frac{1}{2}$	$\frac{1}{2}$

c in general, marginals do not determine joint distribution **d** X and Y not independent in case without replacement **e** The marginal distribution will be discrete uniform over $\{1, 2, \ldots, n\}$ in both the sampling with replacement and the sampling without replacement, but the conditionals and joint distributions will differ because the outcomes with replacement are independent, whereas in sampling without replacement, they are not **5.101 a** 1 **5.103** Make all entries $\frac{1}{8}$; then $P(X^2 = 1, Y^2 = 1) = \frac{4}{9}$ **5.105 a** $f(y|x) = 2(1 - x - y)/(1 - x)^2$ for $0 < y < 1 - x$; $f(x|y)$ is of the same form with x and y interchanged. **b** no; yes **c** $f(x, y) = f_X(x) \cdot f_Y(y) = 3(1 - x)^2 \cdot 3(1 - y)^2$ for $0 < x < 1$ and $0 < y < 1$; $P(X < .5, Y < .5) \approx .77$ **d** .125 **e** .167, .014 **5.109** 4 days **b** yes, any combination such that traversing both requires 90 days

Section 5.7

5.115 a $f(x_1, x_2, x_3, x_4) = f_{X_1}(x_1) \cdot f_{X_2}(x_2) \cdot f_{X_3}(x_3) \cdot f_{X_4}(x_4)$ **b** 0 **c** $f_{X_1}(x_1) =$ $2x_1$ for $0 < x_1 < 1$ **5.117 a** $\frac{1}{6}$ **b** $\frac{3}{4}$ **c** $\frac{1}{2}$ **d** .63 **e** 0 **5.119 a** $p(0|1) = \frac{9}{13}, p(1|1)$ $= \frac{4}{13}$ **b** $\frac{4}{13}$ **c** .213 **d** 1.0 **e** $-.055$ **5.121 a** .767 **b** .185

Section 6.2

6.7 a $F(z) = (1 - e^{-\lambda z})^6$ for $z > 0$ **b** $E(Z) = 4.5/\lambda > 1/\lambda$

6.9 $P(X_1 + X_2 = x) = \sum_{j=0}^{x} P(X_1 = j) \cdot P(X_2 = x - j) = \sum_{j=0}^{x} \frac{e^{-\mu}\mu^j}{j!} \cdot \frac{e^{-\lambda}\lambda^{x-j}}{(x - j)!}$

$$= \frac{e^{-(\mu+\lambda)}}{x!} \sum_{j=0}^{x} \binom{x}{j} \mu^j \lambda^{x-j} = e^{-(\mu+\lambda)}(\mu + \lambda)^x$$

for $x = 0, 1, 2, \ldots$
6.11 a $f(y) = y^2/2$ for $0 < y < 1$; $f(y) = -y^2 + 3y - \frac{3}{2}$ for $1 < y < 2$; $f(y) =$ $(3 - y)^2/2$ for $2 < y < 3$ **b** $f(y) = \frac{1}{2}$ for $1 < y < 2$; $f(y) = \frac{1}{2}(y - 1)^2$ for $y > 2$

6.13 a

u	-6	-5	-3	-2	-1	2
$p(u)$	.1	.2	.2	.3	.1	.1

b

	-6	-5	-3	-2	-1	2
3	.1	.2	0	0	0	0
2	0	0	.2	.1	0	0
1	0	0	0	.2	.1	.1
y/u						

c $p(1|-2) = \frac{2}{3}, p(2|-2) = \frac{1}{3}$

d

	-6	-5	-3	-2	-1	2
5	0	0	0	.1	0	0
3	.1	0	.1	0	0	.1
2	0	.2	0	0	0	0
1	0	0	.1	.2	0	0
0	0	0	0	0	.1	0
v/u						

6.15 $f(r) = re^{-(1/2)r^2}$ for $r > 0$; $E(R) = \sqrt{\pi/4}$ **6.17 a** Distribution 1: $E(X|Y = 0) = 4$; $E(X|Y = 1) = 4.25$; $E(X|Y = 2) = 4.44$; $E[E(X|Y)] = 4(\frac{3}{16}) + 4.25(\frac{4}{16}) + 4.44(\frac{9}{16}) = 4.31 = E(X)$ **b** Distribution 1: $V(X|Y = 0) = .667$; $V(X|Y = 1) = .687$; $V(X|Y = 2) = .731$; $E[V(X|Y)] = .708 \neq V(X) = .736$ **c** Distribution 1: .136

d Distribution 1:

z	3	4	5	6	7
$p(z)$	$\frac{1}{16}$	$\frac{1}{8}$	$\frac{1}{4}$	$\frac{3}{16}$	$\frac{3}{8}$

Section 6.3

6.27 $f(r, \theta) = r/\pi$ for $0 < r < 1$ and $0 < \theta < 2\pi$; since this is $(2r)(\frac{1}{2\pi}) = f_R(r) \cdot f_\Theta(\theta)$ and $R_{R,\Theta}$ is a product set, R and Θ are independent **6.29** Eq. (6.13) is analogous to Eq. (6.5) with J corresponding to $d/dy[g^{-1}(y)]$ **6.31 a** 45 **b** .11 **c** .61

6.33 a

	0	0	.1	0
4	0	0	.1	0
3	0	.3	0	.1
2	.2	0	.1	0
1	.2	0	0	0
y_2/y_1	0	.5	1	2

b no

c

y_1	0	.5	1	2
$p(y_1)$	.4	.3	.2	.1

Section 6.4

6.46 $\sigma_1^2 - \sigma_2^2)/\sqrt{[\sigma_1^2 + \sigma_2^2 + 2\,\text{Cov}(X_1, X_2)][\sigma_1^2 + \sigma_2^2 - 2\,\text{Cov}(X_1, X_2)]}$, $(\sigma_1^2 - \sigma_2^2)/(\sigma_1^2 + \sigma_2^2)$ **6.48 a** 0   **b** .25   **6.50 a** $[P(A \cap B) - P(A)P(B)]/\sqrt{P(A)[1 - P(A)]P(B)[1 - P(B)]}$ **b** $P(\overline{A} \cap \overline{B}) + e^{t_1}P(A \cap \overline{B}) + e^{t_2}P(\overline{A} \cap B) + e^{t_1+t_2}P(A \cap B)$ **6.52 a** If $S = 1$ or 0 according to whether or not the item is satisfactory, then S is Bernoulli (.954)   **b** 91.2   **c** With a single attempt at reworking items above $\mu + 2\sigma$, S is Bernoulli with parameter $p = .954 + .023[\Phi((\mu - \mu_r + 2\sigma)/\sigma_r) - \Phi((\mu - \mu_r - 2\sigma)/\sigma_r)]$; and the mean and variance of the number of satisfactory items are $2000p$ and $2000p(1 - p)$ **6.54 a** $-.289, -.878, .226$ **b** 2   **c** $e^{t_1}/8 + e^{t_2}/16 + e^{t_3}/16 + e^{t_1+t_2}/4 + e^{t_2+t_3}/2$; $E(X_1) = \frac{3}{8}$

Section 6.5

6.67 a .299   **b** .301   **6.69 a** If $|\rho|$ were 1, then X and X^2 would be linearly related with probability 1; but this is a contradiction   **b** $r(x) = E(Y|X = x) = E(X^2|X = x) = X^2$ **6.71** 136 pounds   **6.73 a** $\sqrt{31}/2\pi$ **b** can be expressed in the form of Eq. (6.26) with $\mu_X = \mu_Y = 0$, $\sigma_X^2 = \frac{4}{31}$, $\sigma_Y^2 = \frac{8}{31}$, $\rho = 1/\sqrt{32}$ **6.75 a** $M_{X+Y}(t) = M_{X,Y}(t, t) = \exp[(\mu_X + \mu_Y)t + .5t^2(\sigma_X^2 + \sigma_Y^2 + 2\rho\sigma_X\sigma_Y)]$ by Eq. (6.28); but this is a normal mgf and $\mu_{X+Y} = \mu_X + \mu_Y = 0$.   **b** follows by examination of mgf of part (a) and definition of covariance matrix

Section 6.6

6.87 a .214   **b** .0039   **c** .5   **d** .021   **e** 5.22   **6.89** multinomial (4, .164, .098, .328, .410)   **6.91** 7   **6.93 a** 0   **b** 9   **c** .73

Section 6.7

6.110 because of the stationary-increments property   **6.112 a** .11   **b** 6   **c** 52.3%   **6.114** .43, .21, .85   **6.116** .99; 0.0; .085   **6.118 a** .04   **b** .39

Section 6.8

6.133 Verify that the values in each row sum to 1.   **6.135** $\frac{4}{7}$ **6.137 a** Imagine the states 1, 2, 3, 4 arranged around a circle, and a particle moves one step clockwise if the outcome on the biased coin is H; otherwise, it takes one step counterclockwise.   **b** From any state, the walk may visit either of the two immediate neighbors in one step; it may visit the distant neighbor or return to the given state in two steps.   **c** $p^* = (\frac{1}{4}, \frac{1}{4}, \frac{1}{4}, \frac{1}{4})$ **d** $\frac{3}{4} + (1 + 3p)/2(1 - 2qp)$ **6.139** N is geometric with $p = q\pi$, so $E(N) = 1/q\pi$ and $V(N) = (1 - q\pi)/(q\pi)^2$. These decrease with increasing q and π since $q\pi$ then increases, and it is more likely the fault will be repaired on each trial

Section 7.1

7.14 $f_X(z) = \binom{n}{nz}p^{nz}(1 - p)^{n(1-z)}$ for $z = 0, 1/n, 2/n, \ldots, 1$; $F_{\overline{X}}(z) = F_Y(nz)$; yes   **7.16** $\mu + \sigma\Phi^{-1}(\alpha)/\sqrt{n}$ **7.18** .426   **7.20** 3.58, 4.08   **7.22** .24   **7.24 a** If the proportion of the population favoring the candidate is p, the sample can be considered to be a random sample from a Bernoulli (p) population.   **b** $E(\overline{X}_n) = p$ and $\overline{X}_n$ has small variance for sufficiently large n **c** $\frac{1}{2}$ **d** 41 (using a normal approximation)

Section 7.2

7.32 500 is a conservative estimate, based on the Chebyshev inequality; the normal approximation gives a requirement of about 96. **7.34** by the central limit theorem, $\Sigma_{i=1}^{30} X_i$ is approximately $N(135, 120)$; .914 **7.36 a** No; there is about a .11 likelihood of at least 85% effectiveness in 100 cases even when the new drug is no better than the standard drug. **b** yes, very much stronger **7.38 a** approximately $N(396, 6)$ **b** yes, since the standard deviation of a carton of full boxes is much less than the effect of one or more missing boxes **c** .01 **7.40** .03

Section 7.3

7.57 .20 **7.59 a** $N(0, n)$ **b** $\chi^2_{(n)}$ **c** $N(0, 1/n)$ **d** $\chi^2_{(n-1)}$ **e** $F_{(1,1)}$ **f** $t_{(1)}$ **g** $N(0, 1)$ **h** $F_{(1,1)}$ **i** $t_{(1)}$ **j** $t_{(n-1)}$

Section 7.4

7.77 $n(n - 1)(y_n - y_1)^{n-2}/(b - a)^n$ for $a < y_1 < y_n < b$ **7.79 a** $(-4.0, -.1, .7, 1.2, 2.1, 56)$ **b** .95 **c** .95

Section 7.5

7.93 .57 **7.95** .3394 **7.97 a** $g(x) = c \cdot \ln(x)$; exponential; $c^2 b$ **b** $g(x) = c \int \sqrt{x^2 - x}\, dx$; geometric; $c^2 b$ **7.99 b** event in question is equivalent to $[n(1 - \alpha) < W < n(1 + \alpha)]$, where W is $\chi^2_{(n-1)}$, so search for n is feasible **7.101** .079 **7.103** .99 **7.105** .079

INDEX